Climate Change
Biological and Human Aspects
Second Edition

The second edition of this acclaimed text has been fully updated and substantially expanded to include the considerable developments (since publication of the first edition) in our understanding of the science of climate change, its impacts on biological and human systems, and developments in climate policy. As well as being completely revised throughout, major updates include:

Considerable expansion of the sections on climate impacts on early societies in history, and biological impacts;

Updated data and graphs on energy production and consumption;

Completely new chapter sections on: climate thresholds; the Kyoto II conference; Canadian, Australian and New Zealand energy and climate policy;

A new appendix on 'Further thoughts for consideration' to encourage discussion by students and others.

Written in an accessible style, this book provides a broad review of past, present and likely future climate change from the viewpoints of biology, ecology, human ecology and Earth system science. It has been written to speak across disciplines. It will again prove to be invaluable to a wide range of readers, from students in the life sciences who need a brief overview of the basics of climate science, to atmospheric science, geography, geoscience and environmental science students who need to understand the biological and human ecological implications of climate change. It is also a valuable reference text for those involved in environmental monitoring, conservation and policy-making seeking to appreciate the science underpinning climate change and its implications.

The United Nations Environment Programme (UNEP) cited the first edition as one of the top climate change science books of the 21st century.

Jonathan Cowie has spent many years conveying the views of learned societies in the biological sciences to policy-makers, and in science communication (promotion, publishing, and press liaison). His earlier postgraduate studies related to energy and the environment. He is a former Head of Science Policy and Books at the Institute of Biology (UK). He is also author of *Climate and Human Change: Disaster or Opportunity?* (1998).

Praise for this edition:

"A comprehensive review of the science of climate change, the impacts of climate change on biological and human systems, and their interrelatedness. An excellent contribution to the growing recognition that knowledge of biological and human systems is needed to understand climate change."

Gordon Bonan, National Center for Atmospheric Research

"... readers gain an appreciation of the wide-ranging consequences of climate change with many examples and analogies ... it is a book a climate scientist, or any concerned citizen of the world, should read."

Paul A. Dirmeyer, George Mason University

Praise for the First Edition:

"... a fine treatment of global climate change and interactions with biological systems ... everyone is likely to gain a fresh perspective or learn something new."

EOS

"... reader-friendly, quantitative, authoritative, but above all, stimulating; the pages dare you not to turn them over and read further."

The Biologist

"... measured, informative, balanced, scientifically sound, and as up-to-date as a book can possibly be in these days of rapid information accretion."

Bulletin of the British Ecological Society

"There is so much to gain from Cowie's book ... I know of no other source ... that brings together the breadth and depth of material that this book does. ... the bottom line is that anyone who wants to understand climate change and its impacts ... should buy this book. ... Cowie does a brilliant job of weaving together the evolution of life with the evolution of Earth's climate."

Bioscience

"... an impressive endeavor ... the strength of this contribution is precisely the interdisciplinary approach taken to such a multifaceted challenge."

Global Environmental Politics

Climate Change

Biological and Human Aspects
Second Edition

JONATHAN COWIE

CAMBRIDGE
UNIVERSITY PRESS

32 Avenue of the Americas, New York NY 10013-2473, USA

Cambridge University Press is part of the University of Cambridge.

It furthers the University's mission by disseminating knowledge in the pursuit of education, learning, and research at the highest international levels of excellence.

www.cambridge.org
Information on this title: www.cambridge.org/9781107603561

First published 2007
Second Edition 2013
Reprinted 2013

A catalog record for this publication is available from the British Library.

Library of Congress Cataloging in Publication data
Climate change : biological and human aspects / Jonathan Cowie. – 2nd ed.
  p. cm.
Includes bibliographical references and index.
ISBN 978-1-107-60356-1 (paperback)
1. Climatic changes – History. 2. Paleoclimatology. 3. Climatic changes – Environmental aspects. 4. Human beings – Effect of environment on.
5. Physical anthropology. 6. Mass extinctions. I. Title.
QC903.C69 2013
551.6–dc23   2012013138

ISBN 978-1-107-60356-1 Paperback

In memory of Harry Harrison
Making room

(12th March 1925–15th August 2012)

Contents

Figures

Acknowledgements for the first edition

I would very much like to thank all those in UK bioscience with whom I have interacted in some way or other on climate-change matters. In particular, I should thank a good number who have been on the various Institute of Biology science committees since the 1990s. This also goes to a score or two of my fellow members of the British Ecological Society and the Geological Society of London. A special thank you goes to those who have alerted (and, as often as not, invited) me to workshops and symposia on climate and energy issues as well as on biosphere science. I have found every one useful in at least one way: many provided a number of new insights and all gave me a reality check. Thank you.

This book also owes a lot to some research bodies. In the UK we are quite bad at making data from tax-payer-funded research publicly available (even for education and policy purposes). This is not so in the USA and so I greatly valued the open access that the National Oceanic and Atmospheric Administration give to their palaeoclimate-related data (which I have used to generate a number of the figures). Interested readers can visit their website at www.ncdc.noaa.gov/oa/ncdc.html. I am also extremely appreciative of the UK Environment Agency's current (2006) Chief Executive, without whom Figure 6.5 [Figure 6.9 in the second edition] simply would not have been presented! Then there are the many who sent paper offprints (e-mailed pdf files). There are too many to mention but be assured all are referenced.

Talking of references, as mentioned in the Introduction, as far as possible I have taken either major reports, many of which are available on the internet, or used high-impact-factor journals that can be found in most university libraries (these in turn cite papers in more specialist publications). However, I have also used a number of World Health Organization (WHO) press releases. This comes from my background in science policy, and the WHO have been sending me these for the best part of two decades. You will not find these in university libraries but fortunately you too can seek these out, at www.who.int/mediacentre/news/en.

A mention also has to go to the friendly and helpful librarians of Imperial College London, whose work really is appreciated. Then there are all those who have facilitated my site and field visits in the UK and abroad, be they to power stations (fossil, hydroelectric and nuclear), sites of special scientific interest (in the literal and not just the UK technical sense of the term) and educational institutions.

A thank you also goes to Peter Tyers for the [first edition's] cover picture. This is the second time he has done this for me, but then he is a good photographer.

Finally I must specifically thank Cambridge University Press and freelance copy editor Nik Prowse for work on the manuscript. I like to think that I have long since found my feet with words, but any capability for editorial spit and polish has always eluded me. Nik has also greatly helped standardise the referencing and presentation. I therefore really do value good editors (and so should you) and especially those who appreciate those who try to do things a little differently. With luck you will notice.

Acknowledgements for the second edition

In addition to those who kindly helped with the first (2007) edition – as this book firmly builds on that work – I must thank those who helped me develop this updated and expanded second edition. For permission to use figures and data I am appreciative to the following organisations: the Intergovernmental Panel on Climate Change (IPCC), the Earth Science Research at the Laboratory of the National Oceanic and Atmospheric Administration, and the Met Office UK. For permission to reproduce figures (and a photograph), as well as providing advice on data presentation, I am indebted to Timothy Andrews, Gerd Folberth, Jonathan Patz, Pieter Tans and Jim Zachos. A tip of the hat goes to Ian Spellerberg for facilitating some of the contacts for my Australasian sojourn. Here I am most grateful to David Karoly, Rodney Keenan, Ashok Parbhu, Simon Watts and Jez Weston for being generous in affording time and their briefing on climate change impacts and policy in Australia and New Zealand. I have to confess that in this regard I feel somewhat guilty. I had hoped to give more space to climate change and policy matters in these countries. Alas the sheer volume of new science arising in the past 6 years, and the constraints in fitting this into the allocated word count, meant that I could not include nearly as much as I would have liked. Nonetheless I found their briefings most useful, not to mention fascinating, and I hope that my condensing matters down does not do them a disservice.

At this point I must make the obligatory statement that any errors with the science in this book are my own and not those of the above good folk.

I must also thank the Geological Society and British Ecological Society. Of the 'climate surprises' discussed in this book's first edition, the notion that we might at some stage cross a critical transition and climate threshold somewhat analogous to the initial Eocene carbon isotope excursion (CIE) has gained some traction: it was even identified in the IPCC's 2007 Assessment's Working Group I report (pages 442–3 of that work), although it concluded that there was still 'too much uncertainty'. What was needed was a way to bring the current knowledge on this topic together, and so I proposed to the Geological Society the idea of an international symposium on this topic. This suggestion also gained support from the British Ecological Society. In November 2010 a 2-day symposium on past carbon-induced abrupt climate change and how it might inform us regarding future change was held (the first-ever joint event between the British learned societies for geologists and ecologists). There was also an end-of-symposium evening discussion that attracted governmental policy advisors. The outcomes of this symposium have contributed to the discussion in this second edition. Here appreciation goes to my symposium co-convener Anthony Cohen who was invaluable in identifying some of the speakers and in attracting some further sponsorship, as well as Georgina Worrall of the Geological Society who was the event's organising secretary. Once again, any error in my attempts to convey the science are my own, and not the learned bodies involved or the symposium's speakers.

Finally, as with the first edition, once again I must specifically thank Cambridge University Press staff and freelance copy editor Nik Prowse (www.nikprowse.com) for work on the manuscript. They do what I cannot, and for that I am truly indebted.

Introduction

This book is about biology and human ecology as they relate to climate change. Let's take it as read that climate change is one of the most urgent and fascinating science-related issues of our time and that you are interested in the subject: for if you were not you would not be reading this now. Indeed, there are many books on climate change but nearly all, other than the voluminous UN Intergovernmental Panel on Climate Change (IPCC) reports, tend to focus on a specialist aspect of climate, be it weather, palaeoclimatology, modelling and so forth. Even books relating to biological dimensions of climate change tend to be specialist, with a focus that may relate to agriculture, health or palaeoecology. These are, by and large, excellent value provided that they cover the specialist ground that readers seek. However, the biology of climate change is so broad that the average life sciences student, or specialist seeking a broader context in which to view their own field, has difficulty finding a wide-ranging review of the biology and human ecology of climate change. Non-bioscience specialists with an interest in climate change (geologists, geographers and atmospheric chemists, for example) face a similar problem. This also applies to policy-makers and policy analysts, or those in the energy industries, getting to grips with the relevance of climate change to our own species and its social and economic activities.

In addition, specialist texts refer mainly to specialist journals. Very few libraries in universities or research institutes carry the full range. Fortunately the high-impact-factor and multi-disciplinary journals such as *Science* and *Nature* do publish specialist climate papers (especially those relating to major breakthroughs) and virtually all academic libraries, at least in the Anglophone world, carry these publications. It is therefore possible to obtain a grounding in the biology (in the broadest sense) of climate change science from these journals provided that one is prepared to wade through several years' worth of copies.

This book hopefully scores with its broad biological approach, its tendency to cite the high-impact journals (although some specialist citations are also included) and its level of writing (hopefully appropriate for junior undergraduates and specialists reading outside their field). It should also be accessible to bioscientists as well as those outside of the life sciences. However, here is a quick word of advice. Familiarise yourself with the appendices at the back before you start reading!

Even so, this book can only be an introduction to the biology and human ecology – past, present and future – of climate change. Readers seeking more specialist knowledge on any particular aspect should seek out the references, at least as a starting point.

This book's style is also different to many textbooks. Reading it straight through from start to finish, one may get the feeling that it is a little repetitious. This is only *partly* true. It is true in the sense that there are frequent references to other chapters and subsections. This is for those looking at a specific dimension, be they specialists putting their own work into a broader climate context, students with essays to write or policy analysts and policy-makers looking at a special part of the human–climate interface. In short, this book is written as much, if not more, for those dipping into the topic as it is as a start-to-finish read.

There is another sense in which this book appears repetitious, although in reality it is not. It stems from one particular problem scientists have had in persuading others that human activity really is affecting our global climate. This is that there is no single piece of evidence that by itself proves such a hypothesis conclusively. Consequently those arguing a contrary case have been able to cite seemingly anomalous evidence, such as that a small region of a country has been getting cooler in recent years or that the Earth has been warmer in the past, or that there have been alternating warm and cool periods. All of this may be true individually but none of it represents the current big picture. So, instead of a single, all-powerful fact to place at the heart of the climate change argument, there is a plethora of evidence from wide-ranging sources. For instance, there is a wealth of quite separate geological evidence covering literally millions of years of the Earth's history in many locations across the globe. This itself ranges from ice cores and fossils to isotopic evidence of a number of elements from many types of sediment. There is also a body of biological evidence about how species react to changes in seasons to genetic evidence from when species migrated due to past climate change. Indeed, within this there is the human ecological evidence of how we have been competing with other species for resources and how this relates observed changes in both human and ecological communities with past climate change. This vast mass of evidence all points to the same big picture of how changes in greenhouse gases and/or climate have affected life in the past. Then again, there is the present and the evidence used to build up a likely picture of what could well happen in the future. This evidence seems to be very largely corroborative. Therefore, to readers of this book it can seem as if the same ground is being covered when in fact it is a different perspective being presented each time, which leads to the same concluding picture.

Indeed, because there is so much evidence contributing to the big picture that some may well find that evidence from their own specialist area of work is not included, or is covered only briefly. This is simply because the topic is so huge and not due to a lack of recognition on my part of the importance of any particular aspect of climate change science.

That there are similar themes running through specialist areas of climate change science and the relating biology is in once sense comforting (we seem to be continually improving our understanding and coming to a coherent view) but in another it is frustrating. Over the years I have spoken to a large number of scientists from very disparate disciplines. Part of this has been due to my work (policy analysis and science lobbying for UK learned societies and before that in science journal and book management) and in part because I enjoy going to biosphere science as well as energy-related symposia. (There is nothing quite like looking over the shoulders of a diverse range of scientists and seeing what is happening in the laboratory and being discovered in the field.) The key thing is that these individual specialist, climate-related scientists all tend to say similar things, be they involved with ocean circulation, the cryosphere (ice and ice caps), tropical forests and so forth. They say the same as their colleagues in other specialist areas but equally do not appear to really appreciate that there is such a commonality of conclusion. For example, a common emerging theme is that matters are on the cusp of change. Change is either happening or clearly moving to a point where (frequently dependent on other factors) marked change could well happen. It is perhaps a little disappointing that more often than not such specialists seem to have a limited awareness of how their counterparts in other disciplines view things. (I should point out that, in my view, this has more to do with pressures from how science is undertaken these days rather than the high level of competence these specialists have within their own field. Scientists simply are not afforded the time to take several steps back from their work and view the larger scientific panorama.) That science is so compartmentalised tends to limit wide-ranging discussions, yet these, when properly informed by sound science, can be exceptionally fruitful.

By now you may be beginning to suspect what has been motivating my researching and writing of this book. The question that remains for me is whether this book will have any effect on your own motivations and understanding. As it is quite likely that I will encounter at least some of you over the coming years, I dare say I will find out. Meanwhile, I hope you find this topic as fascinating as I do. Reviews and comments online are positively welcome, if not encouraged, be they in print, on websites, in blogs or on social networking sites. I do read and note any comments that I find and they all helped with the revision and expansion process for this second edition, and will help with any further work I may undertake.

Jonathan Cowie
www.science-com.concatenation.org

1   An introduction to climate change

In most places on this planet's terrestrial surface there are the signs of life. Even in places where there is not much life today, there are frequently signs of past life, be it fossils, coal or chalk. Further, it is almost a rule of thumb that if you do discover signs of past life, either tens of thousands or millions of years ago, then such signs will most likely point to different species than those found there today. Why? There are a number of answers, not least of which is evolution. Yet a key feature of why broad types of species (be they broad-leaved tree species as opposed to ones with narrow, needle-type leaves) live in one place and not another has to do with climate. Climate has a fundamental influence on biology. Consequently, a key factor (among others) as to why different species existed in a particular place 5000, 50 000, 500 000 or even 5 000 000 years ago (to take some arbitrary snapshots in time) is because different climatic regimens existed at that place at those times.

It is also possible to turn this truism on its head and use biology to understand the climate. Biological remains are an aspect of past climates (which we will come to in Chapter 2). Furthermore, biology can influence climate: for example, an expanse of rainforest transpires such a quantity of water, and influences the flow of water through a catchment area, that it can modify the climate from what it otherwise would have been in the absence of living species. Climate and biology are interrelated.

Look at it another way. All living things flourish within a temperature range and have certain temperature tolerances for aspects of their life cycle. Furthermore, all living things require a certain amount of water and the availability of water, terrestrially, is again driven by climate. Given this essential connection of temperature and water to life, it is not difficult to see how important climate is in determining where different species, and assemblages thereof (ecosystems), can be found.

From this we can easily deduce that if climate is so important, then understanding climate change is absolutely critical if we are to predict the likely fate of species in a certain region. As mentioned, it is also possible to use the reverse in an applied sense to note the presence (or past presence) of different species and then use this as an indicator of climate, both in the past and in the present. This interrelationship between life and climate is fundamental. It affects all species, which includes, we sometimes forget, our own: *Homo sapiens*. We also tend to forget that on every continent except Antarctica there are examples of deserted settlements and evidence of long-extinct civilisations. These are societies that once flourished but which have now gone, primarily because of a change in climate (this will be examined in Chapter 5).

If it is not sufficiently significant that living things, including human societies, are subject to the vagaries of climate change, there is now convincing evidence that our modern global society is altering the global climate in a profound way that also has regional, and indeed global, biological implications that will impact heavily on human societies. For these reasons there is currently considerable interest in the way living things interact with the climate, and especially our own species. As we shall see in the course of this book, biology, and the environmental sciences relating to ecology and climate, can provide us with information on past climates and climate change (palaeoclimatology) which in turn can illuminate policy determining our actions affecting future climate. This will be invaluable if we are to begin to manage our future prospects.

1.1 Weather or climate

Any exploration of the biology of climate change needs to clarify what is meant by climate as distinct from weather. In essence, the latter is the day-to-day manifestation of the former. The climate of a region is determined by long-term weather conditions including seasonal changes. The problem is that weather is in its own right a variable phenomenon, which is why it is hard to make accurate long-term forecasts. Consequently, if the climate of a region changes we can only discern this over a long period of time, once we have disentangled possible climate change from weather's natural background variability. An analogy is what physicists and engineers refer to as the signal-to-noise ratio, which applies to electrical currents or an electromagnetic signal, such as a commercial radio broadcast or that from a stellar body. Similarly, with climate change, the problem is to disentangle a small climatic change signal from considerable background weather noise. For example, by itself one very hot summer (or drought, or heavy monsoon or whatever) does not signify climate change. On the other hand, a decade or more of these in succession may well be of climatic significance.

Before we explore climate change and especially current problems, we first need to be aware of some terms and the phenomena driving current global warming.

1.2 The greenhouse effect

The greenhouse effect is not some peripheral phenomenon only of importance to global warming. The greenhouse effect is at the heart of the Earth's natural climatic systems. It is a consequence of having an atmosphere, and of course the atmosphere is where climates are manifest.

The French mathematician Jean-Baptiste Joseph Fourier (not to be confused with the contemporary chemist of the same name) is generally credited with the discovery of the greenhouse effect. He described the phenomenon, in 1824 and then again in a very similar paper in 1827, whereby an atmosphere serves to warm a planet. These papers almost did not get written because Fourier was very nearly guillotined during the French Revolution and only escaped when those who condemned him were ultimately guillotined themselves.

Perhaps the best way to illustrate the greenhouse effect is to consider what it would be like if the Earth had no atmosphere. This is not as difficult as it might first seem. We only have to travel 384 400 km (238 856 miles) to the Moon and see the conditions there. On that airless world (its atmosphere is barely above vacuum at one trillionth $[10^{-12}]$ of the Earth's) the daytime temperature is 390 K (117°C), while at night it drops to 100 K (−173°C), giving a median of some 245 K (−28°C). During the lunar day, sunlight either is reflected off the Moon's rocky surfaces or is absorbed, warming the rocks that then re-radiate the energy. The total amount of incoming radiation equals that outgoing. However, at the Earth's surface the average global temperature is higher, about 288 K (15°C). The Earth's atmosphere keeps the planet warmer than it would otherwise be by some 43 K (43°C). This 43 K warming is due to the Earth's atmospheric greenhouse. It is perfectly natural. This warming effect has (albeit to a varying extent) always existed. It occurs because not all the thermal radiation from the Sun falling on our planet's surface gets reflected back out into space. The atmosphere traps some of it just as on the Moon it is trapped in the rocks that are warmed. However, more is trapped on Earth because the atmosphere is transparent to some frequencies (the higher frequencies) of thermal radiation, while opaque to some other, lower, frequencies. Conversely, rock on the Moon is not at all transparent so only the surface of the rock warms and not the strata deep beneath.

The reason that some of the light reflected from the Earth's surface, or radiated as infrared radiation from the lower atmosphere, becomes trapped is because it has changed from being of the sort to which the atmosphere as a whole is transparent to that to which the atmosphere is opaque. There are different types of light because photons of light can be of different energy. This energy (E) of electromagnetic radiation (light, thermal radiation and other rays) is proportional to its frequency (ν) or colour, with the constant of proportionality being Planck's constant (h, which is estimated to be 6.626×10^{-34} J/s). Therefore, the atmosphere is transparent to some frequencies of light but not others. This transparency mix allows some higher-energy light into the blanket of atmosphere surrounding our planet, but hinders other wavelengths, especially lower-energy infrared (heat-level), from getting out. The exact mathematical relationship between the energy of a photon of light (or any other electromagnetic radiation) was elucidated, long after Fourier, in 1902 by the German physicist Max Planck. It can be expressed in the following simple equation.

$$E = h\nu.$$

E (energy) is measured in joules and ν (frequency) in hertz.

When sunlight or solar radiation is either reflected off dust particles and water droplets in the atmosphere or, alternatively, off the ground, it loses energy. As a result of the above relationship between energy and frequency, this reflected light is now at a lower energy, hence lower frequency. As stated, the atmosphere, although transparent to many higher frequencies, is opaque to many of the lower thermal frequencies. The atmosphere traps these and so warms up. Consequently, the atmosphere acts like a blanket trapping lower-frequency radiation (see Figure 1.1). It functions just as the glass of a greenhouse does by allowing in higher-frequency light, but trapping some of the lower-frequency heat; hence the term greenhouse effect. This

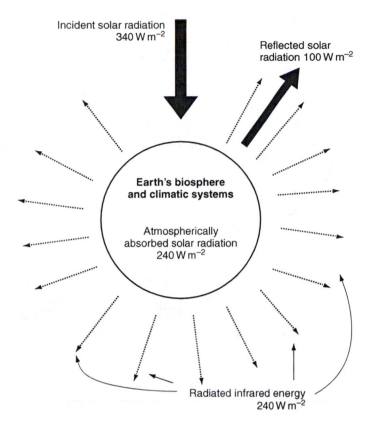

Fig. 1.1 A summary of the principal solar-energy flow and balance in the Earth's atmosphere. Not all the high-energy infrared radiation falling on the Earth is reflected back out into space. Some is converted into lower-energy infrared radiation in the atmosphere. The result is atmospheric warming. Note: the Sun radiates 1370 W m^{-2} to the Earth's distance. However, the Earth is a rotating sphere not a flat surface, so the average energy falling on the Earth's surface is just 340 W m^{-2}.

is why those constituents of the atmosphere that strongly exhibit these properties are called greenhouse gases. The Irish polymath John Tyndall described the greenhouse role of some gases in 1861 and succeeded in quantifying their heat-absorbing properties.

There are a number of greenhouse gases. Many of these occur naturally at concentrations determined by natural, as opposed to human, factors. Water vapour (H_2O) is one, methane (CH_4) another, as is nitrous oxide (N_2O), but the one talked about most frequently is carbon dioxide (CO_2). Others do not occur naturally. For example, halocarbons such as chlorofluorocarbons (CFCs) are completely artificial (human-made), being products from the chemical industry that are used as coolants and in foam blowing. Then again, today there are the naturally occurring greenhouse gases, such as carbon dioxide, the atmospheric concentrations of which are further enhanced by human action.

Tyndall not only recognised that there were greenhouse gases, he also speculated what would happen if their concentration in the atmosphere changed. He considered what it would be like if their warming effect did not take place (as on the Moon).

Indeed, he contemplated that a reduction in greenhouse gases might throw the Earth into another ice age. Strangely though, he never considered what might happen if the concentration of greenhouse gases increased. Consequently, he *never* asked what would happen if human action contributed additional greenhouse gases. In other words, what would happen if there was the addition of an anthropogenic contribution to the natural greenhouse effect?

It is this difference between the natural greenhouse effect and the additional human-generated (anthropogenic) effect that is at the heart of the current issue of global warming. The Swedish chemist and Nobel laureate Svante August Arrhenius first proposed that the human addition of carbon dioxide to the atmosphere would result in warming in 1896, although he himself did not use the term 'greenhouse', but 'hothouse'.

Fourier, Tyndall and Arrhenius are, today, rightly credited with providing the initial grounding science for greenhouse theory. Yet it is often forgotten that in the few decades following 1896 this theory was not high on many scientists' research agendas, and indeed serious doubts arose as to the importance of the increase in atmospheric carbon dioxide in changing the Earth's global climate. However, in 1938 a steam technologist working for the British Electrical and Allied Industries Research Association, one Guy Stewart Callendar, managed to get a paper published in the *Quarterly Journal of the Royal Meteorological Society* in which he noted that humankind had added some 150 000 tons of carbon dioxide to the Earth's atmosphere and that this, he calculated, would have warmed the atmosphere by some $0.003°C$ per year. He also looked at (a limited number of) meteorological records that suggested the climate's temperature had increased at an average rate of $0.005°C$ per year (Callendar refined this last estimate in 1961 using a larger meteorological data set). Callendar's meteorological estimates and greenhouse warming calculation were well within the right order of magnitude and his work, albeit limited, deserves to be remembered in the history of climate change science.

Guy Callendar was not alone. At the end of August 1972 an atmospheric scientist, J. S. Sawyer, estimated that the warming that might be expected with a continued growth in fossil fuel emissions of carbon dioxide to 2000 would be $0.6°C$. This was quite prescient because less than half the total amount of carbon dioxide released into the atmosphere between the Industrial Revolution and 2000 was in the atmosphere by 1970. In addition, as we shall see in Chapter 5, the United Nations (UN) Intergovernmental Panel on Climate Change (IPCC) consider that there was very roughly $1°C$ of warming from the beginning of the Industrial Revolution to 1990, so $0.6°C$ really is not bad for about half the carbon dioxide. What is more, for a couple of decades up to 1970 the global temperature had actually been declining, making Sawyer's prediction particularly brave as it seemingly went against the grain. So he was very close and, with the benefit of hindsight, we can now see that between the 1970s and 2000 the global temperature rose by close to $0.5°C$!

Today the atmosphere is indeed changing, as August Arrhenius thought it might, with the concentration of carbon dioxide increasing in recent times, largely due to the burning of fossil fuels. In 1765, prior to the Industrial Revolution, the Earth's atmosphere contained 280 parts per million by volume (ppmv) of carbon dioxide. By 1990 (which is, as we shall see, a key policy date) it contained 354 parts per million (ppm; either by mass or by volume) and was still rising. By 2005 (when this

Table 1.1 Summary of the principal greenhouse gases (with the exception of tropospheric ozone, O_3, due to a lack of accurate data). Atmospheric lifetime is calculated as content/removal rate

Greenhouse gas...	CO_2	CH_4	CFC-11	CFC-12	N_2O
Atmospheric concentration					
Late 18th century	280 ppm	0.7 ppm	0	0	288 ppb
2010	388 ppm	1.809 ppm	240 ppt	533 ppt	323 ppb
Atmospheric lifetime (years)	50–200	12	45	100	114

ppb, parts per billion; ppm, parts per million; ppt, parts per trillion (all by volume).

book's first edition was written) it had topped 380 ppm and by 2011 (when drafting the second edition) it had reached more than 392 ppm and was still climbing.

If this rise is because of the addition of carbon dioxide to the atmosphere (and it is) then it would be useful to get an idea as to how much carbon is needed to raise the concentration by 1 ppm. Well, the Earth's atmosphere weighs around 5.137×10^{18} kg, which means that a rise of 1 ppm of CO_2 equates to 2.13 Gt of carbon (or 7.81 Gt of CO_2 as carbon dioxide is heavier than carbon).[1]

Over the time since the Industrial Revolution the Earth has also warmed. The warming has not been as regular as the growth in greenhouse gas but, from both biological and abiotic proxies (which I will discuss in Chapter 2) as well as some direct measurements, we can deduce that it has taken place. Furthermore, we now know that Tyndall was right. With less greenhouse gas in the atmosphere the Earth cools and there are ice ages (glacials); as we shall see in Chapter 3 we have found that during the last glacial period, when the Earth was cooler, there was less atmospheric carbon dioxide.

Nonetheless, there has been much public debate as to whether the current rise in atmospheric carbon dioxide has caused the Earth to warm. An alternative view is that the warming has been too erratic and is due to random climate variation. To resolve this issue the United Nations (UN), through the UN Environment Programme (UNEP) and World Meteorological Organization (WMO), established the IPCC. Its four main reports, or assessments (IPCC, 1990, 1995, 2001a, 2007), have concluded that the emissions of greenhouse gases and aerosols due to human activities continue to alter the atmosphere in ways that are expected to affect the climate.

The current rise in atmospheric greenhouse gases (over the past three centuries to date) is well documented and is summarised in Table 1.1.

As we shall see, each of the above greenhouse gases contributes a different proportion to the human-induced (anthropogenic) warming, but of these the single most important gas, in a current anthropogenic sense, is carbon dioxide.

---

[1] You may have noticed that some estimates of CO_2 emissions seem to be about three-and-a-half times as large as others. This is because numbers are sometimes expressed as the mass of CO_2 but are in this book mainly expressed in terms of the mass of the carbon (C). Because carbon cycles through the atmosphere, oceans, plants, fuels, etc., and changes the ways in which it is combined with other elements, it is often easier to keep track only of the flows of carbon. Emissions expressed in units of carbon can be easily converted to emissions in CO_2 units by adjusting for the mass of the attached oxygen atoms; that is, by multiplying by the ratios of the molecular weights of carbon dioxide and carbon respectively, 44/12, or 3.67.

There are two reasons for the different warming contributions each gas makes. First, the concentrations and human additions to the atmosphere of each gas are different. Second, because of the physicochemical properties of each gas, each has a different warming potential.

With regards to post-18th-century changes to the concentrations of the various gases, they were attributable to the post-Industrial Revolution anthropogenic increases in each gas: human influences on the global atmosphere were very different before the Industrial Revolution. The changes in the concentration of these key greenhouse gases have each largely arisen from different sets of human actions. For instance, part of the increase in carbon dioxide comes from the burning of fossil fuels and part from deforestation and changes in land use. Some of the increase in methane comes from paddy fields, whereas part of the rest comes from the fossil fuel industry and biomass (which includes rotting dead plants and animals, and fermentation in animals). We shall examine this in more detail in the next section when looking at the carbon cycle, but other methane increases (or, in the prehistoric past, decreases) are due to more complex factors such as the climate itself, which can serve to globally increase or decrease the area of methane-generating wetlands.

Both carbon dioxide and methane are part of the global carbon cycle (see the following section). Nitrous oxide (N_2O) forms part of the nitrogen cycle and, like carbon dioxide and methane, has both natural and human origins. Naturally, nitrous oxide is given off by the decomposition of organic matter in soils, in particular by tropical forest soils that have high nutrient-cycling activity, as well as by oceans. Human sources include biomass burning and the use of fertilisers. The principal agent removing nitrous oxide from the atmosphere is photolysis – removal by the action of sunlight – ultimately resulting in nitrogen (N_2) and oxygen (O_2).

As to the second factor determining the different warming contribution that each gas makes, each has different physicochemical properties. These are quantified for each gas in what is called its global warming potential (GWP). GWP is a comparative index for a unit mass of a gas measured against the warming potential of a unit mass of carbon dioxide *over a specific period of time*. Carbon dioxide has, therefore, a defined warming potential of 1. A complicating factor is that because different greenhouse gases have different atmospheric residence times (see Table 1.1) GWPs *must* relate to a specific time frame. A GWP expressed without a time frame is nonsense. This can be understood by considering methane, which only has an average atmospheric residence time of a dozen years. Nearly all of a kilogram of methane will still be in the atmosphere after a year. Roughly half of it will be in the atmosphere after 12 years and, assuming exponential decay, a quarter or less after 24 years. This means that the average life time of a typical molecule will be around 12 years.[2] Conversely, nitrous oxide has an average residence time of more than a century. So, clearly, comparing the GWPs of nitrous oxide and methane over a decade will give different warming figures compared with the same comparison over a century. Finally, because of uncertainties, not least with carbon dioxide's own atmospheric residence times, different researchers

[2] Residence times are both estimates and also can alter in different atmospheric conditions such as gas concentration and temperature. So, do not be surprised if you see slightly different figures in the academic literature. Sound advice is to use the most recent as well as authoritative estimates.

Table 1.2 Global warming potentials (GWPs) for some of the principal greenhouse gases over three time frames (IPCC, 2001a)

Gas	Atmospheric lifetime (years)	GWP		
		20 years	100 years	500 years
Carbon dioxide	50–200	1	1	1
Methane	12	62	23	7
Nitrous oxide	114	275	296	156

have different GWP estimates. This can be especially frustrating, as estimates 'improve' with time and – as different theories about the dominating effect of, for example, an aspect of the carbon cycle, come into vogue – it means that GWPs often vary with both research team and time. Even the IPCC's GWP estimates vary a little from report to report. Furthermore, because the IPCC is science by committee – where uncertainty is resolved through consensus of opinion – one cannot simply dismiss one research team's estimates as being completely out of hand. Instead, when looking at a research team's climatic model, you need to see what GWP estimates are used as well as the model itself and then make your own judgement on its results compared to those of another team. Table 1.2 summarises the IPCC's 2001a estimates for GWPs for carbon dioxide, methane and nitrous oxide. Chlorofluorocarbons (CFCs), hydrofluorocarbons (HFCs) and hydrochlorofluorocarbons (HCFCs) are not included because there are so many different ones. However, typically most have GWPs of a few thousand (compared to carbon dioxide's GWP of 1) for time horizons up to 500 years. Fortunately because of their low atmospheric concentration, human-made chemicals such as CFCs and HFCs contribute less than a quarter of current warming (see Figure 1.2).

It is therefore possible to use GWPs to standardise each greenhouse gas into a 'CO_2 equivalent' (CO_2-eq). And so CO_2-equivalent emission is the amount of carbon dioxide emission that would cause the same warming (radiative forcing), over a given time horizon, as an emitted amount of another greenhouse or a mixture of greenhouse gases. CO_2-equivalent emission is obtained by multiplying the emission of a greenhouse gas by its GWP for the given time horizon. For a mix of greenhouse gases it is obtained by summing the CO_2-equivalents of each gas. CO_2-equivalent emission is a standard and useful metric for comparing emissions of different mixes of greenhouse gases, and in policy discussions about reducing warming irrespective of which gas emissions are being reduced. For example, in the latter half of the 2000s, Russia greatly reduced methane leakage from its natural gas industrial infrastructure. Methane is a far stronger greenhouse gas than carbon dioxide: it has a higher GWP even on the 100-year timescale. So the reduction would count for far more than if a similar volume of carbon dioxide emission had been reduced, and the reduction can be calculated in CO_2-equivalents.

There is one other important greenhouse gas that has only briefly been mentioned so far, and that is water vapour. Water vapour is a powerful greenhouse gas, contributing a significant proportion of the natural (as opposed to the human-induced) greenhouse effect. There is sufficient water vapour above the troposphere for it to absorb much of

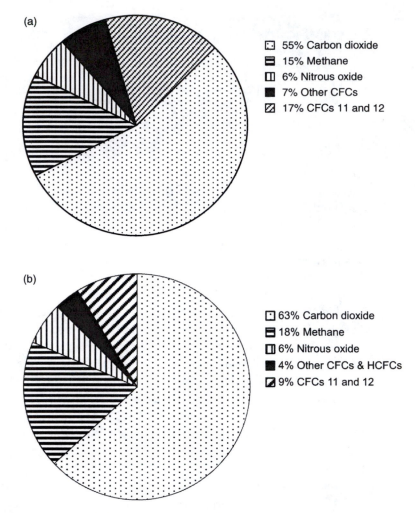

(a)

☐ 55% Carbon dioxide
☐ 15% Methane
☐ 6% Nitrous oxide
■ 7% Other CFCs
☑ 17% CFCs 11 and 12

(b)

☐ 63% Carbon dioxide
☐ 18% Methane
☐ 6% Nitrous oxide
■ 4% Other CFCs & HCFCs
☑ 9% CFCs 11 and 12

Fig. 1.2 The contribution from each of the principal anthropogenic greenhouse gases due to the change in warming (radiative forcing) (a) from 1980 to 1990 (excluding ozone, which may or may not be significant and is difficult to quantify). Data from IPCC (1990). (b) The data for 2005. Data from IPCC (2007).

the infrared radiation at its absorptive frequencies. Indeed, if we were to look at the Earth from space, solely in water-vapour frequencies, our planet would appear as mist-veiled as Venus. This is true even over the dry Sahara Desert. But the concentration of water vapour is not constant throughout the atmospheric column. Tropospheric water vapour, in the atmospheric layer closest to the ground, varies considerably over the surface. In the first 1–2 km of the atmosphere (lower part of the troposphere), the amount of water vapour potentially in a unit volume increases with temperature. In the troposphere above this point, the water-vapour greenhouse effect is most important but is harder to quantify. Nonetheless, current computer models of the global climate account for water-vapour feedback, whereby a warmer world sees more evaporation, hence more water vapour, and this tends to double the warming that one would expect from just a model of fixed water vapour. The ability of current (early 21st century) global climate computer models to reproduce the likely effect of water vapour over a period of warming was given credence in 2005 by a US team of atmospheric scientists led by Brian Soden. They compared satellite observations between 1982 and 2004 at the 6.3 μm wavelength – which is part of water's absorption spectrum and especially useful for measuring its presence in the upper troposphere – with climate models. The satellite measurements and the models showed a good correlation.

Clouds (the suspension of fine water droplets in regions of saturated air) complicate the picture further still. Being reflective, they tend to cool the surface during the day and at night act as an effective greenhouse blanket. However, there are various types of cloud. The picture is complex and our understanding incomplete, hence climate models provide only an approximation of what is going on, but they are revealing approximations nonetheless. However, it has long been thought that biological processes affect cloud formation by releasing volatile organic compounds such as monoterpenes ($C_{10}H_{16}$ compounds such as pinenes, which give pine forests their smell) and sequesterpenes that in turn determine particle formation around which droplets can form, hence clouds. Clouds above a large expanse of boreal forest can affect the local temperature by as much as 5–6°C. Nonetheless it is still early days in developing a clear picture. Indeed, one puzzling feature has been a mismatch between the amount of volatile organic compounds produced by plants and that found in the air above some boreal forests. In 2009 a German-Finnish team led by Astrid Kiendler-Scharr demonstrated, using a plant chamber containing birch (*Betula pendula*), beech (*Fagus sylvatica*), spruce (*Picea abies*) and pine (*Pinus sylvestris*), that isoprene was also released and that some species are high isoprene emitters. One of these species was oak (*Quercus robur*), which the team included in their study. They found that isoprene, with its hydroxyl radical (OH$^\bullet$), inhibits particle formation and so counters the release by other plants of volatile organic compounds which have cloud-forming action. In short, the effect that a forest will have on cloud formation (and hence temperature and water cycling) will depend on the forest's species mix. This work is all the more relevant from a climate change perspective (let alone an ecological management climate perspective) in that the amount of isoprene that vegetation produces is affected by light and temperature. Such studies are now illuminating the construction of global climate computer models, which in the 21st century increasingly have a biological component. One problem is that that there are literally many tens of thousands of organic volatiles produced by plants and teasing out which ones are important from a climate perspective will take time. (We will return to climate change and the water cycle later in this chapter.)

Given that overall the Earth's atmosphere (with its mix of various greenhouse gases) broadly confers a 43°C greenhouse warming effect (because, as we have seen, the airless Moon is cooler), the question remains as to how much warming has been conferred since the Industrial Revolution as a result of the human addition of greenhouse gases. We shall come to this in Chapter 5. Nonetheless, it is worth noting for now that mathematicians Cynthia Kuo and colleagues from the Bell Laboratory in New Jersey, USA, statistically compared instrumentally determined changes in atmospheric CO_2 concentrations between 1958 and 1989 and global temperature

(Kuo et al., 1990). This confirmed that carbon dioxide and global temperature over that period were significantly statistically correlated to over 99.99%. This is to say that were 10 000 alternative copies of the Earth similarly measured, only one would give similar results due to sheer chance and 9999 would give results because there is a link between carbon dioxide concentrations and global temperature. But before we look at how the human addition of carbon dioxide to the atmosphere affects climate we need a better understanding of the natural sources and sinks of atmospheric carbon dioxide. Fundamental to this is the carbon cycle.

1.3 The carbon cycle

Carbon is one of the fundamental elements necessary for life. It is found in virtually all molecules (but not quite every molecule) associated with life. These include all carbohydrates, all proteins and all nucleic acids. As such, carbon is fundamental to biological structures, of both micro- and macro-organisms, including plants and animals; for example, lignin in plants and cartilage and bone in animals. Indeed biomolecules, as we shall see in Chapter 2, can be of great use to palaeoclimatologists because some of them (and hence the remains of species in which they are found) can be used as climatic indicators.

The carbon cycle itself refers to the circulation of carbon in the biosphere. The circulation is driven primarily (but not solely) by biological processes. A planet that does not have any biological processes would have carbon flow through its geosphere[3] driven solely by geophysical processes. On Earth carbon, in the form of carbon dioxide, is fixed by photosynthesis into organic compounds in plants and photosynthetic algae and returned to the atmosphere mainly by the respiration of plants, animals and micro-organisms in the form of carbon dioxide, but also by the decay of organic material in the form of both methane and carbon dioxide. Abiotic drivers include the burning of organic material, be it natural (e.g. forest fires) or through human action (e.g. the burning of firewood or fossil fuels). Another abiotic driver is that of plate tectonic movement. This contributes to the so-called deep carbon cycle operating on a scale of millions of years. Tectonic plates in the process of subduction carry with them organic sediments down into the Earth's mantle. This plate movement results in volcanic activity that in turn converts the organic sediments to carbon-containing gases (again mainly carbon dioxide but also other volatile compounds) that come to the surface via volcanic and related activity. There is a variety of other abiotic processes, including the chemical oxidation of methane in the atmosphere to carbon dioxide as well as processes (which frequently also accompany biotic ones) in organic sediments. An estimate of the principal carbon movements in the carbon cycle is given in Figure 1.3.

The carbon cycle is at the centre of biology's relationship with the global climate (and hence global climate change). More than this, it demonstrates the importance of both the molecular biological sciences and the whole-organism approach to biology.

---

[3] On a lifeless world there is no biosphere, only a geosphere.

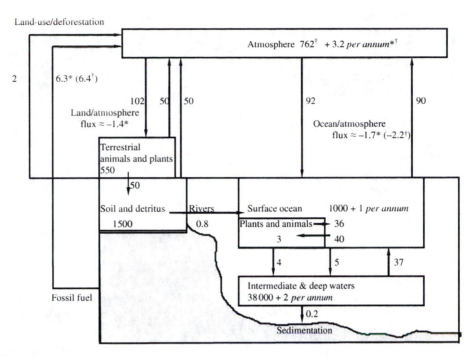

Fig. 1.3 Broad estimates of the principal annual carbon sources and sinks in gigatonnes of carbon (GtC), and approximate annual movements of carbon about the carbon cycle in gigatonnes of carbon *per annum* (GtC year^{-1}) average for the 1990s. It is important to realise that there are a number of uncertainties that are the subject of current research (see text) and also, because we are changing the cycle, it will be different for different decades. To give a sense of how understanding changes and the uncertainties involved, the figure includes 2001 IPCC estimates for annual fluxes over 1990–9, marked *, and 2007 IPCC estimates for the same decade's annual averages, marked †.

A first impression of the biology associated with the carbon cycle might focus on the plant activity sequestering carbon dioxide from the atmosphere. But this is just one aspect, albeit a key one. There is also plant and animal respiration, and the respiration of bacteria and fungi, returning carbon dioxide to the atmosphere. We might also think of biomes (climatically determined regional groups or assemblages of ecosystems) and how they affect the global carbon cycle in terms of marine productivity, or of the carbon in a biomes' biomass, be it in terrestrial tropical rainforests or temperate wetlands, for example. We might also consider the effect of climate change on such ecosystem assemblages and individual ecosystems; and indeed we will do so later in the book. But there is also the biomolecular dimension.

Included in the biomolecular perspective is the role enzymes play. Rubisco (ribulose-1,5-bisphosphate carboxylase oxygenase, which is sometimes portrayed in print as RuBisCO) is the most common enzyme on the planet and is fundamental to photosynthesis. It is therefore probably the most common protein on Earth; it constitutes about half of leaf proteins and is synthesised in chloroplasts. All the carbon dioxide captured by photosynthesis – in algae and multicellular plants – is handled by this one enzyme. That is about 200 billion t (or 200 Gt) of carbon a year! Another important enzyme is carbonic anhydrase, which catalyses the hydration of about a

Table 1.3 Estimated annual carbon emissions in 1980–9, 1990–9 (IPCC, 2001a) and 2000–5 (IPCC, 2007) to the atmosphere and annual transfer from the atmosphere to sinks. This way you can see not only how we think the carbon cycle changes as carbon moves between biosphere reservoirs, but also how understanding/perceptions of the carbon cycle have changed with time. (The latest understanding is given in the 2007 Working Group I assessment in IPCC [2007], table 1.7, p. 516.)

	Carbon transfer (Gt of carbon year^{-1})		
	1980–9	1990–9	2000–5
From fossil fuels to atmosphere	5.4±0.3	6.3±0.4	7.2±0.3
From deforestation and net land-use change to atmosphere	−1.6±1.0	−1.4±0.7	−0.9±0.6
Accumulation in atmosphere	3.4±0.2	3.2±0.1	4.1±0.1
Ocean-to-atmosphere flux	−2.0±0.8	−1.7±0.5	−2.2±0.5
Net imbalance in 1980s estimates	1.6±1.4		

third of the carbon dioxide in plants and in soil water. As we shall see, the way these enzymes handle carbon dioxide molecules with different isotopes of carbon, or oxygen molecules with different isotopes of oxygen, helps us in our understanding of details of the carbon cycle.

It is important to note that there are some uncertainties in the estimates of the rate of flows between the various sources and sinks (Figure 1.3). This is the subject of ongoing research. Part of the problem (other than the cycle's complexity) is that the carbon cycle is not static: there are varying transfers of carbon, so altering the amounts within reservoirs. For example, during cold glacial times (such as 50 000 years ago) the amount of atmospheric carbon (as both carbon dioxide and methane) was less than today. Conversely, currently the atmospheric carbon reservoir is increasing. This dynamism is not just because of human action; it also happens naturally, although human action is the current additional factor critical to what is called global warming.

In practical terms today, carbon cycle uncertainties manifest themselves in a number of ways. Of particular concern is the mystery of where roughly half the carbon dioxide released into the atmosphere by human action (from fuel burning and land-use change) ends up: from measuring the atmospheric concentration of carbon dioxide we know that half of what we burn does not remain in the atmosphere for very long. This imbalance is so significant that there is debate as to whether a major carbon cycle process has been overlooked. Alternatively it could be that the current estimates as to the various flows have a sufficient degree of error that cumulatively manifests itself as this imbalance.

The broad estimates (IPCC, 2001a, 2007; together with estimates of uncertainty) of the contributions of burning fossil fuels and deforestation to atmospheric carbon dioxide and the entry to carbon sinks from the atmosphere for the decades 1980–9 and 1990–9 are listed in Table 1.3. The estimates are derived from computer simulations.

As we can see from Table 1.3, in the 1990 IPCC report (average 1980–9 annual flux column) there was this net imbalance between the estimates of carbon dioxide entering and of those accumulating and leaving the atmosphere. More carbon dioxide was estimated as being released into the atmosphere than was retained by the atmosphere or thought to be absorbed by the oceans. Where was this carbon dioxide going?

This net imbalance was both large and of the same order of magnitude as existing global flows. Subsequent IPCC assessments did not estimate this imbalance in their analogous tables, although there is still an imbalance. This imbalance is why some think that a major route of carbon from the atmosphere has not been identified, or alternatively that one or more of the existing carbon-flow estimates (Figure 1.3) is considerably off the mark, or perhaps a bit of both. As regards the flow of fossil fuel carbon to the atmosphere, this is quite well documented due to the economic attention paid to the fossil fuel industry: we know how many barrels of oil are sold, tonnes of coal are mined, and so forth, so we can be fairly certain that this estimate is broadly accurate (note the smaller plus/minus error).

Looking at the other side of the equation, the accumulation of carbon in the atmosphere is also as accurately charted as it can be, and has been directly monitored over many years from many locations. It is because we know exactly how much extra carbon we release, and have released, from fossil fuels into the atmosphere (again note the smaller plus/minus error), and how much actually stays in the atmosphere, that we can be certain that there is a shortfall and that some part of the carbon cycle has either not been properly quantified or even perhaps not properly identified.

It is uncertainties such as this, and that the global climate-warming signal had to be sufficient to be discernable from the background natural variation (noise), that has helped some argue that global warming is not taking place. As we shall see over the next few chapters, the climate has changed in the past (affecting biology and vice versa) and the atmospheric concentrations of greenhouse gases have played a major part in these changes.

Interestingly, the current year-on-year accumulation in atmospheric carbon dioxide, as measured in either the northern or southern hemisphere, is not smooth. Rather, there is an annual oscillation superimposed on the rising trend. The oscillation occurs because of seasonality outside of the tropics. During winters in the temperate zone there are no leaves on the trees and in the boreal zone too there is little photosynthesis on land or in the sea (in the main by algae). But in the summer there is considerable photosynthesis and so more carbon dioxide is drawn into plants and algae. In winter respiration continues, even though photosynthesis is reduced, and so, on balance, more carbon dioxide is released into the atmosphere. Thus there is an annual cycle of waxing and waning of atmospheric carbon dioxide in the northern and southern hemispheres (see Figure 1.4). Indeed, because while in one hemisphere there is summer and the other winter, the carbon dioxide oscillations in the two hemispheres are opposite and complement each other. However, the seasonal variation of carbon dioxide in the southern hemisphere is not nearly so marked, as that hemisphere is dominated by ocean, which has a strong ability to buffer carbon dioxide. Oxygen, as the other gas concerned with photosynthetic and respiration reactions, also shows a seasonal variation in each of the hemispheres but one that is more marked than that for carbon dioxide (as oxygen is not buffered by the seas). Like carbon dioxide, the seasonal variation of the atmospheric concentration of oxygen is equal and opposite in the northern and southern hemispheres.

The changes in the atmospheric reservoir of carbon are quite well understood (even though our knowledge of other reservoirs is not so complete) because we can measure atmospheric carbon dioxide directly, and because atmospheric mixing

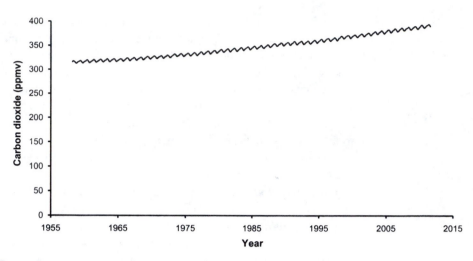

Fig. 1.4 Outline of northern hemispheric seasonality effects of atmospheric carbon dioxide concentration, 1958–2011 through to the early 21st century. Note that in addition to the annual waxing and waning, there is also an overall trend of growth in atmospheric concentration. ppmv, parts per million by volume.

within hemispheres is reasonably thorough. We can therefore be quite certain that our knowledge of this atmospheric part of the carbon cycle is fairly accurate. The problem of the missing carbon seems to be associated with one or more of the other carbon sinks: possibly the accumulation of carbon by terrestrial plants, or alternatively by absorption into the oceans. Nonetheless, the annual waxing and waning of atmospheric carbon dioxide does illustrate the power of the photosynthetic carbon pump (the power of plant and algal photosynthesis to drive the carbon cycle).

Turning to human impacts on terrestrial reservoirs of carbon, estimates of deforestation/land use do have greater error associated with them than changes in atmospheric gas concentrations, which can be measured directly. So it may well be that part of the missing carbon flow in the carbon cycle is associated with this component of the cycle. However, instances of more detailed scrutiny of, for example, deforestation data have revealed that here official estimates are invariably underestimates, and that deforestation probably accounts for a greater contribution to atmospheric carbon dioxide, not less, and so the amount of missing carbon is greater. This means that, if anything, our understanding of the cycle's carbon imbalance is not even as good as we think. Of course, deforestation is but one dimension, of a number, to the changes of the terrestrial reservoirs of carbon.

The importance and power of the photosynthetic pump driving part of the carbon cycle is corroborated by the magnitude of the seasonal oscillation in carbon dioxide. Nonetheless land-use change, along with terrestrial-biome change and ocean accumulation, are key areas of uncertainty (either singly or together) that might account for the missing carbon. It could be that oceans are accumulating more carbon than we think and/or that the increased atmospheric carbon dioxide along with global warming is encouraging terrestrial photosynthesis, drawing down carbon into plants over much of the globe. As stated above, because of the magnitude of seasonal variation in atmospheric carbon dioxide outside of the tropics we can see how powerful the photosynthetic and respiratory carbon pumps truly are. *If* this missing carbon sink is terrestrial (and given that much of the planet's land is in just one, the northern, hemisphere) it appears that the carbon may be being sequestered not in a temporary way by annual plants but in a longer-term way by perennials, and especially temperate and boreal trees. Then there is the carbon stored in soils and detritus. The total carbon store in soils, at 1500 gigatonnes of carbon (GtC), is over twice that stored as biomass and is also more than the atmospheric carbon reservoir. Currently the reservoir of soil carbon is a net sink, although carbon flows to and from it are far less than those to and from either biomass carbon or atmospheric carbon reservoirs. It is important to note that *currently* soil acts as a global net carbon sink. 'Currently' because a world that is just a few degrees warmer could see the soil reservoir act as a net source of carbon as it would be released as carbon dioxide into the atmosphere. In such an instance soils would act to further global warming. (We will return to carbon in soils later in this chapter and again in Chapters 5 and 7.)

Again, as mentioned above, alternatively (or in addition: we just do not know) the oceans could be a greater sink of carbon than we realise, so they could account for the carbon imbalance. Either way, it is almost certain that the driving force behind this missing sequestration of additional atmospheric carbon is photosynthesis, even if we are unable to say whether it is marine, terrestrial or both, let alone where on Earth it is taking place. (Having said that, there is some fascinating progress in the area of satellite monitoring that shows seasonal changes in the photosynthetic pump in various parts of the world, although there is a little way to go yet before we can get meaningful quantitative data.)

With regard to the scale of carbon flow between carbon cycle reservoirs due to photosynthesis, the observed annual natural seasonal variation in atmospheric drawdown and replenishment is considerable, and these are greater than the year-on-year increasing trend in atmospheric carbon dioxide due to the human addition of fossil fuels and land-use change. Just as human action (the action of just one species) is responsible for the current growth in atmospheric carbon dioxide, so one of the most fundamental biological processes – photosynthesis – (through many species) can almost certainly be involved in its amelioration. This re-emphasises that biology and climatology are closely entwined.

Here the ways that photosynthetic and respiratory enzymes handle carbon and oxygen are beginning to illuminate the problem of missing carbon. As stated above, Rubisco is the enzyme globally responsible for fixing atmospheric carbon dioxide as part of the photosynthetic process. However, not all atmospheric carbon is in the form of the ^{12}C isotope. Around 1% is ^{13}C. Rubisco has evolved the greatest specificity for the almost universal ^{12}C and so discriminates against ^{13}C, leaving it behind in the atmosphere. If photosynthetic activity increases (as it does every summer in each hemisphere) then the increase in atmospheric ^{13}C left behind can be measured. This also works if photosynthesis increases due to global warming, because in a warmer world the thermal growing season (TGS) is longer. On the other hand, isotopes of ^{12}C and ^{13}C dissolve more or less equally well in sea water: in fact, if anything ^{13}C dissolves slightly more easily. Consequently, if we detect changes in atmospheric ^{13}C above and beyond the expected seasonal changes in a hemisphere, we can see whether the photosynthetic pump is working harder or not. Similarly, carbonic anhydrase works best on the ^{16}O isotope of oxygen and discriminates against ^{18}O. Isotopic studies are therefore an increasingly important tool in understanding how the carbon cycle works.

Ascertaining details of the carbon cycle is the subject of considerable ongoing research. In terms of addressing the problem of increasing atmospheric carbon dioxide, the solutions will almost inevitably involve modifying carbon flows between reservoirs such that the atmosphere's carbon burden will be reduced. We will return to this towards the end of this section. Before doing so it is important to note that this ongoing carbon cycle research does not just provide extra detail but still turns up major surprises and even false surprises.

One such recent (2006) surprise was the possibility that plants in aerobic conditions (with oxygen available) produce methane. Indeed, it was so surprising that *Nature* ran a small article in its news section entitled 'How could we have missed this?' The discovery was almost fortuitous in that it had been thought that all the principal sources of atmospheric methane had been identified even if their individual quantification needed to be refined. To ensure that it was methane from plants (and not microbes) the researchers attempted to kill off bacteria on the plants with radiation. They also removed methane from the air in the incubators in which the plants were to be grown. Although the amounts of methane detected from individual plants were small, globally it amounts to a significant source. The researchers who made the discovery could only make a very rough estimate (as work has yet to be done on an appropriately representative range of species and conditions) but they thought that the annual atmospheric contribution could be between 60 and 240 million t (Keppler et al., 2006), or between one-twelfth and one-third of the annual amount entering the atmosphere. The result came as a surprise and was important in understanding the carbon cycle (hence for developing global climate models, which increasingly include biological components). However, 4 years later Andrew Rice of Portland State University in Oregon, USA, conducted a study suggesting that trees act like chimneys, transporting methane produced by soil microbes up from the roots through to leaves and that this effect could account for as much as around 10% of methane emissions globally. This itself was a valuable development as it could help explain why methane emissions are higher than expected from wet tropical forest regions. Another advance in carbon cycle understanding came in 2009 by a team led by Eran Hood. They found that dissolved organic matter (rich in carbon) in glacial run-off from 11 Gulf of Alaska watersheds was higher (and more labile) than was thought. This was because most of the previous work done on this aspect of the carbon cycle had been on watersheds where terrestrial plants and soil sources dominate dissolved organic matter. This new insight suggests that glacial run-off is an important source of labile reduced carbon to the planet's oceans. Furthermore, the most glaciated watersheds the team studied were the source of the oldest ($\approx$4000 years old) and most labile (66% bioavailable) dissolved organic carbon. All this is very important when considering a warming world in which glaciers are retreating.

As for the year-on-year growth in atmospheric carbon dioxide since the beginning of the Industrial Revolution (as stated above) this has been carefully charted. It has been done so in two main ways. First, in recent times (since the middle of the

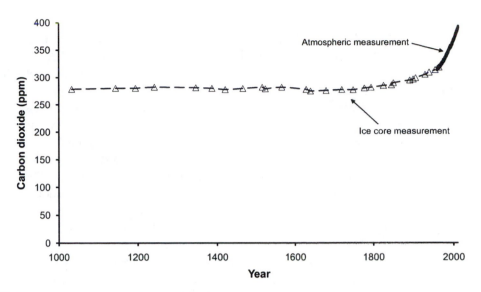

Fig. 1.5 The growth in atmospheric carbon dioxide. The graph shows measurements of air trapped in ice (ice core measurement,), and more recent direct atmospheric measurements taken in Hawaii (solid black line on the right).

20th century) there has been direct measurement of atmospheric carbon dioxide. Second, there has been historic measurement of the bubbles of air trapped in ice caps (mainly from either Greenland or Antarctica) as snow has fallen and then turned to ice. Together these show a continuous growth in the atmospheric concentration from around the time of the Industrial Revolution to the present (see Figure 1.5).

Much (but, remember, not all) of this growth in atmospheric carbon dioxide is due to the burning of fossil fuels. In essence this represents a short-circuiting of the aforementioned deep carbon cycle: the part of the carbon cycle that takes millions of years to complete. For while, as discussed above, much of the annual movement of carbon dioxide is due to respiratory and photosynthetic processes and contributes to the fast carbon cycle, there is also a slower accumulation of about 1 billion t of carbon a year in marine sediments and more being trapped terrestrially in soils and wetlands. After millions of years much of this ultimately ends up as coal and oil, the larger deposits of which we have only recently (in geological terms) mined and burned as sources of energy. We are burning these fossil fuels at a far faster rate, thousands of times faster, than they are currently being formed, for the process of fossil fuel formation continues today. It is as if there are two carbon cycles within the overall carbon cycle. One is driven by photosynthesis, respiration and forest fires and operates over a short period of time. The other is the deep carbon cycle operating in so-called deep time of many millions of years and is driven by the geological formation and entrapment of fossil fuels and the tectonic subduction of carbon-rich sediments at the edge of some plates followed by the emission of carbon from volcanoes.

There is currently much research being done to elucidate the carbon cycle, and to measure or infer carbon sinks (Houghton, 2002). This work might be grouped as follows.

1. Global budgets based on atmospheric data and models. These use data from nearly 100 sites around the Earth of atmospheric carbon dioxide and isotopes.
2. Global budgets based on models of oceanic carbon uptake. These use models of oceanic carbon cycle and chemistry linked to those of terrestrial and atmospheric sources.
3. Regional carbon budgets from forest inventories. Many developed nations have national forest inventories and changes in volumes over time can indicate sources or sinks of carbon. However, we are a long way from making this accurate, either on a national basis or for global coverage, although progress is being made.
4. Direct measurements of carbon dioxide above ecosystems. Using stand towers in forests it is possible to measure changes in carbon dioxide being given off or absorbed. (This is quite different from measuring atmospheric carbon dioxide on top of an Hawaiian volcano – one of the key measurement sites – combined with scores of other direct atmospheric measurements to obtain a hemispheric average.)
5. Earth system science modelling using ecosystem physiology with global models built up from global biome and ecosystem data. One of the big questions here (apart from the accuracy of the size of the ecosystem components) is whether all the important ecosystem processes have been included or properly quantified.
6. Carbon models based on changes in land use. This is related to item 5 and has similar constraints.

Having looked at the broad areas of research into the carbon cycle, this leads us on to the question of whether, because we are already altering one part of the carbon cycle so as to increase atmospheric carbon dioxide, we can alter another part to counteract this effect. Given that carbon dioxide is the principal anthropogenic greenhouse gas, and that the carbon cycle is the determining phenomenon in its atmospheric concentration, it would at least appear logical that we might alter the way carbon is currently cycled so as to offset atmospheric carbon increases. One way would be to increase terrestrial photosynthesis through planting new forests, thus sequestering carbon, and we will return to forests and biofuel options later (see the end of Chapter 7 and also some additional thoughts in Appendix 4). Another might be to increase marine photosynthesis.

Marine photosynthesis is mainly carried out by phytoplankton in the open oceans. The dominant species in the sea are the prokaryotes (organisms without internal structures surrounded by cell membranes) *Prochlorococcus* and *Synechococcus*. Both are cyanobacteria (also known as blue-green algae). In terms of crude numbers, *Prochlorococcus* is probably the most populous species on the planet. In addition to sunlight, carbon dioxide and water, these plankton species also require nitrates, phosphates and small amounts of metals. In the ocean, close to the surface, more than enough sunlight is present to drive photosynthesis, but it has been found (in parts of the Pacific Ocean at least) that raising the concentration of iron to about 4 nM (nanomoles per litre) results in planktonic blooms and associated increased photosynthetic production. The most dramatic of these experiments were the IronEx I (1993) and II (1996) experiments that covered an area of about 70 km^2, although an area larger than this (1000 km^2) had to be surveyed due to the blooms' drift.

This has led to speculation that it may be possible to use oceanic iron fertilisation to sequester atmospheric carbon. However, modifying the base of some of the planet's major ecosystems, such as in this way, may well carry with it unacceptable ecological risks. Furthermore, it is one thing for winds to carry a global load of minerals to fertilise the oceans and quite another for humans to do so. Indeed, it appears that the energy required to distribute the iron over the ocean surface would roughly equate in fossil fuel terms with the carbon assimilated. Even so, it does appear that in the past natural changes in the carbon cycle, almost certainly involving the marine component, have had a major effect on the global climate (Coale et al., 1996).

1.4 Natural changes in the carbon cycle

We know that atmospheric carbon dioxide plays a major role in contributing to the natural greenhouse effect and we also know that this natural greenhouse effect has varied in strength in the past. Perhaps the most pronounced evidence comes from Antarctic ice cores. Snow falls in Antarctica to form ice and in the process tiny bubbles of air from the atmosphere become trapped and sealed within it. As more snow falls, more ice with bubbles builds up. By drilling a core into the ice it is possible to retrieve atmospheric samples of times past. Indeed, cores at one spot, Vostok in eastern Antarctica, have provided an atmospheric record going back well over 100 000 years. This is a long enough time to cover the last glacial–interglacial cycle (and more; a glacial is the cooler part of an ice age, compared with the warmer interglacial, such as the one we are presently in; see Chapter 3). The ice at Vostok is well over 2 km thick and the cores retrieved between the mid-1980s and the present day have shown clearly that concentrations of carbon dioxide and methane were far lower during the cool glacials than they were during the warmer interglacials. Using the ice water's deuterium (^{2}H) concentration these palaeoconcentrations can be directly compared to the estimated difference in temperature between the oceans from which water at that time evaporated and when it fell as snow. This is because it takes more energy (heat) to evaporate water containing the heavier deuterium isotope of hydrogen (^{2}H) than water containing the common isotope of hydrogen (^{1}H). Therefore, a plot of the ice core's deuterium concentration gives an indication of regional temperature. Such temperature changes, it can be seen, closely correlate with carbon dioxide and methane concentrations (see Figure 1.6). Such ice-core evidence suggests that atmospheric carbon – carbon dioxide or methane – really is linked to climate. Because we know from laboratory analysis that these gases are greenhouse gases (absorbing long-wave infrared radiation), we can deduce that this is the mechanism linking them to climate. Equally importantly, because we know that the atmospheric concentrations of both these gases are affected, if not determined, by the carbon cycle, we have a direct link between the carbon cycle and climate. As one of the carbon cycle's key drivers is photosynthesis, we can see that life is clearly linked to the global climate.

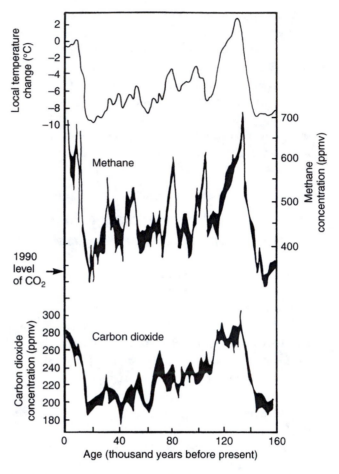

Fig. 1.6 Atmospheric carbon dioxide and methane palaeorecord for the past glacial–interglacial cycle plotted with regional temperature change, as indicated by the ice hydrogen isotope proxy. Adapted from Barnola et al. (1987) and Chappelaz et al. (1990), reproduced from the IPCC (1990) with permission.

1.5 Pacemaker of the glacial–interglacial cycles

There have been 20 or so of these glacial–interglacial cycles over the past 2 million years and, at first sight, they have an apparent regularity. The question then arises as to whether there is anything causal driving this periodicity.

In the 1920s the Serbian mathematician Milutin Milankovitch (or Milanković) – although owing a debt to the work of James Croll in 1864 – suggested that minute variations in the Earth's orbit could affect the sunlight reaching its surface. This happens largely because the Earth's geography is currently asymmetric: the northern hemisphere is land-dominated while the southern is ocean-dominated, so the hemispheres differ in the way they absorb the Sun's heat. Also at the present time the positioning (through plate tectonics) of the North and South American and the

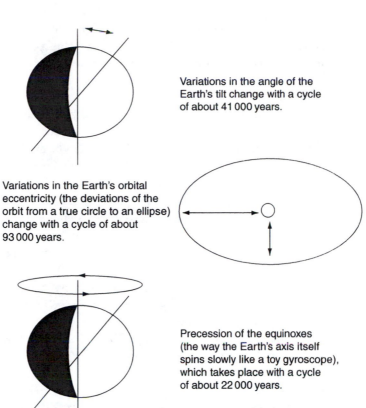

Variations in the angle of the Earth's tilt change with a cycle of about 41 000 years.

Variations in the Earth's orbital eccentricity (the deviations of the orbit from a true circle to an ellipse) change with a cycle of about 93 000 years.

Precession of the equinoxes (the way the Earth's axis itself spins slowly like a toy gyroscope), which takes place with a cycle of about 22 000 years.

Fig. 1.7 Milankovitch orbital parameters of eccentricity, axial tilt and precession of the equinoxes. Each parameter varies through a cycle of different period. Each affects the way the Earth is warmed by the Sun.

Afro-Eurasian continents restricts oceanic currents, and hence heat transport, about the planet. Oceanic currents are further restricted at the poles, in the north by the presence of the North American and Euro-asian continents and in the south by Antarctica. The way heat is accepted by and transported about the Earth under these constraints makes the planet prone to glacials. Milankovitch calculated a theoretical energy curve for changes over time based on the variations of three orbital parameters: angle of axial tilt, or obliquity, which varies between 22 and 24.6° over a 41 000-year cycle; orbital eccentricity (the degree of elliptical deviation from a true circle), which varies every 93–136 000 years (so, approximately 100 000 years) between a true circle (an eccentricity of 0) and an ellipse (of eccentricity 0.05) and back again (there is also a 400 000-year resonance); and precession of the equinoxes (the way the Earth's axis spins slowly like a gyroscope), which takes place with a cycle of roughly 19 000–23 000 years (effectively 22 000 years; see Figure 1.7). The climatic relevance of this last, put simply, means that at a specific time of year, say mid-summer's day, the Earth is closer or not to the Sun than on other mid-summer days. Using these factors Milankovitch could predict when the Earth was likely to experience a cool glacial or a warm interglacial (Milankovitch, 1920).

Milankovitch concluded that a drop in the amount of sunlight falling on the northern hemisphere at the end of both the precession and tilt cycles would make it more likely for there to be a glacial and that when the opposite happened it would coincide with the timing of an interglacial. For many years Milankovitch's theory did not have much currency. This was due to two main things. First, for much of this time there was no real understanding of when in the past glacials and interglacials had actually taken place, so the theory could not be checked. Second, the variations in the solar energy falling on a square metre in the northern hemisphere that Milankovitch was talking about were of the order of 0.7 W m^{-2}; in other words less than one-tenth of 1% of the sunlight (the solar constant is close to 1400 W m^{-2}) bathing the planet. What turned things around was the palaeo-evidence from Antarctica's ice cores in the 1970s and 1980s, showing when glacials and interglacials had taken place. This confirmed Milankovitch's timings and, as we shall see below, there is a considerable body of other biotic evidence corroborating the timing of past climates and climatic change.

If Milankovitch's theory simply provides a glacial–interglacial pacemaker but does not account for sufficient energy changes needed to instigate and terminate glacials, then what is amplifying this signal? The answer lies in the complexity of the global biogeosphere system (from here on simply referred to as the biosphere system). There are numerous factors determining the global climate. Some, such as silicate erosion (see Chapter 3), affect the planet over long timescales. Others, such as the burning of fossil fuels (a major factor), stratospheric and tropospheric ozone (medium factors) and biomass burning, mineral-dust aerosols and variations in the Sun's energy output (very low factors), affect the climate on timescales of far less than a century. Other factors we still know little about and so their climatic effects are hard to quantify (such as aircraft condensation trails, or 'con trails'; see Chapter 5). Complicating matters further still, there are many factors that conspire, or interact synergistically, to affect the climate with positive or negative feedback. These feedbacks either amplify climate change or have a stabilising effect. This text focuses on the biology of climate change but it is important for life scientists interested in climate change to have at least a basic appreciation that such feedbacks exist. Figure 1.8 illustrates three such feedbacks (there are many). Figure 1.8a and 1.8b are physical systems and might operate on a lifeless planet. These are both examples of positive feedback that add to, reinforce or amplify any forcing of the climate. That is to say, if something (be it the release of a greenhouse gas, either human-made or natural) forces the climate to warm up, then these feedbacks will serve to amplify the net warming.

Figure 1.8c represents a biophysical feedback system of a different kind. This is an example of a negative-feedback system that dampens any net change in the climate. We have already referred to iron that can fertilise the oceans, which allows more algae to grow that in turn draws down carbon dioxide from the atmosphere, so reducing the greenhouse effect and cooling the planet. Take this a step further. Consider a world slightly warmer than ours. Being slightly warmer there is more evaporation from the oceans; more evaporation means more rainfall (and/or snow), which in turn means more geological erosion. This increased erosion increases the amounts of iron (eroded from minerals) transported to the oceans that in turn encourages algae, which draws down atmospheric carbon dioxide. Of course, the timescale and

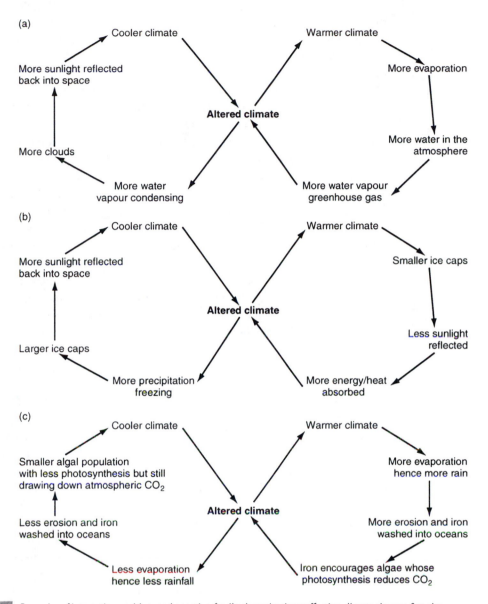

Fig. 1.8 Examples of interacting positive- and negative-feedback mechanisms affecting climate change. See also Figure 4.13 for an example of another feedback cycle: there are many.

magnitude of this natural effect may not be as large as some might wish, hence the discussion as to whether we should deliberately fertilise the oceans with iron or take some other measure (although it would be unwise to tinker with the planet's biosphere mechanisms without a thorough understanding of them). Given that there are feedback processes that amplify change and those that stabilise the climate, it is not surprising that changes in the global climate are not always gentle. For example, we do not see a gentle segue from a glacial to an interglacial. Instead, we see a sharp transition between the two (see Figure 1.6). It is as if positive feedback encourages sharp changes while negative feedback encourages stable states and that either one or other of these two types of feedback dominates at any given time. The combined global climate picture is one of stable (or semi-stable) states between which there is occasional rapid flipping. One of the main pacemakers timing these flips is the combination of Milankovitch's orbital parameters.

Two questions arise from all of this. The first is whether, with current global warming, the Earth is now shifting towards a new feedback system that may encourage further warming. One example of a mechanism that might drive this is soil carbons, especially at high latitudes, and whether it may be released, through warming, into the atmosphere. Such soils include peatlands that are at so high a latitude that they are either frozen for part of the year or are permafrosts. Michelle Mack from the University of Florida and colleagues (Mack et al., 2004), including those from the University of Alaska Fairbanks, have looked at carbon storage in Alaskan tundra. Carbon storage in tundra and boreal soils is thought to be constrained by carbon–nutrient interactions because plant matter is the source of much (nearly all) soil carbon and plant growth is usually nitrogen-limited. Should soils warm in response to climate change, it is thought that nutrient mineralisation from soil organic matter will increase. This should increase plant growth. However, total-ecosystem carbon storage will depend on the balance between plant growth (primary productivity) and decomposition. Experiments at lower latitudes (in temperate and tropical zones) have in the past given variable results (although we will cite a major – albeit rough – assessment of European soil carbon in Chapter 7), but high-latitude ecosystems, because of the large amount of carbon in their soil, show a clearer relationship between productivity and soil carbon storage. In 1981 Mack and colleagues began one of the longest-running nutrient-addition experiments in Alaska by adding 10 g of nitrogen and 5 g of phosphorus m^{-2} $year^{-1}$. This is about five to eight times the natural deposition rate in moist acidic tundra soils. Two decades later poaceden or the graminoid (grass) tundra – which is dominated by the tussock-forming sheathed cottonsedge (*Eriophorum vaginatum*) – had changed to a shrub tundra dominated by the dwarf birch (*Betula nana*). The carbon above ground (in the form of plants and litter) had increased substantially; however, this was more than offset by a decrease in carbon below ground in the soil. This decrease in below-ground carbon was so great that the net result for the ecosystem was a loss of 2000 g of carbon m^{-2} over 20 years. (The ecosystem's nitrogen did not change nearly so much, other than that a greater proportion of it at the end of the experiment was found above ground in the vegetation.) This loss of carbon was approximately 10% of the initial carbon in the ecosystem, and much will have entered the atmosphere as the greenhouse gases methane and carbon dioxide.

Another approach to carbon loss from soils is soil respiration, the flux of microbially and plant-respired carbon dioxide from the soil to the atmosphere. In 2010 Ben Bond-Lamberty and Allison Thomson of Maryland University in the USA conducted a meta-analysis of the literature on soil respiration. They obtained 1434 data points from 439 studies conducted worldwide and matched these with climate data. They found that air temperature was positively correlated with soil respiration. Increasing soil respiration globally with temperature does not necessarily constitute a positive climate feedback as there could be higher inputs of carbon into soil (through plant growth). Nonetheless, their results are consistent with an acceleration of the terrestrial carbon cycle in response to global climate change.

The fear is that as, globally, soils hold 1500 GtC (compared with some 750 GtC in the atmosphere) and that as about a third of soil carbon is in arctic and boreal soils, warming of these soils might make a significant contribution to atmospheric carbon. Should this occur, this carbon contribution would be beyond what is being added to the atmosphere through human actions such as fossil fuel burning and land-use change. It would therefore exacerbate current global warming.

Returning to the aforementioned variable results of carbon-release change with temperature in temperate and tropical soils, matters may be explained by considering soil carbon to be in various forms with different turnover rates. A three-carbon-pool model, with each pool having differing turnover times, was proposed in 2005 (Knorr et al., 2005). This was then applied to data from 13 previously published soil-warming experiments covering tropical and temperate soils that lasted between about 100 days and 2 years. It gave somewhat varying results, but importantly this model was compatible with earlier work. What appears to be happening is that the pools of carbon with a faster turnover mask the effect of pools with slower and larger rates of turnover. This model also suggests that higher carbon release from warmed soils might continue over a number of decades. This last has yet to be tested, but we may get the chance to find out as the Earth continues to warm up.

To put the Alaskan experiment into context of current increases in atmospheric carbon dioxide and existing carbon pools, high-latitude warming could at some stage further increase the current rate of (largely fossil fuel-driven) increase in atmospheric carbon dioxide by between half as much again and double, for a period of two decades or more. However, the exact global effect of climate change on soil carbon is uncertain but (as we shall shortly see) progress is being made. Nonetheless, the key point here is that warming would itself enable the release of more tundra soil carbon, which would fuel further warming, and so on (this is another positive-feedback cycle). The three-pool carbon soil model, applied to real experimental results, suggests that increases in soil carbon release with temperature rises also apply to temperate and tropical soils and not just to high-latitude (and high-carbon) boreal soils. As we shall see in Chapter 7, this could undermine the policy proposals of temperate nations to use soils as carbon sinks to offset atmospheric carbon dioxide increases from fossil fuel burning. For now it is worth noting that soil carbon has a feedback to atmospheric carbon dioxide mediated by temperature change, even if the precise strengths of this feedback have yet to be discerned.

In addition to the issue of whether, with current global warming, the Earth is now shifting towards a new feedback system, the second key question is how much more carbon will be released into the atmosphere with warming? In other words, what is the magnitude of the climate sensitivity of the global carbon cycle?

In 2010 a team of researchers led by David Frank, based in Swiss and German institutes, looked at the global climate together with atmospheric carbon dioxide concentrations between the 11th and 18th centuries. They chose this time period because it was before the Industrial Revolution and so humans were not adding carbon dioxide to the atmosphere through intensive fossil fuel burning, and wholesale clearance of tropical rain forests also had yet to begin. During this time atmospheric carbon dioxide only changed a little (see Figure 1.5) and these small changes (if not anthropogenic) may therefore have been due to global ecosystem responses to temperature. They estimated that the sensitivity of the global carbon cycle was around 7.7 ppm CO_2 per degree Celsius (ppm CO_2 $°C^{-1}$) warming with a likely range of 1.7–21.4 ppm CO_2 $°C^{-1}$ warming. This is a less than previous estimates of around 40 ppm CO_2 $°C^{-1}$ warming. Although we cannot discount the earlier estimates completely until this more recent work has been verified in other ways, it does suggest that amplification of other warming mechanisms is important (possibly ones such as those, among others, given in Figure 1.8).

So much for the broad picture of interacting feedback cycles and flips between semi-stable states. Not surprisingly, climatologists continually monitor research into these positive- and negative-feedback systems. Indeed, climate-modelling research has received huge investments in recent years, especially since the mid-1980s, when computer technology became sufficiently sophisticated to accommodate models of adequate complexity to provide tolerably useful (or at least interesting) output. The models are good in that they do broadly reflect the global climate, but they (currently) lack detail, both spatially and thermally, and there have been problems with some outputs that simply do not tie up with what we know (see Chapter 5). For example, in the 1990s global models were not particularly good at portraying climates at high latitudes (which are warming far faster than models predict), whereas the models of the early 21st century include more developed biological components but still have a long way to go. Another more recent example is that ocean dynamics surprisingly do not seem necessary to model El Niño event timing, although they do seem relevant for El Niño strength (Clement et al., 2011). Furthermore, most climate models erroneously predict the existence of an Intertropical Convergence Zone (ITCZ) – a band near the equator where the prevailing winds from the two hemispheres converge – in the South Pacific, in addition to the real one observed north of the Equator: a problem known as double ITCZ bias. However, computer models are continually getting better as they are built with greater resolution (both vertically and horizontally) and incorporate more of the features and processes operating on Earth. Indeed, as previously mentioned, since the late 1990s programmers have included biological processes in their models, so continuing the trend of being able to increasingly match model outputs with expectations based on reality. Biologists and geologists have much to contribute, for what has taken place climatically is frequently recorded biologically and preserved geologically. Not only do different species live under different climatic regimens but species are affected by climate and these influences can be laid down in ways that are long-lasting (for example, tree rings, to which we will return in Chapter 2).

Prior to the 1980s we had such a poor understanding of the way that the global climate operates that there was great uncertainty as to whether the Earth was warming or cooling. Indeed, when the first ice cores were analysed in the 1970s it was realised that the last glacial period had lasted for roughly 100 000 years, whereas the previous interglacial was just 10 000 years long. Furthermore, the change between the two was sudden and not gradual. Consequently, in the 1970s, given that our current interglacial (the Holocene; the time since the end of the last glacial to the present) has already lasted about 11 700 years, there was for a while genuine concern that a new glacial was imminent and that this global cooling might even threaten civilisation. (This concern was fuelled further by sulphur aerosol pollution in the early and mid-20th century.) More research was required into the strengths of the various factors influencing (that is, forcing) climate. It was realised in the 1980s that, on balance, the factors were warming our planet. Factors such as the current human generation of greenhouse gases were adding to natural warming processes and so were greater than the various cooling factors. The question that remained was how great a warming could we expect from our fossil fuel generation of carbon dioxide and how would this compare against the current range of other climatic factors? This was the subject of the first IPCC report, published in 1990.

1.6 Non-greenhouse influences on climate

Milankovitch variation and the changing sunlight-reflecting properties (called albedo) of ice caps (and indeed the solar-reflecting differences of any surface) demonstrate that non-greenhouse considerations do play a part in climate change. They can, through feedback cycles, either increase or decrease the magnitude of change. They can also superimpose their own variability imprint on climate.

However, it is now accepted by nearly all within the climate community that the anthropogenic addition of greenhouse gases is the key factor determining current global warming. But this does not mean that these other, non-greenhouse, factors are not taking place and have no effect on climate.

One such factor in particular has caused some controversy, although it is generally accepted to have a minor effect compared to current anthropogenic climate change. This is the variation in solar output, or changes in the Sun's intensity, which has had a major part to play over the three billion years or so of evolution of the Earth's biosphere. As we shall see in Chapter 3, the Sun, being a main-sequence star, is growing significantly warmer over a scale of hundreds of million to billions of years. This has considerable implications when it comes to elucidating the evolution of the biosphere. However, on shorter timescales of hundreds or thousands of years it is comparatively (but not entirely) stable. On even shorter timescales the 11-year sunspot cycle has an impact. But this last effect is small, causing a variation in irradiance of only 0.08%, which is too small to have much effect: such changes may affect the global climate by 0.02–0.4°C. Changes in sunspot activity do seem to tie in with similar patterns of change in global temperature, but these are superimposed on larger climate changes determined by other factors (Foukal et al., 2004).

The question of whether the Sun is largely responsible for current global warming came about because the Sun became slightly more active during the 20th century. During much of this time the Earth's temperature rose and fell almost simultaneously with changes in solar activity. There is also the question of the so-called Little Ice Age and solar activity, and we will return to this in Chapters 4 and 5.

Since 1978 we have been able to take space-borne measurements of solar output and correlate them with sunspot activity, for which we have previously only had an observational record going back a few centuries. The relationship is not clear but there *is* a relationship. However, there is also another larger, longer-term (over a timescale of centuries) solar component, the exact historic magnitude of which is a matter for speculation but which is roughly five times that of the solar variation reflected by sunspot activity. Furthermore, there does seem to be a correlation between global temperature and solar output (as we shall see in Chapter 4, which might account for the Little Ice Age) but is this relationship real, partial (with other factors determining climate), weak or just a coincidence?

In 2004 a team of five European researchers from Germany, Finland and Switzerland used the carbon isotope ^{14}C from tree rings going back 11 000 years (Solanki et al., 2004). It is usual to think of ^{14}C as a means of dating objects using radioactivity, because this isotope is produced in the upper atmosphere at *roughly* a constant rate. The ^{14}C isotope is then absorbed into living things and begins to decay slowly at a known exponential rate. So the amount of ^{14}C left compared to the stable ^{12}C enables one to date objects, albeit with a certain amount of error. Yet, ^{14}C is not produced at an *exactly* constant rate. Carbon-14 is produced in the upper atmosphere due to the action of cosmic rays from the Sun on nitrogen and carbon atoms; the level of cosmic rays is an indication of the Sun's output. Counting tree rings from overlapping samples of wood enables each ring to be dated in a different, more accurate, way to carbon dating. It is therefore possible to deduce the amount of ^{14}C produced compared to what would have been produced if solar activity was constant. The research team's ^{14}C-determined calculation of solar output was corroborated by ^{10}Be (a beryllium isotope) from Antarctic and Greenland ice cores, as this isotope also relates to solar output. The researchers found that there was indeed unusually high solar activity at the end of the 20th century and that this would have certainly contributed to some of the global warming experienced then. However, it could not account for it all and was 'unlikely to be the prime cause' (Solanki et al., 2004). Given that it has been so controversial early in the 21st century, should you want more detail a good review of this topic is given by Foukal et al. (2006).

What seems to be happening is this: there are many factors affecting the climate. Some 'force' it in a positive (warming) way and others in a negative way. Furthermore, some climate forcing factors are strong and some weak. Greenhouse gases are strong forcers. Variations in the Sun's output over tens of thousands of years do occur but are comparatively small. Their effects may be superimposed on the climate change that is determined by the sum total of all other forcing agents and as such they may account for small changes in the climate (and possibly even the Little Ice Age). However, it is difficult for small climate forcing (such as the small increase in 20th-century solar output) to account for the large temperature changes measured. Remember, the Milankovitch variations in energy reaching the northern hemisphere in the summer are small. As we shall see in Chapters 3 and 4, these small changes in solar radiation help trigger larger changes in carbon dioxide and methane as the Earth moves between glacial and interglacial climate modes. One should not be surprised that greenhouse gases are not the only positive climate forcing factors contributing to current warming. There are others, both positive and negative, and increases in solar output are but one. These others include volcanic activity, marine release of methane and (with regional effects on climate) oceanic and atmospheric circulation. We shall return to them later in the book. Although these other factors can play an important part in some climate change, they are not the dominant factors of current global warming. Even so, they cannot be easily dismissed because, as we shall see, these factors could make a serious contribution to climate change in the future. In particular, circulation changes can help flip climate regimens between semi-stable states.

1.7 The water cycle, climate change and biology

Something noted above, and to which we will occasionally return later in this book, is that in a warmer world we could reasonably expect more evaporation from the ocean. Again, as noted, this expectation is at least in part corroborated by computer models of atmospheric water vapour as well as satellite observations: so, global warming affects the water cycle. This leads us to another expectation that we would reasonably anticipate from more evaporation, that there would be more precipitation (rain and snow) and thus increased river flow. But how reasonable are these expectations?

The Earth's climatic system is complex. Furthermore, biology is not just affected by a changing climate; biology, as we shall see, plays a key part in affecting the nature of climate change and in affecting some of the consequences of that change. Not all these complications are biological, but the complexities need to be taken into account.

First of all, although a warmer world will lead to more evaporation (other factors, such as the complete solar spectrum, remaining constant), more evaporation does not in itself necessarily mean more precipitation. To take an extreme example, Venus is a far warmer planet than the Earth and its atmosphere imparts a far more power-ful greenhouse effect. Indeed, Venus is so warm that water exists solely in the form of water vapour. There is no rain to soak the ground on Venus and there are no oceans. Second, although a warmer world results in more ocean evapora-tion, if, hypothetically speaking, part of this extra water vapour does return as a partial increase in precipitation this does not mean that river flow would necessarily increase. It could be, hypothetically, that the increased evaporation means that the excess precipitation evaporates before it reaches the rivers (we shall return to this in Chapter 6). Alternatively, it may be that rainfall would increase mainly over the oceans, so leaving river flow unaffected. Third, other factors and indeed biological processes may well play their part. So it is important to identify these factors (or as many as possible) and to try to quantify them.

Climate models and satellite observations both suggest that the total amount of water in the atmosphere will increase at a rate of 7% per degree Celsius (% $°C^{-1}$) of surface warming. Yet climate models have predicted that global precipit-ation will increase at a much slower rate of 1–3% $°C^{-1}$. However, a 2007 analysis by Frank Wentz and colleagues of satellite observations carried out between 1987 and 2006 does not support this prediction of a milder response of precipitation to global warming. Rather, the observations suggest that precipitation and total atmospheric water have increased with warming at about the same rate (around 6% $°C^{-1}$) over the past two decades, even if not all of the extra evaporation falls back as rain (some may stay in the atmosphere). Having said this, Wentz's team were aware that their results were preliminary and that more work is needed in this area.

Time for a quick reality check. First, as already noted, the global climate is an average made up of regional and annual variations. With a changing global climate not only will some climatic components change in some places, and at some times more than others, but it is possible that in some places (and/or times) there may be a change in the direction of the overall trend. (This has been seized upon by some, especially in the 1990s, to argue that global warming is not taking place. Such arguments are based on selective and atypical data, with the fallacious claim that the data are typical.) Second, there are other factors operating that are not related to greenhouse gases. Temperature is not the sole mechanism behind evaporation: direct sunlight (electromagnetic radiation of appropriate frequency) also plays a part. A photon of sunlight can excite a water molecule, causing it to leave its liquid-state companions and become vapour. As we shall see later in this chapter changes in the amount of sunlight reaching the Earth's surface have been happening and have resulted in so-called global dimming due to pollution particles. Another factor affecting water becoming water vapour is biology.

The routes from liquid water back to water vapour are not restricted to straight-forward evaporation but also include plant transpiration, as part of photosynthesis in terrestrial plants. For this reason evapotranspiration (the total water loss from an area through evaporation and vegetation transpiration) is important. Now let us return from theory to reality.

A warmer world due to increased greenhouse gases will, among other things, affect plant physiology. Plants exchange gas and water with the atmosphere through openings on the surface of leaves called stomata. Stomata open and close to reg-ulate photosynthesis in the short term. (In the longer term stomatal densities have been shown to vary on more geological timescales with atmospheric carbon dioxide concentrations; see Chapter 2.) In a warmer, more carbon dioxide-rich world with higher rainfall that serves to enhance photosynthesis, all other things being equal, we might expect plant homeostatic processes (physiological mechanisms acting to keep functions stable) to dampen photosynthesis. If this were happening then we would reasonably expect plant transpiration to decrease. This in turn would lower a plant-covered water catchment's evapotranspiration and so more water would remain in the ground to percolate through to streams. River flow would increase. So much for theoretical expectations. The question then becomes how do the various factors of changed precipitation, warmth, plant physiology and river flow interact?

It is possible to model the individual processes. We do have climate records of temperature and precipitation covering many decades and measurements of solar radiation reaching the Earth's surface, as well as those of river flow. We also know about plant physiology and so can apply broad parameters to plant physiology over a region, and include factors such as deforestation and land use. In short, we know both actual river flow for principal catchments on each of the continents and, broadly, how the various factors that contribute to river flow have changed over the last century. A model of 20th-century continental water run-off has been constructed that reflects actual river-flow measurements. It is then possible to examine the model conducting a 'sensitivity analysis' (or 'optimal fingerprinting' or 'detection and attribution'; the nomenclature varies with research group) to vary just one factor at a time and to see, using statistics, how this causes the predicted run-off to differ from reality. It is known that 20th-century climate change alone is insufficient to account for run-off changes. However, such an analysis of a surface run-off model has indicated that including the suppression of carbon dioxide-induced stomatal closure makes the model's outcome consistent with actual run-off data (Gedney et al., 2006; Matthews, 2006). In short, biology plays a key part in controlling the water cycle to such a degree that without it climate change and other non-biological factors cannot fully explain water-cycle trends.

Yet, as we shall see throughout this book, there is much that is unexpected in the detail of climate and its interactions with the biosphere. For example, as mentioned, in a warmer world we would expect more ocean evaporation, hence more rainfall and more plant evapotranspiration. This is not always so, even if the underlying trend is one of a warmer world leading to more ocean evaporation. In 2010 a large international team involving many research departments, and led by Martin Jung, provided a data-driven estimate of global land evapotranspiration from 1982 to 2008, compiled using a global monitoring network, and meteorological and remote-sensing observations. In addition, they assessed evapotranspiration variations over the same time period using an ensemble of process-based land-surface models. Their results suggest that global annual evapotranspiration increased as expected. On average the increase was by 7.1 ± 1.0 mm per year per decade from 1982 to 1997. All well and good, but after that – coincident with the last major El Niño event in 1998 – the global evapotranspiration increase seems to have ceased until 2008. This change was driven primarily by moisture limitation in the southern hemisphere, particularly Africa and Australia. In these regions, microwave satellite observations indicate that soil moisture decreased from 1998 to 2008. Hence, increasing soil-moisture limitations on evapotranspiration largely explain the recent decline of the global land evapotranspiration trend. Not all the extra ocean evaporation in a warmer world results in increased rain uniformly distributed over the land. So, whether the changing behaviour of evapotranspiration is representative of natural climate variability or reflects a more permanent reorganisation of the land water cycle is a key question for Earth system science.

As we shall see again and again throughout this book, climate and biology are connected: they affect each other. Furthermore, humans, as a biological species that also relies on and which is in turn affected by other biological species, are caught in the middle of this climate–biology dynamic. That human action is also a significant driver of current climate change further complicates matters. We need to understand both how our species affects the global climate and the climate–biology dynamic if our growing population is to survive without a decline in either well-being or environmental quality.

1.8 From theory to reality

In the above review of the causes of climate change it can be seen that climate theories, whether of greenhouse climatic forcing or Milankovitch's orbital effects of incoming solar energy, have to be validated. This is done by comparing theory with real data. What is happening can be measured today, although care is needed. For example, satellite remote-sensing devices need proper calibration and we need to know what their data actually mean. (I cite this example because in the 1990s incorrect assumptions were made about some satellite data.)

The past 10 years have seen tremendous progress in understanding climate change on two fronts. First, there has been a steady improvement in computer models. This improvement continues but there is still a long way to go, both in terms of reducing uncertainty across the globe and in terms of spatial and temporal resolution (see Chapter 5). These inadequacies mean that even the best models can only present a broad-brush picture of a possible likelihood. The 2001 IPCC report provides some good illustrations of the limitations. For example, it presents (p. 10) two global models of projected changes in precipitation run-off from the 1960s to 2050. Both show decreases in run-off in the Amazon basin, much of the rim of Australia, and Central Europe, and run-off increases in south-east Asia, north-west Canada and southern Alaska. At the moment this does seem to tie in with *some* observations (IPCC, 2001b). This similarity of computer-model output with reality (even though we have yet to reach 2050) is not proof of the various models' accuracy but it does lend them credibility: one corroborates the other. However, the same two models differ in that one shows a marked decrease in run-off in the eastern USA and the other an increase. There are also marked contrasting differences in run-off projected for much of the Indian subcontinent and much of north-west Europe. Such problems are not trivial even if progress is continually being made, with models becoming ever more sophisticated and now increasingly including biological dimensions.

Second, there has been a huge growth in understanding of how the natural world reacts to climate change, be it in terms of a single organism or an ecosystem. And so a palaeorecord provides us with a proxy as to what might have really happened in one place at one time under a different climatic regimen. This is, if you will, just one pixel in a bigger picture of the Earth's climate history. But the past decade has seen much building up of records covering many periods of time and places. Such records can help reduce uncertainty and inform us. This is particularly useful because – although it is difficult to build up a picture, as if pixel by pixel, from biological and other indicators of past climates – such a picture reflects something far better than a computer model of the Earth: the Earth itself. Indeed, if at some time the Earth was warmer than today we can use climatic proxies (be they biological or chemical, laid down in the geological record) to determine what the planet was like and use that as an analogue in an attempt to predict what it would be like in a globally warmed future.

However, with this success the past decade has also seen problems. Policy-makers in particular have put much faith in computer models. Some bioscientists on the other hand have found it difficult to relate their work on one or a group of biological palaeoclimate proxies with others. Finally, communication between these modellers and bio- and geoscientists might arguably have been better. A 2001 editorial in *Nature* (v411, p. 1) summed it up thus:

> To get the best return on [climate change research] funding, there must be strong interdisciplinary collaboration between those taking data from sea sediments, ice

and land, and also between those making measurements and those developing models. But above all, the community must deliver more by way of systematic deposition of its data.

In making the most of data, funding agencies, journals and researchers themselves all have a role to play. There seems to be too little awareness by researchers of what is admittedly something of a maze of publicly supported databases. And too often, as researchers will readily complain, trying to extract numbers in a large data set from the originators of published work is like pulling teeth.

This is not to decry much of the very good work that has been, and is being, carried out. The purpose of this book is simply to provide bioscience and other students (not to mention interested members of the public) with a basic, broad-brush picture as to how climate affects biology (and vice versa). For others this book will serve as a background against which they can set their own more specialist work, with reference to human ecology and likely future climate change. However, before this can be done, it will be useful to review some of the key biological indicators of past climate change. This forms the subject of the next chapter.

1.9 References

Arrhenius, S. (1896) On the influence of carbonic acid in the air and upon the temperature of the ground of the absorption and radiation of heat by gases and vapours, and on the physical connection of radiation, absorption and conduction. *Philosophical Magazine*, 41, 237–76.

Barnola, J. M., Raynaud, D., Korotkevitch, Y. S. and Lorius, C. (1987) Vostokice-core provides 160,000-year record of atmospheric CO2. *Nature*, 329, 408–14.

Bond-Lamberty, B. and Thomson, A. (2010) Temperature-associated increases in the global soil respiration record. *Nature*, 464, 579–82.

Callendar, G. S. (1938) The artificial production of carbon dioxide and its influence on temperature. *Quarterly Journal of the Royal Meteorological Society*, 64, 223–37.

Callendar, G. S. (1961) Temperature fluctuations and trends over the Earth. *Quarterly Journal of the Royal Meteorological Society*, 87, 1–12.

Chappelaz, J., Barnola, J. M., Raynaud, D., Korotkevich, Y. S. and Lorius, C. (1990) Ice-core record of atmospheric methane over the past 160,000 years. *Nature*, 345, 127–31.

Clement, A., DiNezio, P. and Deser, C. (2011) Rethinking the ocean's role in the Southern Oscillation. *J. Climate*, 24, 4056–72.

Coale K. H., Johnson, K. S., Fitzwater, S. E. et al. (1996) A massive phytoplankton-bloom induced by an ecosystem-scale iron fertilization experiment in the equatorial Pacific Ocean. *Nature*, 383, 495–508.

Foukal, P., North, G. and Wigley, T. (2004) A stellar view on Solar variations and climate. *Science*, 306, 68–9.

Foukal, P., Fröhlich, C., Spruit, H. and Wigley, T. M. L. (2006) Variations in solar luminosity and their effects on the Earth's climate. *Nature*, 443, 161–6.

Fourier, J. (1824) Remarques générales sur la température du globe terrestre et des espaces planétaires. *Annales de chimie et de physique*, 27, 136–67.

Fourier, J. (1827) Mémoire sur les températures du globe terrestre et des espace planétaires. *Mémoires de l'Académie Royal des Sciences*, 7, 569–604.

Frank, D. C., Esper, J., Raible, C. C. et al. (2010) Ensemble reconstruction constraints on the global carbon cycle sensitivity to climate. *Nature*, 463, 527–30.

Gedney, N., Cox, P. M., Betts, R. A., Boucher, O., Huntingford, C. and Stott, P. A. (2006) Detection of a direct carbon dioxide effect in continental runoff records. *Nature*, 439, 835–8.

Hood, E., Fellman, J., Spencer, G. M. et al. (2009) Glaciers as a source of ancient and labile organic matter to the marine environment. *Nature* 462, 1044–7.

Houghton, R. A. (2002) Terrestrial carbon sinks – uncertain explanations. *Biologist*, 49(4), 155–60.

Intergovernmental Panel on Climate Change (1990) *Climate Change: the IPCC Scientific Assessment*. Cambridge: Cambridge University Press.

Intergovernmental Panel on Climate Change (1995) *Climate Change 1995: the Science of Climate Change – Summary for Policymakers and Technical Summary of the Working Group I Report*. Cambridge: Cambridge University Press.

Intergovernmental Panel on Climate Change (2001a) *Climate Change 2001: the Scientific Basis – Summary for Policymakers and Technical Summary of the Working Group I Report*. Cambridge: Cambridge University Press.

Intergovernmental Panel on Climate Change (2001b) *Climate Change 2001: Impacts, Adaptation, and Vulnerability – Summary for Policymakers and Technical Summary of the Working Group II Report*. Cambridge: Cambridge University Press.

Intergovernmental Panel on Climate Change (2007) *Climate Change 2007: The Physical Science Basis – Working Group I Contribution to the Fourth Assessment Report of the Intergovernmental Panel on Climate Change*. Cambridge: Cambridge University Press.

Jung, M., Reichstein, M., Ciais, P. et al. (2010) Recent decline in the global land evapotranspiration trend due to limited moisture supply. *Nature*, 467, 951–4.

Keppler, F., Hamilton, J. T. G., Brab, M. and Rockmann, T. (2006) Methane emissions from terrestrial plants under aerobic conditions. *Nature*, 439, 187–91.

Kiendler-Scharr, A., Wildt, J. and Maso, M. D. (2009) New particle formation in forests inhibited by isoprene emissions. *Nature*, 461, 381–4.

Knorr, W., Prentice, I. C., House, J. L. and Holland, E. A. (2005) Long-term sensitivity of soil carbon turnover to warming. *Nature*, 433, 298–304.

Kuo, C., Lindberg, C. and Thomson, D. J. (1990) Coherence established between atmospheric carbon dioxide and global temperature. *Nature*, 343, 709–13.

Mack, M. C., Schuur, E. A. G., Bret-Harte, M. S., Shaver, G. R. and Chapin, III, F. S. (2004) Ecosystem carbon storage in arctic tundra reduced by long-term nutrient fertilization. *Nature*, 431, 440–3.

Matthews, D. (2006) The water cycle freshens up. *Nature*, 439, 793–4.

Milankovitch, M. (1920) *Théorie Mathématique des Phénoménes Thermiques Produits par la Radiation Solaire*. Paris: Gauthier-Villars.

Rice, A. (2010) Emission of anaerobically produced methane by trees. *Geophysical Research Letters*, 37, L03807.

Sawyer, J. S. (1972) Man-made carbon dioxide and the greenhouse effect. *Nature*, 239, 23–6.

Soden, B. J., Jackson, D. L., Ramaswamy, V., Schwarzkopf, M. D. and Huang, X. (2005) The radiative signature of upper tropospheric moistening. *Science*, 310, 841–4.

Solanki, S. K., Usoskin, I. G., Kromer, B., Schussier, M. and Beer, J. (2004) Unusual activity of the Sun during recent decades compared to the previous 11,000 years. *Nature*, 431, 1084–7.

Tyndall, J. (1861) On the absorption and radiation of heat by gases and vapours, and on the physical connexion of radiation, absorption and conduction. *The London, Edinburgh, and Dublin Philosophical Magazine and Journal of Science, Series 4*, 22(146), 169–94 and 22(147), 273–85.

Wentz, F. J., Ricciardulli, L., Hilburn, K. and Mears, C. (2007) How much more rain will global warming bring? *Science*, 317, 233–5.

If palaeoclimatologists had their wildest dreams come true then they would undoubtedly travel back in time to make meteorological recordings before humans existed, let alone thermometers. If only life would be so obliging. Instead, preserved indications of past climate have to be identified, calibrated and then used to infer usually just one aspect of the climate, and frequently just at one locality. Such a preserved indication provides us with a single proxy measurement.

Many measurements are needed from many proxies to build up a climatic picture of the past. Indeed, there are also many types of climate proxies and the principal ones will be reviewed in this chapter. The products of proxy analyses are often termed palaeoclimatic data. Palaeoclimatic data commonly come with considerable uncertainty in one or more of the following regards: the inferred temperature, the inferred time and the space to which it applies. It is important to emphasise this and to recognise that most proxy data rarely ever give a direct measure of a single meteorological parameter. For example, tree-ring thickness is affected by numerous factors, not just climate. These include: the aspect in which the tree is found, be it in a valley or on a hillside, and whether it is south- or north-facing; soils, which affect nutrients; and local geology, which affects water availability.

As said, there are many palaeoclimatic proxies. Not only do they each come with their own vagaries as to their relationship with climate, but the passing of time also blurs information. For example, a sediment may contain an isotope or a biomolecule that relates to temperature, but the sedimentation process itself may not have been uniform with time, or the sediment may itself be re-worked by biological and/or geological processes (such as burrowing organisms and/or strata movement, respectively). It is as if this book not only contained information in the form of words on the page, but that the type of paper and even the page order had changed between the time of my writing it to you reading it. An example of this sediment re-working is known as smearing, which is the process of fossils being transported downwards into older sediments (backward smearing) or upwards into younger sediments (forward smearing).

Another example of how external factors can affect the accuracy of proxy dating is exemplified in ^{14}C dating analysis. The radioactive ^{14}C isotope is formed in the upper atmosphere by the action of cosmic rays (mainly from the Sun) on nitrogen and carbon atoms in the atmosphere. Broadly, new ^{14}C atoms are produced at a rate of 2 atoms cm^{-3} s^{-1}. This assumes that cosmic ray (or solar) activity is constant with time. This is not so. However, there is sufficient homogeneity for some meaningful work, and steps have been taken to calibrate palaeo-carbon-14 (see the previous chapter and section 2.1.1, on isotopic dendrochronology).

Radioactive decay dating using different radioactive isotopes from associated geological strata is one common way of pinpointing a climate proxy in time. The first recorded use of radioactivity for dating took place 2 years *before* the discovery of isotopes by Frederick Soddy in 1913, just 15 years after the discovery of radioactivity and 9 years after Ernest Rutherford explained the phenomenon of radioactive decay! Arthur Holmes' classic paper from June 1911 demonstrated how uranium decay could be used to date rocks millions of years old, a far greater timescale than can be used with the aforementioned ^{14}C dating, because ^{14}C has a much shorter decay half-life.

Because widespread direct instrumental climatic measurements have only been available for the past 100 years or so we are dependent on proxy indicators combined with dating techniques for earlier information. To ensure their reliability, reconstructions based on these proxies must be validated by comparison with instrumental records during periods of overlap. Sometimes direct comparison is not possible and so this must be done indirectly by relating to another overlap with another proxy, and so commonly a number of proxies are used simultaneously.

Another problem is that all palaeoclimatic proxies only relate to a certain time period. Glacial terminal moraines largely tell us about the maximum extensions of ice, hence glacial *maximums* and not glacial *beginnings*. Furthermore, the aspect of the climate being indicated may only relate to part of the year. For instance, tree-ring data are only applicable to periods when trees grow and so largely relate to warm-season conditions and do not give an indication as to the severity of winters. (Of course, there are often exceptions and here, to continue with our current example, cold-season tree-ring information is available but largely limited to species from semi-arid or Mediterranean environments.) Such restrictions, in this case seasonal specificity, may give a misleading picture of large-scale temperature changes. For example, the climatic response to volcanic forcing can lead to opposite temperature changes in the summer (with cooling) and winter (warming) over continents.

It is for all these reasons that many proxies are required. Sets of these in turn are used to determine indices against which it may be possible to ascertain large-scale climatic changes. For example, one such set is used to determine what is called the North Atlantic Oscillation (NAO) index (see section 5.1.5, Holocene summary).

However, no matter how each proxy is subsequently used there is considerable need to add to our knowledge of existing proxies and to identify new ones. This is both to establish further palaeoclimatic data and for corroboration.

The good news is that although as many climate proxies as possible are required to build up a picture of past climates, they do all begin to paint the same picture. So, on one hand there is uncertainty, but on the other, by looking at more and more proxies (in both number and type), there is increasing certainty.

Climatic proxies are many but can be divided into two principal groups: biotic proxies (terrestrial and marine) and abiotic proxies (physical and abiotic geological). We will look at a few examples of the principal types of each of these, but because this text concerns the biology of climate change we will focus mainly on those that are in some way biological.

2.1 Terrestrial biotic climatic proxies

2.1.1 Tree-ring analysis (dendrochronology)

Dendrochronology was developed early in the 20th century, not by a biologist but by an astronomer, Andrew Douglass. When he was 27, and working at the Lowell Observatory at Flagstaff, Arizona, Douglass discovered a possible relationship between climate and plant growth. He recorded the annual rings of nearby pines and Douglas firs (*Pseudotsuga* spp.). In 1911 he made corresponding records among trees felled around 50 miles to the south west of the observatory, which prompted his study of sequoias. When viewing a cross-section of a tree, it became clear in certain species that wide rings were produced during wet years, and narrow rings were produced during dry years (Webb, 1983).

Douglass coined the term dendrochronology, meaning tree time study, and worked on exploring the link between tree rings and climate. Between 1919 and 1936 he wrote a three-volume work, *Climate Cycles and Tree Growth*. Collectively, the methods he pioneered are now invaluable for archaeologists to date prehistoric remains. They were first used this way in 1909 when Clark Wissler, of the American Museum of Natural History, put him in touch with the Archer M. Huntington Survey of the Southwest. In 1918, Wissler (who recognized the value of tree rings for dating purposes) arranged for Douglass to receive nine beam sections from Aztec Ruin and Pueblo Bonito so he could begin cross-dating the two ruins. The final missing link to his studies came from an unstable beam that was extracted from the Whipple Ruin in Show Low, Arizona. That beam bridged the gap between living-tree-ring chronology and the archaeological tree-ring chronology that had been established.

The Laboratory of Tree-Ring Research, subsequently initiated by Douglass, has the largest accumulation in any research centre of tree-ring specimens from both living trees and age-old timbers.

Trees respond to seasonal and climatic fluctuations through their annual growth rings. The thickness of each ring reflects the climate of the principal growing season, primarily the summer, in which it is formed. The rings are made of xylem. Pith is found at the centre of the tree stem followed by the xylem, which makes up the majority of the tree's circumference. The outer cambium layer of the tree trunk keeps the xylem separated from the rough bark. Each spring or summer a new layer of xylem is formed, and so the rings are produced, which may be counted as an annual record. More specifically, a tree ring is a layer of wood cells produced by a tree in 1 year, consisting of thin-walled cells formed in the early growing season (called earlywood), and thicker-walled cells that are produced later in the growing season (called latewood). The beginning of the earlywood and the end of the latewood forms one annual ring. In dendrochronology these rings are then counted and their thickness compared. However, because (as discussed in the previous section) other local factors come into play – due to the tree's site, which determines the nutrients, water and sunlight it receives – statistical analysis of a number of tree-ring samples from the local area is required to discern the local climatic factor. Even so, trees are particularly useful climatic proxies when they are near their climatic limits of either rainfall or summer warmth as then it is possible to attribute variability in ring thickness with one, or other, of these climatic parameters. Conversely, trees in the tropics do not exhibit a pronounced annual cycle and so it is not possible to obtain climatic information from their rings.

Where possible, rings not only serve as climatic indicators but are also used in dating. Sections of tree trunks can be compared with others from either slightly younger or older trees in the region, and then the overlap in each tree record can be used to extend the overall ring record backwards or forwards in time, as Andrew Douglass did. Indeed, starting from recent trees in the time of meteorological records it is possible to make a calibration relating tree rings from appropriate samples directly to either rainfall and/or summer warmth back to centuries before meteorological records began. Starting from modern times and using overlapping sequences of ring it is possible to work back in time not just centuries but even, in some cases, a few thousand years. Such is the success of dendrochronology that it has not just been of use to palaeoclimatologists but also to archaeologists wishing to date local wooden artefacts. Palaeoclimatologists who use dendrochronology have created a number of standard sequences against which new samples can be checked. Dendrochronology has proved particularly useful in understanding the palaeoclimate of part of our current interglacial, the Holocene (the time since the end of the last glacial to the present; see Chapter 4), which itself has so far lasted more than 11 700 years (Briffa, 2000).

Further, more detailed information than just the nature of a particular season can be obtained by a closer examination of the wood structure within individual rings. Parameters such as maximum latewood density, minimum earlywood density and width of early- and latewood growth can provide insight into weather fluctuations in the principal growing season.

There are a number of tenets underpinning dendrochronology. The main ones are described here.

> The uniformitarian principle: variability evident in tree-ring growth can be linked to past environmental variability and used to identify environmental change.

> The limiting-factor principle: the fastest rate that plant growth (and other processes such as seed production) can occur is equal to the greatest limiting factor.

> The principle of ecological amplitude: tree species will be most sensitive to environmental flux at the latitudinal and altitudinal limits of their range.

> The principle of aggregate tree growth: any individual tree-ring growth series can be broken down into an aggregate of environmental factors, both natural and human, that have affected the patterns of tree growth over time (and these might include age-related growth trends, the climate that occurred over the course of the year and various other factors that occur within and outside of the forest stand).

> The principle of cross-dating: matching patterns in tree-ring width, density and other characteristics across several tree-ring series allows for the identification of the year in which the growth ring was formed.

> The principle of site selection: sites can be identified based on the criteria that will produce growth-ring series that are most sensitive to the examining factor.

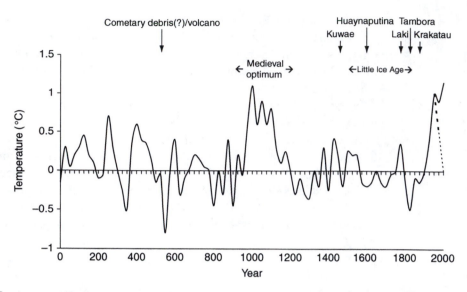

Fig. 2.1 Approximate, smoothed-average (with an ≈70-year filter) northern high-latitude temperature changes over the past 2000 years, based originally on several dendrochronological series and deviating from the 1601–1974 average. Possible cometary debris in one instance, together with five volcanic eruptions, coincide with subsequent lower annual tree-ring growth. See Chapter 3: before 1000 the dendrochronology suggests it was warmer rather than cooler; also shown are the medieval warm period, 900–1200, the Little Ice Age, 1550–1750, and the 20th-century warming. The decline in some boreal dendrochronological series at the end of the 20th century (dashed line) is portrayed separately and is thought to perhaps be due to increased snow fall delaying spring greening or some other environmental factor impeding growth most likely caused by a warming climate. See the section in Chapter 5 on the current biological symptoms of warming. Data in part from Briffa (1998).

It is important to note these underpinning tenets (as it is for all assumptions made in science), for it is easy to assume that they will always hold. This is not so and invariably there are exceptions. Tree-ring series derived from high-latitude growth have provided a solid record for nearly all of the past two millennia. However, ring growth in recent decades, principally since the 1960s, has deviated from the pronounced warming experienced (and instrumentally measured). Growth is less than is expected from temperature alone (see Figure 2.1). The reason for this is the failure of the above principles. There is some evidence, for example (Vaganov et al., 1999), to suggest that what has happened is that our globally warmed world of recent decades has seen increased winter precipitation (snow) in these high-latitude regions. (One would expect more ocean evaporation in a warmer world, and hence, overall, more precipitation. More evidence from high latitudes is covered in Chapter 6.) In these permafrost environments this extra precipitation has delayed snow melt: a greater depth of snow takes longer to melt. As a result the initiation of cambial activity necessary for the formation of wood cells is delayed. Fortunately, dendrochronological problems of such significance are rare.

Today there are a number of long, annual-resolution records of tree growth from temperate regions, determined by the aforementioned method of piecing together overlapping growth-ring patterns from successively older timbers. One of the lengthiest is the oak chronology in central Germany that runs continuously back to 8480 BC. This is even longer than the bristlecone pine chronology from the western USA, which has been available since the 1960s and which provides a record back to 6200 BC. There are some eight chronologies that span the last seven millennia; one from northern Russia, one from western North America and six European ones covering Ireland, England, Germany and Fennoscandia. The rate of development of dendro-chronological databases in the latter half of the 20th century is illustrated by the creation of the Irish oak chronology by researchers at Queen's University Belfast. Here thousands of samples of oak were collected from buildings and archaeological sites. Samples from living trees acted as a chronological anchor. By 1977 a continuous chronology ran back to 885 AD and by 1982 a continuous 7272-year Irish chronology had been determined. By 1984 sufficient correlation between Irish and German dendrochronological records had been ascertained for the announcement of the European oak chronology covering seven millennia.

It is important to remember that greenhouse gases are not the only factors forcing climate. For example, trees are sufficiently sensitive that dendrochronology can identify the years of cooling following major volcanic eruptions (see Figure 2.1). There are notable declines in tree-ring growth associated with large volcanic eruptions, such as Tambora (1815) and Krakatau (1883). The dust from these eruptions entered the stratosphere and resulted in spectacular sunsets for a few years, as well as global cooling and disrupted harvests (see Chapter 5). Acid levels in ice cores (such as the Greenland cores discussed later in this chapter) suggest past super-eruptions and these too have been connected to years of poor tree-ring growth. Such corroborative evidence is important not only to ensure that the climate inference is accurate but also to help provide further detail. For instance, several dendrochronological records suggest a twin climate event in the 6th century: one in the year 536 and something worse in the early 540s. The ice-core record is not nearly so clear; although there is evidence of some volcanic activity at that time there is nothing of the magnitude suggested by dendrochronology. However, one possibility is that the Earth may have accrued cometary debris around that time (Baille, 2003). Nonetheless, despite its limitations (such as its lack of effectiveness in the tropics) dendrochronology remains the only terrestrial biotic record detailing climatic change with annual detail covering the Holocene.

2.1.2 Isotopic dendrochronology

As will be discussed in more detail, isotope analysis can be used to infer climate change. Neutrally charged isotopes of the same element have the same number of protons and electrons as each other but a different number of neutrons. Different isotopes of the same element therefore exhibit the same chemical properties but have different atomic weight and hence different physical properties. Most oxygen in the atmosphere (99.76%) consists of ^{16}O (with an atomic weight of approximately 16) but, of the remaining proportion, the next most common isotope of oxygen is ^{18}O. Because lighter ^{16}O evaporates preferentially to ^{18}O the ratio between the two varies in atmospheric water vapour and in rain water depending on climate. The oxygen in water is then incorporated into wood through photosynthesis and so a record of the isotopic ratio of the rain at the time the wood was growing can be determined. This should give an indication of the climate. However, there are several problems with this form of analysis. These include rain-out effects: the first precipitation from a cloud has a greater composition of the heavier ^{18}O. There are also altitude and ice/snow meltwater effects. Consequently isotopic dendrochronology has only met with limited success and not all such work has borne equivocal results. This form of palaeoclimatic analysis needs some further development. Yet, despite considerable limitations there have been some interesting possibilities. For example in 2006 the Swiss and German team led by Kerstin Treydte reported that a dendrochronological isotope record from the mountains in northern Pakistan correlated well with winter precipitation and that this *suggested* that the 20th century was the wettest of the past 1000 years, with intensification coincident with the onset of industrialization and global warming (Treydte et al., 2006).

Despite the extra caution required with isotopic dendrochronology, as we shall see using $^{16}O{:}^{18}O$ ratios in another biological context, Foraminifera (an order of marine plankton) can be employed as a kind of climate proxy but more exactly as an indication of the global amount of water trapped in ice caps.

Because dendrochronology gives precise annual rings, it has also been used as one of the means to calibrate radioactive ^{14}C dating (as discussed in Chapter 1). In 1947, the US chemist Willard Libby explained how the radioactive isotope ^{14}C was produced, for which he received the Nobel Prize for Chemistry in 1960. ^{14}C, with its half-life of 5730 years, is used for dating biological samples from several decades to thousands of years old beyond the last glacial maximum about 20 000 years ago: the limit of ^{14}C dating is usually considered to be about 45 000 years (and we will return to this when considering human migration during the last glacial in Chapter 4). ^{14}C is continually being generated in the atmosphere due to cosmic rays, mainly from the Sun, hitting nitrogen and carbon atoms. ^{14}C so formed is then incorporated into plants through photosynthesis. The common ^{12}C and the radioactive ^{14}C are in equilibrium with that in the atmosphere up until the plant dies, and at that point the various carbon isotopes become fixed. From then on the fixed reservoir of ^{14}C in the plant begins to decay and so the ratio between ^{12}C and ^{14}C changes. This ratio enables dating.

However, as mentioned, all this assumes that the rate of ^{14}C production in the atmosphere is constant. This is not precisely true as more ^{14}C is produced in the atmosphere at times of, for instance, exceptional solar activity, such as when there are many sunspots. Carbon dating has therefore had to be verified by comparison with other dating methods, including dendrochronology. (In recent times $^{12}C{:}^{14}C$ ratios have been upset by atmospheric tests of atomic bombs in the mid-20th century and by the Chernobyl nuclear power plant disaster.) Notwithstanding these difficulties, ^{14}C decays at an immutable rate of 1% every 83 years. If the rate of decay is measured to obtain an accuracy of, for instance, plus or minus 0.5%, then the observation of around 40 000 decaying carbon atoms would be required. This would, in principle, allow the sample to be dated to an accuracy of ± 40 years and would require a sample of several grams to be measured in about a day. However, detecting radioactive decay is only one way of detecting (inferring) ^{14}C. A mass spectrometer enables direct detection of ^{14}C and ^{12}C and a similar accuracy can be obtained using a spectrometer with a few milligrams of sample in a few hours.

Isotopic dendrochronology has provided a lynchpin for both dendrochronology and ^{14}C dating. First, ^{14}C enables dendrochronologists to date a sample of wood to within a few years or decades. Further accuracy may then be obtained by matching the sample to the corresponding regional dendrochronological record. Second, because ^{14}C production is not constant due to changes in solar radiation it is possible to use the established dendrochronological record as a base against which to determine variations in atmospheric ^{14}C production. Notably, two ^{14}C-calibration curves have been produced: one by researchers from the University of California, La Jolla, using the dendrochronological bristlecone pine data set, and the other by those at Belfast University in Northern Ireland using oak. In particular, the results show that ^{14}C production fluctuates with a periodicity of around 200 years. Overall, for the past 6000 years both curves show that ^{14}C production decreased a little from 6000 years ago to 2500 years ago before a marginal recovery. As ^{14}C production relies largely on solar radiation this helps inform as to alterations in the solar contribution to climate change over time. This can then be added into the mix of the various factors influencing (forcing) climate and so (because we know the atmospheric concentrations of the isotopes) help us estimate the greenhouse component over time.

2.1.3 Leaf shape (morphology)

In 1910 two explorers (Irving Bailey and Edmund Sinnott) noticed that the shapes of leaves appeared to relate to the climate of a given area. The most noticeable trait was that trees that grew in warm areas had large leaves with smooth edges, whereas trees that grew in cold places tended to have smaller leaves with very serrated edges (Bailey and Sinnott, 1916; Ravilious, 2000). This was largely forgotten until the 1970s, when a US researcher, Jack Wolfe, wondered whether fossilised leaves could reveal something of the climate if they were compared with their modern counterparts. He collected leaves from all over North America, recording aspects of their shape and relating them to regional weather records for the previous 30 years. Then, using statistical analysis, he was able to compare fossil leaf assemblages with those of modern leaves and come up with past average temperature and rainfall values for the regions and times in which the fossil leaves once grew. For example, he predicted that the temperature in north-eastern Russia some 90 million years ago (mya) was roughly 9°C warmer than it is today (Wolfe, 1995). Other early work following Bailey and Sinnott, such as by R. Chaney and E. Sanborn in 1933, showed that leaf size and apex shape were also related to climate. Apart from the Wolfe database standard for North America, other databases have been developed. For instance, in the late 1990s Kate Ravilious worked on developing a similar database for the UK. These databases can be used in techniques such as leaf margin analysis (LMA) to infer temperature and leaf area analysis (LAA) to infer precipitation. LMA relates to the aforementioned changes in leaf shape with temperature, whereas LAA relates to changes in leaf area with precipitation level; for instance, plants with access to plenty of water can afford to lose it through transpiration over a large surface area (which also benefits sunlight capture, but which is not restricted by sunlight).

More recently still there has been a further twist in leaf morphology as an environmental indicator. At the turn of the millennium (2001) the UK team of Beerling, Osborne and Chaloner developed a biophysical model that seemed to explain why leaves evolved in the first place. For the first 40 million years of their existence land plants were leafless or had only small, spine-like appendages. It would appear that megaphylls – leaves with broad lamina (leaf blades) – evolved from simple, leafless, photosynthetic branching systems in early land plants, to dissected and eventually laminate leaves. This took some 40 million years. It seems that these transformations might have been governed by falling concentrations of atmospheric carbon dioxide (some 90%) during the Devonian period 410–363 mya (see Figure 3.1). As noted earlier, atmospheric carbon dioxide has a bearing on climate so that the Devonian did represent a period of global climate change. (Although if we are to draw climatic analogies with the more recent [Quaternary] glacial– interglacial climatic cycles of the past 2 million years, we need to bear in mind that the Sun's energy output was slightly less at that time than it is today; see Chapter 3.) This episode again demonstrates that climate change can influence evolution.

Another clue as to why leaf shape helps us elucidate climate is to turn this question on its head: climatic conditions favour certain leaf shapes. This came in 2008 when Brent Helliker and Suzanna Richter discovered that the mean temperature around leaves around midday during the middle of the thermal growing season (TGS) was remarkably constant across a range of latitudes. They found that mean leaf temperature for 39 tree species during the TGS across 50° latitude – from subtropical to boreal biomes – was 21.4 ± 2.2°C. Clearly there is evolutionary pressure for a species to have optimal photosynthesis and leaf shape, as well as leaf albedo, and leaf transpiration (both affected by leaf composition and structure), will affect the leaf temperature. Furthermore, leaf spacing and canopy further affect leaf microclimate. Leaf microclimate itself is a modification of the general climate and the way leaves help modify the general climate is central to their being used as a climatic indicator. Of course, leaf morphology is not the only way different leaf types make the most of different climates. This brings us to leaf physiology.

2.1.4 Leaf physiology

Whereas leaf morphology shows a general relationship with climate, one aspect of leaf physiology in particular has a close relationship with one of the key factors forcing global climate, that of atmospheric carbon dioxide concentration. A leaf's primary function is to enable photosynthesis, a case of form driven by function.

The theory underpinning leaf physiology as an indicator of atmospheric carbon dioxide, and hence global climate, is that the higher the carbon dioxide concentration the fewer stomata – the small openings in plant leaves that facilitate gaseous exchange – are needed to transfer the required amount of carbon dioxide that the plant needs. So, there is an inverse relationship between stomatal densities and atmospheric carbon dioxide concentrations. This relationship was mainly deduced and established in the late 1980s.

Problems with this sort of analysis abound. First, there are all the dating problems commonly associated with terrestrial palaeontology (which we touch on elsewhere in this chapter). Second, the relationship with climate is less direct than other commonly used indicators. The indicator really is atmospheric carbon dioxide concentration and although carbon dioxide does force climate, climate change is known to occur without a major change in carbon dioxide. An example here would be the Little Ice Age (see Chapters 4 and 5), and, of course, vagaries of the weather (the day-to-day manifestation of climate) are independent of changes in carbon dioxide. Third, even the carbon dioxide/leaf stomata relationship produces only a coarse correlation with carbon dioxide concentrations.

Nonetheless, analysis of leaf physiology has on occasion provided some remarkable climatic insights. In 1999 Jenny McElwain and colleagues from Sheffield University in the UK (McElwain et al., 1999) counted stomata densities in fossilised leaves from dozens of species from around the Triassic/Jurassic extinction some 205 mya (see section 3.3.6). They found lower stomata densities after the extinction than before, which suggests that carbon dioxide levels were far higher. They calculated that carbon dioxide levels rose from around 600 parts per million (ppm) before, to between 2100 and 2400 ppm after the extinction. This would probably have forced the global climate to be some 4°C warmer. The resulting climate would have been too warm for large leaves outside of high latitudes to photosynthesise properly and so could go a long way towards accounting for the extinction of 95% of all plant species at that time. The cause of this carbon dioxide pulse is thought to be volcanism during the breaking up of Pangaea. The area of volcanism required to generate the carbon dioxide pulse, the Sheffield team concluded, seems to fit with the likely area of the Pangaean split.

2.1.5 Pollen and spore analysis

Pollen and spores are critical parts of the life cycles of vascular plants. Because different species of plants thrive in different climatic conditions, it is possible to deduce past climates if one has evidence as to which species grew in those past times. Pollen and spore utility as palaeoclimatic indicators arises out of their resilience and durability with time. This is because they have very resistant cell walls, which in turn is due to the outer portion of their cells (the exine) that is composed of a waxy substance known as sporopollenin. Their cell-wall resistance and inert nature allows the preservation of pollen and spores in sediments under a variety of conditions. Pollen and spores typically are the most abundant, most easily identifiable and best preserved plant remains in sediments such as bogs and clays, and in the form of fossils in sedimentary rocks. Together it means that they can be used as a climatic indicator for both recent and geological timescales.

Spores, as referred to here, include the reproductive bodies of lower vascular plants such as club mosses, horsetails and ferns. The earliest occurrences of spores produced by land plants in the fossil record are in Lower Silurian rocks (around 430 mya), slightly preceding the appearance of the first vascular plant fossils. Conversely, pollen grains are the gamete-carrying reproductive bodies of seed plants, including gymnosperms (such as conifers and cycads) and angiosperms (the flowering plants). Fossilised pollen first occurs in Upper Devonian rocks (about 370 mya), corresponding to the occurrence of the earliest fossil seeds (*Archeosperma*). Pollen and spore walls of each plant species have a distinctive shape with characteristic apertures and these can be used to identify the types of plants represented by pollen in a sample from a given site. A microscope must be used for such identification because pollen and spores are small, typically between 10 and 200 μm. First, though, pollen and spores must be isolated from sediments and rocks using both chemical and physical means, before they are mounted on microscope slides for examination and identification. For most geological and environmental applications of pollen and spore analysis scientists count and identify grains from each sample using a microscope and generate pollen diagrams of the relative (percentage) and absolute abundance of pollen in samples from a site's stratigraphic sequence (the geological age column): it is not just the presence or absence of a species used in this form of palaeoclimatology that is important, but its abundance. Typically, the results of spores and pollen analysis from several species are pulled together to establish a picture of the changing plant ecology, and hence palaeoclimate, of a site.

Whereas dendrochronology provides an annual indicator of climate, pollen and spore analysis does not have such a high resolution. In some circumstances, specifically in well-dated sediment cores of high resolution, it is possible to obtain high-resolution records of botanical change at a decadal scale or less (in some cases to within a year) and also to document community changes over the last few centuries and even millennia. However, this is rare. Typically the resolution achieved is of several years or even a few decades. Even so, it is important to bear in mind that the presence or absence of pollen or spores is not a hard-and-fast identifier of climate, as other ecological factors need to be considered.

For example, if the climate warms it will take a little while for a new species to establish itself, especially if the climate change is marked and sharp, so often necessitating species migration over a long distance. Indeed, there may also be other factors impeding migration. Suppose climatic cooling affects a northern boreal forest; even if the cooling is not marked enough to kill any trees, some species, such as black spruce, may stop growing cones and producing pollen. Black spruce pollen will then be absent from the record even though the species is as abundant as it was before, and it may even multiply by the growth of new roots where drooping branches touch the ground; so the trees will not die out even if they cannot seed. Yet, simply by being there they are impeding the arrival of new species whose pollen might be a better indication of the climate.

Aside from resolution and ecological problems, other difficulties with pollen and spore analysis as climate indicators include those of human land use. Humans tend to clear land for various purposes, from intensive urban landscapes to semi-natural systems such as agriculture. Either way, the natural community is no longer there. Consequently spore and pollen analysis to elucidate past climates is best from sites in areas where there has been minimal human interference. (Although, of course, this particular problem can be turned on its head with pollen and spore analysis being used for archaeological purposes as an indicator of human activity.)

One also needs to recognise the problem of pollen decay. Pollen of some species are less resilient than others. For example, white pine pollen is resilient to breakdown whereas larch (*Larix* spp.), aspen (*Populus* spp.) and poplar (also *Populus* spp.) pollen

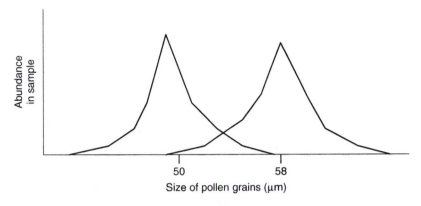

Fig. 2.2 The variation in size of pollen grains from two different species of the same genus may overlap, impeding identification. Shown are data for lodgepole pine (*Pinus contorta*), with an average pollen-grain size of 50 μm, and western white pine (*Pinus montecola*), of average size 58 μm. Statistical analysis of the difference of means of a sample against those from a known species may be required for identification purposes.

decays quickly. So, a sample of fossil pollen dominant in spruce (*Picea* spp.) could have come from a pure spruce forest or a mixed forest of spruce and poplar; there is no way of knowing. This means that pollen analysis rarely, if ever, provides a complete picture of a past ecosystem. Furthermore, some sites are better for conserving pollen and spores than others. The cold, acidic, anaerobic (oxygen-deficient) water of a peat bog is a better environment for preservation than warm, eutrophic (nutrient-rich) waters where bacteria thrive.

Considerable expertise by those who study pollen, palynologists, is needed to correctly identify which species particular pollen comes from. For example, the pollen grains of junipers (*Juniperus* spp.) and arborvitaes (or cedars, *Thuja* spp.) are indistinguishable. In other cases, although it may be possible to identify the genus of a pollen grain, it may be difficult to tell the species. Sometimes the grains from different species belonging to the same genus are identical except for a slight difference in size. Furthermore, pollen grains from a single species do have natural variability in size and if the range from a single species overlaps with the range of another species in the same genus, say lodgepole pine (*Pinus contorta*) and western white pine (*Pinus montecola*), the identity of the grain becomes uncertain. In fact these two species have average pollen-grain lengths (excluding bladders) of 50 and 58 μm, respectively (error, ±8 μm). In difficult instances where there is doubt a statistical analysis of the difference of means is required (see Figure 2.2).

Consequently pollen and spores as palaeoclimatological indicators really only come into their own as a means of confirming (but not necessarily disproving) the picture portrayed by other palaeoclimatological proxies, be they local or hemispheric, and in this sense can assist in helping ascertain not only whether there was genuine climatic change of such-and-such a degree but whether it was a local variation or part of a more global climatic change. It is arguably better (as with species; see below) to think of pollen-and-spore evidence not in terms of climatic proxies but as evidence contributing to a picture of the effects of climatic change. We shall return to this later when discussing past ecosystems as being palaeoclimatic ecological analogues of possible future climatic change. (For an example of a typical summary pollen diagram see Figure 4.8.)

2.1.6 Species as climate proxies

It is well known that different fauna and flora live under different climatic conditions and we have seen how the adaptations of leaves to different climate conditions help with their use as indicators of climate. Such adaptations are not just restricted to leaf morphology but (as we have also seen) also physiology, and many other ways. (Indeed, we are beginning to see these in genetic codes. For example, the expression of the gene *Hsp70* in plants varies linearly with temperature; Kumar and Wigge, 2010.) The bottom line is that plants and animals evolve to live under certain (or a limited range of) environmental conditions. So, as with pollen and spores, it might seem obvious to use species as climate proxies. Aside from the apparently obvious connection with climate, another attraction of considering species as palaeoclimatic indicators is that fossils are frequently all about us in unaccounted millions. Although large fossils are comparatively rare, small fossils, of a few millimetres or less in size, are frequently abundant. They come in two principal groups, macrofossils and microfossils, depending on whether a microscope is needed to identify them.

There is a considerable variety of small macrofossils. Usually only the hard parts of a plant or animal become fossilised. The soft parts decay quite rapidly after death. Animals are therefore often represented by the bones and teeth of vertebrates, the shells of molluscs (snails and clams) and the hard chitinous parts of arthropods (insects, spiders, etc.). These parts are rarely found intact but are fragmented to some degree. Among the commonest insect fossils are bits of beetles (Coleoptera), especially their upper wings (the elytra) and the shield covering most of their thorax (the pronotum).

Plants are commonly represented by fossil leaves, seeds, twigs and other wood fragments. Fossil charcoal gives evidence of past forest fires. As with animals, the soft parts of plants usually decay. However, fossil leaves are found if only because a single tree produces so many leaves, year after year, and that they can be carried some limited distance. One therefore only needs a small fraction of these to be buried in a way that maximises the chance of fossilisation for there to be some samples surviving until today.

Microfossils are far more abundant. The most numerous are fossil spores and pollen, discussed in the previous section, but also there are diatoms (a class of eukaryotic algae of the phylum Bacillariophyta), Foraminifera (which belong to another class of eukaryotic algae), dinoflagellates, ostracods (a group of small bivalved crustaceans), chironomids (a family of midge) and stoneworts (part of the Charophyta algae). These are all important to micro-palaeontologists. Microfossils can be found terrestrially and in the marine environment (see section 2.2).

As with pollen and spore analysis, there are many problems in using species as indicators of past climates and past climate change. Not least are problems of dating. Microfossils are capable of being washed from one place to lie in the sediment in another. Then there are ecological problems. Species do not migrate and become established instantaneously. Indeed, taking beetles as an example, a specialist plant-eating beetle needs for its food plants to be already established before it can migrate into an area and, if it is in turn preyed upon by a carnivorous beetle, then a population of the plant-eating beetle needs to established before the carnivorous predator can be sustained. All this takes time; meanwhile, the climate has already changed. Consequently the remains of species are not ideally suited as palaeothermometers (except in a very broad-brush sense). Instead the remains of species are better viewed as indicators of the ecological impact of climate change. We shall return to this later when discussing the impact of climate change (Chapters 4, 5 and 6) and especially the likely effects of current global warming (Chapters 7 and 8) for which in some instances there are palaeo-analogues.

Of course, the above techniques only work if we have exact knowledge of species' climatic preferences: that is to say, that we can study live members of the species today. When trying to ascertain the climate many millions of years ago (as opposed to hundreds of thousands of years ago) we have to increasingly rely on fossils of species that are no longer with us. Consequently the next best thing that can be done is to look at the nearest living relatives of the fossil species. Even if this does not give us as exact an idea of the past climate it still gives us a good idea and can provide corroborative evidence alongside other palaeoclimatic indicators. The fundamental assumption behind analysis of nearest living relatives is that the climatic tolerances of living taxa can be extrapolated without modification to ancestral forms. The problem here is that this oxymoronically implies evolutionary stasis. However, confidence in nearest-living-relative species analysis can be increased if making a number of (hopefully reasonable) assumptions and adopting operational practices. Namely: (1) that the systematic and evolutionary relationship between the extinct species and the living one is close; (2) being able to look at as large a number of fossil and nearest-living-relative pairs as possible, assuming that this will rule out individual species' idiosyncrasies; (3) using living relatives that belong to a diverse and widespread higher taxon as this increases the chance of finding a closer match with the fossil; and (4) using plant groups that have uniformly anatomical or physiological features that are constrained by climatic factors. Nearest-living-relative analysis can therefore provide us with a broad-brush picture of climate change many millions of years ago. (For examples see Allen et al., 1993.) However, as this book is largely driven by the present need for science to underpin policy concerns over current global warming, much of this text's focus is on more recent change in the past 300 000 years or so (the past two glacial–interglacial cycles) and likely future change due to anthropogenic global warming. Over this recent and short time horizon, many of the species that were around then are still with us. (Of course, there are always exceptions and many large animal species, or megafauna, outside of Africa have become extinct in the last quarter of a million years due to human, and combined human and climatic, pressures. African large animals were more resilient as they co-evolved with humans.)

Naturally, individual species are not alone in responding to climate change and it is often necessary to use more than one species to ascertain the timing and degree of the same climate change. For example, Post and Forchhammer (2002) published research on caribou and musk oxen on opposite sides of Greenland, demonstrating that populations of different species may respond synchronously to global climate change over large regions.

But terrestrial species are not the only ones to be used as palaeoclimatic indicators. Marine species are too (see section 2.2) and even those that can live both in the sea and on land. In 2004 it was reported that the migration of fish that can live for extended periods on land can be used as a palaeoclimate indicator, when Madelaine Bohme from the University of Munich used fossil snakeheads, a group of predatory fish (Channidae) native to Africa and south-east Asia, from the Miocene to Pliocene (23–5 mya; Böhme, 2004). The species provides a good indicator of summer precipitation. Its present-day distribution is limited to climates with at least 1 month of rainfall of 150 mm and a mean temperature of 20°C. Bohme's study suggests that the most extended migration events (17.5 mya and between 8 and 4 mya) must have been linked to changes in northern-hemisphere air circulation, resulting in formerly dry summers having higher rainfall.

2.2 Marine biotic climatic proxies

Marine proxies for palaeoclimates have several advantages over many terrestrial indicators, especially biological ones. The sea surface is subject to fewer variables affecting biotic proxies compared to their terrestrial counterparts. Surface topography is constant and so there is no need to worry about whether the proxy originated on a long-since-gone slope. Vagaries in water availability for the proxy are also not a problem. Finally, the sea temperature (because of the sheer volume of water being affected by the climate) is more applicable to the regional climate (albeit affected by heat transport by ocean currents) than the local climatic factors that affect terrestrial biotic indicators. Marine biotic indicators are therefore particularly important when attempting to ascertain palaeoclimates. There are a number of marine biotic indicators of climatic proxies. These include terrestrial counterpart methods such as the use of species. However, of all the methods used, one in particular has generated much information as to climate change, both generally over several tens of millions of years and more recently in greater detail over hundreds of thousands of years. This method employs ^{18}O analysis.

2.2.1 ^{18}O Isotope analysis of forams and corals

For decades now, the concentrations of the heavy isotope ^{18}O in the microshells of Foraminifera – an order of plankton – and also corals have been widely used to reconstruct the temperature profiles of ancient seas. Originally it was in 1947 that Harold Urey (who previously won a Nobel Prize for Chemistry for his discovery of deuterium) pointed out that the oxygen isotope composition of fossil seashells could serve as a palaeothermometer of past sea-surface temperatures. The stable heavy isotope ^{18}O accounts for about 0.2% of the oxygen in sea water H_2O compared to the most common ^{16}O isotope. The greater proportion of ^{18}O that a shell incorporated, Urey showed, the colder the water in which it was formed. The slightly heavier nucleus of ^{18}O also affects evaporation and precipitation, but here it is crystal formation that is important. Urey argued, on thermodynamic grounds, that the formation of crystalline $CaCO_3$ (calcite) – with dissolved calcium in sea water and which is taken up by forams and corals – should involve some isotopic fractionation. The idea was that the fraction of ^{18}O incorporated into the calcite should decrease with increasing water temperature, essentially because the heavier nucleus slows down the vibrational modes of molecules with ^{18}O- compared to ^{16}O-containing water molecules, which then have a thermodynamic advantage in the process of calcite formation. However, to exploit this isotopic effect for palaeothermometry one has to correct for the global extent of ice caps and glaciers at the time in question. This is because water molecules with the normal, lighter ^{16}O preferentially evaporate compared to ^{18}O, which requires more energy that goes with higher temperatures; it is this last that underpins the theory behind ^{18}O dendrochronology mentioned above. So, using ^{18}O in shells for palaeothermometry is only really effective when there is little ice on the planet. When there is a lot of ice (such as now, with Antarctica and Greenland) ^{18}O in shells is a better indicator of the amount of ice worldwide than global temperature. (We shall return to this ^{18}O ice issue shortly.)

So, because ice caps take water molecules with a mix of oxygen isotopes out of the oceanic pool of water molecules, ice caps complicate things. However, from the late Cretaceous (65 mya) to the late Miocene (5 mya), the Earth had negligible ice caps and terrestrial ice (such as on mountain tops). Consequently ^{18}O palaeothermometry is an applicable palaeoclimatic tool for this geological window. There is a complication in that different plankton species, although they all float in the top few hundred metres of the ocean, live at different depths below the surface (hence possibly at different temperatures), and they often change habitat depth with the seasons. Some species, for example, keep to the upper layers because they live symbiotically with photosynthesising algae. Not knowing the detailed life cycles of all the long-extinct creatures means that some inferences need to be made, with some inadvertent erroneous assumptions. Consequently, in these palaeoclimatic studies researchers customarily deduce surface temperature for a given multi-species sample from the species that yields the lowest ^{18}O fraction, presumably corresponding to the habitat closest to the surface.

Since the mid-1980s, however, models of carbon dioxide greenhouse warming have confronted ^{18}O data from fossil planktonic (floating) Foraminifera with the so-called cool-tropics paradox. In stark contrast to many global climate models, the planktonic ^{18}O data seemed to suggest that 50 mya tropical ocean surfaces were about 10°C *cooler* than they are now, even though this was a time when the carbon dioxide level was almost certainly much higher than it is today, and the Arctic was warm enough for crocodiles and giant monitor lizards.

The controversy was resolved largely due to an analysis of planktonic Foraminifera from the late Cretaceous to the late Eocene (67–35 mya). This was conducted by Paul Pearson (at the University of Bristol in the UK) and co-workers in 2001, and does much to lay this troubling cool-tropics paradox to rest. Because planktonic Foraminifera, while alive, float at or near the surface, researchers had assumed that their ^{18}O concentration reflects the temperature of the sea surface. But Pearson and company, doing isotopic analyses of unusually well-preserved samples of pristine shells selected with the help of electron microscopy, concluded that the surprisingly high ^{18}O level of traditional samples is a misleading consequence of extensive recrystallisation of the fossil shells in the much colder waters at the bottom of the sea (Pearson et al., 2001). Daniel Schrag (Harvard University) reached much the same conclusion in 1999 by means of a mathematical model of the recrystallisation of these very porous microshells buried in the sea floor (Schrag, 1999). At higher latitudes, where there is less temperature contrast between the surface and the bottom of the sea, this recrystallisation issue is less of a problem for ^{18}O palaeothermometry.

There are other problems with this form of ^{18}O analysis, especially with regards to determining climates prior to the last glacial–interglacial cycle (of the order of a hundred thousand years) and going back millions of years. These relate to dating benthic (deep-sea) forams. First, more ancient deep-sea sediments are increasingly disturbed due to a variety of factors over millions of years, including plate tectonic movement and the weight of fresher sediments being superimposed on top. Second, ^{14}C, as used in dendrochronology (see section 2.1.1), is not applicable over such long timescales and so radionucleotides with appropriately longer half-lives are required. However, this reduces the resolution of sediment age determination. Consequently, whereas we might *in theory* (see below) be able to use various techniques to determine regional and global palaeoclimates down to virtually an annual level over the past glacial–interglacial cycle (of the order of 100 000 years), when attempting to go back tens of millions of years the palaeoclimatic resolution tends to be of the order of several centuries, if not millennia, at best. However, it is possible to get a broad-brush (low-resolution) picture as to how the temperature of the planet has changed over such timescales (see Figure 2.3).

There are only a few isotopes with long half-lives, including those relating to the uranium-235 (^{235}U) decay series, that can be used to date remains further back in time than ^{14}C. And among these there are a number of radioisotopes that are not incorporated into biotic remains but with the surrounding geological strata instead. This obviously begs the assumption that the biotic proxy and the surrounding geological strata originated at the same time. Nonetheless, radioactive isotope dating methods can even be used with extremely long timescales that take us back to before life on Earth arose. Rubidium-87 (^{87}Rb), with a half-life of 4.8×10^{10} years, is used by geologists for dating back to the formation of the Earth.

Another problem is the resolution of change discernable from ^{18}O analysis. The problem is that the climate can sometimes change faster than ^{18}O concentrations. For ^{18}O concentration to properly reflect the climate there needs to be an appropriate amount of mixing of the oceans' waters and new balances struck with ^{18}O in its various guises in the atmosphere and ocean. This takes time. Broadly speaking, ^{18}O analyses are not particularly useful in resolving climatic episodes of much less than 1000 years in duration.

In short, ^{18}O Foraminifera and coral palaeothermometry is of value (1) but only for periods of time on the planet when there were virtually no ice caps, (2) if one concentrates on such proxies from high latitudes away from the tropics so as to minimise recrystallisation problems and (3) if the climate being elucidated is one that lasts for roughly 1000 years or more.

However, there is another value in ^{18}O analysis when there were or are large ice caps on the planet, such as now in Antarctica and Greenland, even in our warm (Holocene)

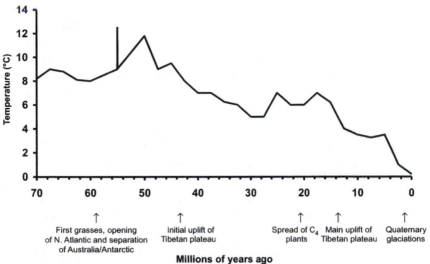

Fig. 2.3 Approximate (and with coarse resolution) Cenozoic–present global temperature trends compared to 1990, as estimated from [18]O analysis of composite benthic foram records from the Atlantic Deep Sea Drilling Program sites. Based on data from Miller et al. (1987) and simplified from Willis and Niklas (2004) and Intergovernmental Panel on Climate Change (2007). Indicated are some of the factors affecting the global climate. Grasses and C_4 plants enabled enhanced photosynthetic sequestration of atmospheric carbon. The raising of the Tibetan plateau not only affected the continental Asian climate but also created a considerable body of mid-latitude ice that affected planetary albedo. (Milankovitch effects occur on too short a timescale to be deduced in this way and resolved over this period of time.)

interglacial, and especially in glacials with their substantive ice caps. This applies to all the Quaternary period of the past 2 million years. As noted, ^{18}O-containing water molecules require more energy to evaporate from the oceans than [16]O water. The principal thing here is that as the lighter [16]O evaporates preferentially from the sea it tends to leave behind [18]O and so the sea becomes enriched in [18]O as the ice caps grow. This is reflected in [18]O in forams and corals. Consequently, whereas [18]O may be a reasonable palaeothermometer between the Eocene and Cretaceous (when there were no ice caps), when there are ice caps this isotopic partitioning between marine and terrestrial ice environments comes to the fore. At such times [18]O in marine calcite is an indicator of terrestrial ice and, hence, global sea levels (as opposed to temperature). The higher the concentration of marine calcite [18]O the greater the extent of terrestrial ice and the lower the sea level. (As terrestrial ice caps – rich in lighter isotopes – grow, accounting for more water in the biosphere, so the amount left for the oceans decreases.) Intuition suggests that this indicator of ice volume is sort of a palaeothermometer in itself, as one might reasonably suspect that when the planet has large ice caps it must be cooler than when it does not. But this is not strictly true. Ice caps do not come and go overnight. We see substantial ice caps today in Antarctica and Greenland, on which snow falls even now in the middle of a warm interglacial (even if around the margins of these caps there is melting). So, at times when our planet (or in all probability any other Earth-like planet) does have ice caps,

^{18}O in marine calcite is an indicator of the extent of these terrestrial ice caps, and, commensurately, sea level, but not nearly so much of temperature.

Because ^{18}O concentration in forams and corals is determined primarily by two factors – temperature in the absence of terrestrial ice and sea level in the presence of terrestrial ice – a number of global palaeotemperature graphs have two temperature scales, one relating to before 35 mya, when Antarctic glaciers began forming, and one after, when terrestrial ice became a major factor. This is because of the problems discussed above about this form of palaeoclimatic determination over periods of many million years. For this reason Figure 2.3 should actually be considered to be the merging of two ^{18}O graphs relating to temperature, one covering the period before 35 mya and one after. Alternatively, it could have been a single ^{18}O concentration curve but necessitating two temperature-related scales, again one before 35 mya and one after. Figure 2.3 was drawn the way it was to clarify matters simply.

But forams are not the only biological sources of ^{18}O that can be analysed. Coral analyses have certain advantages over foram analysis. Unlike forams, which are more subject to the vagaries of water movement, coral reefs are fixed. Consequently, problems with changes in temperature with depth are far less important.

Corals are animals in the cnidarian (coelenterate) phylum, order Scleractinia, class Anthozoa. Their basic body plan consists of a polyp containing unicellular endosymbiotic algae known as zooxanthellae overlaying a calcium carbonate exoskeleton and it is this calcium carbonate (calcite) that is used for ^{18}O analysis. Of particular value is that many corals outside of the tropics also have annual growth rings: actually they have daily rings that are further apart in the summer and narrower in the winter and during their monthly (tide-driven) breeding cycle. (This last has been used to determine the distance between the Earth and the Moon millions of years ago but that is a separate story.)

2.2.2 Alkenone analysis

Long-chain (C_{37}–C_{39}) alkenones are cell membrane lipids that are organic molecular 'fossils' from phytoplankton and represent about 8% of total phytoplankton carbon. They can be used to indicate the sea-surface temperature. The principle behind this is not exactly known, although the correlation between the unsaturated proportion (that with double and triple bonds in the chain) and temperature is undeniable. It is speculated that as temperature changes so does the rigidity, or some other physical property, of phytoplankton cell membranes. What is known is that, in culture, living algae respond to changing temperature conditions in a matter of days. Because of this they are particularly good for recording rapid temperature change subject to accurate dating. Also, alkenones are robust molecules that are capable of being preserved for many hundreds of thousands of years and so are useful in charting more than our current glacial–interglacial cycle.

One of the longest established alkenone-based proxies was devised at the University of Bristol in 1985. Known as the U^{K}_{37} index, it is based on the number of double bonds in alkenones with 37 carbon atoms that are found in many unicellular eukaryotic marine algae, including *Emiliania huxleyi*, one of the most common species of such algae.

Table 2.1 Isotopic species of water	
Composition	Natural abundance (%)
$^1H^1H^{16}O$	99.732
$^1H^1H^{18}O$	0.200
$^1H^1H^{17}O$	0.038
$^1H^2H^{16}O$	0.030
$^2H^2H^{16}O$	0.020
$^1H^2H^{18}O$	0.006
$^1H^2H^{17}O$	0.001
$^2H^2H^{18}O$	0.0004
$^2H^2H^{17}O$	0.00008

In 1986 another robust index was established, called the TEX 86 index, named after the Dutch island of Texel on which it was developed. TEX 86 is based on molecules that have 86 carbons. However, these molecules come from far less common single-celled Archaea, but because such Archaea have been found in sediments spanning the past 100 million years they represent a powerful complement to the U^K_{37} index.

2.3 Non-biotic indicators

Whereas (formerly) living organisms are most useful as an aid to estimating palaeo-climates because they either live solely within usefully narrow climatic restrictions or because in the course of their lives they somehow incorporate a palaeobiotic indicator into their bodies, there are other indicators of past climates that operate completely independently of living creatures. Of these, one of the most useful in terms of helping determine climate change hemispherically (if not globally) in the past few hundred thousand years has been the isotopic analysis of water ice.

2.3.1 Isotopic analysis of water

Water is not just H_2O or HHO. As discussed in preceding sections, oxygen exists as two principal isotopes, ^{16}O and ^{18}O, together with another, ^{17}O. Further, there is also a second stable isotope of hydrogen in addition to the common ^1H and this is ^2H, or deuterium (sometimes portrayed as D). Only a tiny proportion of hydrogen and oxygen is not of the respective principal isotopic forms, but nonetheless these rarer isotopes do exist. Consequently water is not just HHO, as in $^1H^1H^{16}O$, but has a variety of isotopic variations (see Table 2.1).

As has previously been noted the take-up by photosynthetic organisms of water containing ^{18}O for photosynthesis is affected by temperature and is complicated by preferential evaporation of water containing lighter oxygen isotopes. Yet such preferential evaporation of lighter isotopes with temperature can also be of use as a climate indicator. Deuterium-containing water, because of hydrogen's smaller size, is not so discriminated against by living things, so the preferential evaporation and condensing of water containing isotopes of deuterium is more solely temperature-dependent.[1] More accurately, the deuterium concentration of ice in terrestrial ice caps such as Antarctica and Greenland reflects the temperature difference of the water (primarily the ocean) from which it evaporated and the air from which it froze out as snow. Deuterium concentration in such ice caps therefore reflects part-hemispheric more than global climate change. Yet, as we shall see in the next chapter, there is considerable agreement between the deuterium records of the northern hemisphere's Greenland cores and those from the southern hemisphere's Antarctic cores. However, there are also some interesting differences that will be discussed.

The problem with the isotopic analysis of water or, to be specific, ice as a determinant of palaeoclimate is that it is only possible where ice has existed undisturbed as far back as the climate you wish to study. This restricts such palaeoclimatic studies, both in space and time; space because the places where ice exists undisturbed with longevity are rare, and time also because where it has lain undisturbed has not been long: even a stable ice cap will slowly build up so that the early lower layers come under increasing pressure, with an increasing likelihood of deformation. However, both the Greenland and Antarctic ice sheets have enabled us to piece together the global climate in the past couple of glacial–interglacial cycles. This has been a key validation of the Milankovitch theory of orbital climatic forcing, although we are aware from other evidence that Milankovitch climatic forcing has existed for many millions of years and so presumably for as long as the Earth has had a troposphere with discernable weather (as opposed to other factors affecting conditions on the planet's surface such as planetesimal accretion).

The great utility of the isotopic analysis of ice is that when it does lay undisturbed it is possible to discern annual layers of snowfall that subsequently have become compacted to form ice. But there are only a few places where ice is so undisturbed. As mentioned, one is at the top of the aforementioned Greenland ice cap where there is minimal lateral movement of ice, and another is the ice cap over the lake at Vostok in Antarctica. The cores from these places enable us to discern isotopic proportions of ice with resolutions of a few years, yet the records cover hundreds of thousands of years. Even so, it just takes one layer to be less clear for the counting of subsequent layers to be out by one; there is a huge difference between much of a core having a high resolution and the entire core being so clear. Consequently, although a core of a thousand years of age may exhibit a high resolution, the precise dating of the layers through physical counting of the entire length of the core to the present may only be accurate to within a decade. Nonetheless, this means that it is possible to discern a particularly detailed picture of the palaeoclimate going back to more than two glacial–interglacial cycles and a highly detailed picture emerges from the end of the last glacial maximum some 20 000 years ago. As we shall see this has provided evidence that the hemispheric climate has 'flickered' in the past, changing rapidly in just a few years (within a decade) or a decade or two. As we shall see in the next chapter, climate change is not always as gradual as sometimes portrayed.

---

[1] Those interested in physical chemistry can look up the term Clausius–Clapeyron relation.

2.3.2 Boreholes

As the climate warms and cools, so varying amounts of heat are absorbed into the underlying geology. Consequently, it is theoretically possible to drill a deep borehole to measure the temperature at varying depths as an indication of past climate. Depths of borehole are typically of the order of a kilometre or more. Assumptions based on geological knowledge of the area, especially those related to geothermal heat, need to be made.

Problems with using boreholes as indicators of palaeoclimate include that the resolution of the thermal record is low and decreases with depth. However, in 2000 Shaopeng Huang and colleagues from Michigan University, USA, and the Western Ontario University, Canada, analysed temperatures from 616 boreholes from all continents, except Antarctica, to reconstruct century-long trends in the past 500 years (Huang et al., 2000). The results confirmed the unusual warming of the 20th century, which was greater than for the previous four centuries. Their results also corroborated the consensus as to past climatic change over this period from other palaeoindicators.

2.3.3 Carbon dioxide and methane records as palaeoclimatic forcing agents

In addition to indicators of past climates there are also records of palaeoclimatic forcing agents. Here the principal examples are the ice-core records of the past atmosphere. As snow falls on ice caps so bubbles of air become trapped and encased in ice. Bubbles in ice cores from Vostok in Antarctica, and Greenland provide a record of atmospheric composition going back hundreds of millennia. A history of both the key greenhouse gases, carbon dioxide (CO_2) and methane (CH_4; see Chapter 1), can be ascertained. These strongly reflect hemispheric climatic change as elucidated by the deuterium isotope record and so demonstrate the strong link these forcing agents have with climate. The methane record largely reflects the area of wetlands globally that produce the gas. Its concentration is higher in the warmer, hence wetter (due to increased ocean evaporation), interglacial times. Consequently, being a greenhouse gas, its atmospheric concentration change is another example of one of the many feedback systems operating in the biosphere. The warmer the global climate the more wetlands there are generating methane, which being a greenhouse gas further warms the climate, and so on. Conversely, the cooler the climate the less the wetland area, hence less methane is produced, and less climate warming from atmospheric methane.

Of course wetlands are not the only source of methane. Oceanic methane hydrates (clathrates) have been known to release pulses of methane into the atmosphere, providing a warming burst.

The problem with carbon dioxide and methane gases as a record of climate forcing agents is that other forcing factors operate as well and so the record cannot be said to properly reflect climate. Indeed, one brief climatic return to glacial conditions (the Younger Dryas) prior to our current Holocene interglacial took place in the northern hemisphere but does not seem to have taken place to anything like the same extent in the southern hemisphere. It is also less prominently reflected in the Antarctic atmospheric carbon dioxide record. Furthermore, as we shall see in Chapters 4 and 5, the so-called Little Ice Age, in the 15th to 17th centuries, took place but was not reflected in changes in atmospheric carbon dioxide concentrations.

Another problem with relating atmospheric carbon dioxide to climate is that if the climate is changing and plants are absorbing more carbon dioxide, or rotting and producing more, or it is getting wetter and wetlands are expanding, thus pushing up methane concentrations, then it takes time for this to be reflected to any great degree in the atmospheric concentration. True, atmospheric carbon dioxide varies by 5 parts per million by volume (ppmv) every 6 months outside of the central tropics due to seasonal vegetation changes, but carbon dioxide concentration changes associated with glacial–interglacial transitions are of the order of 150 ppmv. If the hemispheric carbon pump can only cope (optimistically) annually with less than 10 ppmv then climate changes associated with greater carbon dioxide change than this fall outside of this window. In part related to this, there is the problem that the atmospheric residence time of a carbon dioxide molecule is of the order of a century or two. Consequently, climate change episodes not primarily driven by atmospheric carbon dioxide (such as those driven by changes in ocean circulation) and which occur quickly, cannot easily, if at all, be discerned from the atmospheric gas record. This is yet another reason why it is best to use as many as possible indicators of climate so as to understand what is really going on.

2.3.4 Dust as an indicator of dry–wet hemispheric climates

Dust blown many hundreds of miles is an indicator of the dryness of the climate. In a wet climate rain is more likely to wash dust out of the air than in a dry climate. Indeed, we know that warm interglacials are globally far wetter than the cool glacials. As noted above, this is because in a warmer world there is more oceanic evaporation, hence precipitation, than in a cooler one. Of course, to discern this indicator one needs a recording location that is hundreds of miles from the nearest large source of dust. Such locations include the middle of ice caps and, as mentioned, one in particular – Greenland – lends itself to the building up of ice layer by layer each year and has ice strata going back hundreds of thousands of years. Measuring the dust trapped in the ice can be done by measuring the ice's electrical conductivity. (Conductivity being the reciprocal of resistance.) The more dust there is in the ice then the less conductive it becomes as the dust neutralises the acidic ions that facilitate conductivity. As with dendrochronology, providing dating can be accurately achieved, where there is clear layering it is similarly possible to discern conductivity levels on close to an annual basis. As we shall see in Chapter 4, electrical conductivity measurement analysis of the Greenland ice core has revealed dry–wet (hence presumably cold–warm) climatic flickers of roughly a year or two, and certainly far less than a decade (see Figure 4.12b).

2.4 Other indicators

The above sections deal with many of the key indicators (with a focus on the biologically related ones) used to determine past climates. However, it is important to note that there are others. Planktonic species abundance in oceanic cores (analogous to pollen analysis in peat bogs), lake-level reconstructions, lake sediments and glacial moraines are among a number of other palaeo-indicators of climate. Together all indicators provide a valuable insight into past climates, over a variety of scales, both spatially and temporally.

2.5 Interpreting indicators

Because proxy indicators are not *actual* instrumental measurements of climate but a *reflection* of climate it is important to recognise their limitations. One might consider the bringing together of several proxy data sets to form a record of a region's climate as superimposing photocopies of different parts of a picture, with some pictures having parts that are different to others and others the same. Such superimpositions are perfectly readable but they are still blurred. Yet even statistical analyses of palaeo-records, even if they reflect when times were warmer or cooler, may not necessarily accurately reflect by exactly how much warmer or cooler past times actually were. Palaeo-indicators such as tree rings are, as previously noted, themselves subject to a number of non-climatic factors such as whether the tree grew on the north or southern side of a hill, soil suitability and so forth. The data from such proxies are therefore said to be noisy.

In 2004 a European team, led by Hans von Storch, attempted to see whether past climate reconstructions from noisy proxies reflected climate by working backwards from a climate model that had already been shown to reasonably accurately reflect the northern hemisphere's climate for the past 1000 years. They used this model to generate data as if from climate proxies and then added noise. They then analysed the resulting proxy data to reconstruct the northern hemisphere's climate just as had already been done with real proxy data. They found that their simulated noisy proxy data did not accurately reflect the magnitude of the (simulated) past climate change. This suggests that some (actual) palaeoclimate reconstructions (including those used by the UN Intergovernmental Panel on Climate Change, IPCC) may only reflect half the actual change that has historically taken place since the Industrial Revolution (von Storch et al., 2004). If this were so it means that the IPCC's forecasts up to 2007 may not have fully reflected the degree of climatic change we may actually experience (see Chapter 5 and the subsection on uncertainties and the IPCC's conclusions).

2.6 Conclusions

Each technique for ascertaining an aspect of past climates has its limitations in terms of: climatic dimension being indicated (or reflected), be it temperature or aridity; resolution in time and space; and finally the sensitivity of the proxy. These limitations have to be borne in mind when attempting to reconstruct past climates and any interpretation made with reference to other climatic proxies. However, palaeo-indicators of climate have made it possible to begin to establish a picture of how the climate has changed in the past, even if this picture is more detailed in some places than others. Nonetheless, we now have a sufficiently good picture of climate change on the global scale going back the past few glacial–interglacial cycles at the decadal-to-century scale to begin to inform us about possible future climate change.

At the regional level in many places we also have a detailed understanding of the palaeoclimate with resolutions of a year or less going back many thousands of years. On timescales of back many millions of years we have a coarse understanding at the mid-latitude continental level and with resolutions of fractions of millions of years. Several palaeoclimatic proxies taken together can prove to be a powerful tool. One such multiple proxy index is the North Atlantic Oscillation index and such indices can be used to reconstruct large-scale hemispheric changes in temperature. Consequently not only is a picture of global climate change beginning to emerge but also – given that carbon dioxide is a principal forcer of climate – an appreciation of how the carbon cycle has broadly changed over the past few hundred years is beginning to emerge: although this, and palaeoclimatology itself, is still the focus of a considerable research effort in many countries.

2.7 References

Allen, J. R. L., Hoskins, B. J., Sellwood, B. W., Spicer, R. A. and Valdes, P. J. (eds) (1993) Palaeoclimates and their modelling with special reference to the Mesozoic Era. *Philosophical Transactions of the Royal Society of London Series B, Biological Sciences*, 341, 203–343.

Bailey, I. W. and Sinnott, E. W. (1916) The climatic distribution of certain types of angiosperm leaves. *American Journal of Botany*, 3, 24–39.

Baille, M. (2003) Dendrochronology and volcanoes. *Geoscientist*, 13(3), 4–10.

Beerling D. J., Osborne, C. P. and Chaloner, W. G. (2001) Evolution of leaf-form in land plants linked to atmospheric CO_2 decline in the Late Palaeozoic era. *Nature*, 410, 352–4.

Böhme, M. (2004) Migration history of air-breathing fishes reveals Neogene atmospheric circulation patterns. *Geology*, 32, 393–6.

Briffa, K. R. (1998) Influence of volcanic eruptions on northern hemisphere summer temperatures over the past 600 years. *Nature*, 393, 450–8.

Briffa, K. R. (2000) Annual climate variability in the Holocene: interpreting the message of ancient trees. *Quaternary Science Reviews*, 19, 87–105.

Helliker, B. R. and Richter S. L. (2008) Subtropical to boreal convergence of tree-leaf temperatures. *Nature*, 454, 511–14.

Holmes, A. (1911) The association of lead with uranium in rock minerals, and its application to the measurement of geological time. *Proceedings of the Royal Society of London Series A*, 85(578), 248–56.

Huang, S., Pollack, H. N. and Shen, P. (2000) Temperature trends over the past five centuries reconstructed from borehole temperatures. *Nature*, 403, 756–8.

Intergovernmental Panel on Climate Change (2007) *Climate Change 2007: The Physical Science Basis – Working Group I Contribution to the Fourth Assessment Report of the Intergovernmental Panel on Climate Change.* Cambridge: Cambridge University Press.

Kumar, S. V. and Wigge, P. A. (2010) H2A.Z-Containing nucleosomes mediate the thermosensory response in *Arabidopsis. Cell*, 140(1), 136–47.

McElwain, J. C., Beerling, D. J. and Woodward, F. I. (1999) Fossil plants and global warming at the Triassic-Jurassic boundary. *Science*, 285, 1386–9.

Miller, K. G., Fairbanks R. G. and Mountain G. S. (1987) Tertiary oxygen isotope synthesis, sea-level history and continental margin erosion. *Paleoceanography*, 2, 1–19.

Pearson, P. N., Ditchfield, P. W., Singano, J. et al. (2001) Warm tropical sea surface temperatures in the late Cretaceous and Eocene epochs. *Nature*, 413, 481–7.

Post, E. and Forchhammer, M. C. (2002) Synchronization of animal population dynamics by large-scale climate. *Nature*, 420, 168–71.

Ravilious, K. (2000) Leaf shapes: climate clues. NERC News, Spring, 6–7.

Schrag, D. (1999) Effects of diagenesis on the isotopic record of the late Paleogene tropical sea surface temperatures. *Chemical Geology*, 161, 215.

Treydte, K. S., Schleser, G. H., Helle, G. et al. (2006) The twentieth century was the wettest period in northern Pakistan over the past millennium. *Nature*, 440, 1179–82.

Vaganov, E. A., Hughes, M. K., Kirdyanov, A. V., Schweingruber, F. H. and Silkin, P. P. (1999) Influence of snowfall and melt timing on tree growth in subarctic Eurasia. *Nature*, 400, 149–51.

von Storch, H., Zorita, E., Jones, J. M. et al. (2004) Reconstructing past climate from noisy data. *Science*, 306, 679–82.

Webb, G. E. (1983) *Tree Rings and Telescopes.* Tucson, AZ: University of Arizona Press.

Willis, K. J. and Niklas, K. J. (2004) The role of the Ice Ages in plant macroevolution. *Philosophical Transactions of the Royal Society of London Series B, Biological Sciences*, 359, 159–72.

Wolfe, J. A. (1995) Palaeoclimatic estimates of Tertiary leaf assemblages. *Annual Review of Earth and Planetary Sciences*, 23, 119–42.

3

Past climate change

Post-industrial global warming is not the Earth's first major climate change event. If we are to appreciate the significance and relevance to biology of current climate change, it is important to be aware of past climate events, at least the significant ones. This chapter summarises some of the major episodes between the Earth's formation and the beginning of the current, Quaternary, ice age. When reading this chapter you may also want to refer to Appendix 2.

3.1 Early biology and climate of the Hadean and Archaean eons (4.6–2.5 bya)

3.1.1 The pre-biotic Earth (4.6–3.8 bya)

The Earth and the Solar System formed some 4.6 billion years ago (bya), give or take a few hundred million years. The Earth formed with the Solar System (containing 'Sol', our sun) accreting out of a dust and gas cloud. The dust, ice (composed of not just water but various volatile compounds) and rocks were not all small and particulate but themselves had accreted into small and large asteroid-sized bodies that ranged in size up to, and including, small planets. One of these, a Mars-sized planetoid (Thea) is thought to have had a glancing blow with the proto-Earth, exchanged material, and formed the Moon (Luna) 4.5 bya. This is not irrelevant to the nature of the Earth's climate. The lunar/Earth ratio of mass of 1:81.3 is much greater than any satellite/planet mass ratio for any other planet in the Solar System. Taking the Copernican principle, that there is nothing cosmologically special about the Earth as a life-bearing planet, this begs the question as to whether our large moon is a necessary factor facilitating a biosphere, or at least a biosphere with longevity. There is a suggestion that the Earth–lunar system is one that confers some axial stability to the Earth (affecting the variation in its angle of tilt), and hence climate stability, so enabling complex ecosystems to form. Indeed, in Chapter 1 we discussed axial tilt as a dimension of Milankovitch forcing of climate, but planets without large moons are prone to larger axial tilting and this means that good portions of such worlds spend half the year in sunlight and half the year in darkness. On Earth this only takes place within the polar circles, which form a minor proportion of the planet.

Aside from the formation of the Moon and Earth's axial stability, the other likely result of the Thea impact is that it affected the Earth's atmosphere. This suggestion arises from the concentrations of inert gas in the Earth's atmosphere compared to those of Venus (Genda and Abe, 2005). However, the notion has not become firmly established with any definitive link as to biosphere formation.

As for what was effectively the 'climate' immediately following the Thea–Earth collision, a proportion of the Earth (and presumably Thea) was vaporised and the Earth would have been surrounded by a tenuous photosphere with a temperature of around 2300 K. Subsequent cooling to allow liquid rock to form on the surface took the order of 1000 years and another 10 million years or so were needed for a thin, solid crust to form. After that there would have been some tectonic cycling of rocks, which would have drawn down some of the great excess of carbon dioxide thought to exist at that time. This too would have taken millions of years, possibly a hundred million.

Following the creation of the Moon and cooling on Earth, the oceans formed from water arising from the proto-Earth's rocks and from the comets, asteroids and planetesimals still bombarding it. At the time the Earth's atmosphere was very different from today, being largely oxygen-free and dominated not by nitrogen but by carbon dioxide and possibly methane. There is also some debate as to whether there was a significant amount of hydrogen in the very early mix. Either way, the early Earth's atmosphere must have conferred a powerful greenhouse effect. This was needed at that time as the early Sun (as a typical main-sequence star) was thought to be only 30% as bright as it is today. Studies of other main-sequence stars have made it clear that young main-sequence stars do not generate as much energy as when they are older. Given, from the previous chapter, that there is currently a natural greenhouse effect, the early Earth under a fainter Sun would have needed a far more powerful natural greenhouse effect if liquid water (which would greatly facilitate life) was common on the planet: this is known as the faint young Sun paradox. So, as we shall see, greenhouse gases were vitally important to ensure that conditions on the Earth's surface were warm enough for life. Furthermore, as the Earth's atmosphere evolved so a different mix of greenhouse gases were required to maintain temperatures. Life was fundamental to this process.

3.1.2 The early biotic Earth (3.8–2.3 bya)

Life arose fairly quickly in planetary geological terms, if suggestive isotopic evidence (possible microfossils containing a high proportion of ^{12}C) is to be believed. It is thought to have begun somewhere between 3.5 and 4.2 bya, which is about when the heavy bombardment of the Earth ceased (somewhere around, or shortly after, 4 bya). However, both the Earth and the Moon were still subject to hits by large planetesimals, as the 1000 km-wide impact basins visible on the Moon testify, although the frequency of such hits gradually waned.

There is reasonable evidence that water existed on the Earth at this time and there is a growing body of literature on the pre-biotic chemistry that led to the building blocks for life. Work by Oleg Abramov and Stephen Mojzsis in 2009 indicated that early microbial life, had it arisen before the late heavy bombardment, could have survived this period. This period would have seen the Earth's crust fractured, possibly allowing more hydrothermals, but irrespectively the energy of such impacts would have added heat to the environment (albeit locally/regionally). At around the same time, in 2008, work by Eric Gaucher, Sridhar Govindarajan and Omjoy Ganesh elucidated theoretical temperature-sensitive proteins that early bacteria may have used, suggesting an environment of around 65–73°C and that this bacterial environment cooled over the next billion years or so to 51–53°C. Then a French and French Canadian team led by Bastien Boussau looked at ribosomal RNA and protein sequences (as did Gaucher and colleagues) from evolutionarily ancient but extant bacteria and, seeing how these differed between species, created a model taking evolution backwards and deducing likely early proteins (Boussau et al., 2008). This work suggests that the earliest forms of living bacterial cells may have thrived in an 'optimal growth temperature' of up to 50°C. A second thermophilic phase of evolution then may have taken place in which the precursors to today's bacteria and then later Archaea and Eukaryota thrived in a thermophilic environment of 50–80°C. This tentative theory has the benefit of reconciling previous, seemingly contradictory, views.

Of relevance to the question of biology and climate change, temperature appears to be a factor in life arising in the first place. Having said that, the common notion that life began around deep-sea hydrothermal vents is not as cut and dried as much popular science would have it. If many parts (if not most) of the Earth's seas and oceans were far warmer than today then this would have increased the chance of life arising compared to localised hydrothermals. Furthermore, with the thin crust of the early Earth, there were many locations other than deep-sea hydrothermal vents where magma was close to and in contact with sea water. Having said that, it is important to be aware of nuance. There is a reasonable case that serpentinite hydrothermal vents provided a chemical energy source for proto-life and/or early life as well as iron-, nickel- and sulphur-bearing minerals of catalytic value that early life could have exploited. (Serpentine is a magnesium siliceous mineral mix with, among other things, magnesium and iron hydroxides, and which is formed by the reaction of olivine-rich rocks with its magnesium and iron silicates.) This vent environment, in addition to being contained, had a pH gradient that could have been exploited by life for energy (Sleep et al., 2011). Even so, it is a leap to say that this environment is where life actually started and even if hydrothermal vents are required for the chemical energy gradient they may not have been *deep-sea* vents. Today we see Archaea around deep-sea hydrothermals possibly because this is the refuge to which they retreated as the Earth cooled: it is not because they necessarily evolved there.

It is not known whether life based on RNA and DNA (and its RNA world precursor) was the very first bioclade[1]; it may well have been that some other system did arise first (with a preference for ^{12}C that we may even eventually isotopically detect), but if it did this non-RNA/-DNA life failed to become established. Equally, it may be (we simply do not know) that large impacts capable of sterilising large areas of the Earth's surface, and even the smaller impacts that continued up to 3.8 bya, made the surface too perilous for sustained life, a view that was firmer prior to the work of Oleg Abramov and Stephen Mojzsis (2009) mentioned above. Conversely, microbes sheltering deep in the Earth's crust would have survived: perhaps this is why the genes of living microbes suggest that the ancestor of all life had much in common with the hyperthermophiles that thrive in hot springs of 113°C or more. It has even been

---

[1] A bioclade is a related ensemble of life arising from a common ancestor. As noted in the text, the accepted view is that life on Earth arose once from a single common ancestor (although simple pre-biotic chemical precursors could have formed independently many times).

contemplated that life arose on Mars, which had a far more benign environment at that time and presented a small target for asteroids, as well as having a lower gravitational well attracting them. Indeed, even if Mars had had an impact large enough to affect life (which it frequently did) the early Martian seas were so much shallower than Earth's that the damage would have been less, with the resulting steam condensing out in just decades. Life would have had a better chance of arising on Mars first and then hitching a ride to Earth; ejecta from an asteroid impact on Mars could have carried life to seed Earth. (Again, Mars' shallower gravitational well would have more easily facilitated material leaving it, compared with the Earth's.) Conversely, others engaged in this scientific whimsy argue that as we know that life got off the mark quickly in geological terms, so the Martian hypothesis (albeit theoretically possible) is not necessary, and in any case would only have given a few hundred million years' (about half a billion years at the most) advantage over any life arising on Earth. Furthermore, life needed to arise somewhere, be it Mars or Earth, and so the point is moot (unless one considers a non-planetary origin for life such as panspermia theories). Either way, as we shall see, consideration of the climate in which life arose remains central to discussion of its immediate subsequent evolution.

The exact temperature of the environment in which early live arose is not known, other than it was much warmer than the temperature of tropical seas. In 2006 French researchers François Robert and Marc Chaussidon, using $^{18}O/^{16}O$ and $^{30}Si/^{28}Si$ isotope analysis of cherts (siliceous geological strata often including the remains of siliceous organisms), found that the ambient temperature of sea water then was roughly 70°C and that this temperature cooled to around 20°C about 800 mya. So it does seem as though there was a time when sea water was around 70°C, and this fits with the Boussau team's idea that there was an early life phase living in an environment of 50–80°C. Nonetheless, the evidence is being challenged – as in all good science – and work to develop a firm picture continues. For example, one challenge came in 2009 when the US team of Hren, Tice and Chamberlain used a combination of deuterium and oxygen isotopes to analyse South African Buck Reef Cherts from 3.42 bya in an attempt to get around some of the assumptions made by many earlier workers (including that the early Earth was essentially ice-free). They concluded that the sea temperature was 40°C or less. I mention this because there is (rightly) debate in the literature, although of course it is important not to make the 'It rained on Mongo' mistake[2] of assuming that the sea temperature everywhere in the primordial Earth was the same.

What is known is that our RNA/DNA bioclade arose in a biosphere very different from the one we know today. As stated above, 3.8 bya the Sun was dimmer than it is today: this is an important point, as atmospheric greenhouse gases have played a major role in helping to stabilise the planet's surface temperature within a window enabling life to survive if not flourish. If the Sun was dimmer then something had to be making the early seas warmer than today for thermo- and mesophilic (warmth-loving) early organisms to thrive. (Even if the very early Earth had a thinner crust and bombardment, which would have caused warming, there were still billions of

---

[2] This is writer and space engineer Jerry Pournelle's way of explaining the oversimplification that science fiction authors sometimes make of treating planets where local or regional conditions apply as happening to the whole world. It is a lesson to be kept in mind when considering climate and environmental proxies.

years after that in which temperatures had to remain high despite a dimmer Sun.) In addition, also as noted above, the atmosphere of the primordial Earth was markedly different from today's: it had little if any oxygen, far below the concentration of 21% that we see today. So, what were the greenhouse gas considerations back then?

Well, at that time carbon dioxide was possibly the major atmospheric constituent and a major greenhouse gas along with water vapour and methane. Consequently, anaerobic prokaryotes (bacteria and some algae) ruled the biosphere: aerobes came later (see the next section) followed by eukaryotic life (cells with internal membranes, including one around the genetic material) around 1.7 bya or earlier. The fossilised remains of some colonies of these anaerobic prokaryotes can be found today as stromatolites (clumped, fossilised, laminated sedimentary structures; although it is important to note that some stromatolites are thought to have non-biological origins). Structures thought to be stromatolites have been discovered dating from 3.2 bya (Noffke et al., 2006).

A key question about the anaerobic Earth 3–3.5 bya is whether carbon dioxide (and water vapour) alone would have been enough to stop the Earth freezing over under a far dimmer Sun: a resolution to the faint young Sun paradox. If carbon dioxide had been acting alone as the principal greenhouse gas then concentrations some 300–1000 times higher than atmospheric levels today would have been required for sufficient greenhouse warming to counter the fainter primordial Sun! But at such high concentrations the gas would have combined with iron to form iron carbonate (siderite, $FeCO_3$) in fossilised soil strata (palaeosols) and there is simply no evidence for this (although it has been found in ancient marine sediments). So, already the primordial Earth gives us a greenhouse conundrum.

One view, from US cosmologists Carl Sagan and Christopher Chyba in the 1990s, is that ammonia (NH_3) could have contributed to anaerobic Earth's greenhouse effect. The problem with this suggestion is that ammonia is sensitive to sunlight and so would have required a (very) high-altitude reflective methane haze for protection, but that haze itself would have reflected warming sunlight at that high altitude, so cooling the Earth's surface as much as the ammonia might have warmed it, and indeed if there was too warm an Earth then there would have been little methane haze! (Note: methane as a haze at that height would be reflective, as are water-vapour clouds, even though methane and water vapour as gases both have strong greenhouse properties.)

So, how was the early anaerobic life-bearing Earth kept warm? Possibly life itself could have provided a solution, in the form of methanogenic bacteria. Methanogen ancestors of the bacteria that now produce methane from marshes, river bottoms, landfills and sewage works could have produced methane with levels 1000 times higher than today's concentration of 1.7 ppmv. Methane gas lower in the atmosphere is a greenhouse gas and has the opposite effect of high-altitude methane haze. Such a great methane concentration could only have happened in an anaerobic Earth, because if there had been any oxygen (as in an aerobic Earth) then it would have chemically mopped up the methane. Indeed, the average atmospheric residence time of a methane molecule today (in an aerobic atmosphere of about 21% oxygen) is about a decade, but back in the anaerobic Earth, without free oxygen in the atmosphere, a methane molecule could have survived the order of 20 000 years. With the greenhouse effect of methane gas combined with that of carbon dioxide, the Earth could well have been warm enough not to necessitate a very high-altitude methane haze to protect photodegradable ammonia. Nonetheless there still would have been a methane haze in the lower atmosphere and there could also have been a bio-feedback-regulating effect keeping the haze just right: not too much haze to cause more reflection, and not too little to mean that the methane content of the atmosphere was so low that its (non-haze) greenhouse warming properties did not keep the Earth warm enough for life. It might have worked something like this: a warming Earth would see more methanogenic bacterial action producing more methane, and so more haze that would be reflective and hence be cooling. Conversely, a cooling Earth would see less haze and so be less cooling. This early Earth would have been shrouded in a brown, photochemical sooty haze much like Saturn's moon Titan is today. A visitor to the Solar System back then would not see a blue planet Earth as today, but a brown one.

Even so, the above synopsis needs to be treated with caution, as we simply do not yet have enough evidence for a firm view of conditions on primordial Earth. One contrary view is that there were trace amounts of oxygen in the anaerobic Earth's atmosphere and that this was enough to prevent siderite forming in ancient soils, although it would have done under water. Yet, if this was the case then there could not have been appreciable quantities of methane in the atmosphere, although siderites have been found in a number of ancient marine sediments. Yet, if there was no methane then the level of atmospheric carbon dioxide would need to be at the very least roughly 100 times that of today to maintain a liquid ocean before 2.2 bya (Ohmoto et al., 2004). Even so, despite which view ultimately prevails, we do know that a strong natural greenhouse effect (one far stronger than today) was required early in the Earth's history and that carbon dioxide played a part.

Eventually, when Calvin-cycle photosynthetic metabolic pathways evolved in the primordial anaerobic Earth, a new energy source was tapped. This was to change the very nature of the biosphere with the introduction of copious quantities of oxygen.

However, there is much debate as to the early biotic Earth's chemistry and bio-chemistry. Some say that deep-sea hydrothermal vents were responsible for some early carbonaceous deposits and not life. Others say that anaerobic life was responsible. Indeed, recently there has been the suggestion that 3400 million-year-old carbon-aceous strata in South Africa appear to have been formed by an algal mat in a shallow sea, and that the algal mat was photosynthesising, but in a way that did not involve oxygen production (Tice and Lowe, 2004). Certainly photosynthesis, whatever its form, could not take place in the dark environment of hydrothermal vents. The debate about early life continues.

3.2 Major bio-climatic events of the Proterozoic eon (2.5–0.542 bya)

3.2.1 Earth in the anaerobic–aerobic transition (2.6–1.7 bya)

Evolving robust metabolic pathways in a single cell takes time. Oxygenic photosyn-thetic organisms (as opposed to earlier anoxygenic photosynthetic organisms that did not generate oxygen) evolved well before 2–2.5 bya, around a billion years after life first arose and when the Earth's atmosphere suddenly had an appreciable amount of oxygen; although not nearly as much as today. Oxygenic photosynthesis could possibly have evolved as early as 3.5–3.8 bya (Buick, 2008; Lenton and Watson, 2011) but if so did not have any appreciable effect on the atmosphere until 2–2.5 bya. To put this into some temporal context, this means that for between one-third and one-half of the time of life on Earth it existed only in a simple unicellular form in an atmosphere that would be poisonous to nearly all of today's metazoans (multicellular animals). If the 3.8 billion years of life on Earth were represented as a single day then the anaerobic Earth would have lasted until close to midday.

These new photosynthetic organisms (prokaryotic algae) slowly began to change the atmosphere by consuming carbon dioxide and releasing vast amounts of oxygen. The oxygenation of the atmosphere was due to the photosynthetic reaction (see below) accompanied by the burial in marine sediments of some of the carbohydrate, so leaving behind an oxygen surplus (the same process that was to happen on a large scale in the Carboniferous period and which is examined later in this chapter):

$$6CO_2 + 6H_2O \rightarrow C_6H_{12}O_6 + 6O_2$$

The aerobic Earth (with an atmosphere that includes free oxygen) did not come into being overnight. Evidence exists for the first major rise in atmospheric oxygen some 2.23 bya (Bekker et al., 2004), but the oxygen level would still have been a lot less than today. Over the next half a billion years or so (until 1 bya) the atmosphere's oxygen content continued to rise (notwithstanding the Carboniferous excess discussed below) until it was close to today's value (Holland, 1990). (The amount of oxygen in the atmosphere before 2.4 billion years is debatable. Although it is known that quantities were not significant compared to today's value, exactly how small they were, and when free oxygen first exactly could be found in the atmosphere, is still a subject of debate; Knauth, 2006.)

It is important to note that the generation of oxygen by primitive algal mats did not immediately result in a commensurate increase in free oxygen in the atmosphere. Just as today, there were buffering systems in the biogeosphere (hereafter referred to as the biosphere). These buffers soaked up any free oxygen that was released and in the process the abilities of these buffers themselves changed. One of the first greenhouse casualties of the release of free oxygen would have been the primordial abundance of the greenhouse gas methane, irrespective of whether it was just high or extremely high. As previously noted, methane is quite easily oxidised by free oxygen and today has an atmospheric residence time of just 12 years (see Chapter 1). However, over 2 bya, when oxygen was beginning to be released into the atmosphere, initially it would be soaked up by the methane and any other agent (such as iron) capable of being oxidised. Indeed, iron oxide strata were laid down from the oxidation of iron in the primordial oceans to form characteristic banded formations (called banded iron formations or BIFs). At some point, as oxygen continued to be released, much of the methane was used up, so lowering its atmospheric concentration, bringing it closer to today's low value: this meant that the strong methane greenhouse warming of the anaerobic Earth had gone.

We know that micro-organisms have flourished in the oceans since at least 3.8 bya, but there has been much debate as to when land was first colonised. Whereas some 2.7 billion-year-old stromatolite fossil algal mats have been occasionally suggested as being of terrestrial origin, there has been doubt over this idea because algal mats are more likely have been living in shallow marine or freshwater environments. However, in 2000 work by Yumiko Watanabe from Japan and colleagues in the USA and South Africa announced that elemental analysis of 2.6 billion-year-old palaeosols (fossilised soils) revealed ratios of carbon, hydrogen, nitrogen and phosphorus consistent with an organic origin. This is possible geochemical evidence for early terrestrial ecosystems (Watanabe et al., 2000).

With regards to Earth's early, primordial climate, we simply do not know exactly what it was like during the very early part of the oxygenating phase. This phase lasted approximately 1 billion years. Now, while oxygen (aerobic) metabolism does confer an evolutionary advantage (higher energies are involved), there is a flip side to this: anaerobic metabolism has its own advantage in that because it involves lower energies, fewer photons are needed in anaerobic photosynthesis to process a molecule of carbon dioxide. This means that in this environment much of the surplus oxygen generated by the early oxygenic photosynthesisers was mopped up by iron and anything capable of being oxidised, although some would have been immediately used for metabolism by the organism and/or its symbionts (partners in an association in which both somehow benefit) and/or commensals (one partner benefits and the other not harmed). So for many hundreds of millions of years oxygen levels were not high. This meant that anaerobes could continue to live, and live alongside aerobes. Indeed, because they did not need high energies to photosynthesise they would survive as long as oxygen levels stayed low. Then, once all the iron and other oxidisable material had been oxidised, the mopping up of oxygen in the biosphere stopped and oxygen started building up in the atmosphere. This atmospheric oxygen began to erode the methane greenhouse warming effect and this would have caused cooling.

Now, whether or not other factors of significance were at play we simply do not know. One possibility is that the oxygen metabolisers in the absence of the anaerobes (now killed off by the greater free oxygen in the atmosphere) would have been able to expand into the anaerobes' former ecological niches. With their higher energy metabolism it is possible that they globally processed more carbon than their anaerobic predecessors and this extra carbon would have come from the carbon dioxide in the atmosphere. Therefore, not only was any previous methane greenhouse effect reduced (of which we are certain, irrespective of its strength), it is possible that the carbon dioxide greenhouse effect was also reduced.

At this point, despite the Earth being subject to a far greater carbon dioxide greenhouse effect than today, in the absence of the additional methane greenhouse effect (which previously existed in the absence of oxygen) and possibly a reduction in the carbon dioxide greenhouse effect, there was a major glaciation, very roughly 2.2 bya (possibly 2.4–2.3 bya). This was the so-called Snowball Earth I (from which you may deduce that there is at least one other Snowball Earth to come).

There is some discussion about how early life altering the atmosphere's greenhouse properties may have initiated (or at least contributed to) this major glaciation, as I present above. Until the early 21st century little attention was given to Snowball Earth I and alternative explanations have flaws in them that themselves require explanation. Conversely, this life-evolving hypothesis is self-contained, fits the evidence and does not require patching to cover flaws. Since the brief coverage of this topic in the first edition of this book (2007), British Earth system scientists Tim Lenton and Andrew Watson have looked at this major (near-) global glaciation together with early evolution in some detail. They have produced a coherent narrative about how life caused major changes in the Earth system as opposed to the earlier view that major changes in the Earth system enabled evolutionary change. This they recount in their book, *Revolutions that made the Earth* (Lenton and Watson, 2011) and science students interested in this aspect of biosphere (or Earth systems) science might want to refer to this text.

The 2.2 bya Snowball Earth I glaciation was extreme, far more so than the geologically recent series of glacials of the Quaternary ice age, which have occurred in the past 2 million years. Even so, the Earth was obviously not too cold for life to continue, because you are reading this today, but there is much uncertainty regarding the details of Snowball Earth I (and II). It is likely that climatic feedback systems were operating differently then and/or operating at a different magnitude to those in the present, due to the lower solar output. Nonetheless, Snowball Earth I was a cold place, for there is evidence of glaciation at tropical latitudes. What we do not know is whether the Earth was completely covered in ice, right to the equator. An argument against this would be that life in the surface of the sea would surely have been wiped out. The arguments for complete ice coverage include that once ice had reached deep into the tropics then there would have been so much sunlight reflected back off the planet that the Earth would have cooled further still, in a runaway feedback effect. This would occur until ice had reached the equator, with pockets of life surviving only near areas of geothermal activity. Although there is some debate as to the complete extent of ice coverage in Snowball Earth I, we have evidence of glaciation as close as 6 and 16 degrees latitude to the equator (Evans et al., 1997). The Earth did not remain a snowball forever (after all, we are here today). It is thought that volcanoes continued to steadily pump out geological carbon dioxide that boosted the atmosphere's greenhouse effect, which warmed the planet once more (Kirschvink, 1992). This geological carbon dioxide ended Snowball Earth I.

Meanwhile, as geological time passed, the Sun itself continued to warm. It is estimated that about 1.4 bya the Sun's luminosity was only 88% of its present value, and so its output had grown considerably since the Earth's pre-biotic days 2.4 billion years previously (Clough, 1981). It has been shown that the solar 'constant', as a percentage of its current value of 1373 W m^{-2}, can be described as:

$$\text{Percentage change in solar constant} = [1 + 0.4(1 - t/4.7)]^{-1}$$

Here, t is the time in billions of years since the creation of the Solar System (taken for this purpose as being 4.7 bya). So, in another 4.7 billion years (when $t = 9.4$), the Sun will be about 67% hotter in terms of its outgoing radiation (as distinct from its core temperature).

3.2.2 The aerobic Earth (from 1.7 bya)

With the demise of an anaerobic atmosphere, the Earth had experienced its first pollution catastrophe: by comparison, the current anthropogenic (human-generated) release of greenhouse gases is a mere hiccup in the biosphere's cycling processes. Referring to the creation of an oxygen atmosphere as 'the worst atmospheric pollution incident that this planet has ever known' (Lovelock, 1979) may seem somewhat of an odd statement but at the time it foreshadowed great change for the then living species. The anaerobic species that had dominated the Earth previously could not exist in the presence of oxygen and so became confined to anaerobic muds and other oxygen-free environments, where they are still found today.

The availability of oxygen, albeit in low concentrations compared to today but far more than before Snowball Earth I, enabled life to exploit new thermodynamically advantageous metabolic pathways, so giving early oxygen-using species an evolutionary edge in energy terms. Hence they had the prospect of diversifying more than their anaerobic cousins had. Indeed, looking at life on Earth in a very simplistic way, writ large over deep time, diversification and carbon processing have increased and carbon dioxide in the atmosphere has decreased (although of course in the detail there have been ups and downs along the way).

So, very broadly speaking, the past 2 billion years have seen an overall trend of decreasing atmospheric concentration of carbon dioxide balanced by increased solar output. I say 'broadly speaking' because there have been deviations from this trend and periods in time when the poles were ice-free and other times (glacials) when there was considerable ice reaching as far as today's temperate latitudes. (We will return to the question of the overall decline in atmospheric carbon over the past billion years or so shortly.)

Some 1.4 bya water vapour (which has been in the Earth's atmosphere virtually since the planet was formed) and carbon dioxide would have been two of the principal greenhouse gases. Had there not been considerably more carbon dioxide in the atmosphere then than there is today, given the Sun's lower luminosity back then, there would have been continued major glaciation. Given, 1.4 bya, much of the primordial atmosphere's methane had been soaked up by the rising oxygen levels, and that, 'broadly speaking', water vapour was as it is today, then the question arises of just how much more carbon dioxide was necessary in the Earth's early aerobic atmosphere to keep the planet ice-free.

In 2002 two US researchers, Alan Kaufman and Shuhai Xiao from Maryland University and Tulane University, respectively, made ingenious use of biological processes to estimate the 1.4 billion-year-old atmosphere's carbon dioxide concentration (Kaufman and Xiao, 2003). Of carbon's five isotopes only the common ^{12}C isotope and the rare ^{13}C isotope are stable (we examined the utility of the radioactive – non-stable – ^{14}C as a dating tool in palaeoclimatology in the previous chapter). Biological processes, including photosynthesis and the Calvin cycle, have evolved in an overwhelmingly ^{12}C-dominated environment. These processes therefore preferentially use this ^{12}C isotope over ^{13}C. (Hence the reference in section 3.1.2 to isotopic evidence for the possibility that some microfossils were biotic.) The degree of this preferential biological fractionation of carbon isotopes depends on two things: the proportion of ^{13}C in carbon dioxide and the partial pressure of carbon dioxide (in other words, its atmospheric concentration). Kaufman and Xiao (2003) determined the level of fractionation 1.4 bya by isotopic analysis of microfossils of the acritiarch species *Dictyosphaera delicata* compared with carbonate produced by inorganic processes. This in turn enabled them to estimate the atmosphere's carbon dioxide concentration 1.4 bya. Their figures suggest elevated levels of between 10 and 200 times the present carbon dioxide concentrations.

Brief summary of $\delta^{13}C/^{12}C$ analysis

Carbon consists of two stable isotopes with atomic masses of 12 and 13. The heavier isotope makes up 1.1% of naturally occurring carbon (Nier 1950). Variations in the ratio of ^{13}C to ^{12}C are usually expressed as per mil (parts per thousand or ‰) deviations from the ratio in the Vienna PeeDee Belemnite (V-PDB) standard. The original sample was used up some years ago, but a Vienna-based laboratory calibrated a new reference sample to the original, giving rise to the widespread use of the term Vienna PeeDee Belemnite standard, abbreviated to V-PDB. The change in $^{13}C/^{12}C$ is given by the equation:

$$\delta^{13}C/^{12}C = \frac{[(^{13}C/^{12}C)_{Sample} - (^{13}C/^{12}C)_{VPDB}]}{(^{13}C/^{12}C)_{VPDB}} \times 1000‰$$

The $(^{13}C/^{12}C)_{V\text{-}PDB}$ standard always has a value of zero on the $\delta^{13}C/^{12}C$ scale. Samples enriched in ^{13}C relative to the standard have positive values and those depleted in ^{13}C have negative values (for example, see Figure 3.4, below). In terrestrial ecosystems, atmospheric CO_2 is used in photosynthesis, and terrestrial C_3 photosynthetic carbon (see section 3.3.11) has a $\delta^{13}C$ value of around $-27‰$. Conversely, terrestrial C_4 photosynthetic carbon has a $\delta^{13}C$ value of around $-13‰$. The marine situation is more complex.

The V-PDB standard is a specific geological formation that contains the remains of cephalopod mollusc species of the subclass Belemnoidea from the Mesozoic era. The strata come from the PeeDee Formation in South Carolina, USA. This material has a high absolute $^{13}C/^{12}C$ ratio (0.0112372), and was established as a ^{13}C (V-PDB) value of zero. Use of this standard gives most natural material a negative $\delta^{13}C$ value. Because most $\delta^{13}C$ (V-PDB) values are negative (although some, like in the late Ordovician, are positive), the equation researchers sometimes use is:

$$\delta^{13}C/^{12}C = \left(\frac{[(^{13}C/^{12}C)_{Sample} - (^{13}C/^{12}C)_{VPDB}]}{(^{13}C/^{12}C)_{VPDB}}\right) - 1 \times 1000‰$$

The existence of this alternative equation means that it is important, if you are conducting your own $\delta^{13}C$ analysis and relating those results to those of others, to read the Methods section of academic papers (or the Methods appendix) to see how the researchers have used $\delta^{13}C$ analysis. Fortunately for more casual readers of academic papers, most simply show a graph denoting a change in $\delta^{13}C$ across some period of geological time.

The isotopic compositions of carbonate carbon and organic carbon leaving the oceans by burial in sediments are enriched and depleted, respectively, in ^{13}C compared with most carbon in the carbon cycle. On long time scales the sizes and isotopic compositions of these major pools in the carbon cycle must balance to match the bulk carbon isotopic composition. On shorter time scales, however, there can be readjustment of isotopic compositions that are a response to changes in the ratio in which organic and carbonate forms are buried. Essentially, if the rate of burial of (^{13}C-depleted) organic carbon increases then ^{12}C is removed from the oceans disproportionately quickly compared with ^{13}C. In response to this, ^{13}C will accumulate in all ocean carbon pools (i.e. their $\delta^{13}C$ value will increase) until either the rate of organic carbon burial falls or a new isotopic equilibrium is reached (with a higher rate of organic carbon burial but having a higher $\delta^{13}C$). Consequently, changes in the $\delta^{13}C$ values of marine carbonates and organic carbon over geological time can be used to estimate the fractions of carbon that are buried as carbon ultimately from an organic source (photosynthetically derived, such as carbon from decaying plants, or carbonate from carbon dioxide from the same, or fossil fuel, which itself is of photosynthetic origin) and carbonate forms with carbon from inorganic sources.

This summary is all too brief. For further information, and information on other isotopes as indicators of past environmental change see Newton and Bottrell (2007), from which this text box is partly taken.

As a greenhouse gas, carbon dioxide has retained its importance ever since. But what brought us the oxygen in the first place? What were the early photosynthetic species? Well, the earliest oxygenic photosynthesisers were most probably organisms like cyanobacteria (also known as blue-green algae). These early microbial species were then incorporated into other cells so that there were then, if you like, cells within cells (this is the endosymbiotic theory). The absorbed cells became organelles such as chloroplasts (found in plant cells doing photosynthesis) and mitochondria (for energy-releasing metabolism). These more complex cells with specialised organelles are called eukaryotes. Fossils have been found of one of the first early eukaryotic aerobic species, dating from 2100 mya. It was similar to a species of alga living today, *Grypania spiralis* (Han and Runnegar, 1992). *G. spiralis* grows in long, spiral filaments about 1–2 mm wide. Another good candidate for a modern descendant of an early eukaryotic oxygenic photosynthesiser has been presented by a large international team of researchers in 2008 led by Robert Moore and Miroslav Obornik. It is a microbial marine brown ball related to present-day apicomplexian parasites (one of which you may know: *Plasmodium falciparum*, the malaria parasite). This brown ball is called *Chromera velia* and its discovery came as a surprise, for although it is related to the apicomplexians, it is fully photosynthetic.[3]

Subsequent to Snowball Earth I, over 2 bya, the anaerobic species were to lose their domination of the planet to the aerobes that generated and/or used oxygen. Meanwhile, the huge carbon dioxide greenhouse blanket declined over the next billion years while the Sun continued to grow stronger. If this greenhouse effect had not declined then the Earth may have seen a runaway greenhouse effect. In this, warming of the Earth would have caused more water to evaporate from the oceans. Water vapour is a greenhouse gas and would have generated further warming, so resulting in a positive-feedback loop (opposite to the albedo feedback loop that would have cooled the Snowball Earth).

Where did the primordial carbon dioxide go? The carbon dioxide was chemically transformed by biological processes into carbonates that were laid down as great beds of chalk, and by abiotic, purely chemical, processes.

By 800 mya life had diversified so much that multicellular organisms had evolved, including animals. Indeed, the eon 542 mya to the present day is known as the Phanerozoic eon, which means the age of visible life. The significance of multicelled living things was profound for the biosphere. More than one cell means that it is possible to have a more advanced structure than the simple cellular repetition seen in algal colonies. A more advanced structure in turn enables specialization of cells for processing food, respiration and other metabolic functions. This meant that carbon

---

[3] Patrick Keeling wrote a short article explaining this organism's evolutionary significance for *Nature* in 2008.

could be processed in greater quantities, and more quickly. Greater carbon processing could have led to a dip in atmospheric carbon dioxide. This would explain the global cooling thought to have taken place between 850 and 635 mya. We know this cooling took place because it was so marked that large ice caps formed, leaving geological traces that can be seen today. The large ice caps reflected sunlight. This further cooled the Earth, so enlarging the ice caps, just as happened in Snowball Earth I, and this led to a second Snowball Earth, Snowball Earth II in the aptly named Cryogenian period (meaning ice-generating). It should be pointed out that this was not one single event but a clutch of at least three main events taking place within the Cryogenian period.

An interesting possibility as to one aspect to the extra biological carbon processing taking place prior to Snowball Earth II is that life colonised the land around 850 mya. US researchers Paul Knauth and Martin Kennedy in 2009 used $^{13}C/^{12}C$ isotope ratios from many old sediment samples originally deposited by freshwater run-off from continents. The results suggested that the continental run-off contained carbon that was biological in origin, which in turn implies terrestrial life. In short, they suggest that the continents were covered by a thin green carpet in many places before Snowball Earth II. However, the problem with this theory, compelling as it is (it fits in with the new theories of evolution and the Earth system and its climate), is that the evidence is somewhat circumstantial. Algae covering the land, unlike larger vascular plants with good structure such as leaves, do not leave much in the way of fossils. Nonetheless, something must have altered the $^{13}C/^{12}C$ isotope ratios and terrestrial life is arguably the best candidate. However, there is some fossil evidence for very early terrestrial life prior to Snowball Earth II from a small team of British and US researchers – Paul Strother, Leila Battison, Martin D. Brasier and Charles Wellman – in 2011. They reported the discovery of large populations of diverse microfossils with cell walls that contained organic compounds. These assemblages contained multicellular structures, some approaching 1 mm in diameter (which is large). The fossils offer direct evidence of eukaryotes living in freshwater aquatic and sub-aerially exposed habitats during the Proterozoic era. The apparent dominance of eukaryotes in non-marine settings by 1 bya indicates that eukaryotic evolution on land may have commenced far earlier than previously thought.

As with Snowball Earth I, there is considerable debate as to whether Snowball Earth II saw heavy glaciation across the equator. One idea (Pollard and Kasting, 2005) is that the Snowball Earth came about slowly and at the equator ice may have formed at rates as low as 7 mm a year and reached the equator, at least in part, by glaciers flowing from higher latitudes. In this case, the researchers speculate, the sea glacier ice formed would have been far more transparent than we find in many of today's glaciers. Such transparent ice may have allowed photosynthesis to take place in the water below. Whatever the case, there is little doubt that Snowball Earth I and II were extremely large events for the Earth's biosphere.

As with Snowball Earth I, it is thought that volcanic carbon dioxide came to the rescue of Snowball Earth II by building up over several million years to atmospheric concentrations a few hundred times today's levels. As with the first Snowball event, this led to considerable global warming that ended the massive glaciation. However, there is one final detail worth mentioning, because we will see similar events later. There is evidence to suggest that the end of Snowball Earth II was accelerated by methane that was released by frozen biogenic carbon. As ecosystems prior to Snowball Earth II drew down carbon dioxide, so this carbon became trapped. At the end of Snowball Earth II, warming of the Earth released this carbon back into the atmosphere, an event that further accelerated the warming. Martin Kennedy, David Mrofka and Chris von der Borch, who published this work in 2008, suggest that this carbon came from permafrosts at low latitudes (where the tropics are today).

Then, with the ice caps gone and a super greenhouse effect in operation, the average global temperature was far warmer than today and might even have been as much as 50°C.

As with much palaeontology, especially relating to anything before the beginning of the Cambrian period (542 mya; see Appendix 2), it is difficult if not impossible to derive a detailed picture of any exactness. We do not know how long this warm period lasted. What we do know is that at some stage this super carbon dioxide greenhouse effect declined, as carbon in the form of carbonate layers was deposited over the glacial strata from Snowball Earth II. We also know that once again, as followed Snowball Earth I, carbon-cycling feedback systems re-organised and stabilised matters and the climate.

Whereas global carbon cycling, through the evolution of photosynthesis and then photosynthetic plants with more structure, is now one of the principal theories posited to explain Snowball Earth II (which is presented here because the text deals with biology and climate), there are other theories, just as there are for Snowball I. For example, ocean circulation plays an important role in transporting heat and it has been proposed that major global cooling only occurs when high-latitude oceanic gateways open and low-latitude ones close (Smith and Pickering, 2003). Geologists supportive of this gateway theory have been quick to point out that even if gateways are important in regulating heat flow from the equator, they cannot be too critical a factor due to the lack of glacial deposits 1–2 bya, a time when oceanic gateways came and went.

Another theory has it that Snowball Earths only occur when supercontinents break up (the zipper-rift model), but then an explanation is needed as to why there was no glaciation 1.7 bya when the the Nena-Columbia supercontinent broke up.

But Snowball Earth II did take place and then it ended. Into this subsequent warm world came the surviving multicelled plant and animal (metazoan) species (and, of course, the surviving older single-celled ones, both prokaryotic and eukaryotic). Beyond their Snowball refugia they had nearly a whole planet to colonise. Ancient primitive fossils are scarce and their interpretation is difficult, but from fossil evidence it is likely that lichen-like symbiosis between coccoidal cyanobacteria (a group of spherical, photosynthetic, nitrogen-fixing bacteria) or algae and fungi took place some 600 mya before the evolution of vascular plants (Yuan et al., 2005). The animal metazoans had all the basic body plans we see today, such as diploblastic species (two multicellular layers) and triploblastic coelomate species (three-layered with a body cavity). The time just prior to the Cambrian was a period of considerable speciation, which not surprisingly is known as the Precambrian boom. The first animals had no backbones. These invertebrate species included trilobites, clams and snails. Between then and now species in terrestrial ecosystems have continued to evolve. But just as the time between an anaerobic Earth and an aerobic Earth (with its multicellular creatures) saw periods of considerable biosphere re-organisation, the road between the Cambrian and now was not a steady one, as is evident from times of speciation and times of extinction, both beyond and within background rates. It is not possible to review even a summary of the extensive palaeobiological literature here (for which readers should seek out other texts such as Briggs and Crowther, 2001), but it is worth citing a few of the major climate-related biological events. Then we can put recent changes of the past million years or so, not to mention likely future changes over the next thousand years (see Chapter 8), into a broader context.

3.3 Major bio-climatic events of the pre-Quaternary Phanerozoic (542–2 mya)

3.3.1 Late-Ordovician extinction (455–435 mya)

In the Ordovician period life was mainly marine. During most of this period life was flourishing in the seas and oceans, and terrestrial life was largely restricted to algae in splash zones at the edge of the sea and possibly in brackish environments. But by the late Ordovician life was beginning to get an enhanced foothold on land. Yet at this time life was threatened when globally about 22% of all families of species became extinct, which makes the late Ordovician the time of one of the largest episodes of mass extinction in the Earth's history. The extinction is thought to have happened in two principal phases about 0.5–2 million years apart that wiped out nearly all major benthic and planktic groups. Among the molluscs, 125 genera of nautiloids were reduced to 25, the 88 gastropod genera went down to 55 and bivalve genera were down from 84 to 32. The two phases of the late Ordovician were associated with the growth and decay of a large Gondwanan ice cap when the southern continents drifted over the South Pole. Sea-water temperatures fell by as much as 8°C, and a similar degree of cooling even took place in the tropics, which caused the loss of many tropical species. The sea level fell by between 50 and 100 m depending on regional eustatic conditions. Changes in ^{13}C of about 7‰ compared to ^{12}C in sediments indicate a major change in carbon cycling (see section 1.3 and the text box in section 3.2.2). The rapid global cooling and growth of the ice caps contrasted with the greenhouse conditions prevalent before and after the event. The two extinction phases are thought to have related first to the cooling (caused by the weathering of recently formed mountains ranges: see the silicate carbon dioxide equation in section 3.3.3) and then second to the warming: species surviving the first change were then faced with the second. What appears to be characteristic is that both events were sudden, in geological terms. It is perhaps worth reflecting, as we shall see, that recent (Quaternary) glacials exhibit periods of rapid, rather than gradual, climatic change. (We will return to this when discussing climate thresholds in section 6.6.8.) As we shall see with some future major extinctions, there was a fungal spike associated with each event, suggesting a world of rotting biomass. With the late-Ordovician extinction there was another spike: a spike of microbialites (including stromatolites, thrombolites and dendrolites).

The above account is the standard story. More recently the notion has arisen that the early terrestrial colonising vegetation, such as mosses (Bryophyta), could have caused silicate drawdown of carbon dioxide, much like the later silicate drawdown that occurred with the rise of vascular plants (see section 3.3.3). The idea is that, even though mosses do not have the root system of vascular plants, they could have enhanced the erosion of rock, albeit far more modestly that vascular plants: after all, mosses have to access mineral nutrients too. Not only would this (as we shall shortly see) result in carbon dioxide drawdown, causing cooling, but this erosion of rock (the surfaces of which were previously devoid of such life) released nutrients. This would have released phosphorous, which then got washed into the oceans, so enhancing marine biological productivity that further drew carbon dioxide to the point where glaciation took place. In 2012 British researchers Tim Lenton, Michael Crouch and colleagues supported this idea with experiments determining the effect of mosses on rock and then applied the results to a model of the carbon cycle to see the likely effect on atmospheric carbon dioxide. This idea certainly fits the ^{13}C evidence of a change in global carbon cycling; however, further corroborating work is needed.

Microbialites (including stromatolites, thrombolites and dendrolites) were formed by microbial mats that were common in the late anaerobic Earth. They made a resurgence following the late-Ordovician extinction and in one sense this is analogous to the opportunistic fungi at the end-Permian and Cretaceous–Tertiary extinctions. However, in this case the opportunity came both in the form of a ready food supply (the seas containing dead marine life would have provided nutrients for algae, which would release phosphorous in the way that Lenton et al. explored) and reduced pressure from grazing animals. The microbialite resurgence lasted some 5 million years (Sheehan and Harris, 2004). There is also some evidence for a similar stromatolite resurgence associated with the later end-Permian extinction (see section 3.3.5).

3.3.2 Late-Devonian extinction (365–363.5 mya)

The late-Devonian extinction appears to have been a succession of extinction pulses. Many cephalopods were decimated, as were the armoured fishes that are associated with much of the Devonian. Many coral reef species and crinoids (sea lilies) were also lost. Both terrestrial and marine systems were affected, the tropics more so than the poles. There is much debate as to the cause of this cooling but one of the preferred theories is multiple asteroid strikes (possibly five significant ones; there is evidence from Morocco). Another cause is likely to have been the spread (through changes in ocean circulation) of anoxic (oxygen-deficient) deep water. Of course, there could have been more than one factor operating.

3.3.3 Vascular plants and the atmospheric depletion of carbon dioxide (350–275 mya)

Vascular plants (the pteridophytes and seed plants) have a vascular system of xylem and phloem that enables them to have a free-standing structure, and hence they would have gained height on the land compared to the more primitive plants. Equally importantly they have extensive roots that helps to break up bedrock. Vascular plants evolved after some 400 mya and from around 350 mya larger vascular plants arose and started to spread across much of the terrestrial globe (as today, where conditions permit). These plants contained new types of biomass, such as lignin, which is important for imparting structural integrity to those plant species that have it. Some of these structural compounds were resistant to the then available pathways of organic degradation. This in turn increased their chance of burial. (Dead plants today tend to rot on the ground before they are buried.) Chances of burial were also further enhanced by the Permo–Carboniferous glaciation, 330–250 mya (see section 3.3.4), with the repeated waxing and waning of its ice sheet causing the sea level to fluctuate with periods suggestive of Milankovitch influences. Burying this tremendous quantity of carbon led to a residual excess of atmospheric oxygen.

This oxygen excess can be easily understood when considering the biospheric equilibrium between photosynthesis and respiration: carbon dioxide and water combining through photosynthesis to form carbohydrate and oxygen one way, and the respiratory reaction the other (see the equation below). With the biosphere in a steady state these reactions are equal and opposite (regional and seasonal variations notwithstanding). However, remove some of the carbohydrate by burial (shifting the equilibrium to the right) and an oxygen surplus results:

$$6CO_2 + 6H_2O \rightleftharpoons C_6H_{12}O_6 + 6O_2$$

$\downarrow$ (Part burial leaves an
atmospheric oxygen surplus)

Geologically isolated organic
compounds $C_6H_{12}O_6$

The forests of the Earth 400 mya were hugely different from those of today, or even those emerging 350 mya. These early forests of the Devonian were largely composed of moss-like species with primitive arthropods living among them (some of which were quite large by comparison with their present counterparts, growing to a metre or more). The Devonian forests were only a few centimetres high, forming green mats on the land. However, the plants that followed had thick-walled cells to allow water to be conducted up their stems, but, even so, none of these late-Devonian plants were more than a metre or two in height.

It was the evolution of vascular plants in the Carboniferous period that allowed there to be forests of heights comparable with those today, even if the dominating species were different. Vascular plant roots penetrated further into the substratum than the plants that preceded them. Vascular plants also allowed forests to have a higher biomass per unit area and, because of their roots, allowed a greater area of the land to be covered with plant life. Globally, therefore, the terrestrial biomass increased, but equally some of this biomass became buried, as described above. This buried material gave rise to the vast Carboniferous and Permian coalfields. With so much organic material taken out of the carbon cycle there was a significant atmospheric oxygen surplus. It is thought that oxygen concentrations started to rise after 400 mya and reached a peak 300 mya of 35–40% before returning to close to today's concentration of 21% over 200 mya. The high-oxygen environment would have facilitated forest fires and this would have helped raise carbon dioxide levels around

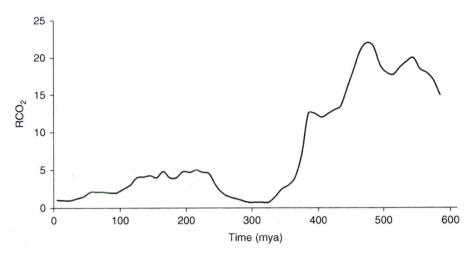

Fig. 3.1 A broad outline of Phanerozoic atmospheric carbon dioxide (shown in low resolution that excludes sharp spikes and short-lived fluctuations). RCO_2 means the carbon dioxide relative to mid-20th-century carbon dioxide levels of 300 ppmv. The best-estimate curve shown is summarised from Berner (1998). Berner's original work also includes upper and lower estimates and relates this curve to fossil stomata and isotope data which broadly relate to the key features, namely: that for the past 500 million years carbon dioxide has declined to present levels with a dip 300 mya due to the rise of vascular plants.

250 mya (see Figure 3.1), so helping end the Permo–Carboniferous glaciation (see section 3.3.4).

However, biological interactions with atmospheric carbon dioxide are a little more complex than this. They also depend on geological carbonate and geological silicate (summarily represented as SiO_2 below). Indeed, silicates previously trapped in the substratum could be more easily accessed by the root systems of the new vascular plants. Geological magnesium and calcium silicates have a chemical relationship with atmospheric carbon dioxide that may be summarised succinctly as follows:

$$CO_2 + CaSiO_3 \rightleftharpoons CaCO_3 + SiO_2.$$

Bringing all these factors together, computer models (such as GEOCARB II) have been used to suggest that the role of plants can greatly affect the level of atmospheric carbon dioxide through a stabilising feedback loop. The evolution of plants with strong root systems enhanced the weathering of rocks, hence the release of calcium and magnesium silicates (driving the above equilibrium to the right) and so more atmospheric carbon dioxide became drawn down as calcium carbonate over the period around 350 mya. This would have been doubly so because plants themselves have biomass so that carbon dioxide would also have been sequestered as organic plant material. With carbon dioxide drawn down the planet became cooler. This would have resulted in less evaporation from the oceans, and hence less rainfall. A cooler world with less carbon dioxide and lower rainfall is one in which water erosion and plant growth are suppressed. Less water erosion and less plant-root erosion reduce silicate-bearing rock weathering. Of course, all this assumes that other contributions to atmospheric carbon dioxide (such as volcanic activity) remained constant (Berner, 1998): as we shall see below, this was not always so. However, the lesson to be learned here is that life does have an effect on the biosphere and plays an important part in the long-term carbon cycle. Some argue – for example Britain's Andy Watson (2004) – that the effect of life on climate is largely a stabilising one other than in periods of evolutionary innovation, during which there is some re-organization of the carbon cycle. As discussed in Chapter 1, it is the short-circuiting of this long-term or deep carbon cycle that is causing much of current global warming.

3.3.4 Permo–Carboniferous glaciation (330–250 mya)

During the Carboniferous period the temperature dropped for two reasons. First, global climatic forcing from atmospheric carbon dioxide was low, for the reasons discussed above: vascular plants were becoming established, so increasing mineral weathering, and more biomass was being buried. Second, many of the Earth's land-masses came together to form the supercontinent Gondwana. This stretched from the equator to the South Pole and a large part of it sat over the pole. With oceanic currents unable to transport warmth from the equator, a terrestrial ice sheet formed, covering much of Gondwana (similar to the one on Antarctica today). The rest of Gondwana would have been ice-free but somewhat arid, since on a cool Earth there is less oceanic evaporation and hence low global precipitation levels.

Back then, the present-day North America and Eurasia lay together on the equator, and a steamy coal-forming belt stretched in a continuous band from Kentucky to the Urals that was then the warmest place on the planet. The coal formed due to the reasons, discussed in the previous subsection, of enhanced carbon cycling and carbon burial, especially in swampy delta-like areas. The burial rates were further increased by the Permo–Carboniferous glaciation due to the repeated flooding of continental margins arising from the waxing and waning of Gondwana's south polar ice sheet. It is now well established that many of these Carboniferous sea-level cycles had an approximate range of 75 m (this is only slightly less than the sea-level cycles associated with the current series of glacial and interglacials, which are closer to 120 m). It is also well established that each cycle saw the sea slowly decline over tens of thousands of years yet rapidly increase within a thousand years or so; again this is similar to sea-level change in our current glacial–interglacial series. There is a strong suspicion that the waxing and waning of the Gondwana ice sheet in the late Carboniferous was driven by Milankovitch parameters (see Chapter 1).

The ecosystems that were largely to become coal-bearing strata were dominated by two tree-sized club mosses (Lycophyta): *Lepidodendron* and *Lepidophoois*. Mean-while, inland and away from the poles, during the glacials much of what are today temperate latitudes were cooler and dryer. There is evidence that during the warm Permo–Carboniferous interglacials these inland dry zones became far wetter than during the glacials. Because Gondwana was so large, huge drainage systems were formed through erosion. Often these carried material that covered the organic mat-ter in the continental margin swamplands and this again enhanced organic burial. Meanwhile, the resulting atmosphere's oxygen excess enabled giant flying insects to evolve, as flying requires a fast metabolism and hence effective oxygen transition, which is greatly facilitated by the gas having a higher atmospheric partial pressure. Metabolic models of these insects have been constructed. We know their size from fossils and can reasonably assume a similar density to modern-day insects; hence we can deduce their mass. We can therefore estimate the energy they would have required for flight and so, in turn, their oxygen requirements. This gives us another way of estimating the oxygen excess 300 mya: such estimates seem to fit those derived from the carbon model.

3.3.5 End-Permian extinction (251 mya)

The largest of all the mass extinctions of the past 600 million years was the end-Permian or Permo–Triassic event, which occurred 251 mya. The marked turnover of species – fauna and flora, marine and terrestrial – is used to mark the boundary between the Palaeozoic and Mesozoic eras. At that time more than 60% of animal species, both marine and terrestrial, disappeared. Some 90% of ocean plant and animal species vanished, as did 70% of vertebrate families on land.

The tremendous volcanic activity of the Siberian Traps at the end of the Permian undoubtedly played a major part in this extinction. The Siberian Traps are a large igneous province that formed as a result of a mantle plume. They are centred around the present-day Siberian city of Tura and also encompass Yakutsk, Noril'sk and Irkutsk. This was a continental-flood basalt event. The Traps area, including associated pyroclastics (rocks formed by materials thrown out by explosive volcanic eruptions), is just fewer than 2 million km^2 and greater than that of present-day Central and Western Europe combined. Estimates of the original volume of the Traps range from 1 to 4 million km^3 (3 million km^3 is often cited). The Siberian Traps eruptions lasted at full intensity for about 1 million years, which coincides with the end-Permian extinction. The most accurate dating method available at the moment is argon-argon radiometric dating which still has too many uncertainties to conclusively prove the exact timing. The eruptions ejected volcanic ash and sulphur aerosols into the atmosphere. This would have caused global cooling that probably lasted for decades or even centuries after the main eruptions ceased. To put this into some kind of present-day context, the largest eruption of the twentieth century, that of Mount Pinatubo, was tiny compared to the Siberian Traps but even that caused a 0.5°C drop in global temperatures in the following year. The largest eruption in historic memory occurred in Iceland, 1783–4, and generated some 12 km^3 of lava on the island: the Siberian Traps by comparison released 1–4 million km^3. The poisonous gases given out by the Icelandic event are recorded as killing most of the island's crops and foliage and lowering global temperatures by about 1°C. It is therefore most likely that the end-Permian event caused a short, sharp period of cooling that killed off much life. In addition to ash and sulphur aerosols there would have been the release of considerable quantities of greenhouse gases such as carbon dioxide, methane and water vapour. Carbon dioxide stays in the atmosphere much longer than dust and sulphur aerosols, especially if the Earth's photosynthetic pump is disabled, so its climate-changing effects last longer. The warming from any major carbon dioxide release would have become apparent only once the cooling dust and aerosols had been removed from the atmosphere.

If the Siberian Traps volcanic activity were not enough, it has been suggested that the resulting global warming from volcanic carbon dioxide may have been enhanced by large releases of methane from the oceans. Methane hydrates (or clathrates) exist in considerable quantities off the continental shelf (see section 6.6.4 on methane hydrates, and the Toarcian and Eocene warming events below). Methane hydrate stability depends on pressure and temperature: reduce the former or increase the latter and they quickly dissociate, releasing methane gas, which (as noted in Chapter 1) is a powerful greenhouse gas. At the end of the Permian if volcanic carbon dioxide (or carbon dioxide or methane from magma-warmed, organic-rich sediments) had warmed the planet then some of this heat would have been conveyed to oceanic deep waters and possibly destabilised methane hydrates. The methane would have further enhanced any global warming, so further stressing ecosystems globally.

As if this methane enhancement were not enough there is (somewhat speculative) evidence to suggest that much of the extinction 'could have occurred in a single bad day' (Ward et al., 2000). Geological strata, where the Karoo Basin subsequently formed in South Africa, imply that a forest was literally stripped away in a very short time (Ward et al., 2000). Could it be, in addition to the Siberian Traps vulcanism generating global cooling and then warming, and the consequential oceanic methane release, that there was an asteroidal or cometary impact (Becker et al., 2001; Jin et al., 2000)? In 2004 an Australian/US team of researchers reported their discovery of an impact crater that they say may have been responsible for the end-Permian extinction. Located off the north-west Australian coast (due west of Roebuck Bay and north of De Grey) the crater may be about 100 km in diameter but this estimate is tentative due to erosion (Becker et al., 2004). Even so, it is the same order of magnitude as the Chicxulub crater in Yucatan, Mexico, which is thought to have played a part in the Cretaceous–Tertiary dinosaur extinction (see section 3.3.8).

Of course, even without evoking climate change, there were other, minor (non-global) effects of the Siberian Traps vulcanism. Being in or around Siberia at the time would not have been healthy simply because of the widespread volcanic activity. Another minor effect (comparatively speaking) would have been the destruction of the ozone layer, caused by gas emissions. Chlorine and fluorine gases are released by almost all volcanic eruptions and they destroy the ozone layer. Without the ozone layer, harmful ultraviolet rays can kill exposed terrestrial organisms, so contributing to the mass extinction.

For many years the cause of the end-Permian mass extinction appeared to involve 'a tangled web rather than a single mechanism' (Erwin, 1994). Yet we know that much of life was simply wiped out, but not all life. Indeed, for a brief time it appears that fungi flourished. There is a fungal spike in the geological record around the onset of the Siberian Traps eruption, supposedly due to fungi living on dead and rotting vegetation. (A somewhat similar fungal spike has been identified in strata associated with the Cretaceous–Tertiary extinction: see below; only 'somewhat' because that fungal spike did not take place at the onset of the vulcanism that led up to that extinction, but is associated with an asteroid strike.)

Three papers published in 2011 provide some clues for a single principal cause. The first was in *Nature Geoscience* and written by Stephen Grasby, Hamed Sanei and Benoit Beauchamp. They discovered that many end-Permian strata contained char that looked remarkably like fly ash from modern coal-fired power stations. The rocks were from the Canadian High Arctic and dated to a time immediately before the mass extinction. Based on the geochemistry and petrology of the char, they propose that the char was derived from the combustion of Siberian coal and organic-rich sediments by flood basalts, which was then dispersed globally. The second paper, by a western European team led by Micha Ruhl, got around the problem that even a long, drawn-out period of volcanic emission from the Siberian Traps would not have imposed a volcanic emission burden sufficient to result in global mass extinction. They used geological carbon isotope data that indicated a strong ^{13}C depletion in the atmosphere in the end-Triassic, within only 10 000–20 000 years. The magnitude and rate of this carbon cycle disruption can be explained by the injection of at least 12×10^3 Gt of isotopically depleted carbon as methane into the atmosphere. Concurrent vegetation changes reflect strong warming and an enhanced hydrological cycle. Hence, end-Triassic events are robustly linked to methane-derived massive carbon release and associated climate change. The third paper was by an international European team led by Stephan Sobolev and Alexander Sobolev. They came up with a mechanism, a model and field evidence suggesting that recycled oceanic crust rich in volatiles caused a massive gaseous eruption (particularly a massive degassing of CO_2 and HCl). The volatiles were released in a short period of time even in the period of volcanic activity that lasted a long while. This short but considerable release of volatiles could have been enough to cause the short, sharp extinction we see in the geological record. Taking these three papers together it seems likely that the extinction was caused by a geologically very short period of intense, explosive activity involving organic-rich geological strata, probably involving the Siberian Traps, that added considerable greenhouse gases to the atmosphere, causing major abrupt climate change.

Just after the end-Permian extinction there is also some evidence of a stromatolite resurgence, similar to that of the late-Ordovician extinction. This is thought to be due to a decline in marine animals that would feed off of algal mats. Isotopic evidence using ^{13}C suggests disrupted global carbon cycle activity for 3 million years following the extinction event. Although the exact reasons for this remain unclear, it suggests a relationship between biodiversity and Earth system (biosphere) function (Payne et al., 2004). Almost certainly part of the change in $^{13}C/^{12}C$ is due to the decay of vegetation releasing carbon rich in ^{12}C, but some may possibly be due to the release of organic material in geological strata due to volcanic action of the Siberian Traps. This ^{13}C evidence comes in the form of a carbon isotope excursion (CIE) and we will return to CIEs shortly.

3.3.6 End-Triassic extinction (205 mya)

The end-Triassic extinction has long been recognised by the loss of most ammonoids (a now-extinct group of single-shelled gastropod molluscs, the same group as present-day snails), many families of brachiopods (a shrimp subclass of the Crustacea), Lamellibranchiata (bivalve molluscs), conodonts (a now-extinct group of fishes) and marine reptiles. There were also terrestrial extinctions. The cause of this extinction is unclear although asteroidal impacts have been suggested and there is some evidence suggesting a possible multiple strike, with a 40 km-diameter crater in Saint Martin, western Canada, a crater 100 km wide in Manicouagan in eastern Canada and a 25 km crater in Rochechouart, France. The arguments against this are that the craters have different magnetic fields. (As a crater cools so its magnetic field aligns with the Earth's, but the Earth's field takes a thousand years or more to re-set, which suggests that the craters were formed at least that far apart and not simultaneously.) The evidence against this argument is that the largest crater had such a volume of molten rock that it would have taken more than a thousand years to cool to the Curie point (580°C for pure magnetite), when the Earth's magnetic field would have become embedded in the rock (Vent and Spray, 1998).

An alternative explanation for the end-Triassic extinction is volcanic action due to the break up of the old continent of Pangea. This would have injected carbon dioxide (warming) and/or sulphates (cooling) into the atmosphere, causing a climate blip that resulted in the extinction. Another explanation is that both these scenarios happened simultaneously. What is known is that the geological strata of the volcanic Chon Aike province of South America and Antarctica were created by a super volcano 170 mya, as was the Karoo-Ferrar province 183 mya. These eruptions lasted millions of years and would have had an impact on the global climate. So, from the end of the Triassic through to the first half of the Jurassic there were a number of global climate-related extinction events of varying degrees.

3.3.7 Toarcian extinction (183 mya)

As noted above, the first half of the Triassic saw a number of events but there has been particular recent interest in the Toarcian (early [late lower] Jurassic) extinction due to both its magnitude and growing evidence for its similarity to that of the Eocene thermal maximum, or Palaeocene–Eocene Thermal Maximum (see section 3.3.9). For a brief time (120 000 years), in the geological sense, during the Toarcian (183 mya), many of the Earth's ecosystems experienced severe disruption. Carbon laid down in Toarcian strata reveals a pronounced decline of approximately 5–7‰ in the proportion of ^{13}C. This suggests that a pulse of ^{12}C diluted the carbon pool. The source of this ^{12}C includes carbon from marine carbonates, marine organic matter and terrestrial plant material (Cohen et al., 2007).

This decline in ^{13}C is referred to as a carbon isotope excursion (CIE). As discussed in Chapters 1 and 2 the majority (98.9%) of stable carbon exists as the ^{12}C isotope but some exists as other isotopes and radioactive (hence unstable) ^{14}C, which is used in carbon dating. Around 1.1% of stable carbon exists as ^{13}C, but the Rubisco enzyme in photosynthesis has a higher affinity for ^{12}C. This means that organic matter (photo-synthetic organisms and the animals that live off of them) has proportionally far more ^{12}C than ^{13}C compared to non-organic geological strata. This relates to the isotopic balance in photosynthetically fixed carbon (even if the carbon is from animal remains further up the food chain or on a higher trophic level). However, there is also carbon in carbonate ions that is not photosynthetically fixed ^{12}C and here the isotopic balance is subject to the environmental sources (of which previously photosynthesised carbon is still one) and sinks of the various carbon isotopes. This carbonate is present in creatures such as the Foraminifera which have shells containing carbonate. Analysis of the carbon isotopes in such species provides an indication of the isotopic car-bon balance in their environment. The implication for the Toarcian CIE (and for the end-Permian and the Eocene CIE) of a decrease in ^{13}C is that at the time the sediments were formed there must have been comparatively far more ^{12}C in the atmospheric carbon pool, which was then fixed by living creatures into the carbonate in the strata being laid down. As organic matter is the major ^{12}C-rich source then it is likely that there was an injection of organically sourced carbon (be it fast or deep) into parts of the fast carbon cycle from which the deep carbon cycle sediments were formed and which we analyse today. This carbon conceivably could have come from forests (although this is unlikely). (If it had all come from forests then woodlands planet-wide would have needed to contribute the carbon, but we do not see such an extinction of tree species.) Alternatively, the ^{12}C could have come from organic carbon that had originally been buried in soils, or the combustion of carbon-rich sediments (such as coal) or methane hydrates in the ocean. This last is discussed further in section 3.3.9 with respect to the Eocene.

The Toarcian CIE (as we will also see with the Eocene CIE) also suggests that the time was one of elevated atmospheric methane and carbon dioxide. (Note: even if initially there was an increase in atmospheric methane, much would then have become oxidised to carbon dioxide.) Of course, both methane and carbon dioxide are greenhouse gases and there is evidence of a simultaneous rise in global temperatures. The Toarcian CIE was also a time of a 400–800% increase in global levels of erosion (suggesting a marked increase in global precipitation) and elevated levels of carbon burial (which again one would expect in a warmer and wetter greenhouse world). There was also a terrestrial and marine mass extinction, and strata from present-day Yorkshire in the UK reveal a decline in marine invertebrate species of greater than 50%. Although the duration of the most significant climate change part of the event lasted 120 000 years, overall the event lasted some 200 000 years and is coincident with the geological formation of the Karoo-Ferrar large igneous province in South Africa and Antarctica. Here, there is slowly mounting evidence suggesting that the relationship between these two is causal. It could be that magma from the Karoo-Ferrar province intruded into Gondwana coal seams, so igniting or driving off methane (McElwain et al., 2005). It could be that the subsequent warming destabilised marine methane hydrates, so releasing more ^{12}C into the atmosphere. Much depends on the estimated size of the methane hydrate reservoir at the time and whether there was enough to account for the Toarcian CIE. There is also a notion that matters within the CIE event might have been further timed (or exacerbated) by Milankovitch astronomical pacing (Kemp et al., 2005), but this last is fine detail (although it is likely to be important when considering critical thresholds in biosphere stability; see section 6.6.8).

In short, what is known is that the Toarcian extinction was caused by a perturbation in the global carbon cycle that resulted in a period of greenhouse warming (above that of the average greenhouse warming of the age) and a perturbation in ocean chemistry. We will return to the question of CIEs in sections 3.3.9 and 6.6.4.

3.3.8 Cretaceous–Tertiary extinction (65.5 mya)

Evidence abounds that the Earth during the mid-Cretaceous was warmer than it is today. For example, the breadfruit tree (*Artocarpus dicksoni*) has present-day relatives that flourish in a tropical 15–38°C, and fossil remains of this plant from the period have been found in western Greenland, and remains of large crocodile-like champosaurs have been found in the Canadian Arctic archipelago. This suggests that in the Arctic the mean ambient temperature exceeded 14°C. This has been supported by biomolecular climate proxy records based on the composition of membrane lipids that exist in the common Crenarchaeota marine plankton. (A warm Arctic, however, contradicts computer climate models that to date do not appear to reflect high-latitude temperatures nearly as well as they do mid-latitudes; see also Chapters 5 and 6.) Indeed, Arctic sea-surface temperatures are thought to have at times reached 20°C (Jenkyns et al., 2004).

The plate-tectonic distribution of the continents was different in the Cretaceous than today and so, therefore, were the Earth's oceanic and atmospheric circulation systems transporting heat and moisture about the planet. However, while warm for much of this time, a stage of the Cretaceous (known as the Aptian) was significantly cooler. Looking primarily at ^{18}O isotope levels in marine bivalve molluscs from latitudes between 8° and 31°N, Thomas Steuber and colleagues (2005) have shown that when the Cretaceous was warm, high summer temperatures reached approximately 35–37°C but with low seasonal variability (<12°C) at latitudes 20–30°N. In contrast, when cool (cooler by about 6–7°C), seasonal sea-surface temperature variability increased to up to 18°C. To put this last into context, it is comparable with the seasonal range found today. This raises the question as to whether a warm greenhouse world (of the kind we are moving towards in the 21st century and beyond) will be one of warmth but low seasonality. Although one still has to remember that planetary circulation was different in the Cretaceous, current computer modelling (of both modern and palaeo-Earth) is not sufficiently developed to accurately reflect palaeoglobal climates such that the results can be viewed with confidence.

A warm Cretaceous Earth was ideal for reptiles but the Earth was too cool for them during the latter half of the Cretaceous period. The Arctic appears to have gradually cooled by 10°C over 20 million years (Jenkyns et al., 2004). Nonetheless, it was not this cooling (at least not by itself) that caused the extinction event.

The Cretaceous–Tertiary, or end-Cretaceous, extinction is probably the most famous mass extinction event of all as it saw the demise of the dinosaurs. Between 75 and 80% of terrestrial species became extinct, with possibly somewhat fewer at higher latitudes. It is now well established, due to an iridium layer in rock strata, that this was due to an asteroidal impact. An iridium layer is a reasonably reliable indicator of an asteroidal impact due to the way the Earth's crust was formed. In the beginning, when the Earth was molten, heavy metals like iridium sank to its core, leaving the Earth's mantle depleted of such metals in comparison with asteroids. In the late 1970s Alveraz and his team in the USA detected significant levels of iridium in the Cretaceous–Tertiary layer, which pointed to an asteroid impact (Alveraz et al., 1980). This discovery spawned investigations into possible extraterrestrial causes of other mass extinctions.

The Cretaceous–Tertiary impact is thought by many to have taken place in central America at Chicxulub, Yucatan, in Mexico (there is some debate about this). Whether or not this was the impact that caused the extinction, there was an impact in middle America somewhere around the extinction time and this caused ejecta that reached north, well into the USA. Ejecta dust undoubtedly entered the stratosphere, resulting in diminished photosynthesis and cooling on a global scale for several years (not to mention spectacular sunsets). As with the end-Permian extinction, a fungal and algal spike has been identified (Vajda and McLoughlin, 2004). This was followed by a recovery led by pteridophytes and gymnosperms (ferns and seed-bearing/woody plants). The Cretaceous–Tertiary and end-Permian vegetation recoveries represent similar responses to terrestrial ecosystem collapse, although the end-Permian bio-diversity crisis was much more prolonged. Whether or not there was a subsequent warming period depends on how much carbon dioxide was released. Either way, a change in the climate, albeit for a few years, would have been the main cause of the dinosaur extinction worldwide, although on a continental basis there would have been shock events and raining ejecta, causing life to be wiped out. Here endeth the summary story . . . well, not quite.

There are still some who say that an asteroid did not wipe out the dinosaurs. As a point of pedantry this view has much merit. To begin with, only some genera of dinosaurs became truly extinct. Other genera only suffered a pseudo-extinction: they did not die out as such but new species evolved. For example, birds evolved. Second, the dinosaurs were already in decline prior to the asteroid impact. The 60-million-year-long Cretaceous period was the third period (the Triassic and Jurassic being the other two) in the Mesozoic era of the dinosaurs, beginning 251 mya. The Mesozoic saw considerable evolution in the dinosaurs, and some dinosaur species that were around in the Triassic (at the beginning of the Mesozoic) were not around in the Cretaceous. Conversely, some dinosaurs did not evolve until the Cretaceous. Indeed, the *Tyrannosaurus rex* star of the book (and film) *Jurassic Park* is associated with the Cretaceous, not the Jurassic. Overall, many dinosaur genera were in decline well before the end of the Cretaceous and its asteroid impact, and so the asteroid cannot have been responsible. The possible culprit behind the – if you will – pre-asteroid dinosaur decline may well be the considerable volcanic activity associated with the Deccan Traps in north-west India. The Traps represent the remains of a continental flood basalt event (like the end-Permian Siberian Traps) which is thought to have been active (albeit to varying degrees) for 8 or 9 million years somewhere around 60–65 mya, although there are some associated geological events that occurred earlier than this (70 mya). The continental flood basalt event would have released cooling dust and ash as well as warming carbon dioxide and methane. Either way, a global impact of the continental flood basalt event was virtually unavoidable. As such, this combination of volcanic traps and an asteroidal impact is possibly similar to the end-Permian extinction, although the blame for the end-Permian event lies more firmly with the vulcanism of the Siberian Traps. What may have happened in the Cretaceous is that the Deccan Traps flood basalt event, which lasted many millions of years, altered the biosphere sufficiently to stress dinosaurs, and then the end-Cretaceous asteroid impact provided the *coup de grâce*.

Then again, the dinosaurs were not the only life on Earth at that time: indeed, there were other newcomers. The late Cretaceous and the early Tertiary were periods of vegetation change due to the appearance, and then increasing dominance, of flowering plants (angiosperms). There is also evidence of a fungal spike similar to that found associated with the end-Permian extinction. In all likelihood fungi would have flourished in a world of much dead and rotting biomass.

It is well established that in the Cretaceous the large supercontinent of Pangea continued to break up and that there was considerable volcanic activity (such as the Deccan Traps in north-west India, discussed above). New oceanic currents were established, and old ones diminished or ceased. The late Cretaceous was a time of major sea-level transgressions and regressions and in particular no fewer than six such events have been identified. Globally, oceanic isotopic data (such as forams; see Chapter 2) from the tropics and also belemnites (a small, extinct cephalopod or octopus-type mollusc) from higher latitudes suggest a consistent picture of at least two periods of warming and cooling (Frakes et al., 1992). Add to this the regional climatic changes due to the aforementioned circulation, and times were hard, and especially so for those creatures more wedded to a specific biome. Considered this way, it is not surprising that the dinosaurs were in decline. Whether or not an asteroidal *coup de grâce* was required to terminate the Mesozoic age of the dinosaurs we may never know for certain. That such an event was necessary for so complete a termination is, however, likely. For some the debate will continue, as it did with an exchange of lucid articles and letters over many issues of *Geoscientist* in 2003–4.

The end-Cretaceous clearing of many species meant that when ecosystems recovered there were vacant ecological niches. This, and the recovery period of environmental change, facilitated speciation of surviving fauna. Of relevance to this book's readers, among the new species to arise were the placental mammals from Eutheria, the forebears of the mammals. Such an early Placentalian was *Maelestes gobiensis* (Wible et al., 2007). Environmental change is an important factor in biological evolution.

3.3.9 The Eocene (55–34 mya) and the Initial Eocene Thermal Maximum (~55 mya)

Following the Cretaceous–Tertiary dinosaur extinction 65 mya, the archaic mammals began to diversify, but further change was to come. Prior to 55 mya, although the temperature globally was warmer than today, there is a case that the terrestrial climate was more seasonal than it previously had been. Many of the major landmasses had already tectonically moved towards the northern high latitudes and so globally there may have been a more deciduous type of vegetation. Certainly, folivorous (leaf-eating) mammals were not apparent until the Eocene.

So, the late Palaeocene and the Eocene was already a warm time. The Eocene itself began 55.8 mya and ended 33.9 mya. A few million years into the Eocene was warmest sustained part of this time but at the very beginning of the Eocene there was a short (geologically) and abrupt warm event at 55.8 mya that was, for a brief while (a hundred thousand years), even warmer than the later, more sustained, warmest part of the Eocene (see Figure 3.2): the Eocene (Climatic) Maximum[4], which occurred 50–52 mya. This longer period of warmth in the already warm Eocene was towards the end of the first quarter of the Eocene. During this warm time there was even a period of some 800 000 years when it appears that Arctic waters were much fresher

---

[4] Note that Eocene (Climatic) Maximum is not the same as Initial Eocene Thermal Maximum (IETM)/Palaeocene–Eocene Thermal Maximum (PETM).

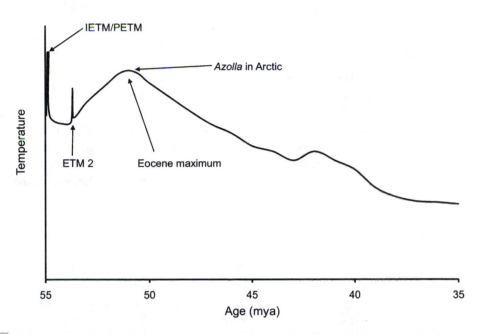

Fig. 3.2 A sketch outline of the principal temperature changes of the Eocene (55–34 mya). More accurately, it reflects the deep-sea temperature changes. The temperature axis is not quantified due to uncertainty about the precise temperature proxies. Despite insufficient data points globally and throughout this period, and different proxies and proxy interpretations giving slightly different temperature estimates, the overall picture of warm and cool periods, as well as the Initial Eocene Thermal Maximum (IETM) (or Palaeocene–Eocene Thermal Maximum, PETM) climate blip, does consistently show up in all relevant geological records. Note there were hyperthermals other than the IETM and Eocene Thermal Maximum 2 (ETM2), but that these were smaller.

(more brackish) and not as saline, as the remains of freshwater ferns, *Azolla*, have been found. *Azolla* is a genus of aquatic floating ferns that get their nitrogen from symbiotic nitrogen-fixing blue-green algae. Today *Azolla* generally flourishes only in waters with less than 0.55% salt, and some species of *Azolla* only thrive in waters less than 0.2% saline (for comparison, average sea-water salinity today is 3.47%), so why was the Arctic sea so fresh? Two factors played their role. The first is that during the Eocene the Arctic was (due to continental drift) smaller and far more enclosed by land, and hence more cut off from the rest of the Earth's oceans. This more isolated Arctic water was fed by fresh water from rivers in what would become the north of North America, Greenland, Europe and Asia. Second, the warmer Eocene and its greater rainfall would have further contributed to this supply of fresh water.

The climatic background prior to the Eocene was one of an already warm Earth as the result of a number of factors. The Palaeocene (the first epoch in the Tertiary period, beginning 65.5 mya) was a time of higher atmospheric carbon dioxide concentration than the present (although it had been higher beforehand; see Figure 3.1). So, there was more carbon dioxide greenhouse forcing than now, but not as much as in earlier times (240–66 mya). At the time of the Eocene ocean circulation, transporting heat about the planet, was substantially different from today, and heat transfer from the tropics more efficient, although the distribution of the continents on the Earth's surface was beginning to become recognisable as the present layout. Back then, a number of tropical oceanic gateways were open, such as the seaway between Indonesia and Papua New Guinea, and that between North and South America. Conversely, the high-latitude southern-hemisphere gateways of the Tasmanian Straits and the Drake Passage between South America and Antarctica were more constrained than today. These factors together helped transport heat away from the tropics. More heat then went to higher latitudes and less heat was re-radiated back into space from the tropics. With less heat leaving the planet the result was an overall warmer Earth, but it was to get even warmer, with the Initial Eocene Thermal Maximum.

This event is sometimes known as the Initial Eocene Thermal Maximum (IETM), or alternatively the Palaeocene–Eocene Thermal Maximum (PETM).[5] The very beginning of the IETM/PETM for geologists actually defines the start of the Eocene epoch and technically the start of the Lower Eocene stage, or Ypresian, as it is sometimes known.

Interest in the IETM/PETM event and other CIEs (such as the aforementioned Toarcian CIE 183 mya) began towards the end of the 20th century but, although palaeoclimatically interesting and possibly important, it largely remained a little-known curiosity to the broader scientific community. Indeed, as we shall see in Chapter 5, the UN Intergovernmental Panel on Climate Change's (IPCC's) 2007 scientific assessment (Working Group I) ran to nearly 1000 pages, and only a couple of pages were devoted to the IETM/PETM. Then, in 2010, largely as a result of working on the first edition of this book, I suggested that the UK Geological Society run a 2-day international symposium on CIEs and their climate and ecological importance. This event was the first Geological Society symposium to garner support from the British Ecological Society (even though both bodies had been members of Britain's 'Science Council' of learned bodies for many years). What follows in the next two paragraphs is a summary largely taken from this symposium's abstract booklet (Cowie and Cohen, 2010).

CIEs themselves have a number of broadly similar aspects that include severe global warming, acidification and deoxygenation of the oceans, extinction of marine and terrestrial species, and sudden shifts in the Earth's climate and its hydrological cycle.

The IETM/PETM CIE event itself took place at the beginning of the Eocene ($\sim$55 mya). The sea-surface temperatures rose in a few centuries (a couple of thousand years at the very most) by some 5°C and regionally temperatures were approximately

---

[5] When doing a literature search it is best to search for IETM and/or PETM. Some papers use one term and some use the other. PETM is currently (2010 onwards) more in vogue with western European and North American researchers, although some (and others elsewhere) still use IETM. Why the difference? The term Initial Eocene TM suggests that the event started at the beginning of the Eocene (and indeed the International Commission on Stratigraphy uses IETM/PETM strata as a Global Standard Section and Point marking the start – the initiation – of the Eocene). Conversely, Palaeocene–Eocene TM suggests that this geologically brief climate event straddled the Palaeocene–Eocene boundary. If you like, it is a bit like arguing whether the no-man's land between North and South Korea belongs to either North or South Korea. The IETM or PETM is also sometimes called the Eocene Thermal Maximum 1 because there were other hyperthermals (short, brief, distinct warm periods) in the Eocene (of which the IETM/PETM is by far the greatest). Fortunately the term Late Palaeocene Thermal Maximum (LPTM), yet another name for the same event, never gained much currency.

5–9°C higher at mid to high latitudes than before, with polar temperatures rising by as much as 8 or 9°C. This rise lasted for tens of thousands of years, but probably less than 100 000 years as an extreme event (although less significant warming may have lasted longer than 100 000 years). Evidence also suggests that temperatures in the mid-latitudes of 45° were some 18°C warmer than they had been at the end of the Cretaceous (the extinction event and aftermath notwithstanding), and about 30°C warmer than the same latitude is today. In short, the IETM was the most pronounced transient warming of the last 100 million years.

In palaeo-Great Britain, the geological record from Cobham in the county of Kent documents a major vegetation change. A pre-IETM/PETM herbaceous fern and woody angiosperm community of low-diversity, fire-prone fauna is replaced at the onset of the CIE: ferns are lost, fires cease, wetland plants increase (including swamp conifers and water ferns like *Azolla* and *Salvinia*) and a wider variety of flowering plants, including palms, arise. With regards to the marine environment, CIEs are marked by deep-sea acidification, but evidence for acidification at the ocean surface (deformation of calcifying phytoplankton) is not universally accepted. Microfossil and geochemical evidence indicates widespread low-oxygen conditions during the IETM/PETM and possibly other CIEs in coastal regions and epicontinental basins (New Jersey, New Zealand, North Sea and parts of Tethys), but also at mid-depths in the south-eastern Atlantic, where the Oxygen Minimum Zone expanded. The geographic and bathyal (deep-sea) extent of this lack of oxygen (hypoxia) is not well defined. Eutrophication (nutrient enrichment) may have increased in marginal and continental edge basins, whereas oligotrophy (nutrient depletion) increased in open ocean environments, possibly linked to increasing ocean heat stratification. There are comparisons between this climate threshold (tipping point) event and modern-day global warming (see section 6.6.8).

So how did interest in all this begin? One of the earliest papers was published in 1990 by marine scientists James Kennett and Lowell Stott, both then at the University of California in Santa Barbara. It reported an analysis of marine sediments from the time of the IETM/PETM showing that, not only had the surface of the Eocene Antarctic ocean heated up, but the entirety of the ocean column had warmed to some degree and that its chemistry had changed markedly. Such complete ocean warming would be expected only with a warming event lasting at least tens of thousands of years. Subsequent work, such as by Tripati and Elderfield (2005), showed that Atlantic and Pacific tropical and subtropical bottom waters warmed by some 4–5°C. Again, this was indicative that the warming had lasted for thousands of years; indeed, it would have taken this length of time for the oceans to warm this much from the climate, such are ocean mixing times. Further, the volume of the Earth's oceans having been warmed, the ocean mass provided an additional thermal buffer, prolonging the warming. The warming caused profound environmental change and ecosystem dislocation, comparable with, if not greater than, that associated with more recent (Quaternary) glacial–interglacial transitions. Indeed, above the Arctic Circle the very warm climates of this period permitted the growth of moderately diverse and productive deciduous forests. With regards to the ocean's chemical changes there was severely reduced oxygen in deep-sea waters (warmer waters tend not to hold dissolved gases so easily), and 30–40% of deep-sea calcareous Foraminifera abruptly

Fig. 3.3 A marine core from the early Eocene. The abrupt depletion of white calcareous forams – the dark band – is clearly seen. Picture copyright Jim Zachos and reproduced with permission.

became extinct, this last in no small part due to increased ocean acidity. Wherever in the world cores have been taken of early Eocene marine sediments this depletion of calcareous forams is seen (see Figure 3.3).

But what caused this abrupt warming and, indeed, the increased ocean acidity? The picture of what probably happened has been built up over a number of decades of research and even now there is considerable debate as to the various factors involved and how they interacted. Yet, as we shall see, a broad picture is emerging and one that increasingly appears to be relevant in the 21st century.

Analyses of both marine carbonate fossils and terrestrial fossils from this time reveal a sharp decline in the normal (low) ratio of ^{13}C to the common ^{12}C (see Figure 3.4). The logical explanation for this CIE is that a reservoir of ^{12}C flooded the active biosphere (i.e. its living components). As noted with the Toarcian event, because photosynthetic enzymes have evolved a stonger affinity for ^{12}C than ^{13}C it is likely that the flooding carbon was ultimately biogenic (generated from life). However, such was the amount of ^{12}C flooding the biosphere, it must have come from a ^{12}C-dominated reservoir not usually associated with the short-term carbon cycle, but from the deep carbon cycle. Carbon-12 reservoirs include the living part of the biosphere itself (short-term or fast carbon cycle), fossil fuels of the deep carbon cycle (or fossil fuel-type sources, since no one was refining fuel back then), methane hydrates (clathrates) in submarine continental slopes and some sea floor sediments (both deep carbon cycle), and volcanic gases (which ultimately owe much to the tectonic recycling of organic sediments and which are again part of the deep carbon cycle and so can be ^{12}C-rich).

With regard to the first of these options, if all the carbon associated with the living terrestrial component of the biosphere was turned to carbon dioxide (for example, by some sort of cataclysm causing, say, global forest fires) then this still would not have been enough to provide the amount of ^{12}C released. More than three times the amount found planet-wide in terrestrial plants and animals would have been required. Soils and detritus globally contain sufficient carbon (see Chapter 1) but releasing *all* of this would have necessitated an extinction event of unprecedented severity and

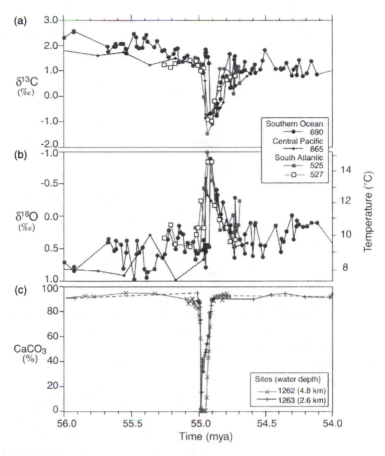

Fig. 3.4 (a) The IETM/PETM CIE as revealed by the analysis of benthic (bottom-dwelling) foraminifera (see Zachos et al., 2003 for details). (b) ^{18}O climate proxy temperature analysis of IETM/PETM strata (see section 2.2.1). (c) Carbonate ($CaCO_3$) content of sediments in two cores from the south Atlantic (Zachos et al., 2004, 2005). All three panels reproduced with permission from *Climate Change 2007: The Physical Science Basis. Working Group I Contribution to the Fourth Assessment Report of the Intergovernmental Panel on Climate Change*, figure 6.2, Cambridge University Press.

this is also not seen. Alternatively, the burning of a massive peat deposit could have caused the CIE. This last could be a possibility but the source of this deposit has not been found. Deep (abyssal) marine sediments could have provided more than enough carbon but disruption of these on a planetary scale would have provided too much, and in any case would have also disrupted continental shelf methane hydrates. Other proposals include a volatile-rich asteroid impact, but again there is no evidence for this, nor an iridium layer that is associated with other large, global climate-changing asteroid-impact events. This leaves fossil fuel-type sources, marine methane hydrates (clathrates) or volcanic gases as source candidates.

In 1995 Gerry Dickens and colleagues from the University of Michigan proposed that only geological methane gas had enough light carbon (^{12}C) to produce the early Eocene CIE. They proposed that a belch of methane escaped from methane hydrates (methane clathrates) in continental slope sediments as the Earth warmed during the late Palaeocene (Dickens et al., 1995). But if this occurred, what triggered the release? As we shall see in the next chapter, methane 'burps' from marine landslides on the continental slopes have affected the climate during the last glacial, but not on the IETM scale.

Nonetheless, in 1999 US geologists Miriam Katz and Dorothy Pak, together with Gerald Dickens working in Australia, reported the discovery of a submarine landslide off the Florida coast dating from the time of the Eocene CIE, and ^{18}O analysis of the benthic forams suggested that bottom waters rose in temperature by at least some 6°C. The landslide was just the sort one would expect with methane hydrate dissociation. The problem was that at least some 1000–2000 Gt of ^{12}C would need to be released to generate such a pronounced CIE and this was not possible from this single submarine avalanche alone. The Katz team therefore predicted that if their 'smoking gun' hypothesis was true then there must have been other Eocene submarine landslides, the remains of some of which should still be evident today. Certainly, globally there should have been enough methane hydrate to account for a significant proportion of the CIE.

Then in 2004, Henrik Svensen and fellow Norwegian colleagues announced they had found evidence of literally thousands of hydrothermal vents in a complex off the More and Voring basins in the Norwegian sea, dating from the Palaeocene–Eocene boundary. Could the intrusion of mantle-derived melts in the carbon-rich sedimentary strata in the north-east Atlantic have caused an explosive release of carbon? Such a structure on land is associated with the Siberian Traps that resulted in the end-Permian extinction approximately 250 mya (see section 3.3.5). Meanwhile, the complex of the More and Voring basins is part of a greater North Atlantic volcanic province. Work by Michael Storey and colleagues in 2007, using $^{40}Ar/^{39}Ar$ age determinations, shows that the Danish Ash-17 deposit, which overlies the IETM/PETM by about 450 000 years in the Atlantic, and the Skraenterne Formation Tuff, representing the end of 1 ± 0.5 million years of massive volcanism in East Greenland, stem from around the same time. The relative age of Danish Ash-17 therefore places the IETM/PETM onset after the beginning of massive flood basalt volcanism at 56.1 ± 0.4 mya but within error of the estimated continental break-up time of 55.5 ± 0.3 mya, marked by the eruption of basalt-like flows in the mid-ocean ridge. These correlations support the view that the IETM/PETM was triggered by greenhouse gas release during magma interaction with basin-filling carbon-rich sedimentary rocks proximal to the embryonic plate boundary between Greenland and Europe.

An additional slant on this has been proposed by a European and American team of geologists led by the Dutch (Lourens et al., 2005). They have found another smaller, subsequent warming event, the Eocene Thermal Maximum 2 (ETM2), and the timing of this and the IETM are possibly suggestive that long-term Earth orbital factors may have played a part, just as Milankovitch orbital variations pace Quaternary glacials (see Chapters 1 and 4). (There is a similar suggestion concerning the Toarcian event described earlier in this chapter.) However, whether or not this orbital role was causal in (rather than tweaked the timing of) the dissociation of methane hydrates remains to be seen, although the evidence for the timing option is gathering: there is a lot to discover about the IETM/PETM. Nonetheless, orbital factors could well have played a part in the pacing of climate detail within the event. Anything that increased the temperatures of intermediate sea waters would increase the chance (and hence the pacing) of methane hydrate dissociation.

Carbon dioxide, be it from part of the volcanic event, which itself might have been methane-dominated, or as a result of the oxidation of released methane, would easily dissolve in the oceans, so increasing their acidity. This in turn would increase the dissolution of calcite shells of microplankton which are the dominant component of sea-floor sediments, leaving behind only non-soluble clays; hence the foram-extinction event. Other species with calcareous shells are also likely to have been affected (see the ocean acidity subsection in Chapter 6). It is well documented that a change in the colours of sediments, from bright white carbonate to deep red clays, marks the Palaeocene–Eocene event (see Figure 3.3). Normal deposition of microscopic carbonate foram shells on the deeper reaches of the sea floor did not resume for at least 50 000 years, and the total recovery time to a so-called normal state took as long as 100 000 years. Finally, ocean sediments also reveal a boom in the number of planktic Foraminifera species a few million years after the IETM/PETM (Emilliani, 1992). This might be expected if the ocean contained an abundance of material from which to make carbonate. However, whereas this recovery begins immediately after the IETM event, a more gradual rise in foram biodiversity (as opposed to population numbers, which, as said, declined during the IETM) began prior to the IETM, so making it difficult to claim with certainty that the foram biodiversity peak and IETM were connected, even if this is arguably likely.

It is worth noting that Svensen's team estimate that the *rate* of carbon released as methane during this Norwegian sea event – over a period of between 30 and 360 years – is comparable to an average year's worth of carbon release from the burning of fossil fuels as carbon dioxide in the 1990s. This rate is low (as 1000–2000 Gt of ^{12}C in total is needed for the CIE), nonetheless the suggestion is that our current burning of fossil fuels will eventually warm the planet and hence, as in the Eocene, eventually the oceans, possibly enough to disrupt methane hydrates and so trigger further greenhouse warming above and beyond that. The relevance to today is that we are releasing ^{12}C into the atmosphere by burning fossil fuels and clearing forests; could we be about to trigger an event analogous to the IETM/PETM?

In addition to the Svensen team's estimate of the *rate* of carbon release from these vents, they also estimate that the *total* carbon release over the period of the CIE might be 1500–15 000 GtC. Given that the CIE is thought to involve around 2000 GtC or more, *if* these estimates are accurate the vents themselves may have been responsible for the CIE on purely carbon-volume terms (Svensen et al., 2004). However, this alone is unlikely. We know the CIE happened and it resulted in global warming that affected ocean deep-water temperatures, and so thermal dissociation of methane hydrates would have been virtually inevitable. In addition, with warming, carbon was also likely to have been released from soils. The question then becomes one of whether the North Atlantic volcanic province was the trigger for a separate, subsequent release of methane from marine clathrates and also, in all likelihood, other terrestrial carbon pools.

As for comparisons with current carbon releases driving present-day global warming, it is worth highlighting a few figures. The release of at least some 1000–2000 GtC was needed to generate the Eocene CIE. This compares with some 6.1 GtC released into the atmosphere in 1990 alone through human fossil fuel burning and industry (such as cement works), and since then human carbon release has increased. Meanwhile, the IPCC (2001a) forecast that the total 21st-century fossil carbon release is likely to be between 1000 and 2100 GtC. In short we could be on our way to creating a climatic event analogous to the IETM.

One key question that follows is what was the atmospheric concentration of carbon dioxide during the Eocene peak? There are a number of ways to calculate this, notwithstanding the volume of carbon needed to cause the ^{12}C isotope excursion. Another is to use (non-biological) chemical estimates such as that by UK geologists Paul Pearson and Martin Palmer in 2000. They looked at boron-11 (^{11}B) concentrations in marine calcium carbonate sediments, because boron in solution occurs as $B(OH)_3$ and $B(OH)_4^-$, the equilibrium between which is pH-dependent, and marine pH is largely governed by atmospheric carbon dioxide concentration. They give an IETM atmospheric carbon dioxide concentration peak of some 3800 ppm (with a lower error limit of 2800 ppm and a higher limit of 4600 ppm). On the one hand this is far higher than the IPCC consider in their stabilisation scenarios of current carbon emissions: the highest being 1000 ppm (IPCC, 2001b). On the other hand it needs to be remembered (from Figure 3.1) that this peak was superimposed on the existing higher-end Palaeocene and early Eocene atmospheric carbon dioxide concentrations. Consequently the early Eocene *increase* in atmospheric carbon dioxide above pre-Eocene levels (already high by today's standards) was only around 1000 ppm or more. So again this nearly chimes with the current carbon pulse generating the global warming the IPCC is examining and the levels of atmospheric carbon dioxide that might occur beyond the end of the 21st century with current trends.

In contemporary greenhouse terms, assuming a similar CIE today (which is neither an insignificant nor an unrealistic working assumption)[6], this carbon would need to be added to the pre-industrial atmospheric concentration of 280 ppm, so making a total of 1280 ppm or more. This is clearly higher than the highest stabilisation scenario the IPCC considered in their 2001 report (IPCC, 2001b), but is touched upon in possible high-end scenarios beyond the 21st century in its 2007 report. Of course if we are beginning to disrupt natural biosphere pools of carbon (be they in soils or as marine methane clathrates), and we do then trigger an analogous CIE to that of the early Eocene, then this natural carbon would be in addition to that being released through fossil fuel combustion and land-use change: we would not need all the carbon from fossil fuels and land-use change the IPCC consider to result in an extra 1000 ppm of carbon dioxide.

Returning to the Eocene event, consider its duration. Another matter that springs to mind from Chapter 1 is that if carbon dioxide has an atmospheric residence time of just a century or two, why was the approximately 100 000-year or longer IETM so protracted?

---

[6] We do need to note that even though the temperature *rise* of the Eocene CIE was of several degrees and so broadly similar (albeit greater) to the warming we might expect at the end of the 21st century or sometime in the 22nd, the climate *baseline* temperature for the Eocene was already warmer than today. The continents (hence ocean circulation and heat transport) were also positioned differently.

The answer to this is that matters are more complex than dealing with a simple carbon pulse into the atmosphere. In 2004 an Anglo-American research team, led by Gabriel Bowen and David Beerling, suggested that terrestrial and, importantly separately, oceanic carbon cycling may have changed significantly as a result of the Eocene carbon release and that this in turn would have a consequential regional climate change effect (as opposed to the average global increase in temperature expected). Turnover of soil organic matter therefore probably increased dramatically, suggesting vigorous plant growth in a number of terrestrial ecosystems (due to the extra warmth and precipitation, and increased carbon dioxide from the slow oxidation of methane), so doubling terrestrial carbon cycling. The picture they construct is that there would need to be an increase in soil moisture and humidity of at least 20%. Water vapour is itself a greenhouse gas and this would have helped further the IETM, during which the Earth appears to have switched into a new climatic mode. However, water vapour concentration is a response to warming and would not itself maintain it, as we perceive from the geological record, but reinforce something else that was going on. The Bowen–Beerling team could not state with confidence what this factor (or factors) might have been but did make a number of suggestions. These included changes in atmospheric and/or ocean circulation and heat transport, higher atmospheric carbon dioxide due to circulation and ocean chemistry changes, or higher methane due to increased wetlands arising from the higher precipitation, humidity and temperature as well as some of the carbon release from soils once global temperatures had started to rise (Bowen et al., 2004).

All the above notions are likely factors but in themselves do not fully explain the duration of the Eocene CIE. Yet, pulling together the evidence published over the past decade, the following scenario is one well worth considering.

This is what may have happened. An event such as magma intrusion into biogenic carbon rich sediments, as reported by Svenson et al. (2004), could (if the conclusions of the analysis are to be believed) have resulted easily in the release of more than enough ^{12}C to generate the Eocene CIE. (Not surprisingly, you will recall, major volcanic activity also took place at the time of the Toarcian CIE.) Yet, even if this were so, the raising of deep-ocean water temperature by 4–5°C was very likely to have destabilised those marine methane hydrate deposits that were already close to dissociation. After all, the degree of methane hydrate stability is a function of both temperature and pressure so that those deposits that were closer to the surface would be more likely to dissociate before than those deeper. These latter deposits would only dissociate with further abyssal warming and that would take time. Furthermore, there is evidence suggesting that there was an abrupt change in ocean circulation during the Eocene warming event, from a more temperature-driven thermohaline circulation (see Figure 4.1 and associated text) to a more salinity-driven halothermal circulation. Foram benthic (deep-water) carbon-isotope evidence from four ocean basins dating from the time of the IETM/PETM event also suggests a switching of ocean circulation during the climate event itself (Nunes and Norris, 2006). This would have exposed other methane hydrate strata to thermal destabilisation that had previously been protected by cooler currents. In short, methane hydrate deposits dissociated not simultaneously but gradually, as the oceans slowly warmed and as the ocean circulation changed. Further, once the planet began to warm up, other carbon reservoirs may have seen changes in carbon flow: soil carbon being but one likely example. In short, a pulse of light carbon (^{12}C) probably cascaded from one biospheric carbon pool to another, to another and so on, largely via the atmosphere which consequently was greenhouse-warmed. One might speculate that fossil carbon triggered warming that initiated the release of ocean hydrate carbon that lengthened, and enhanced the CIE and warming, which in turn released high-latitude soil carbon, and that atmospheric carbon was in part absorbed by the oceans (causing acidity) and by terrestrial ecosystems that released it slowly via rivers and so forth (in addition to the atmospheric route). Then there would be a recovery period during which the carbon pools were (largely) restored. So each carbon cascade between biosphere reservoirs of carbon (both ecological and inorganic) could have been quite short, but taking all the cascades together we then get the 100 000–200 000 year CIE duration.

Whereas after the IETM/PETM the global climate returned to what it had been beforehand, it is likely that the biosphere's reservoirs remained sufficiently altered so that only a small trigger (be it Milankovitch pacing or perhaps a smaller volcanic intrusion) forced the Earth system through a critical transition and across a subsequent climate threshold, such as the Eocene Thermal Maximum 2 (ETM2). If the first quarter of the Eocene saw the Earth system continually close to a critical transition then this would have resulted in further more minor hyperthermal events. Indeed, these we see (Sexton et al., 2011) and ETM2 was of some significance (see Figure 3.2).

Unfortunately at the moment computer models are rather limited in being able to reproduce the Eocene CIE and warming. This is because ecosystem and marine hydrate climate and sufficiently detailed biosphere carbon pool responses to warming are not included in current models. (Biome cover in the albedo sense has been included in models since the late 1990s and biome carbon tentatively included since the early 2000s: I say 'tentatively' because the biology of carbon flows has not been elucidated in nearly sufficient detail. This last is a current focus of research [see Chapter 7].)

The above IETM/PETM story, which has developed since the 1990s, has been a bit of a marathon but it is important given its relevance to current anthropogenic warming. The astute student may have a couple of questions. First, if the above explanation has currency then why was Arctic *Azolla*, which was in such abundance during the Eocene maximum, not present in the IETM/PETM? Second, if the Eocene did see the Earth system close to a critical transition, then why are hyperthermals mainly seen before and not after the Eocene maximum? I must stress at this point that the next paragraph is speculation (albeit informed) on my part and so students and non-palaeoclimatology readers should treat it with caution. It nonetheless fits all the currently known facts.

There are three possible reasons, or a combination thereof, why *Azolla* does not appear (at least in the quantity) in the IETM/PETM as it does later during 800 000 years of the Eocene maximum. You might expect *Azolla* to be present in the Arctic in the IETM/PETM as continental run-off was greater, hence the palaeo-Arctic sea was fresher at this time, just as it was during the later Eocene maximum: *Azolla* needs fresher water. First, it could be that ecotoxicological effects on the marine environment from the North Atlantic and related volcanism could have inhibited *Azolla* growth during the IETM/PETM. (For example, we know that there were ash deposits in what are now Greenland and Denmark.) Second, cores of strata may simply not have picked up *Azolla* fossils from the IETM/PETM, which was of shorter duration than the *Azolla* event in the Eocene maximum, and also sedimentation processes may have been different in the two events. Third, there were differences in ocean circulation that enhanced Arctic sea-water freshening in the Eocene maximum beyond that experienced in the IETM/PETM. Indeed, the depth of the North Atlantic changed considerably during the lower Eocene (Ypresian) including to the point where there was some land in the middle of the ocean.

As for why smaller hyperthermals (of which ETM2 was the largest) are seen after the IETM/PETM and before the Eocene maximum *Azolla* event, but not after the *Azolla* event, it could be that the *Azolla* event took sufficient carbon out of the system so that the biosphere was no longer near a critical transition leading to a climate threshold event. In addition, after the geologically short-term IETM/PETM ended, the subsequent long-term early Eocene temperature continued to rise (Figure 3.2). This would have slowly brought the Earth system close to a critical transition point, and then exceeded it, triggering a hyperthermal with the addition of periodic Milankovitch orbital forcing. This extra warming naturally affected the oceans and, if already near-saturated with carbon dioxide, would have caused them to vent the gas into the atmosphere causing the ETM2 hyperthermal (Sexton et al., 2011). Conversely, after the Eocene maximum the longer-term trend for the rest of the Eocene was one of cooling: cooler oceans are able to store more carbon dioxide (taking the Earth away from a potential critical transition) and so hyperthermals would not take place after the Eocene maximum even though Milankovitch orbital forcing continued.

How does the IETM/PETM relate to current climate change concerns today? In terms of modern anthropogenic global warming, the suggestion that lends itself from the whole IETM episode is that if humanity continues to release carbon, as it has recently, then the resulting warming might eventually disrupt present-day methane hydrate sediments and/or enhance methane generation from wetlands and other sources, so *further* increasing the atmospheric carbon burden. However, this extra burden would not be as carbon dioxide but (initially at least) as the more greenhouse-potent methane. This would be an event analogous to the IETM itself. Broadly this means that should we experience a repeat of this Palaeocene–Eocene methane (and other light ^{12}C) event then the current IPCC warming forecasts of 1990–2007 (see Chapter 5) would be greatly exceeded. There may also be a switch (following a critical transition) to a new climatic mode (a climate threshold would have been crossed), or at least a very warm interval, lasting many tens of thousands of years. Either way, if current fossil fuel emissions of carbon do trigger a methane hydrate and other light carbon release, then the total atmospheric greenhouse burden from carbon may well exceed that of the Eocene. This is because we are already set to release broadly 2000 GtC this century from fossil fuel (broadly similar to all the carbon involved in the Eocene CIE), so that any methane hydrate release from the oceans would be extra, and this does not even include likely carbon releases from boreal soils or further carbon-pool cascade effects.

The aforementioned notion, if it came to pass, should not be considered an undue surprise. The IPCC clearly warn that their forecasts do not include unforeseen events and emphasise that we should be aware of 'climate surprises' (see section 6.6.9). As such, the IETM might represent a possible quasi-palaeoanalogue (perhaps even only a modest one) to future anthropogenic warming should we continue to pump fossil carbon into the atmosphere (see Chapter 8). 'Quasi' because if the scenario in the previous paragraph did take place then it would be a more severe climatic event than the IETM and one that would have a far faster onset: at most the Eocene CIE took possibly some 30 000–40 000 years from its beginning to the start of its plateau-like peak, whereas we are set to release 2000 GtC in just 100 years.

The impact on the marine environment, especially some calcareous species, is noted above, but what of terrestrial biological impacts? It is generally accepted today that the warmer (and wetter) the environment the greater the terrestrial species diversity. So what was species diversity like during the IETM/PETM? Tropical South America has the highest plant diversity of any region today. In 2003 a team of US palaeobiologists led by Peter Wilf reported an analysis of fossil flora in sediments from a caldera lake in Patagonia (which today is far south of the tropical zone). They recorded 102 species of dicotyledons with proper leaves ('leaf species'; the most diverse group had 88 leaf species at the site), monocotyledons, conifers, ginkgophytes (one of the five main divisions of seed-bearing plants but with only one living genus, *Ginkgo*), cycads (another of the five main seed-bearing plant divisions) and ferns (Wilf et al., 2003). This, adjusted for sample size, correlates with a diversity that exceeds that of any other (non-thermal maximum) Eocene leaf diversity in that area. So parts of the Earth during the IETM had sufficient rainfall to maintain phototranspiration and plant diversity at levels seen today in tropical rainforests in parts of northern Brazil and elsewhere. In the early Eocene, however, this was at palaeolatitude (to allow for a little continental drift) approximately 47°S, in the south of South America. What we do not (yet) know is exactly how much of the planet's land area sustained plant diversity and how much did not. Nor do we know the detail of plant–animal interactions across the planet.

We do know a little, however, and indeed some of Peter Wilf's earlier work has shown that back in the Eocene, where today's south-western Wyoming is located, fossil leaves exhibit greater signs of insect herbivory (both in terms of frequency and types of damage) than just 1 or 2 million years earlier in the late Palaeocene (Wilf and Labandeira, 1999). Today we see greater insect herbivory in warmer tropical forests than in temperate ones so although this discovery is not unexpected it is an important corroboration of the broader ecological changes that one might expect with an episode of global warming.

If tropical species migrated to what are today temperate latitudes, what happened nearer the poles? In 2006 a clutch of papers was published in *Nature* of an international investigation of the palaeo-Arctic environment during the Eocene. It was summarised in the same issue by Heather Stoll, a geologist at the University of Oviedo, Spain (Stoll, 2006). The teams, mainly European and US but including some Japanese, analysed marine sediment cores beneath the Arctic. This was no mean feat as in the Arctic sea ice is constantly on the move, so a new technique was tried to ensure that drilling could continue for 9 days over one spot. While the drill ship maintained a fixed position, it was protected by two other ships that broke up the ice around it, pushing the larger pieces away. The cores revealed that the surface of Arctic waters during the IETM rose to around 18°C in the summer, which is comparable to that in

Brittany, France, today. (Although again it needs to be remembered that seasonality was probably higher, with a greater difference between Eocene summer and winter temperatures compared to lower latitudes.)

Even though we do not know much of the detail of ecological change that took place during the IETM/PETM, what we do know is that many ecosystems saw great disruption. For example, a study of plant fossils in present-day USA showed that initial Eocene maximum floras were a mixture of native and migrant lineages and that plant shifts were large and rapid (occurring within 10 000 years). Leaf margin analysis (LMA) of 23 dicotyledon species suggested that temperatures in the present-day USA area rose by approximately 5°C during the initial Eocene maximum compared to LMA of fossils 250 000 years beforehand and is consistent with isotope-derived estimates. Leaf area analysis (LAA) suggests increased precipitation but some local reductions and a reduction (approximately 40%) early on in the event. This overall increase from LAA is again in line with isotopic evidence elsewhere on the planet suggesting an increase in precipitation during the IETM/PETM (Wing et al., 2005).

It is likely that a similar event today, with a global human population of several billion fragmenting the natural landscape, would not see such successful species migration as happened in the Eocene. Furthermore, if the event were faster (due to our 21st-century pulse of a similar amount of carbon being more rapid) there would be considerable dislocation of largely human-managed ecosystems (such as agricultural land). Consequently, it is likely that there would be much human suffering (and economic dislocation). Whether it would have significant long-term evolutionary implications is simply not known; even without climate change, humans are affecting many ecosystems globally and causing biodiversity loss, which itself is significant.

In terms of the IETM's long-term biological impact, as a result of the early Eocene climatic-maximum ecosystem disruption the older groups of mammals began to decline and were replaced by a diversification of the newer, modern orders of mammals. This diversification favoured smaller creatures. Smaller creatures have a higher surface-area-to-volume ratio and so tend to lose heat faster than larger animals, which would have been an advantage in a hot climate. In 2012 Ross Secord, Jonathan Bloch, Stephen Chester and colleagues reported a high-resolution analysis of the fossil record across the IETM/PETM. It showed a decrease in the size of equids (horse family) of around 30% across this CIE followed by around a 76% increase in size after it. Carnivorous mammals also diversified across the IETM/PETM but at first were more generalist and they did not assume the role of larger predators until the late Eocene (Janis, 2001). The Eocene also saw the order Primates begin to diversify, having first appeared in basal (early) Eocene deposits of North America and Europe and slightly later in Asia.

A question that has arisen is which was more important to mammal evolution: the IETM climate change or the end-Cretaceous extinction? Clearly both affected mammalian evolution but different researchers emphasise the importance of different events. An international research team led by Robert Meredith of the University of California in 2011 constructed a molecular supermatrix for mammalian families and examined genome data (using relaxed clock analyses). Their results suggest that it was the end-Cretaceous event that provided more of a spur to the diversification of mammals and the rise of modern mammals, even if the IETM/PETM facilitated the further diversification of modern mammals.

Meanwhile, leaving the IETM/PETM behind, another major extinction was to follow. But before looking at that it is worth reminding ourselves that there were other CIEs than this initial Eocene event: remember the Toarcian extinction (183 mya) in the Jurassic. The reason why more is known about the Eocene event is that there are more geological strata (and hence biological palaeorecords) to access, with more recent deposits (in this case 55 million years old) than for older events. Consequently we know more about the Eocene event. Nonetheless, other CIEs have taken place and, as we shall see, today's anthropogenic warming is itself a CIE.

3.3.10 Eocene–Oligocene extinction (approximately 35 mya; or 33.9 mya?)

The Eocene–Oligocene boundary has recently been re-dated (hence the two dates given in this subsection's title, with the latter, younger date being the one appearing in some recent literature). It was a cooler period, with palaeotemperatures at a latitude of 45° dipping to within a few degrees of today. Antarctica, which was moving to cover the South Pole, also saw the beginnings of an ice cap on some of its mountain ranges, including the Gamberstev mountains that are currently under the centre of East Antarctica's Dome A ice sheet (Bo et al., 2009). It was a time noted for one of the biggest extinctions since the Cretaceous–Tertiary extinction. In Europe this extinction was known as the Grand Coupure, although extinctions have been noted on the other continents and the full explanation is not entirely clear. Nonetheless, approximately half the mammal genera became extinct within about a million years (Hallam, 2004).

But, was this extinction caused by cooling due to indigenous planetary climate forcing (be it greenhouse or atmosphere/ocean circulation changes) or some other factor? Evidence emerged in the mid-1990s of an asteroid impact aged 35.5–35.2 mya at Chesapeake Bay off the shore of Virginia, USA. This is a structure of 85 km in diameter and the probable source of the huge field of tektites (characteristic glassy nodules) covering mainly Georgia some 1100 km away and microtektites in Barbados 2250 km away. This was clearly a significant impact, albeit one that was smaller than that associated with the end-Cretaceous event.

If this were not enough, soon after the Chesapeake discovery there was the dating of a second crater, the 100 km-diameter structure in Popigai, northern Siberia, which is therefore possibly the fifth largest impact crater ever found. (The Cretaceous–Tertiary structure at Chicxulub, Yucatan, is 180–310 km in diameter.) The Popigai structure contains some 2000 km^3 of impact melt rocks, and an iridium anomaly has been detected in Italy (over 3200 km away) exactly matching the Popigai dates. The structure had been known for many years but its date of origin was ambiguous, with estimates ranging between 5 and 65 mya. But in 1997 Canadian and Russian researchers announced that they had managed to refine the dating to 35.7 ± 0.2 mya (Bottomley et al., 1997). This suggests two possibilities. Either the Earth was subject to two major hits within a few hundred thousand years, or it was subject to two major ones simultaneously. Both are possible. Given that major hits are rare events that usually occur a few tens of millions of years apart one might at first think that the two were independent and not contemporaneous: that sheer chance meant that two just happened in fairly quick succession, in geological terms. However, many asteroids are known to exist as binaries orbiting each other. Also, it is well known that an asteroid passing within the Roche (tidal stress) limit of a large gravity well (say, that of the gas giant Jupiter) could be disrupted, so ending up as two or more asteroids on a similar trajectory, which in this case intercepted the Earth's orbit.

Either way, both the impacts would have been profound enough to have resulted in a mass extinction of the magnitude we see at the Eocene–Oligocene boundary with an initial physical disruption of large regions of two continents, climatic cooling (and certainly photosynthetic disruption) for a number of years due to stratospheric dust, and possibly longer-term warming if there was a carbon dioxide pulse. Further, there is evidence to support a short, sharp cooling event, and Antarctic ice first appeared, as it did in some other parts of the planet, in a 400 000-year glacial (Zachos et al., 2001). This is another time when the Earth system was approaching a critical transition point with the potential to cross a climate threshold. As soon as ice appears then albedo effects come into play, with more sunlight being reflected, hence more cooling and more ice. An Earth with appreciable ice at the poles would have behaved in a different thermodynamic way (see Figure 1.8). Such a transition point would see ecological stress in addition to that caused by an asteroid strike.

Having said this, it is important to re-emphasise that there is some debate surrounding the Grand Coupure. Tony Hallam of the University of Birmingham, whose career has in no small part focused on mass extinctions, does recognise the cooling taking place at the time. But he also notes – as glaciers grew on the land and the sea level fell – that land bridges formed around that time between North America and Asia as well as North and South America, and that immigrant species contributed to a number of post-Grand Coupure regional biomes (Hallam, 2004). Whether or not species migration causally contributed to the extinction event is an interesting point.

The Earth emerged from the Eocene–Oligocene extinction warmer than today but the global temperature was 2–3°C cooler than it had been prior to the extinction event (i.e. 3 or 4°C warmer than today) and possibly there was also greater seasonality with cooler winters and warmer summers. This relatively cool period lasted for roughly 10 million years until about 24 mya, before it warmed up, returning to close to end-Eocene temperatures. Although, once Antarctic ice appeared it never went away, even if it did reduce at the end of the aforementioned cool period. After this the planetary climate slowly cooled 3 or 4°C from then to the late Miocene, 5.3 mya, after which this overall slow cooling trend continued.

3.3.11 Late-Miocene expansion of C_4 grasses (14–9 mya)

Before recounting the major biosphere changes of the Miocene (25 mya) we need to briefly look at two types of photosynthesis in plants. Modern grasses use two different photosynthetic pathways, the more ancient C_3 pathways and the more recent C_4 pathway, where C_3 and C_4 refer to the number of carbon atoms in the first product during carbon dioxide assimilation. The C_4 pathway includes a carbon dioxide-enrichment mechanism that enables it to photosynthesise more efficiently than C_3 plants at lower carbon dioxide partial pressures (which is effectively similar to atmospheric concentration). Indeed, it is worth noting now, for we will only briefly return to this point later, that the photosynthetically efficient C_4 grasses – being ideally suited to a low-carbon dioxide atmosphere – will undoubtedly play a significant role in the biosphere's continued evolution on the scale of hundreds of millions of years. Now let's return to the Miocene.

Conventional wisdom has it that the general decline in atmospheric carbon dioxide (as seen in Figure 3.1), hence long-term cooling of the planet, was enhanced from the Miocene due to the photosynthetically efficient C_4 plants and this efficiency enabled the C_4 photosynthetic grasses to expand considerably in the late Miocene (approximately 8–4 mya). While C_4 grassland biomes had a significant presence on the Miocene Earth, the grasses did evolve earlier and grass seeds have been found in dinosaur coprolites (fossilised excreta). The idea is that not only did the C_4 grasses help drive atmospheric carbon dioxide concentrations down, at lower carbon dioxide levels they had more of an evolutionary advantage (such plants' other evolutionary factors notwithstanding) over those plants with a lower photosynthetic efficiency. This is the commonly held conventional wisdom, but as nearly always there are details and it is prudent not to take too simplistic a view.

Other environmental factors facilitating the rise of grasses was that the late Miocene was also the time of orogenesis (mountains building), which resulted in formation of the Rocky Mountains, Andes and Himalayas and in turn in regional climatic effects from rain shadows and rainy seasons. In addition, of course, the uplifted areas themselves were cooler.

Whereas it is true that carbon dioxide levels were considerably lower in the Miocene than compared with the previous couple of hundred million years, and that this gave C_4 plants and grasses a competitive advantage, there is some evidence to suggest that carbon dioxide levels had largely already stabilised (recent Quaternary glacials notwithstanding) by around 9 mya. As we shall see in the next chapter, although the evolution of C_4 plants was earlier, their actual global spread did not take place until from about 8 mya (when there was a major expansion) and there was not a decisive lowering of carbon dioxide at that time to act as a trigger for this grassland expansion. So it is likely that C_4 expansion was driven by additional factors such as tectonically related episodes, such as orogenesis or vulcanism (Pagani et al., 1999), which caused cooling climate change that disrupted ecosystems, so enabling the grasses to capitalise on their advantage in the already lower carbon dioxide levels.

This distinction may seem trivial and the point a subtle one, but it is nonetheless worth noting. As we have seen with climate change itself, and as we shall see in later chapters with regards to ecological change in response to climate change, the resulting biological change is frequently a result of the necessary interplay of a number of factors.

The importance of such evolutionary subtlety is also relevant to the broader change in mix of global plant species. For example, it had long been recognised that ferns are an old group of species and that one type, the leptosporanginates (which feature the majority of contemporary or extant ferns), is more than 250 million years old and today number over 10 000 extant species. However, it used to be thought that

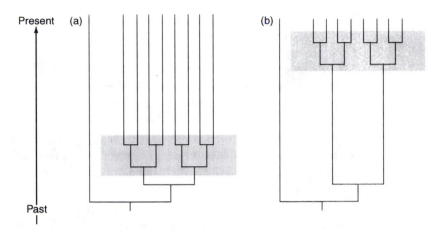

Fig. 3.5 The wide diversification of ferns did not solely take place when they dominated many terrestrial biomes (as represented in panel (a)) but extant ferns diversified later (more as represented in panel (b)), filling the new environmental niches created by the angiosperms.

the golden age of a high number of fern species was prior to the Cretaceous (135 mya). It was previously thought that this time before the Cretaceous, when fern-dominated biomes were far more common, was when the great diversity of ferns arose. It was similarly thought that there were fewer species of fern evolving actually in the Cretaceous due to the rise and increase in angiosperms (flowering plants). Supposedly, the rise and diversification of one offset the diversification of the other, as they were competitors. Of course, some fern species – indeed many – have died out over the past couple of hundred million years, but there is evidence that the richness of fern species we see today does not appear to have stemmed from before the Cretaceous. Conversely, genetic and systematic analysis (of species' evolutionary relatedness) suggests later development of the range of modern ferns that coincided with – not preceded – the rise of angiosperms in the Cretaceous.

This may at first seem counterintuitive. It would seem likely that there were more of today's species early on when the ferns dominated many terrestrial ecosystems, not later, with only a few surviving extinction episodes and competition from the rise of more advanced plants. However, what actually appears to have happened is that the new angiosperm communities established in the Cretaceous also created new environmental niches, and of course the ferns were already there, poised to exploit them (Schneider et al., 2004). The rise of angiosperms was therefore coincident with, and not subsequent to, the diversification of the fern species we see today and the fern diversification was in fact a response to the rise in angiosperms (Figure 3.5).

The relevance of this to the late-Miocene expansion of C_4 grasses (14–9 mya) is that the expansion of grasses also provided new ecological niches, hence the evolutionary opportunity for speciation (the formation of new species). One speciation event that began at this time, and which arguably has had the greatest impact on the planet, was that of the Hominoidea, the apes. This branch of speciation continued ultimately to include modern humans, *Homo sapiens*, but began back in the late Miocene. One of the early Hominoidea speciations was that of the sivapithecines, whose remains have been found in Turkey, India and Pakistan. They are thought to have survived up to about 8 mya and are evolutionary cousins (as opposed to grandparents) of *H. sapiens*.

There is other more tangential relevance of the drivers behind the earlier Cretaceous diversity of ferns to the Miocene expansion of grasses. We have more recent evidence that other factors were at work than the decline in atmospheric carbon dioxide, with which C_4 plants are better able to cope than C_3 flora. Isotopic studies of microfossils from the end of the last glacial 13 000–21 000 years ago from two Central-American lake beds show that C_4 plants began to flourish around one in Guatemala but not the other, in Northern Mexico, even though the atmospheric carbon dioxide would have been the same for both. The reason was a climatic change at the end of the last glacial and local environmental conditions pertaining to each lake. Consequently, while earlier in the Miocene low carbon dioxide may have facilitated C_4 grass expansion, other factors such as regional climate were critical in determining where this happened. Where the conditions were right, C_4 species flourished (Huang et al., 2001). This is exactly analogous to the Cretaceous fern diversification (in niches created by the rise of angiosperms). In short, major evolutionary drivers are rarely the result of single factors and a holistic approach to understanding the development of ecosystems is required.

We will return to this theme of a holistic approach in later chapters when looking at the biological consequences of future climate change, which is not just dependent on the climate change itself but also the nature and distribution of ecosystems in the landscape.

Other evolutionary developments associated with the rise of grasses and grasslands is that equids (the Equidae, the family of horses, asses and zebras) and other mammals developed hypsodont (high-crowned) teeth in the early Miocene (beginning 25 mya) in North America. This perspective is supported in that fossilised grasses and grass pollen appear in abundance in the Miocene and become increasingly common to the present. Grasses are far tougher to eat (see below) compared to the berries and leaves of tropical forests and woodlands that equids used to inhabit. However, since the late 1990s it has been noted that the increase in hypsodonty in mammals occurred in the Oligocene (beginning 35 mya) in South America. It is now thought that grasses were significantly present a little earlier in South America and maybe also in Africa. Unfortunately for palaeontologists, grasses are most abundant in oxidising and well-drained environments that do not encourage fossilisation. So, field evidence is literally thin on the ground.

The Miocene increase in open grasslands in favour of closed forests meant that equids faced new pressures. In forests, being short helped movement through the undergrowth, where manoeuvrability was the key to escaping from predators. Conversely, in open grasslands it was speed that was essential. Those individuals with the longest feet and the ability to use their toes to give them a little extra speed tended to survive while slower individuals did not. Some equids became taller and evolved into (hooved) ungulates.

Grasses belong to the family Poaceae (formerly known as the Graminae) of the order Poales (which in turn is sometimes referred to as the Glumiflorae), in the monocotyledon class (Monocotyledonae) of angiosperms (flowering plants). They include the C_4 species of maize, wheat and rice and so today are of central importance to agrarian and technological humans. (Remember, C_4 plants form carbohydrates through photosynthesis using a different metabolic carbon pathway from C_3 plants.)

Grasses, particularly C_4 grasses, have several properties marking them as different from C_3 dicotyledons. (Dicotyledons have two cotyledons – leaves emerging from the seed – as opposed to the one produced by grass monocotyledons.) First, grasses usually have a lower nitrogen content, which makes them intrinsically less nutritious. (Note: as food plants humans use the seed part of monocotyledons.) Second, C_4 and C_3 grasses rarely have secondary toxins that conversely are common to many C_3 dicotyledons. Third, protein in C_4 plants is protected by a bundle of sheath cells making them harder to digest (hence the agricultural pre-processing of such crops). And finally, grasses have a higher silica content than dicotyledons. This makes them more abrasive and this abrasiveness during mastication provides the link with hypsodonty (teeth with high crowns good for grazing) that enabled mammals to take advantage of the new food source.

Herbivorous animals' food can be determined one way through ^{13}C analysis. C_3 plants (including some grasses) have ^{13}C values of 26‰ whereas C_4 plants (which are mostly grasses) have ^{13}C values of 12‰; consequently fossil teeth provide a record of the proportion of C_4 and C_3 plants in herbivore diets and palaeosol (fossilised soil) carbonates. However, there is a problem: C_3 (monocotyledonous) grasses have the same ^{13}C content as C_4 dicotyledonous plants. This means that ^{13}C analysis only reveals the expansion in C_4 grasses. Conversely, most C_4 plants are grasses and the C_4 grasses are the dominant grasses at low latitudes (around the tropics). C_4 plants can have a carbon dioxide photosynthetic advantage over their C_3 counterparts. Carbon dioxide (on an annual basis) is effectively the same the planet over and so at low latitudes C_4 plants are better placed to make use of the higher-intensity insolation (the solar radiation received at the Earth's surface) and temperature. C_4 grasses are therefore the dominant grasses in the tropics whereas C_3 grasses are found at higher latitudes (or altitudes, if in the tropics).

The ^{13}C record of North American equids shows an abrupt increase starting about 7 mya, although Pacific coast and Canadian equids' ^{13}C values suggest that this increase occurred later. A similar change was noted at the end of the Miocene (approximately 7–5 mya) in Pakistan, Kenya and South America.

Just as ferns did in the Cretaceous, over the Tertiary period (65.5–1.8 mya) mammals diversified at the end of the Miocene. Back in the Mesozoic era (the age of the dinosaurs) the largest mammal was the size of a modern house cat. In the Tertiary this was to change. Here again, equids are illustrative of general mammal diversification. The early horses (such as Eohippus) were small and well adapted to the global tropical-like forests of the Tertiary period's early Eocene epoch (55.8–33.9 mya). Their later diversification into increasingly larger forms reflects their being more suited to a cooler climate (increased size has a lower surface-area-to-volume ratio, and hence reduced heat loss) and to eating nutritionally poor grasses (which necessitates gastrointestinal specialisation that takes up room inside the body cavity). And, as noted, these larger forms were more suited to grassland plains than the smaller forms which were more suited to moving among the tangle of forests. (Once again, note the interplay of factors.)

During this time North America was geographically separated from South America (there was no connecting isthmus). Consequently the smaller early horses could not migrate south towards the equator. If they had been able to then the modern horses may have evolved allopatrically (in a population in a different geographic area), with the more ancient forms surviving today, somewhat in the way the mouse deer (tragulids) did in the Old World tropics.

Of course, and perhaps of special interest to readers, as noted above, the precursors to the Hominoidea began diversifying 14–9 mya and this continued with the evolution of the australopithecines that lived 5–1.2 mya.

From the end of the Miocene, and throughout the Pliocene (2–5 mya), the Earth cooled. This trend culminated in our current (Quaternary) ice age, consisting of a series of glacials interspersed by warmer interglacials such as the one we presently experience. This will be examined next.

3.4 Summary

A key feature of the biosphere's evolution (including its abiotic components of geology and atmosphere) up to the current (Quaternary) ice age has been the interrelation of biology and climate. Key evolutionary steps in biological evolution – such as the developments of photosynthesis, multicellular organisms, vascular plants and new photosynthetic processes – resulted in changes in both the atmosphere and climate: the Snowball Earths I and II. A number of these evolutionary steps have had consequences for the re-organisation of elemental cycling, especially carbon, its various reservoirs and in relations between them. Remember, life on Earth is carbon-based. So it should not be too great an intellectual leap to appreciate that the evolution of life will affect the global carbon cycle. So it is perhaps a little surprising that the notion of evolving life's key evolutionary stages – the development of photosynthesis and then the cell differentiation of eukaryotes and hence rise of multicelled animals (metazoans) and plants – caused the Snowball Earths only began to be discussed seriously at the beginning of the 21st century.

Changes in the carbon cycle, be they major as in Snowball Earths or lesser as in CIEs of the Phanerozoic eon (past 542 million years), implicitly affected atmospheric carbon (both carbon dioxide and methane) and therefore changes in the greenhouse forcing of climate, hence climate itself. Taking all these factors together provided further evolutionary opportunities that included those that (much later) allowed for the rise of *H. sapiens*.

In addition, the biosphere's evolution has been punctuated by a number of extinction events imposed independently of biological evolution but which have affected evolution (or ended it). These events have principally been vulcanism-related or asteroidal. Of the former some have been associated with (or at least helped to trigger) carbon releases and both have had global climate effects. However, even asteroidal biological extinctions, although not directly related to the biosphere's evolution per se, are necessary components of a cosmology that enables Earth-like planets in general, and the Earth specifically, to exist. Asteroids and planetesimals are necessary in a proto-planetary system for planets to form. In contrast, vulcanism is a manifestation of the energy that drives plate tectonics, hence the deep carbon cycle, and so is a major factor enabling allopatric speciation and biosphere stability (see the discussion on ice-world recovery in section 3.2.2). Some of the main causes of mass extinctions therefore relate to factors that enable life. Others, such as methane releases from submarine methane hydrate (clathrates), wetlands and soils, ultimately arise out of life itself. Life and mass extinction events are intricately connected well beyond the obvious symmetry of one virtually being an antithesis of the other, and you cannot have an extinction without life.

With regard to our current anthropogenic global warming and climate change, there is much we can learn from the biosphere's history. There are perhaps two key points of particular relevance. First, re-organisation of the carbon cycle, as we are now doing by liberating formerly geologically isolated carbon into the atmosphere, results in major environmental change. (A point that may seem obvious to many now, but which was not so a decade or two ago; see the IPCC's first assessment, 1990.) Second, such environmental change affects the number and range of species extant on the planet both positively and negatively. This should be of considerable human concern, for not only are we an animal species in our own right, but we rely on many species of plants and animals for our daily survival. We rely on an even greater population of species through what ecologists call ecosystem function, beyond the daily time frame, to maintain environmental quality. We will return to these dimensions in later chapters.

3.5 References

Abramov, O. and Mojzsis, S. J. (2009) Microbial habitability of the Hadean Earth during the late heavy bombardment. *Nature*, 459, 419–22.

Alveraz, L. W., Alveraz, W., Asaro, F. and Michel, H. V. (1980) Extraterrestrial cause for the Cretaceous-Tertiary extinction – experimental results and theoretical implications. *Science*, 208, 1095–1108.

Becker, L., Poreda, R. J., Hunt, A. G., Bunch, T. E. and Rampino, M. (2001) Impact event at the Permian-Triassic boundary: evidence from extraterrestrial noble gases in fullerines. *Science*, 291, 1530–3.

Becker, L., Poreda, R. J., Basu, A. R. et al. (2004) Bedout: a possible end-Permian impact crater offshore of northwestern Australia. *Science*, 304, 1469–75.

Bekker, A., Holland, H. D., Wang, P.-L. et al. (2004) Dating the rise of atmospheric oxygen. *Nature*, 427, 117–20.

Berner, R. A. (1998) The carbon cycle and CO_2 over Phanerozoic time: the role of land plants. *Philosophical Transactions of the Royal Society of London Series B*, 353, 75–82.

Bo, S., Siegert, M. J., Mudd, S. M., Sugden, D. et al. (2009) The Gamburtsev mountains and the origin and early evolution of the Antarctic ice sheet. *Nature*, 459, 690–3.

Bottomley, R., Grieve, R., York, D. and Masaitis, V. (1997) The age of the Popigai impact event and its relation to events at the Eocene/Oligocene boundary. *Nature*, 388, 365–8.

Boussau, B., Blanquart, S., Necsulea, A., Lartillot, N. and Gouy, M. (2008) Parallel adaptations to high temperatures in the Archaean eon. *Nature*, 456, 943–5.

Bowen, G. J., Beerling, D. J., Koch, P. L., Zachos, J. C. and Quattlebaum, T. (2004) A humid climate state during the Palaeocene/Eocene thermal maximum. *Nature*, 432, 495–9.

Briggs, D. E. G. and Crowther, P. R. (eds). (2001) *Palaeobiology II*. Oxford: Blackwell Science.

Buick, R. (2008) Early evolution of oxygenic photosynthesis. *Philosophical Transactions of the Royal Society of London Series B*, 363, 2731–43.

Clough, D. O. (1981) Solar interior structure and luminosity variations. *Solar Physics*, 74, 21–34.

Cohen, A. S., Coe, A. L. and Kemp, D. B. (2007) The Late Palaeocene Early Eocene and Toarcian (Early Jurassic) carbon isotope excursions: a comparison of their time scales, associated environmental changes, causes and consequences. *Journal of the Geological Society*, 164, 1093–1108.

Cowie, J. and Cohen, A. S. (eds). (2010) *Past Carbon Isotopic Events and Future Ecologies Abstract Book*. London: Geological Society (www.geolsoc.org or www.science-com.concatenation.org/archive/cie_symposium_2010.html together with other supporting material).

Dickens, G. R., O'Neil, J. R., Rea, D. K. and Owen, R. M. (1995) Dissociation of methane hydrate as a cause of isotope excursion at the end of the Paleocene. *Paleoceanography*, 10, 965–71.

Emilliani, C. (1992) *Planet Earth: Cosmology, Geology and the Evolution of Life and the Environment*. Cambridge: Cambridge University Press.

Erwin, D. H. (1994) The Permo-Triassic extinction. *Nature*, 367, 231–6.

Evans, D. A., Beukes, N. J. and Kirschvinck, J. L. (1997) Low-latitude glaciation in the Palaeoproterozoic era. *Nature*, 386, 262–6.

Frakes, L. A., Francis, J. E. and Syktus, J. I. (1992) *Climate Modes of the Phanerozoic*. Cambridge: Cambridge University Press.

Gaucher, E. A., Govindarajan, S. and Ganesh, O. K. (2008) Palaeotemperature trend for Precambrian life inferred from resurrected proteins. *Nature*, 451, 704–7.

Genda, H. and Abe, Y. (2005) Enhanced atmospheric loss on protoplanets at the giant impact phase in the presence of oceans. *Nature*, 433, 842–4.

Grasby, S., Sanei, H. and Beauchamp, B. (2011) Catastrophic dispersion of coal fly ash into oceans during the latest Permian extinction. *Nature Geoscience*, 4, 896–7.

Hallam, T. (2004) *Catastrophes and Lesser Calamities: The Causes of Mass Extinctions*. Oxford: Oxford University Press.

Han, T. M. and Runnegar, B. (1992) Megascopic eukaryotic algae from the 2.1 billion year-old Neguanne iron formation. *Science*, 257, 232–5.

Holland, H. D. (1990) Origins of breathable air. *Nature*, 347, 17.

Hren, M. T., Tice, M. M. and Chamberlain, C. P. (2009) Oxygen and hydrogen isotope evidence for a temperate climate 3.42 billion years ago. *Nature*, 462, 205–8.

Huang, Y., Street-Perrott, F. A., Metcalfe, S. E. et al. (2001) Climate change as the dominant control on glacial-interglacial variations in C3 and C4 plant abundance. *Science*, 293, 1647–51.

Intergovernmental Panel on Climate Change (1990) *Climate Change: the IPCC Scientific Assessment*. Cambridge: Cambridge University Press.

Intergovernmental Panel on Climate Change (2001a) *Climate Change 2001: Mitigation – A Report of Working Group III*. Cambridge: Cambridge University Press.

Intergovernmental Panel on Climate Change (2001b) *Climate Change 2001: Impacts, Adaptation, and Vulnerability – Summary for Policymakers and Technical Summary of the Working Group II: Report*. Cambridge: Cambridge University Press.

Intergovernmental Panel on Climate Change (2007) *Climate Change 2007: The Physical Science Basis – Working Group I Contribution to the Fourth Assessment Report of the Intergovernmental Panel on Climate Change*. Cambridge: Cambridge University Press.

Janis, C. M. (2001) Radiation of Tertiary mammals. In Briggs, D. E. G. and Crowther, P. R., eds., *Palaeobiology II*, pp. 109–12. Oxford: Blackwell Science.

Jenkyns, H. C., Forster, A., Schouten, S. and Damsté, S. (2004) High temperatures in the late Cretaceous Arctic ocean. *Nature*, 432, 888–92.

Jin, Y. G., Wang, Y., Wang, W. et al. (2000) Pattern of marine mass extinction near the Permian-Triassic boundary in South China. *Science*, 289, 432–6.

Katz, M. E., Pak, D. K., Dickens, G. R. and Miller, K. G. (1999) The source and fate of massive carbon input during the Late Palaeocene Thermal Maximum. *Science*, 286, 1531–3.

Kaufman, A. J. and Xiao, S. (2003) High CO_2 levels in the Proterozoic atmosphere estimated from the analyses of individual microfossils. *Nature*, 425, 279–81.

Keeling, P. J. (2008) Bridge over troubled plastids. *Nature*, 451, 896–7.

Kemp, D. A., Coe, A. L., Cohen, A. S. and Lorenz, S. (2005) Astronomical pacing of methane release in the early Jurassic period. *Nature*, 437, 396–9.

Kennedy, M., Mrofka, D. and von der Borch, C. (2008) Snowball Earth termination by destabilization of equatorial permafrost methane clathrate. *Nature*, 453, 642–5.

Kennett, J. P. and Stott, L. D. (1991) Abrupt deep sea warming. palaeoceanographic-changes and benthic extinctions at the end of the Palaeocene. *Nature*, 353, 225–9.

Kirschvink, J. L. (1992) A palaeogeographic model for vendian and Cambrian time. In Schopf, J. W. and Klein, C., eds., *The Proterozoic Biosphere*, pp. 567–82. Cambridge: Cambridge University Press.

Knauth, L. P. (2006) Signature required. *Nature*, 442, 873–4.

Knauth, L. P. and Kennedy, M. (2009) The late Precambrian greening of the Earth. *Nature*, 460, 728–32.

Lenton, T. and Watson A. (2011) *Revolutions That Made the Earth*. Oxford: Oxford University Press.

Lenton, T. M., Crouch, M., Johnson, M., Pires, N. and Dolan, L. (2012) First plants cooled the Ordovician. *Nature Geoscience*, 5, 86–9.

Lourens, L. J., Sluijs, A., Kroon, D. et al. (2005) Astronomical pacing of late Palaeocene to early Eocene global warming events. *Nature*, 435, 1083–7.

Lovelock, J. E. (1979) *Gaia: A New Look at Life on Earth*. Oxford: Oxford University Press.

McElwain, J. C., Wade-Murphy, J. and Hesselbo, S. P. (2005) Changes in carbon dioxide during an oceanic anoxic event linked to intrusion into Gondwana coals. *Nature*, 435, 479–82.

Meredith, R. W., Janečka, J. E., Gatesy, J. et al. (2011) Impacts of the Cretaceous terrestrial revolution and KPg extinction on mammal diversification. *Science*, 334, 521–4.

Moore, R. B., Obornik, M., Janouškovec, J. et al. (2008) A photosynthetic alveolate closely related to apicomplexian parasites. *Nature*, 451, 959–63.

Newton, R. and Bottrell, S. (2007) Stable isotopes of carbon and sulphur as indicators of environmental change: past and present. *Journal of the Geological Society*, 164, 691–708.

Nier, A. O. (1950) A Redetermination of the relative abundances of the isotopes of neon, krypton, rubidium, xenon, and mercury. *Physical Review*, 79, 450–4.

Noffke, N., Eriksson, K. A., Hazen, R. M. and Simpson, E. L. (2006) A new window into Early Archean life: Microbial mats in Earth's oldest siliciclastic tidal deposits (3.2 Ga Moodies Group, South Africa). *Geology*, 34, 253–6.

Nunes, F. and Norris, R. D. (2006) Abrupt reversal in ocean overturning during the Palaeocene/Eocene warm period. *Nature*, 439, 60–3.

Ohmoto, H., Watanabe, Y. and Kumazawa, K. (2004) Evidence from massive siderite beds for a CO_2-rich atmosphere before $\sim$1.8 billion years ago. *Nature*, 429, 395–9.

Pagani, M., Freeman, K. H. and Arthur, M. A. (1999) Late Miocene atmospheric CO_2 concentrations and the expansion of C4 grasses. *Science*, 285, 876–9.

Payne, J. L., Lehrmann, D. J., Wei, J. et al. (2004) Large perturbations of the carbon cycle during recovery from the end-Permian extinction. *Science*, 305, 506–9.

Pearson, P. N. and Palmer, M. R. (2000) Atmospheric carbon dioxide concentrations over the past 60 million years. *Nature*, 406, 695–9.

Pollard, D. and Kasting, J. F. (2005) Snowball Earth: a thin-ice solution with flowing glaciers. *Journal of Geophysical Research* 110, C07010.

Robert, F. and Chaussidon, M. (2006) A palaeotemperature curve for the Precambrian oceans based on silicon isotopes in cherts. *Nature*, 443, 969–72.

Ruhl, M., Bonis, N. R., Reichart, G.-J., Damsté, J. S. S. and Kürschner, W. M. (2011) Atmospheric carbon injection linked to end-Triassic mass extinction. *Science*, 333, 430–4.

Schneider, H., Schuettpelz, E., Pryer, K. M. et al. (2004) Ferns diversified in the shadow of angiosperms. *Nature*, 428, 553–7.

Secord, R., Bloch, J. I., Chester, S. G. B., Boyer, D. M. et al. (2012) Evolution of the earliest horses driven by climate change in the Paleocene-Eocene Thermal Maximum. *Science*, 335, 959–62.

Sexton, P. F., Norris, R. D., Wilson, P. A., Palike, H. et al. (2011) Eocene global warming events driven by ventilation of oceanic dissolved organic carbon. *Nature*, 471, 349–53.

Sheehan, P. M. and Harris, M. T. (2004) Microbolite resurgence after the late Ordovician extinction. *Nature*, 430, 75–7.

Sleep, N. H., Bird, D. K. and Pope, E. C. (2011) Serpentinite and the dawn of life. *Philosophical Transactions of the Royal Society of London Series B*, 366, 2857–69.

Smith, A. G. and Pickering, K. T. (2003) Oceanic gateways as a critical factor to initiate icehouse Earth. *Journal of the Geological Society London*, 160, 337–40.

Sobolev, S. V., Sobolev, A. V., Dmitry V. Kuzmin, D. V. et al. (2011) Linking mantle plumes, large igneous provinces and environmental catastrophes. Nature, 477, 312–16.

Steuber, T., Rauch, M., Masse, J.-P., Graaf, J. and Malkoĉ, M. (2005) Low-latitude seasonality of Cretaceous temperatures in warm and cold episodes. *Nature*, 437, 1341–4.

Stoll, H. M. (2006) The Arctic tells its story. *Nature*, 441, 579–80.

Storey, M., Duncan, R. A. and Swisher III, C. C. (2007) Paleocene-Eocene Thermal Maximum and the opening of the Northeast Atlantic. *Science*, 316, 587–9.

Strother, P. K., Battison, L., Brasier, M. D. and Wellman, C. H. (2011) Earth's earliest non-marine eukaryotes. *Nature*, 473, 505–9.

Svensen H., Planke, S., Malthe-Sørenssen, A. et al. (2004) Release of methane from a volcanic basin as a mechanism for initial Eocene global warming. *Nature*, 429, 542–5.

Tice, M. M. and Lowe, D. R. (2004) Photo synthetic microbial mats in 3,416-Myr-old ocean. *Nature*, 431, 549–52.

Tripati, A. and Elderfield, H. (2005) Deep-sea temperature and circulation changes at the Paleocene-Eocene Thermal Maximum. *Science*, 308, 1894–8.

Vajda, V. and McLoughlin, S. (2004) Fungal proliferation at the Cretaceous-Tertiary boundary. *Science*, 303, 1489.

Vent, D. V. and Spray, J. G. (1998) Impacts on the Earth in the Late Triassic (separate letters). *Nature*, 395, 126.

Ward, P. D., Montgomery, D. R. and Smith, R. (2000) Altered river morphology in South Africa related to the Permian-Triassic extinction. *Science*, 289, 1740–3.

Watanabe, Y., Martini, J. E. J. and Ohmoto, H. (2000) Geochemical evidence for terrestrial ecosystems 2.6 billion years ago. *Nature*, 408, 574–7.

Watson, A. (2004) Gaia and the observer self-selection. In Schneider, S. H., Miller, J. R., Crist, E. and Boston, P. J., eds., *Scientists Debate Gaia, The Next Century*, pp. 201–10. Cambridge, MA: MIT Press.

Wible, J. R., Rougier, G. W., Novacek, M. J. and Asher, R. J. (2007) Cretaceous eutherians and Laurasian origin for placental mammals near the K/T boundary. *Nature*, 447, 1003–6.

Wilf, P. and Labandeira, C. C. (1999) Response of plant-insect associations to Paleocene-Eocene warming. *Science*, 284, 2153–6.

Wilf, P., Cuneo, N. R., Johnson, K. R. et al. (2003) High plant diversity in Eocene South America: Evidence from Patagonia. *Science*, 300, 122–5.

Wing, S. L., Harrington, G. J., Smith, F. A. et al. (2005) Transient floral change and rapid global warming at the Palaeocene-Eocene boundary. *Science*, 310, 993–5.

Yuan, X., Xiao, S. and Taylor, T. N. (2005) Lichen-like symbiosis 600 million years ago. *Science*, 308, 1017–34.

Zachos, J., Pagani, M., Sloan, L., Thomas, E. and Billups, K. (2001) Trends, rhythms and aberrations in global climate 65 MA to present. *Science*, 292, 686–93.

Zachos, J. C. et al. (2003) A transient rise in tropical sea surface temperature during the Paleocene-Eocene Thermal Maximum. *Science*, 302, 1551–4.

Zachos, J. C. et al. (2004) *Early Cenozoic Extreme Climates: The Walvis Ridge Transect, Sites 1262-1267*. Proceedings of the Ocean Drilling Program, Initial Reports Vol. 208, Ocean Drilling Program, College Station, TX.

Zachos, J. C. et al. (2005) Rapid acidification of the ocean during the Paleocene-Eocene thermal maximum. *Science*, 308, 1611–15.

4

The Oligocene to the Quaternary: climate and biology

We are currently in the middle of an ice age! This ice age is known as the Quaternary ice age, and it began roughly 2 million years ago (mya) (I say 'roughly' because how much ice do you need on the planet to say that it is an ice age?). We might not think we are in an ice age and this is because we are in a warm part, called an interglacial. As we shall see, there have been a number of glacials and interglacials in our Quaternary ice age. However, this ice age did not just start by itself but arose out of a number of factors that became relevant earlier, in the Oligocene (34–23 mya) and Miocene (23–5.3 mya) epochs, well before the beginning of our ice age and the Pliocene and Pleistocene glaciations (Zachos et al., 2001). These glaciations actually had their beginnings some 5.3 mya. To understand how our Quaternary ice age came about we will need to briefly re-cap part of the previous chapter and note some other material to provide a more biological perspective while leaving out the extinction events.

4.1 The Oligocene (33.9–23.03 mya)

Between 35 and 15 mya the Earth's temperature was roughly 3–4°C warmer than today and atmospheric carbon dioxide concentrations were twice as high. However, climate forcing factors were coming into play that were to cool the planet. Carbon dioxide levels were falling and, as noted in the previous chapter, this fall could only have been furthered by the new C_4 plants (even though their period of major expansion was not to take place until 8 mya: see below).

A major forcer of global climate came about by the collision of the Indian and Asian tectonic plates. These plates first made contact with each other around 55 mya (coincidentally, roughly around the time of the Eocene climatic event) and they continued to compact. This resulted in the subsequent uplift of what was to become the Tibetan Plateau. This gradually rose and is still rising today.

The plateau increasingly affected global climate in three ways. First, it forced the re-organisation of atmospheric circulation, with the creation of the East Asian monsoon and associated climatic patterns. Second, the uplift caused ice to form in the mountains and notwithstanding this there were other albedo effects (due to altitude and changing nature of the land's surface), so reflecting more of the Sun's energy. This is one of the main reasons why Milankovitch curves are frequently calculated on the solar energy received in northern latitudes between 28° and 40°N. Third, it increased erosion, so increasing the amounts of silicate, and hence calcium silicates that could react with carbon dioxide, precipitating it in solid form as calcium carbonate and so encouraging the further drawdown of atmospheric carbon dioxide (see Chapter 3).

All these factors, together with those discussed in section 3.3.11, helped to switch the biogeosphere from being more stable in a greenhouse mode (the Eocene and earlier), to greater stability in a cooler mode (the Oligocene and later: 33.9 mya and onwards).

$\delta^{18}O$ Analysis of forams and ocean sediments to ascertain past calcite compensation depths (sometimes called CCDs) – the water depth at which the calcium carbonate rain is balanced by the dissolution rate due to ocean acidity – suggests that the first indications of the aforementioned shift in stability took place around 42 mya in the Eocene. A more definite shift then took place somewhere around 34 mya at the Eocene–Oligocene boundary (Tripati et al., 2005). (This of course is distinct from the comparatively transient [in geological terms] change in ocean chemistry associated with the Eocene thermal maximum event; see Chapter 3.)

Earlier, in 2003, US palaeoclimatologists Robert DeConto and David Pollard demonstrated through a general-circulation computer model coupled with atmosphere and ice-sheet components, and incorporating palaeogeography of the time, that a combination of declining Cenozoic carbon dioxide (from four times to just double the pre-industrial level) and the thermal isolation of Antarctica enabled an ice sheet in the east of the continent to begin to form 34 mya. This is supported by evidence from geological sediments. The thermal isolation was caused by the opening of oceanic passages between South America and Antarctica (the Drake Passage) and Australia and Antarctica (the Tasmanian Passage). This enabled the Antarctic Circumpolar Current (or ACC as it is sometimes abbreviated) to form. However, in this instance of long-term global cooling, ocean circulation change does not appear to have been as important early on as the decline in atmospheric carbon dioxide. This may be because a depth threshold, to allow a critical volume of water to flow, needs to be reached. But aside from carbon dioxide, it appears that an ice albedo positive-feedback effect (see Chapter 1) also played a major role. Here the expanding ice sheet reflected more sunlight and so cooled the continent, allowing the ice sheet to expand further still, which reflected even more sunlight, and so on.

The first Antarctic ice sheets were sensitive to Milankovitch forcing (see Chapter 1), much as the northern hemisphere ice sheets were during the Quaternary glacial–interglacial series of waxing and waning ice sheets (which we will examine shortly). It has been suggested that orbital frequencies of 40 000 years (obliquity) and 125 000 years (eccentricity) dominated the way the ice margins behaved at that time.

This time was one of critical transition as the Earth system crossed a climate threshold (see section 6.6.8). In addition to the circulation and albedo climate forcing factors already mentioned, there is strong evidence to suggest that the switch to a cooler climate was amplified by changes in greenhouse gas concentrations. In 2009 Paul Pearson, Gavin Foster and Bridget Wade used boron ($^{11}B/^{10}B$) isotope analysis on Foraminifera shells from the sea surface and the remains of which are found in marine sediments. The change in the ratio of these isotopes ($\delta^{11}B$) is correlated with pH of the sea surface and pH in turn is determined by the amount of carbon dioxide dissolved in the sea (as carbonic acid), and hence by implication the carbon dioxide in the atmosphere above. Their results suggest a dip in carbon dioxide concentrations at

36.6 mya (just after the Eocene–Oligocene boundary) to about 760 ppmv. Following this there was a rebound to just over 11 000 ppm over the next 100 000 years, and then a decline again. The modelled threshold for Antarctic glaciation is between 700 and 850 ppm. What exactly caused this temporary dip is not clear: there are a number of possibilities and combinations thereof. Irrespective of our lack of knowledge, after this the Earth continued to cool and Antarctica's ice sheets grew. As we shall shortly see, the expansion of grasses and C_4 plants, ocean circulation changes due to continental shift and orogenesis (building of mountains such as the Himalayas; hence silicate weathering to form carbonate; see section 3.3.3) are all implicated in this long-term decline in carbon dioxide over many millions of years.

4.2 The end Miocene (9–5.3 mya)

As mentioned in section 3.3.11 the new C_4 photosynthetic plants were better able to take advantage of the atmosphere's lower carbon dioxide levels compared to plants that evolved earlier, in the Tertiary (prior to 20 mya), and marginally help with carbon dioxide drawdown. This ability to thrive in the then carbon dioxide-poorer atmosphere gave them an advantage over C_3 plants where other conditions (such as soil, temperature and moisture) were favourable. As noted in the last chapter, C_4 plants originated well before they became globally established: global expansion takes time. Indeed, it is thought that atmospheric carbon dioxide had already declined to the equivalent of Quaternary levels (of the past 2 million years or so) by 15 mya, and if anything atmospheric carbon dioxide is thought to have been on a minor increase between 15 and 5 mya. So, although the lowering of atmospheric carbon dioxide over much of the Tertiary facilitated speciation and the rise of C_4 plants, lowering atmospheric carbon dioxide by itself cannot explain their global spread at the end of the Miocene. It may be that tectonic activity, changes in seasonal precipitation and some other factor(s) or combinations thereof facilitated matters (Pagani et al., 1999). Indeed, as mentioned in the previous section there was the opening of the high-latitude southern hemisphere oceanic passages and, of course, uplift, forming the Tibetan Plateau. By 10 mya the east-Antarctic ice sheet had grown (but was still smaller than today) from its beginnings 35 mya, and 10 mya one had already formed on Greenland and was itself slowly growing. Superimposed on this overall trend of a cooling Earth and increasing ice was Milankovitch-driven ice-sheet waxing and waning. Climate change was taking place.

As noted in the last chapter, the spread of C_4 plants during this time has been charted in a number of ways. The US geophysicist Thure Cerling and colleagues published in 1993 the results of isotope analysis of soil and tooth remains: metabolic pathways in C_3 and C_4 plants have different preferences for (or fractionation of) the ^{13}C and ^{12}C carbon isotopes. Initially there was some discussion that this interpretation might be undermined if animals migrated from where they fed to where they died. However, in 1997 Cerling and colleagues published the results of carbon analysis of tooth remains from more than 500 hypsodont equids from Asia, Africa, Europe and North and South America. In addition, they analysed the teeth of fossil proboscideans (elephants and allied species) and notoungulates (an extinct order of South American mammals). The advantages of looking at teeth, as opposed to soils, are that animals' selective feeding enhances the isotope signal, and tooth identification and dating is comparatively easy, whereas soil contains a mix of ^{13}C and ^{12}C originating from both C_3 and C_4 remains. The results of Cerling's team show that large mammals before 8 mya from around the world all had diets either dominated by C_3 plants or which consisted purely of C_3 plants. In contrast, by 6 mya equids and some other large mammals from many regions in latitudes below $37°$ (the subtropics and tropics) had C_4 diets.

During this time the Drake and Tasmanian Passages widened further, so increasing Antarctica's thermal isolation and facilitating growth of its ice sheets. By 6 mya ice-sheet growth in west Antarctica had become significant. This ice sheet and the one already existing in east Antarctica were subject to positive albedo/temperature-feedback effects and so ice slowly grew to fill the Antarctic continent.

4.3 The Pliocene (5.3–2.6 mya)

The Pliocene was the last epoch in the Tertiary period. It was a time of global transition from warm conditions with global surface temperatures about $3°C$ warmer than today, smaller ice sheets and higher sea levels compared with current cooler conditions. In 2004 a team led by Ana Ravelo from the University of California and Boise State University published an analysis of several palaeoclimate proxy records. These included an ^{18}O isotope record from benthic Foraminifera, an alkenone record, a biogenic carbonate record, an organic carbon record and a biogenic isotopic carbon-sediment record. Together these records and those of other researchers provide a reasonably clear picture of how the climate changed, although of course the detail is limited.

Relative to today (the dawn of the 21st century) the approximately $3°C$ warmer early Pliocene Earth had a sea level some 10–20 m higher than now and an enhanced thermohaline circulation (to which we will return later). Atmospheric and ocean-circulation patterns were in places significantly different from today. It was thought that during the Pliocene the Pacific Ocean was subject to permanent El Niño-like conditions and different ocean- and atmosphere-circulation patterns (Wara et al., 2005). (Today the El Niño Southern Oscillation [ENSO] is a semi-regular event the effects of which have repercussions a hemisphere away; see Chapter 5 and the Holocene summary as well Chapter 6.) However, more recently evidence has been gathering that El Niño events were probably taking place in past warm times much as they do today. For example, in 2011 Japanese researchers led by Tsuyoshi Watanabe reported that the idea of a permanent El Niño mode during the Pliocene warm period was not supported by coral evidence. Indeed, in even warmer times such as the Eocene maximum the ENSO still seems to be present in the same year. In the same year, 2011, Linda Ivany at Syracuse University in New York and her team analysed Antarctic fossils from the Eocene, 56–34 mya, when the average global temperature exceeded today's by at least $10°C$. They measured the width of growth bands in the fossils of wood and bivalves (a class of mollusc) as indicators of annual growth rate.

They found growth rates that varied with similar frequency to the present-day El Niño cycle, suggesting that it was occurring at that time.

The answer to this question is important to ascertain because the early Pliocene can be considered a palaeoclimatic analogue of a possible late-21st-century globally warmed Earth. This means that Pliocene research is of interest to those concerned with present-day climate change.

Antarctica's ice sheets were growing throughout the Pliocene, but were still smaller than they are today. Similarly, there was ice in the northern hemisphere, but again this was greatly reduced compared to today. Atmospheric carbon dioxide, although declining, was also slightly (approximately 30%) higher later in the Pliocene than recent pre-industrial levels, at around approximately 360 ppm. This compares with carbon dioxide levels in the 1990s, but the reason that the Pliocene Earth was warmer than the 1990s is because its oceans had already been warmed, its Pliocene ice caps were smaller than today (reflecting less sunlight) and there were also some atmosphere- and ocean-circulation differences.

Associated with this last, tectonic movement of the continents also continued through the Pliocene and that also impacted on the global climate. The aforementioned Drake and Tasmanian Passages continued to widen but there was also the closing of a key tropical gateway linking the Pacific and Atlantic Oceans, with the joining of North and South America. This affected heat transportation from the tropics as global ocean circulation re-organised.

In addition to ocean circulation, which is important for heat distribution about the planet, there were the Milankovitch orbital factors affecting the amount of energy the Earth receives between hemispheres (before albedo reflection considerations) and at different times of the year. As stated, the rise of the Tibetan Plateau was central to this. Evidence from late-Pliocene temperate pollen analysis suggests a temperate-zone climate periodicity of 124 000 years, with a non-linear response of the climate system to Milankovitch forcing.

Not only did some temperate ecosystems wax and wane in extent and location due to climatic changes, but plant populations responded to the new selection pressures. The rise of C_4 plants was a major evolutionary step, but there were also more modest, but still important, speciations to exploit the cooler Earth. Much Arctic flora is thought to have originated some 3 mya, towards the end of the Pliocene. The flora evolved to survive both the cold and the climatic waxing and waning. The likely pattern, as exemplified by Richard Abbott et al. (2000), is that an Arctic species evolved in one location (different locations for one or more different species) and then spread in a ring around the pole. As the Pliocene Earth warmed and cooled these species extended from, and returned to, various ecological refugia. Abbott and his colleagues demonstrated this by looking at the chloroplast DNA in species of saxifrage (an order of dicotyledon trees and shrubs), and specifically *Saxifraga oppositifolia*. As time passes mutations arise in the DNA and these increase additively with each generation. Furthermore, if a population becomes increasingly isolated, such as when confined to small refugia due to adverse climate regimes, the population retains its characteristic set of mutations. It is therefore possible, through genetic analysis, to ascertain to which refuge population a plant belongs, and indeed the primary location where it originally evolved. The evidence of Abbott et al. suggests that in the case of

S. oppositifolia the species originated in Beringia, part of which was the now sub-merged land bridge between present-day Siberia and Alaska. It then spread around the pole to Greenland, Iceland, the northern British Isles, Scandinavia, Siberia and Canada, and then relied on some 14 refugia when times changed and were harsh (during glacials). Refugia are ecologically important in climate change terms, as we shall see later.

It is worth noting now that in terms of polar ecology Beringia was of hemispheric significance, and we will mention it again later. Beringia, when an exposed land area, was covered with vegetation even in the coolest of times (including more recently in the depth of the last glacial some 24 000 years ago, as then the land-bridge area was substantive due to the sea level being 100 m lower than today). The Pliocene Beringia vegetation supported populations of large mammals including mammoths, horses and bison. This Pliocene vegetation was unlike that found in modern Arctic tundra. Both the Pliocene and modern vegetation communities can sustain a relatively low population of mammals, but the Pliocene Beringian community was a productive dry grassland ecosystem that resembled modern sub-Arctic steppe (dry and treeless) communities. It is likely that Beringia consisted ecologically of a mosaic patchwork with wet tundra in poorly drained lowlands and marked ecological differences between the central and its eastern-most regions (Zazula et al., 2003). This added to the habitat diversity: it was of an appropriate quality for an ecological refugium, enabling it to provide habitats for a range of species.

For millions of years up to somewhere between 4.8 and 7.4 mya Beringia was important in terms of both global climate and ecology. Ecologically it formed a bridge between the North American and East Asian landmasses (enabling animals to cross between the two) and, in terms of marine ecology and hemispheric climate, it was a barrier between the Pacific and Arctic oceans (preventing the exchange of marine life). The tectonic separation of North America and East Asia, between 4.8 and 7.4 mya, split Beringia and allowed water – at times of high sea level – to flow between the Arctic and the Pacific while hindering terrestrial fauna and flora travelling between the two continents. This dating of this significant opening was estimated by the presence of a formerly Arctic and North Atlantic bivalve mollusc, *Ascarte* (Marincovich and Gladenkov, 1999). The separation facilitated the high-latitude re-organisation of ocean circulation including the way heat entered and left the north polar region. It was yet another factor forcing global climate. As we shall see, this land bridge continued to form and close as the sea level rose and fell with the major glacial cycles throughout the Quaternary to the present. One consequence of this sea-level change was that Beringia's role as a refuge was insecure. Because it did not flood uniformly in the warm interglacials of the past million years, its highest hills tended to survive as islands, and do so even today.

Between 5.3 and 2.7 mya the Earth gradually cooled due to the aforesaid combination of factors of tectonic movement and changes in air and ocean circulation. It was around this time (4.4 mya) that an early ancestor to humans lived, *Ardipithecus ramidus* in Africa. It is an evolutionary connection between later small-brained, small-canined, upright-walking *Australopithecus* and the earlier last common ancestor that we shared with chimpanzees and the bonobo some time before 6 mya. Although fossil remains of *A. ramidus* were found in 1992, research providing important details about the find was not published until 2009 with a clutch of papers in the journal *Science*. These included one on its evolutionary significance by Tim D. White, Berhane Asfaw, Yonas Beyene, Yohannes Haile-Selassie and colleagues. The relevance of this discovery to the biological and human aspects of climate change is that the precursors to humans evolved in a world that had not been so cool for scores of millions of years. Indeed, as we shall see (section 4.6.4), modern humans themselves evolved into the even cooler Quaternary ice age (2 mya–present) of glacials and interglacials. Although Milankovitch orbital variations were not triggering the very cold glacials of the Quaternary, they would still have been taking place in the Pliocene (the Earth has been wobbling about the Sun since its formation). While, in the Pliocene, Milankovitch factors were not triggering glacials and interglacials, there would still have been other climate manifestations such as changes in circulation patterns, and hence regional rainfall and so forth (these have been discerned in geological strata). Indeed, as there was a small but significant amount of terrestrial ice on the planet at this time, these Milankovitch factors would not have been insignificant and changes in, for example, Pliocene monsoon strength have been discerned[1]. The notion arising from this is that these Milankovitch-paced climatic changes resulted in local/regional environmental change, and hence evolutionary pressure, the manifestation of one aspect of which was the evolutionary line that led to modern humans (*Homo sapiens sapiens*[2]; hereafter often referred to as just *H. sapiens*).

There is debate as to the exact nature of the evolution of human ancestors but considerable evidence again points to an African origin for hablines (*Homo habilis*). Indeed, in 2011 a largely US team headed by Brenna M. Henn and Christopher R. Gignoux presented genetic evidence for a South African origin for modern humans. (Older than the hablines, earlier sivapithecine remains have been found in Turkey and India from 14–8 mya; sivapithecines are more closely related to orang-utans than chimpanzees, gorillas and humans.) The exact relationship (and as mentioned there is little doubt that at least in part there is one) between climate and human evolution has yet to be clearly elucidated. This is because terrestrial biogeological and geological records of East African environmental change are rare, geographically dispersed and incomplete. Atlantic Ocean and (perhaps more importantly) Indian Ocean sediment records have been used to reconstruct climatic changes in the region. In 2005 Martin Trauth and colleagues reconstructed the East African climate 0.5–3 mya from 10 lake-basin records in the area. They identified three possible hominid periods: 2.7–2.5, 1.9–1.7 and 1.1–0.9 mya. They note that these could well have had an important impact on the speciation and dispersal of mammals and hominids at that time.

At about 2.75 mya a sharp change in this global cooling took place, in all likelihood due to the interaction of climate-influencing factors that prompted new ocean- and/or atmospheric-circulation patterns, and not a single threshold event. Notably,

---

[1] The manifestation of Milankovitch variations in geological strata over the past quarter of a million years was, in 1998, the subject of an excellent Royal Society discussion meeting, the proceedings of which appeared in a special edition of the *Philosophical Transactions of the Royal Society Series B* (Shackleton et al. 1999).

[2] Our species is (supposedly) so wise that they named it twice.

however, North and South America had joined by this time, which would have had a pronounced effect on ocean circulation. What is known, from alkenone analysis and oxygen isotopes from diatoms, is that at this time late-summer sea-surface temperatures in the sub-Arctic Pacific rose. At first it may seem paradoxical that a sea-surface temperature rise is associated with glaciation. What seems to have happened is that at this time, with the re-organisation of ocean circulation, there was increased water-column stratification (temperature difference), and hence the surface water failed to sink. Meanwhile, winter sub-Arctic Pacific sea-surface temperatures did cool. Consequently, the late-summer warm water increased precipitation in the autumn so encouraging the building up of snow and ice over neighbouring Canada which, with the cooler winters, was preserved for longer into the following year (Haug et al., 2005). The northern hemisphere ice sheets grew: ice sheets can only grow if the preceding year's snow and ice survives the summer into the subsequent winter. In this way the North American Laurentide ice cap began to expand to complement the Fennoscandian ice cap over northern Europe and Russia. The Fennoscandian ice sheet obtained much of its water from prevailing winds running above the North Atlantic Drift (commonly called the Gulf Stream). The overall result was a step-like progression in the Earth's cooling that marked the onset of northern hemisphere glaciation. This is reflected in a number of palaeoclimatic indicators including high-latitude benthic ^{18}O, the mass accumulation rate (or MAR) of some northern hemisphere oceanic biogenic opal (this almost ceased), the magnetic susceptibility record of the sub-Arctic Pacific (an indicator of ice-rafted debris which increased sharply having previously been at a low level) and other isotopic mass accumulation rate evidence of foram species (Ravelo et al., 2004).

With the onset of northern hemisphere glaciation approximately 2.75 mya, which complemented the already existing Antarctic ice, the current Quaternary (and Pliocene) ice age could be said to have begun. Having said this, the key ice deposits that, according to the International Commission on Stratigraphy, officially mark the start of the Quaternary (and Pleistocene) date from 2.588 mya. As is often the case, the trigger for climate change came first and the key geological strata (used for key dating) came later. Similarly, today we already have had significantly elevated levels of atmospheric carbon dioxide for the best part of a century (depending on how you quantify a significant elevation) but only recently have we begun to discern the climate change fingerprint both biologically and from meteorological records (see Chapter 6). This biological fingerprint is almost certain to become more pronounced and then leave behind remains that may be discerned in the future as indicators of environmental change.

4.4 The current ice age

Before reviewing the current ice age it is worth noting the nomenclature appearing in the academic literature as variations in usage can confuse. The depth of the current Quaternary ice age is defined broadly in the nomenclature of geological timescales in terms of two parts or epochs. First, there was a series of glacial and interglacial cycles that comprise the Pleistocene. Second, there is the Holocene: this is just the current interglacial. Confusingly the Pleistocene (2.558–0.0115 mya) is sometimes referred to as an ice age in its own right, which conjures up a picture of a single cold spell for the planet, as opposed to oscillations between glacial and interglacial times. (Indeed, some parts of past interglacials were slightly warmer than today.) Similarly, the end of the Pleistocene is sometimes described as the end of the ice age when in fact it is just the end of the last glacial. Notwithstanding human greenhouse considerations, our Holocene interglacial will eventually end with a return to glacial conditions. This confusion in terminology is simply a result of the historical way in which we have uncovered the past. It reflects the way our understanding has changed since the term ice age was first coined by the Swiss naturalist Louis Agassiz in 1837 (after whom the large last glacial maximum [LGM] lake was named; see later). This ambiguity of nomenclature is just something of which the student of palaeoclimatology has to be aware. (Such anomalies are also common in other areas of science, such as the use of the term fertility; see Chapter 7.)

Recently, aside from the Holocene and Pleistocene, a new term, the Anthropocene, has begun to be used. This denotes the time from which humans had a discernable impact on the (global) biosphere. Even here there is some confusion. Some say that this has only happened since the Industrial Revolution, but others, notably Bill Ruddiman of the University of Virginia, interestingly – but controversially – say that humans have been affecting the global climate for some 8000 years (Ruddiman, 2003) and that the spread of agriculture 5000 years ago released methane (Ruddiman and Thomson, 2001)[3]. Irrespective of the exact time when humans began to have a discernable effect on the biosphere – specifically global cycling systems (such as for carbon, nitrogen and phosphorus) and physically (such as climate) – humans today are certainly imparting such an impact. Today, globally some 15% of ice-free land is dominated by human use (such as urban areas and roads) whereas 55% is part-dominated by human use (such as agriculturally dominated landscapes and canalised rivers). This human influence on land is so much so that it is estimated that just under a third of terrestrial primary productivity (the amount of energy, carbon or biomass) fixed by photosynthesis is influenced by humans (Imhoff et al., 2004), so that saying that we are in the Anthropocene is not inappropriate. However, notwithstanding the vagaries of terminology, the onset of 'major' glaciation in both hemispheres can be said to mark the onset of the current Quaternary ice age of the past 2 million years.

The previous gradual cooling that had been taking place over a few million years before the onset of significant northern hemisphere glaciation was not entirely smooth (the Earth system was going through a critical transition and passing a climate threshold) but consisted of temperature oscillations with a period of a few tens of thousands of years. After the onset of the northern hemisphere glaciation these oscillations quickly (within a hundred thousand years) became more marked and at that

---

[3] However, a back-of-the-envelope calculation suggests that human agriculture 5000 years ago was unlikely to have had an impact on the atmosphere because the global human population and hence the likely area farmed at the time would have been too small. We will return to the question of the Anthropocene and when it started in section 4.6.3.

time had a period approximating 41 000 years. Although this cannot be attributed to a single event (again rather a combination of circumstances) it is perhaps useful to re-emphasise that the Isthmus of Panama formed before some 2.5 mya, physically separating the Pacific and Atlantic Oceans. Even though the isthmus formed then, as the two continents drew closer, the water flow between the two oceans would have gradually declined over previous millennia and this factor might have become, in association with others, more climatically significant at some point prior to the continents' physical joining. The 41 000-year periodicity subsequent to the onset of northern hemisphere glaciation reflects Milankovitch obliquity and it dominated global climate variations 1.75–1 mya (see Chapter 1). However, the Earth system went through a climate threshold called the Mid-Pleistocene Transition (MPT) around a million years ago. After this the Milankovitch eccentricity cycle of roughly 100 000 years began to dominate global climate fluctuations. As to what caused this switch, there is some evidence from modelling to suggest that as each successive glacial got deeper a point was reached when the two components of the Laurentide glacial ice sheet in North America joined together. Prior to 1 mya the ice sheet had two distinct components. This joining would have conferred a regional albedo effect. This meant that the overall North American ice sheet (with the now joined components) could outride the obliquity component of Milankovitch solar variation and so the eccentricity component dominated. This it has done through to the present. (Even in the depth of the last glacial the overall shape of the North American ice sheet exhibits the echo of its two separate components as they were in the depth of pre-MPT glacials: see Fig. 4.3.)

In addition to the changes in surface ocean circulation, which helped cool the higher latitudes, there were amplifying feedbacks. One of these came to play a significant part once the Earth had undergone some initial cooling and so encouraged further cooling.

Carbon resides in a number of reservoirs in the biogeosphere (see Chapter 1). This includes reservoirs in the atmosphere (of climate significance) and both in the surface and deep waters of the oceans. Both these latter reservoirs are larger than the atmospheric reservoir and so flows to and from the oceans have a major impact on the atmospheric reservoir; indeed, the size of the ocean deep-water carbon reservoir is over 45 times that of the atmosphere's (see Figure 1.3). In fact, the flow of carbon between these three reservoirs exceeds that of the flow to the atmosphere by human fossil fuel burning by at least an order of magnitude and so these natural flows are important determinants of the global climate. For carbon dioxide to return to the atmosphere from the deep ocean there needs to be mixing with surface waters. Sigman et al. (2004) suggested that before about 2.7 mya, at high latitudes the oceans were sufficiently warm that, by contrast, the sea in winters would have had a cool surface with a temperature not far from the deep-water temperature (which was warmer than that today). This would allow surface water to sink and for deep water to surface elsewhere and return carbon to the atmosphere. However, after 2.7 mya the Earth's deep oceans became so cold that vertical mixing was reduced. This is because warm water, when cooled by a degree, becomes proportionally more heavy than cool water cooled a degree. In other words, the relationship between sea-water density and temperature is not linear (see Figure 4.1). It is an idea that seems to fit in with

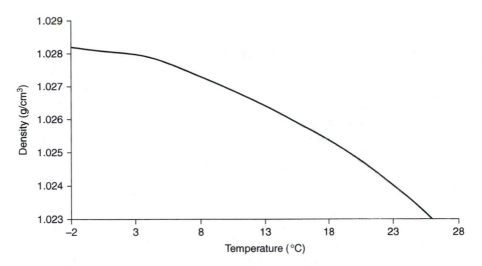

Sea-water density and temperature. As sea water cools it gets more dense but as the temperature approaches freezing point it gets denser at a slower rate. Therefore, a degree-cooler water in cool seas has less negative buoyancy than a degree-cooler water in warm seas, which will have a greater tendency to sink.

the aforementioned sub-Arctic North Pacific scenario. Meanwhile, in today's world there is less deep water returning to the surface bringing with it carbon dioxide. More carbon dioxide remains trapped within ocean deep water. This tendency for lower atmospheric carbon dioxide after 2.7 mya added an additional cooling forcing factor to the already cooling global climate.

As we shall see later in this chapter when discussing the Broecker thermohaline circulation (section 4.5.3), today there is some climatically important mixing of surface and deep waters but this takes place only in special places where the sea is both more salty (and hence more dense) and cool. It is this combination of salt (halinity) and temperature (thermality) affecting density that drives the Broecker thermohaline circulation, and not temperature and density by themselves.

The idea that changes in sea-water temperature/density relationships changed 2.7 mya is substantiated by biological evidence other than that from the sub-Arctic North Pacific (Sigman et al., 2004). This is exemplified by two examples that relate to both the northern and southern hemispheres. First, opal is a mineral composed of amorphous silica and forms a constituent of the cell wall of a group of phytoplankton. It is a useful palaeo-indicator because it is comparatively resistant to decomposition. Second, living organisms prefer to take up the nitrogen isotope ^{14}N to ^{15}N from the fixed biosphere pool of nitrogen. So, when there are more living organisms consuming nitrogen the ratio between the two isotopes changes. Looking at sediments around Antarctica and the Arctic, Sigman et al. found that both the rate of opal production and the nitrogen isotope proportions changed around 2.7 mya. This signifies a change in nutrient mixing, hence surface- and deep-water mixing necessary to support a large phytoplankton population. Finally, this scenario ties in with ocean sediments that suggest that ice rafting also commenced about this time. Therefore, it would appear that ice rafting began around a time of changing surface- and deep-water mixing and a possible causal relationship between climate and surface-/deep-water mixing.

The high-latitude changes in ocean circulation and the onset of significant northern hemisphere glaciation were not only interrelated; the high-latitude cooling (hence the polar ice) also resulted in the subtropical regions beginning to cool. There is a clear and obvious synchronicity between the two and so we know that the cooling experienced then was a planet-wide event. The tropics themselves, though, did not see any fundamental changes, although undoubtedly there were changes in tropical biome extent: there was sufficient environmental similarity to warmer times in some parts of the tropics to provide a refuge for many species. This lends support to the view that the rapid climate transition about 2.75 mya must have primarily involved extra-tropical processes such as ocean (thermohaline) circulation, ice albedo, biogeochemical processes and so forth that came together in an ice-sheet-determined and/or some combination of tectonic and greenhouse threshold.

As the planet continued to cool a new threshold was reached around 2 mya with new modern tropical circulations. These included the development of what is known as the Walker Circulation and cool subtropical temperatures.

The term Walker Circulation was first introduced in 1969 by Professor Jacob Bjerknes for two circulation cells in the equatorial atmosphere, one over the Pacific Ocean and one over the Indian Ocean. Schematically these are longitudinal cells where, on one side of the ocean, convection and the associated release of latent heat in the air above lifts isobaric surfaces upward within the upper troposphere and creates a high-pressure region there (Gilbert Walker was an early 20th-century British climatologist who studied air circulations over the Pacific Ocean). This modern atmospheric circulation, combined with the previous re-organisation of the northern hemisphere's ocean circulation, again served to cool the Earth further.

Following the climate threshold, from about 0.75 mya, the glacial–interglacial cycles became more pronounced and more clearly took place with an increase in periodicity from 41 000 years (reflecting the Milankovitch obliquity cycle) to approximately 100 000 years (reflecting a 93 000-year eccentricity). This 100 000-year periodicity and degree of glacial–interglacial temperature swing has persisted to the present day (Ravelo et al., 2004); our present day, of course, is a warm interglacial.

The switch between the 41 000-year global climate cycle to 100 000 years appears to be more driven by a change in ocean circulation than the gradual overall decline in atmospheric carbon dioxide across glacial–interglacial cycles. Oxygen isotope analysis (see Chapter 2) of planktonic Foraminifera reveals changes in Pacific circulation at this time (de Garidel-Thoron et al., 2005). This is not to dismiss atmospheric carbon dioxide as a driver of climate change, but other factors also play their part and may be the dominant (as opposed to sole) factors initiating a change in the Earth's climate system at some threshold point.

One might summarise the Quaternary and Pliocene trend in global climate as one that had three characteristics: first, overall long-term cooling; second, Milankovitch-driven oscillations between warm and cold; and third, an increase at a transition point in both the period of climate oscillations and their degree.

In addition to isotopic analysis of biological remains in sediments, the fossil record and other techniques (involving biology and/or geology) used to discern global climate change over the past few million years, ice cores from Antarctica have been useful in ascertaining regional climate change in the southern hemisphere over the past few glacial–interglacial cycles. Greenland ice cores have provided similar information, particularly relating to the last glacial–interglacial cycle. (Ice cores can be analysed for variations in deuterium, dust and greenhouse gases as discussed in section 2.3.1.) The longest ice core to date comes from Dome C in Antarctica. It provides a continuous record of chemical climate proxy data covering 740 000 years and several glacial–interglacial cycles. The core's physical length is about 3 km. We will shortly return to ice cores, but for now note that the Dome C record does strongly suggest that iron-dust deposition was higher in the cool parts of glacials (when globally there was less rain, and hence it was drier on land) and that subsequently the glacials reached their coldest phase. It is thought that iron fertilisation of the oceans encouraged algal blooms that further drew down carbon dioxide into the oceans, so causing the coldest periods of glacials (Wolff et al., 2006).

One question to arise is the nature of glacial pacemaking. It is clear from so much evidence that the Milankovitch orbital parameters drive the glacial–interglacial cycles. However, do the Milankovitch orbital variations *lead* the driving, or forcing, of the climate? Or, in other words, does global climatic change *lag* some time behind the parameters, as opposed to the parameters having a more or less instantaneous effect? Finally, do the waxing and waning of the ice sheets that amplify such forcing do so in time with either global climatic change and/or Milankovitch forcing? As noted in section 1.5, Milankovitch factors provide the pacemaker for glacial–interglacial cycles, but are the two (orbital and glacial timing) and the changes in atmospheric greenhouse gases perfectly synchronised?

Nicholas Shackleton of Cambridge University's Godwin Laboratory addressed these questions in 2000. After all, Milankovitch forcing is weak. Although we are considering changes of about 15% of 440 W m^{-2} on the Earth's surface at 65°, this is countered by opposite changes in the southern hemisphere. The actual difference taken globally is less than 1 W m^{-2}, or less than 0.1% of the solar constant (the energy received per square metre perpendicular to the Sun in the Earth's orbit). This is due to the solar energy the entire planet receives and absorbs, which in turn arises from the summation of a variety of factors: its albedo at various locales about the planetary sphere, the time of year these locales receive the energy and the distance from the Sun, again integrated across the planetary sphere. So how does this weak signal magnify into a glacial–interglacial transition? Shackleton shed some light on this when comparing the timing of changes in benthic ^{18}O foram palaeorecords with the ice core and trapped gas Antarctica records in addition to June insolation at 65°N (another often-used latitude for Milankovitch forcing curves that crosses both the former Laurentide and Fennoscandian ice sheets). He found that while the *extremes* in Milankovitch forcing are in time with changes both in global climate and atmospheric carbon dioxide, ice volume lags behind by a few thousand years. This might suggest that northern hemisphere summer sunlight (or perhaps the northern hemisphere thermal growing season?) affects the carbon cycle so as to drive atmospheric carbon dioxide concentrations and that this in turn has an immediate affect on temperature. This, while initially plausible, is unlikely because

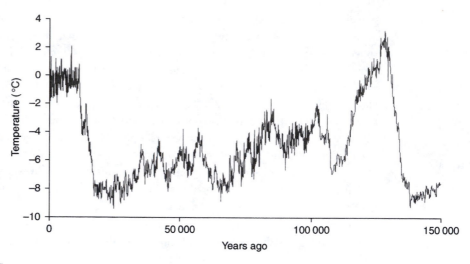

Fig. 4.2   Temperature changes around Antarctica as revealed by deuterium analysis of ice cores. Based on data from the National Oceanic and Atmospheric Administration (see Acknowledgements).

the annual variation in atmospheric carbon dioxide is too small to have a significant climate effect. On the other hand, although ice albedo effects may possibly magnify temperature changes (see Figure 1.8) the thermal inertia in the vast ice caps of the Antarctic and Greenland means that they take longer to respond and so only respond to sustained year-on-year climate forcing factors as opposed to factors that take place within a year.

As we shall see in subsequent chapters, this is somewhat relevant to our current anthropogenic release of carbon dioxide by fossil fuel combustion. The anthropogenic future will be determined not so much by the weak Milankovitch forcing but our effects on the carbon cycle on a year-on-year (as opposed to within-year) basis, as well as albedo changes as glaciers and ice caps retreat. This makes policy decisions that determine how much carbon dioxide we release, such as energy policy, of fundamental importance should we be concerned about future climate change.

4.5 The last glacial

4.5.1 Overview of temperature, carbon dioxide and timing

The last glacial began about 115 000 years ago with temperatures that oscillated towards ever-colder conditions. It reached its climax (the last glacial maximum; LGM) some 18 000–22 000 years ago and ended (after a false start to the current interglacial 14 000 years ago) approximately 12 000 years ago (see Figure 4.2).

The importance of understanding the last glacial–interglacial cycle is that it illuminates the planet's current climatic mode. It is also important because, being closest to us in time, it has been the easiest era for which to gather data. Nonetheless, although the last glacial–interglacial cycle is fairly typical of the series for the past 700 000 years (in the 100 000-year Milankovitch-dominated Earth; see the preceding section), the fourth cycle ago features an interglacial that may be the best analogy to the current interglacial in its natural state (without anthropogenic carbon dioxide). Milankovitch forcing changes that took place four cycles ago were similar to those taking place now. However, as the last cycle is the easiest to study, and indeed in the history of palaeoclimatic research it was the one that was first studied closely, it is appropriate to look at it.

The big difference in global climate between glacial and interglacial times relates to global precipitation and, obviously, temperature. In glacial times there is less rainfall globally simply due to the fact that there is less evaporation from the oceans. With regards to temperature, the glacial Earth is broadly about 4°C cooler than its interglacial counterpart and possibly at least 5°C cooler during the LGM, 18 000 years ago. However, this represents a global average which, as we shall see, is the same order of magnitude (i.e. single-figure degrees Celsius) as is anticipated for global average anthropogenic warming over the next century. However, at high latitudes, near and on the ice sheets, temperatures during glacials were far cooler and more like 12–14°C cooler than now.

The tropics were less affected by glacials but they did not entirely escape the cooling effects. Analysis of forams from the Pacific, and other palaeoclimatic indicators, suggests that the tropical Pacific sea surface was about 3°C cooler during the LGM than today. The Pacific, being such a vast reservoir of heat, is an important component of the Earth's climatic system. If the high latitudes were to experience a greater glacial–interglacial temperature difference than the 5°C global average then one might expect that major parts of the tropical zone would see a smaller deviation to maintain the average. However, this is less true on land than in the sea. On land the effects of mountains ranges on climate and other factors come into play. Some parts of the tropics were clearly less affected by the glacial as tropical species survived not just the last glacial but the previous several that took place in the Quaternary. Yet this was not so everywhere. For example, some parts of Brazil were over 5°C cooler during the LGM than today (Stute et al., 1995).

There is much evidence for our understanding of at least the broad glacial–interglacial temperature changes and timing. There is even a considerable body of evidence exploring many detailed aspects; alas there is not the space to review it all here. But the work of Timothy Herbert and his team reported in *Science* in 2010 serves to illustrate one of the sorts of evidence available. They determined the timing and amplitude of tropical sea-surface temperature (or SST) change as it is an important part of solving the puzzle of the Plio-Pleistocene ice ages. They used alkenone-based tropical sea-surface temperature records (see section 2.2.2) from the major ocean basins to show coherent glacial–interglacial tropical temperature changes of 1–3°C that align with (but slightly lead) global changes in ice volume and deep-ocean temperature over the past 3.5 million years. They showed that tropical temperatures became tightly coupled with benthic $\delta^{18}O$ (an indicator of global ice volume in a time of ice ages, such as the Quaternary) and Milankovitch orbital forcing after 2.7 million years. In addition to providing a chart of temperature and timing, they concluded that there was an inception of a strong carbon dioxide/greenhouse gas feedback and amplification of orbital forcing at approximately 2.7 mya that has connected the fate of northern hemisphere ice sheets with global ocean temperatures since that time. This marginal sea-surface temperature leading global changes in ice volume and deep-ocean temperature over the past 3.5 million years is exactly what would be expected with Milankovitch pacemaking of glacial and interglacials. Indeed, as ice cores indicate, surface temperatures lead carbon dioxide change[4]. Building on the discussion in section 4.4, what is happening is this: By the Quaternary period, continental plate tectonic movement had placed most of the Earth's land area in the northern hemisphere (see section 1.5). This and the rise of the Tibetan Plateau in the northern hemisphere and its reflective waxing and waning ice (section 4.1) meant that the solar energy received by the Earth at these latitudes in the summer (which determines how much of the previous winter's ice melts) is indicative of Milankovitch pacemaking. Depending on whether warming or cooling (more or less energy is received) is taking place will determine how much carbon is released or sequestered from wetlands, soils and other terrestrial biosphere carbon pools and so alter atmospheric carbon dioxide and methane, which in turn amplifies the said warming or cooling. So Milankovitch pacing slightly leads carbon dioxide change. The subsequent amplified warming or cooling will determine global ice volume. Finally, it takes time for temperature change on the surface of the Earth to be carried by ocean currents to deeper waters, and so deep-ocean temperatures also lag behind Milankovitch pacing.

Whereas glacial–interglacial temperature changes were smaller in the tropics compared to the poles, climate-related events in high latitudes near the north polar circle did affect the tropics. Biological productivity in the tropics off Venezuela appears to have increased and decreased, reflecting periods of iceberg discharge from the large glacial Laurentide ice sheet over northern North America, as represented in the climate change recorded in the Greenland ice sheet. Chemical change in the sediments off Venezuela bears an even closer correlation with the Greenland ice-core record for the ^{18}O isotope, which more than adequately reflects regional climate change during the glacial (Peterson et al., 2000). Having said this, the Venezuela sediment record may relate more to deep-ocean ventilation than the tropical climate per se because, as we shall shortly see, global ocean circulation helps transport heat hemispherically and link the oceans globally and transport heat away from the tropics. Changes in ocean circulation also manifested themselves elsewhere in the tropics. For example, the Kalahari Desert in Africa saw significant arid events approximately 41 000–46 000, 20 000–26 000 and 9 000–16 000 years ago, which are thought to be times of circulation change (Stokes et al., 1997).

The detail required to understand and climatically model the planet is far greater than we have so far acquired, as we shall see in Chapters 5 and 6. So there is a

---

[4] Some climate change sceptics use the argument that initial (Milankovitch-paced) glacial–interglacial temperature change leads carbon dioxide change to say that changes in atmospheric carbon dioxide concentration are not very important with regards to global climate change. This example in the public climate debate shows how easy it is to mislead those with little understanding of the science with a specious argument.

clear need to obtain as much palaeoclimatic information as possible from wherever possible if we are to accurately determine how the palaeoclimatic systems worked. Much research is still required today. As there remains some uncertainty as to the exact conditions in the past, the validation of climate computer models cannot be undertaken with the exactness that policy-makers require or in a number of instances realise (see also section 5.3.1).

If some tropical palaeoclimatic records correlate with polar records, it should not be surprising to find that many current temperate-zone palaeorecords exhibit similar correlations. For example, rapid environmental changes similar to those indicated by Greenland ice cores are discernable in southern European palaeorecords from Italy covering the last glacial (Allen et al., 1999). So, even though one line of evidence – be it a drill-core alkenone-determined tropical sea-surface temperature record or whatever – may not have the resolution or be as representative as we would like, we can still place this evidence alongside that of other palaeodata and begin to see a coherent picture emerge. Such is the amount of disparate evidence that, despite shortcomings in any area, this broader coherent picture is quite clear.

4.5.2 Ice and sea level

During the LGM, and other recent glacial maximums, ice sheets of up to 3 km thick covered much of northern North America down to the Great Lakes (the Laurentide sheet, together with the Cordilleran sheet; Figure 4.3) and also much of northern Europe, Scandinavia and the northern British Isles (the Fennoscandian sheet; Figure 4.4). Meanwhile, in the southern hemisphere, much of New Zealand, Argentina and Chile sported ice caps, as did mountains in South Africa and southern Australia. With so much of the Earth's water trapped on land as ice sheets, the sea level fell by about 120 m, so exposing much of the continental shelves.

4.5.3 Temperature changes within the glacial

The temperature change in the last glacial in the regions surrounding Antarctica is depicted in detail in Figure 4.2, which is based on a deuterium isotopic analysis (see Chapter 2) of the Vostok core. Note that most of the current (up to but not including the incomplete present) late-Quaternary interglacials were short, of the order of 10 000 years or less, whereas glacials were about 90 000 years in duration. Also note that the beginning and end of the glacials were comparatively sharp (critical transitions and climate thresholds are mentioned later in the book, e.g. section 6.6.8), as were some of the less dramatic changes in temperature during the glacials themselves. What the glacial–interglacial cycles are not are gently sinusoidal temperature oscillations.

The temperature graph from Figure 4.2 is shown on Figure 4.5 and added to it is the ice-core record of past atmospheric carbon dioxide and methane concentrations covering additional glacial–interglacial cycles. It can be seen that carbon dioxide and temperature effectively parallel each other, which is further evidence as to the importance of greenhouse gases in determining global climate. Methane too, as

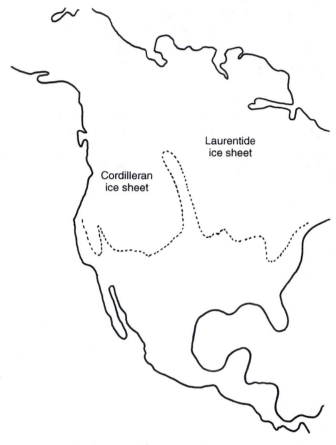

Fig. 4.3 Approximate southern extent of ice over northern North America during the deepest part of late-Quaternary glacials. Note that the present-day coastline is portrayed whereas during glacials the lower sea levels would have meant a larger land area.

we shall shortly see, shows a broadly similar pattern of high concentration during interglacials and low during glacials, with the lowest concentration during the LGM. Atmospheric methane reflects the global extent of wetlands (including, in modern times, rice paddies). As previously mentioned, when the Earth is warmer the greater ocean evaporation leads to greater rainfall which in turn adds to the extent of wetlands. Conversely, when the planet is cooler and drier there is less methane produced. So methane concentration can also be loosely said to track temperature and indeed, being a powerful greenhouse gas (see Chapter 1), amplify temperature changes.

Importantly, the temperature and gaseous changes recorded in the Vostok cores in Antarctica are reflected in their northern hemisphere counterparts. The ^{18}O isotope records of Vostok and Greenland cores broadly, though not nearly exactly, follow each other. Figure 4.6a shows a low-resolution graph (with 25 000-year filter) of how ^{18}O has changed over the past 90 000 years from the present to early in the last glacial. The LGM transition to the current Holocene interglacial can be clearly seen.

The reason for using a filter on the above data is that a high-resolution analysis of northern and southern hemisphere ocean temperatures during the last glacial

Fennoscandian
ice sheet

Fig. 4.4 Approximate glacial ice extent over Europe during the deepest part of late-Quaternary glacials. Extending over all of northern Europe, this ice sheet fixed a considerable body of water and was in places more than 1.5 km thick. Note that the present-day coastline is portrayed whereas during glacials the lower sea levels would have meant a larger land area.

reveals a time-lagged exchange of interglacial and glacial heat change between the hemispheres. The temperatures around Greenland and Antarctica vary by about 4–6°C with a period of about 2500 years, each being out of phase with the other (Knutti et al., 2004). Heat is thought to be transferred from one hemisphere to the other through the Broecker thermohaline conveyor (see Figure 4.7; and see below) of surface and submarine currents that connect all the oceans. The result is that there is a time delay during a glacial between when one hemisphere feels the effects of warmth and when the other hemisphere does. Having said that – which is regards to major glacial and interglacial heat change and exchange between hemispheres via the Broecker ocean circulation – there is also more immediate atmospheric heat exchange. For this reason, could it be that some northern hemisphere climate events are reflected in the southern hemisphere Antarctic record? (See Figure 4.12a, below.)

Climate change and biosphere science arguably have more than their fair share of seeming contradictions, which is one reason why climate sceptics can present (specious) arguments that gain traction with some of the public. The northern and

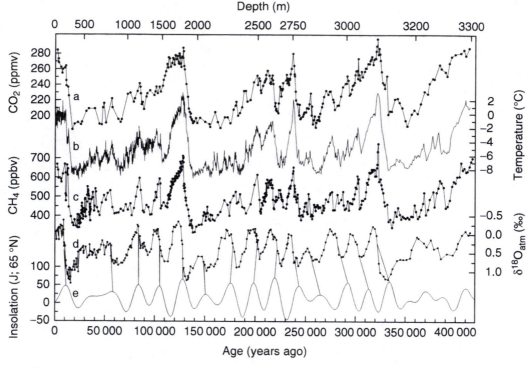

Fig. 4.5 Carbon dioxide and methane relate to temperature. (a) Carbon dioxide, (b) temperature (as derived from isotopic composition of the ice) and (c) methane. The patterns for all three are broadly similar. (d) Track changes in the ice's ^{18}O isotope. (e) Milankovitch insolation (or solar energy received) as calculated for June at 65°N. This shows how peaks and troughs in Milankovitch climate forcing act as a pacemaker of glacial–interglacial cycles. Reproduced with permission from Pétit et al. (1999).

southern hemisphere glacial–interglacial transition being out of phase (or rather, not exactly in phase) is one such seeming contradiction. If the global Broecker thermohaline slows then heat ceases to be transported to the North Atlantic from the southern oceans. This in turn means that the southern oceans warm and the northern oceans cool. At the end of the last glacial we have the apparent paradox of glaciers in Patagonia, South America, growing while those in Europe were retreating. This trading of hot and cold between hemispheres is called the bipolar see-saw.

In 2009 a British and North American team led by Stephen Barker and Paula Diz of Cardiff University, and including Wallace Broecker of global thermohaline circulation renown, looked at both Antarctica and Greenland ice core records together with marine sediment cores (for ^{18}O analysis from forams) from off the US east coast, off Spain, off Chile and in the South Atlantic off South Africa. The results showed that northern hemisphere ice sheets began to dwindle some 21 000–19 000 years ago (in agreement with Milankovitch theory), yet Antarctic temperatures and carbon dioxide only started rising 18 000 years ago, and more abruptly (not quite as more gradually as Milankovitch theory would suggest). So far so good with respect to the bipolar see-saw. But was change in the strength of the Broecker thermohaline conveyor involved

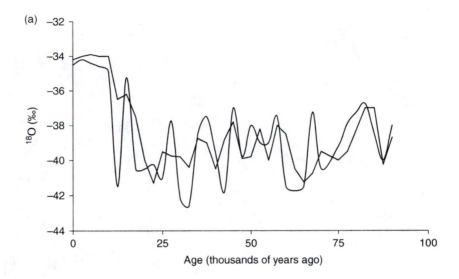

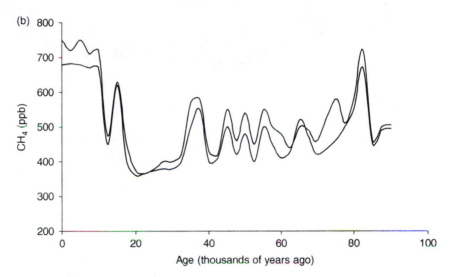

Fig. 4.6 (a) ^{18}O Isotope and (b) methane records from the northern and southern hemispheres (Greenland and Antarctica), smoothed (with a 5000-year filter).

in generating this lag? The marine cores from the northern and southern Atlantic and East Pacific indicate that the global thermohaline circulation was reduced (presumably due to fresh meltwater in the North Atlantic from the melting northern ice sheets, as we shall shortly see). The southern ocean warmed. About 300 years later atmospheric carbon dioxide levels started to rise as the deep ocean warmed, releasing the gas. This amplified the Milankovitch warming of the Earth. The thermohaline circulation then switched back to its 'warm north' mode 14 700 years ago (or perhaps nearer 15 000 years ago). At this point with the thermohaline circulation switched on, increased Milankovitch June solar energy and finally increased atmospheric carbon dioxide, the fate of the northern hemisphere ice sheets was sealed. One other interesting aspect

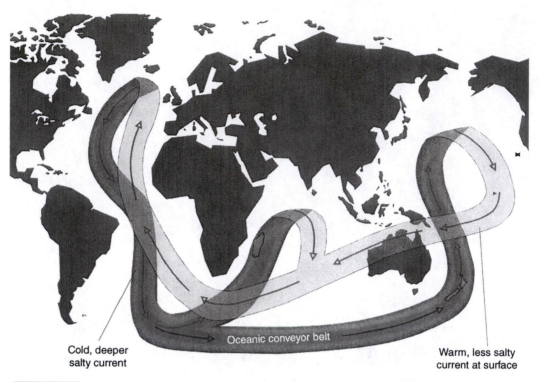

Cold, deeper
salty current

Oceanic conveyor belt

Warm, less salty
current at surface

Fig. 4.7 A simplified representation of the Broecker conveyor (reproduced from Cowie, 1998).

of this work is that it demonstrated the importance of the deep ocean as an increased reservoir of carbon dioxide during cold glacials[5].

However, were that life was this simple (if indeed the above is 'simple'). While the above, teased from the complexities, is a fair generalisation of the global situation, researchers gleaning evidence elsewhere in the field may receive a more muddled picture. Whereas there is the aforementioned ocean temperature evidence, there is other evidence suggesting that terrestrial temperatures at mid latitudes within glacials differ between hemispheres to the ocean time lag suggested above. Local insolation at mid latitudes of the southern hemisphere is almost completely out of phase with that of high northern latitudes. Mid-latitude long-term terrestrial records going back one or more complete glacial–interglacial cycles are rare but one such exists in the form of a 10 m-deep peat bog in Okarito Pakihi in southern New Zealand that was outside the limits of the ice during the LGM (see Figure 4.8). Palaeovegetation analysis of bog peat cores has provided an indication of climate change going back two glacial–interglacial cycles. This revealed an earlier onset and longer duration of the LGM. The explanation for this is likely to relate to the Milankovitch-driven levels of insolation at that latitude (Vandergoes et al., 2005).

---

[5] See also a very readable 2009 review article by Jeffrey Severinghaus on the work of Barker et al. (2009). For a comprehensive review of ocean circulation changes and glacial and interglacial timing see also Sigman et al. (2010).

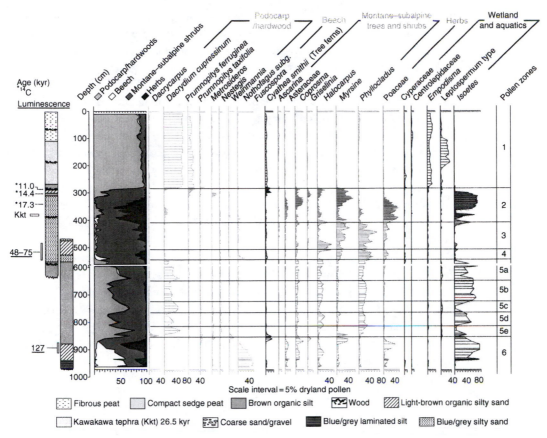

Fig. 4.8 Summary pollen diagram from Okarito Pakihi, New Zealand. Note that the timescale gets more condensed as one travels the further into the past. Also, the current interglacial (Holocene), LGM and early glacial vegetation types show up clearly. The last interglacial vegetation does not, and local insolation was low at that time. Reproduced with permission from Vandergoes et al. (2005). kyr, thousand years.

The atmospheric gas composition of bubbles of northern and southern hemisphere ice cores track more exactly, which is not surprising given that global mixing of the planet's atmosphere takes a little over a year or two. Figure 4.6b is a low-resolution portrayal of methane from both hemispheres over the entirety of the past glacial–interglacial cycle and demonstrates synchrony.

The cores clearly demonstrate that glacials and interglacials are not just hemispheric events but planet-wide in nature. Further, they show that some trends and features within the glacial are also global (even if hemispherically they are not exactly in phase).

The core records also demonstrate the broad structure of the glacial and interglacial cycle. For example, the LGM occurred at the end of the glacial as opposed to the middle, which would be likely if glacial–interglacial temperature cycles were uniformly sinusoidal. However, cores have shown that other features that are pronounced in one hemisphere are only weakly reflected in the other. The end-glacial hiccup return to glacial conditions (the Allerød, Bølling and Younger Dryas; see section 4.6.2), as the Earth was beginning to emerge into the current interglacial (Holocene), is strongly represented in the Greenland ice cores but only weakly in its Antarctic counterparts. This suggests that some climatic-forcing events were driven by factors located in one hemisphere rather than the other, even though glacials and interglacials overall are planetary in nature.

Looking at ice cores in more detail reveals that the temperature changes are not smooth (the indication of crossing a 'critical transition' and a climate threshold). There are a number of factors causing this. Two of the main ones are changes in ocean circulation and also, not surprisingly, greenhouse gas concentrations.

First, as previously noted, ocean circulation is a very important mechanism for transporting heat away from the tropics and so it is worth examining aspects of this in a little more detail. In recent geological times, in the land-dominated northern hemisphere (which loses heat quickly), one of the major currents transferring heat from the tropics has been the North Atlantic Drift (or Gulf Stream). This runs from east of the Gulf of Mexico, up the south-east coast of the USA and north east across the Atlantic up to Iceland as well as the British Isles and the western coasts of Scandinavia. In winter months this warms the western edge of northern Europe so that it is 10–15°C warmer there at latitudes of 40–60°N compared with Russia and much of Canada: the heat transported from the tropics is considerable. On its journey north evaporation takes place so that by the time the water has reached Icelandic latitudes the current is more concentrated in salts than other parts of the ocean. All this time, as well as evaporating, the current is shedding heat (and hence moderating winters in north-west Europe) and so cooling down. At Icelandic latitudes the North Atlantic Drift is cooler and saltier and so it sinks. But the current does not end there. This action of dense (cool and saltier) water sinking is one of the main drivers of a global current that extends from the surface to beneath the oceans. It forms a global conveyor driven by thermohaline gradients (*thermo* being Greek for heat and *haline* meaning salt). The conveyor runs either on the surface or beneath the major oceans and importantly connects the North and South Atlantic with the Pacific (see Figure 4.7).

In 1992 Wallace Broecker of Columbia University argued that the global thermohaline conveyor might be disrupted if one or more of its driving components were disabled. One way to disrupt the conveyor would be to dilute the saltier sea water at points where it sank. Adding fresh water at these points could do this. If this were done in the North Atlantic then the water would not sink and indeed block further water coming up from the south. Of course, the North Atlantic gyre would not cease, as Coriolis forces from the Earth's rotation would remain. Instead, the northern part of the North Atlantic Drift would be displaced south and so cease to warm north-western Europe in winter but further warm Spain and North Africa. Broecker suggested that the switching off of the conveyor would result in a sudden cooling of north-western European in winters.

Briefly jumping forward to 2010, the European (Spanish, British and German) team of researchers led by César Negre and Rainer Zahn solved a conundrum of what had been seemingly contradictory evidence by using isotopic data that they obtained from the South Atlantic and comparing it with data from the North Atlantic. They showed that the abyssal currently southerly current reversed during glacial times. (Figure 4.7 is a rough sketch as to how the Broecker thermohaline circulation works in our warm interglacial.)

Meanwhile, back in 1993, the year after Wallace Broecker presented his hypothesis of the importance of thermohaline circulation change to glacials, fellow Columbia University researchers, with Gerard Bond and colleagues, presented evidence and a picture of what might have been happening during the last glacial. It had been known since the late 1980s that at several distinct times during the last glacial icebergs had carried moraine (rocks and geological detritus) into the North Atlantic. There the icebergs melted, releasing the moraine; this can be seen in ocean sediment cores. These events were summarised in 1988 by Hartmut Heinrich. What Bond and colleagues did was to tie these Heinrich events into a class of glacial temperature changes within the last glacial (known as Dansgaard–Oeschger cycles) as revealed by Greenland ice cores. They used a palaeoclimatic bio-indicator, the abundance of planktic foram *Neogloboquadrina pachyderma*, in North Atlantic sediment cores. The emerging theory was that in the northern hemisphere, as the glacial progressed, the ice sheets grew. As the one over North America (the Laurentide sheet) became thicker it pressed down on the underlying geology and provided increasing thermal insulation. Geothermal heat and the ice's weight caused basal melting and the sheet slipped, fracturing to release icebergs into the North Atlantic. This iceberg armada freshened the Atlantic sea water, so shutting down the Broecker conveyor and in turn cooling the North Atlantic. (We now know, from the work of Negre et al. [2010], that the southerly abyssal current actually reversed during glacial times in addition to – if not as part of – changes in North Atlantic circulation, hence considerable change in the North Atlantic down-welling (or meridional overturning circulation [MOC], as it is sometimes called in the academic literature). This weakening of the MOC then facilitated ice-sheet build-up (and with iceberg flow temporarily reduced) until the process repeated, while the thermohaline circulation slowly recovered. There is also evidence to suggest that the presence of a large ice sheet over northern North America affected air circulation but that ice-sheet collapse may have allowed the air circulation to return to a pattern similar to today's, which would have helped the restoration of the conveyor. Meanwhile, with each cycle the ice sheet grew bigger and although its oceanic margins could see iceberg loss, its terrestrial margins would not. Similarly the Fennoscandian ice sheet bordering the sea over northern Europe also discharged icebergs while its terrestrial-bounded edges obviously did not.

Whereas the Heinrich events and associated climate changes are largely instigated by northern hemisphere factors, only the larger of such events of the last glacial are reflected in the southern hemisphere's Antarctic ice-core record; the weaker ones are not.

Nonetheless, whereas Heinrich events are a major part of northern hemisphere climatic events and only weakly reflected at the South Pole, they do have a pronounced effect in the tropics in terms of both climate and biology. The Intertropical Convergence Zone (ITCZ) is a band of rainfall that spans the Atlantic. It migrates north–south with the seasons. The effect of the ITCZ in our current interglacial is that it makes a band across Brazil wet and a parallel area to the south dry. This arid zone of eastern South America currently separates two tropical rainforests, the Amazonian and the Atlantic forests. In 2004 researchers, led by Xianfeng Wang, reported on a geological climate record of speleothem (secondary calcium carbonate deposits in caves formed by running water) and travertine deposits (again, calcium carbonate

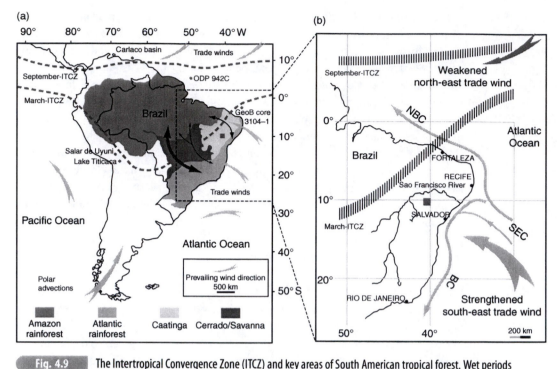

Fig. 4.9 The Intertropical Convergence Zone (ITCZ) and key areas of South American tropical forest. Wet periods coincident with Heinrich events enabled the Amazon and Atlantic forests to periodically join up in the depth of glacials. Reproduced with permission from Wang et al. (2004).

carried by water in a cave but also sometimes associated with hot springs). Layers in the deposits were accurately dated using uranium and thorium isotopes and this showed periods of time when there was considerable rain (pluvials) and other periods when it was dry. They found that these pluvials coincided with Heinrich events. During these pluvials the two tropical rainforests, Amazonian and Atlantic, were able to merge. This merging of the forests enabled species (and gene) transfer between the two zones (Figure 4.9).

Changes in the Broecker thermohaline conveyor are not just important in that it helps explain the drunken walk of global climate change during the last glacial but, as we shall see later (Chapter 6), it is of relevance to present-day concerns about anthropogenic climate change.

The second major factor affecting temperatures during a glacial arises from the concentration of greenhouse gases. As noted, methane effectively tracks glacial–interglacial temperature largely due to the global extent of wetlands. These decline as the planet cools and the atmosphere becomes less moist. Atmospheric moisture (water vapour) itself is a greenhouse gas and so a cooler, hence drier, atmosphere holds less moisture and also exerts less greenhouse forcing of climate. Both atmospheric moisture and the extent of wetlands (affecting methane release into the atmosphere) are further examples of positive-feedback phenomena, noted in section 1.5.

Another source of methane comes from methane hydrates (or methane clathrates as they are sometimes known) that exist in considerable quantities in permafrosts and in the seabed on the slopes of the continental margins (see section 3.3.9 on the Eocene). Estimates of the total quantity of methane stored as hydrates vary mainly between 1×10^{19} and 2×10^{19} g. This is about 4000 times the amount in the atmosphere today. Because methane hydrates contain a smaller proportion of the ^{13}C isotope compared to atmospheric carbon dioxide – you will recall that the enzymes driving photosynthesis prefer ^{12}C to ^{13}C – these clathrates are biogenic in origin (see section 3.2.2). As noted in the last chapter when discussing the end-Permian extinction, the physicochemical properties of methane hydrates are that they are meta-stable and will easily dissociate into hydrogen and water with an increase in temperature or a decrease in pressure.

In 2000 James Kennett and his team, from the Universities of California and California State, discovered oscillation in carbon isotopes in forams that could be explained by methane releases from hydrates. The team discovered four major releases including one just before the beginning of the Holocene interglacial, and these took place at the same time as four major Heinrich events at the beginning of Dansgaard–Oeschger cycles.

In 1998 a team led by Guy Rothwell of Britain's Southampton Oceanography Centre suggested that the biggest climate change over the last glacial–interglacial cycle was possibly attributable to methane release from marine methane hydrates. Seismic studies of the Balearic abyssal plain to the west of Sardinia revealed an extensive ancient submarine landslide three times the size of Wales involving 500 km^3 of mud. Rothwell's team radiocarbon-dated sediments above and below the landslide and got an estimated age of 22 000 years. This date ties in with the LGM. Could it be that the landslide was the result of a massive methane release well in excess of half a billion tonnes? This would have represented a very significant input of gas to the Earth's atmosphere at a time when it contained just 0.8–1.0 billion t. As a greenhouse gas methane is far more powerful, molecule for molecule, than carbon dioxide, but lasts for less time in the atmosphere (see section 1.2). Even so, such a methane release could easily have given the Earth's climate system a jolt. A decade or so of warming would trigger other changes. For example, continuing with methane, warming could have released more of the gas (and other forms of carbon) from permafrosts, and there would be implications for atmospheric and ocean circulation. What is known for certain is that there was a short period of warming about this time, in the depth of the last glacial, even though temperatures did not return to anywhere near interglacial levels.

But what triggered the massive landslide? It could have been coincidence as such landslides happen by chance. Whereas chance is certainly a factor in marine landslides, in that they are fairly common, what made such a large area collapse all at once? The suggestion is that as the glacial progressed and ice caps grew, sea level dropped. As the sea level dropped the pressure of water stabilising the meta-stable hydrates declined. Hydrates in this depth zone risked dissociating and all it took was one trigger event (an initial slide) to trigger others. Of course, sea levels did not just decline in the Mediterranean, so if this climate change mechanism has validity one would expect other major slides worldwide but perhaps spread out over a few centuries. Interestingly, Antarctica and Greenland ice cores reveal that there was a major methane overshoot prior to the beginning of the previous interglacial, and this might signify a number of major landslides occurring more or less simultaneously, geologically speaking (within a century or two).

Before we leave methane it is worth mentioning here that the Intergovernmental Panel on Climate Change (IPCC) have warned policy-makers to be wary of surprises and not to rely solely on climatic forecasts (we shall return to this in Chapter 6). However, one such surprise we may encounter as the Earth continues to warm might be a major release of methane from marine hydrates as temperature destabilises them. Of course, since in the 20th century sea levels globally have been rising marine methane hydrates are under more pressure rather than less and this will confer greater stability. Yet if the temperature of the water around the hydrates were to increase by about 5°C then there may be a massive release of methane, as was likely back in the Eocene. One recent estimate of the methane that could be released this way globally was provided by a geological team led by Matthew Hornbach (2004). They suggest that a release of some 2000 Gt of methane is possible. To place this into the context of present-day anthropogenic warming from carbon dioxide releases, this is roughly 30 times the mass of carbon that was released as carbon dioxide from fossil fuels in the final decade of the 20th century. Furthermore, from Chapter 1, bearing in mind methane's higher global warming potential (GWP) then the warming possible over two decades of such methane release would be far greater. However, methane in the atmosphere over several years gets oxidised to carbon dioxide, which is a comparatively weaker greenhouse gas but which stays in the atmosphere for a century or two. In short, should we experience a theoretical release of methane today from some hypothetical dissociation of all methane clathrates, then we would see a very considerable initial warming followed a longer period when it would be less warm than during the initial burst. Fortunately, a 5°C rise in abyssal waters – which is necessary to destabilise methane hydrates – is highly unlikely this century (although we should heed the IPCC's warning of surprises and other destabilising mechanisms). Nonetheless, such a risk becomes increasingly relevant should anthropogenic warming and fossil fuel consumption continue to grow and especially if it does into the next century. What we do not want to happen is to see an event analogous to the Initial Eocene Thermal Maximum/Palaeocene–Eocene Thermal Maximum (IETM/PETM; see section 3.3.9).

Turning away from the marine environment, it is also worth emphasising that wetlands are not the only terrestrial methane source. Peatlands represent a huge store of carbon. High-latitude peatlands, with between 180 and 455 Gt of carbon, represent up to about a third of the global soil carbon pool. Most have been formed since the LGM and so represent a major terrestrial carbon sink with 70 Gt of carbon being sequestered since that time. Much of this initial peatland formation took place in Siberia's western lowland before their counterparts in North America: all told, Russia's peatlands contain perhaps half of the planet's peat. Peatlands are, in interglacial times, an atmospheric methane source but much is still unknown as to how exactly they fit into the dynamic carbon cycle over glacial–interglacial cycles (Smith et al., 2004). Indeed, another carbon cycle unknown is the degree to which these high-latitude peatlands and permafrost soils might release their tremendous store of carbon if warmed (see Chapters 1 and 7). These are just two more reasons why climate computer models are not complete (see section 5.3.1).

Another reason given for the suddenness of the warming after the LGM is that changes in the Broecker thermohaline conveyor played a significant part. But were the changes in the planet's surface and deep-ocean circulation initiated in the northern hemisphere? There has been considerable attention given to the circulation's North Atlantic pump, but there are other possibilities (for example, see the discussion of the MOC earlier in this section). One, according to a Canadian team led by Andrew Weaver (2003), is that a pulse of comparatively warm water could have originated from Antarctica, warmed due to Milankovitch and methane factors. Had a southern hemisphere event affected the global conveyor then – because the conveyor is global – this would have affected the northern hemisphere. The jury is still out, but we should not dismiss the possibility that factors (note the plural) in both hemispheres played their part.

Finally in this subsection on temperature changes during the glacial we should look at annual events within the timeframe of several years. There are a number of climate cycles in this scale that impact on meteorology (more related to weather than climate). Of these, the one of greatest significance today is the ENSO: it determines things like droughts and moist times on the east coat of South America and the strength of the monsoon in Asia, among other climate-related/meteorological phenomena. You will remember the discussion of the ENSO in the warmer Pliocene and how ENSO state might be an analogue for what we might expect in a future warmer Earth (see section 4.3), so what was happening to the ENSO in cooler glacial times? To cut to the chase, it seems that in cooler glacial times ENSO events were slightly less common and, when they did occur, led to less variable meteorological events. This is perhaps not that surprising: after all, in cooler glacials there is less energy in the biosphere for ocean evaporation and the generation of temperature gradients. The evidence for this includes that by researchers led by Christian Wolff and Gerald H. Haug in 2011. As said, the ENSO affects many parts of the globe and indeed interannual rainfall variations in equatorial East Africa are tightly linked to the ENSO, with more rain and flooding during El Niño and droughts in La Niña years, both having severe impacts on human habitation and food security (see parts of sections 7.3.1 through to 7.4.2). The Wolff-Haug team reported evidence from an annually laminated lake sediment record from south-eastern Kenya for interannual to centennial-scale changes in ENSO-related rainfall variability during the last three millennia suggesting that there were reductions in both the mean rate and the variability of rainfall in East Africa during the last glacial period. Furthermore, climate model simulations support forward extrapolation from these lake sediment data that future warming will intensify the interannual variability of East Africa's rainfall. In effect, it would seem that if the frequency of ENSO events in a given number of years, and their intensity, were lower in cooler glacial times, then they would be more frequent and more intense in a future anthropogenically warmed world.

4.5.4 Biological and environmental impacts of the last glacial

The depth of the last glacial saw increased aridity over much of the planet. With ice caps extended and the sea level lowered, there was a considerable migration of species. The former induced migration to lower latitudes and the latter facilitated migration by exposing land bridges. Because there were a number of glacial–interglacial cycles in the Pleistocene, flora and fauna would have been subject to evolutionary selection pressures favouring those that were capable of migrating from one place to another, often in suboptimal conditions. Many species (but not all) that survived from one climatic cycle to another did so in ecological refugia. Such refugia might be latitudinal – that is, closer or further away from the equator – or alternatively altitudinal, in mountain ranges with vertical migration upwards in response to warm interglacials and downwards in cooler glacial times.

Genetic analyses provide insights into the dynamic ecology of refugia over time. The southern part of South America provides a good illustration of how such insights can be gained. There the Andean mountains chain, separating Chile and Argentina, was glaciated during the LGM while the lowlands on either side were ice-free. Pollen analysis reveals that there were trees there throughout the LGM, suggesting that these lowlands were refugia for temperate forest species. The question then arises as to whether there was just one refuge during the LGM on one side of the Andes or two distinct refugia, one on each side. If the former, then an analysis would reveal that the two populations were genetically similar. Conversely, if there were two separate refugia then populations would exhibit significant genetic differences. In 2000 Andrea Premoli from Patagonia and researchers from the University of Colorado looked at a tree species, the Patagonian cypress (*Fitzroya cupressoides*). Today it can be found substantially spread out in lowland Chile but only in scattered populations on the Argentinean side of the Andes. They examined 11 enzyme systems associated with 21 genetic loci in this species and found that spatially distinct populations on both sides of the Andes did have distinct genetic differences, indicating that there were two separate refugia.

Were life as simple as the above suggests. It would be facile to conclude that just because one species reached out from, and retreated to, two refugia that temperate forest species all behaved in a similar way. *Fitzroya* is a species that exhibits considerable longevity, with individuals living in excess of 3000 years, but it has poor regenerative powers. This indicates that it is not best suited to migrating. However, other temperate forest species are and they may be able to relocate further afield. In short, component species of the same ecosystem do not necessarily respond uniformly to climate change; indeed, responses are varied.

There are many examples of genetic analysis illuminating how a population outrode the last glacial. One European study, by the French botanist Rémy Pétit, examined the chloroplast DNA of 22 widespread species of European trees and shrubs (all woody angiosperm species) from the same 25 European forests (Pétit et al., 2003). The advantage of using chloroplast DNA is that it accompanies seeds only (not pollen, which can travel considerably further). The forests were located in Sweden, Scotland, Germany, England, Austria, Hungary, Italy, France, Greece, Slovakia, Croatia, Romania and Spain. During the last glacial European forests were considerably more restricted than those in North and South America, or for that matter South Africa, due to the Mediterranean Sea forming a barrier to latitudinal migration. Nonetheless, one might expect the greatest genetic diversity to be found in those trees and shrubs in the south, as the glacial refugia would be more likely to be located there. The reality is more complicated. True, the southern European latitudes were important for providing refugia during the LGM and the Iberian, Italian and Balkan peninsulas were all important locations. But pollen and fossil analysis also indicates that several tree species were localised in small, favourable spots at the southern edge of the then cold, dry steppe-tundra in eastern, central and south-western Europe. With post-glacial warming some of the surviving populations expanded (horizontally), some migrated altitudinally (vertically up mountains) and some unable to migrate in time became extinct. One would expect that those populations that had been able to migrate and persist would show the greatest genetic diversity. One might also anticipate a gradual decrease in diversity away from source populations. Again the reality was more complex.

Pétit and colleagues (2003) analysed their data in two ways. First they tested the genetic diversity of species within populations, and second they looked at the genetic divergence of populations in forests from all remaining populations in other forests. This last is an indication of how strongly the separate populations are related to each other.

It transpires that for the most part there is a greater genetic diversity in southern forests which suggests that their population persisted within the region throughout the last glacial. There was, though, one exception and that was on the Iberian peninsula, where populations exhibited low genetic diversity. This may well be because that peninsula seems to have been exposed to particularly severe climatic episodes of cold and aridity, and because the Pyrenees formed a particularly formidable barrier. (The lower LGM sea level did not provide extensive land bridges there, as there were between the British Isles and mainland Europe or Italy and the Balkan peninsula.)

The second thing discovered was that the genetic divergence of populations from all remaining populations was highest not in the low-latitude refugia but across France, Germany, Austria and Hungary.

What appears to have happened is that species were constrained to southern refugia during the glacial, and a few fortunate higher-latitude refugia. However, after the glacial each refuge made its own contribution to the genetic diversity of the current interglacial higher-latitude forests on land in between the refugia. Interglacial populations of trees close to glacial refugia retained the genetic make-up of the refuge populations. Populations further away and equidistant from refugia tended to share source-refugia population genes. However, with the rapid transition from glacial to interglacial conditions and back it was the populations of trees that were closest to refugia that were able to migrate back while the mixed-gene populations furthest away tended to die out. Consequently, the refuge populations tended to retain their genetic identity.

Of course, as with any study, one has to be aware of limitations and this analysis only looked at diversity within species (intraspecific diversity) and not interspecific biodiversity. Furthermore, whereas this scenario applies in this circumstance and with these species, it does not necessarily apply to other species or circumstances, such as with the refuge (or refugia) being further away.

Nonetheless, genetic analysis does provide some insights. One such possible inference applied to humans came in 2009 from work by Jeremy Searle and colleagues published in the *Proceedings of the Royal Society of London Series B*. There are two competing views as to the origin of the Celts, whose descendents today inhabit Cornwall, Ireland and Scotland in the British Isles and who have some cultural connections with some in Brittany in France. The traditional view is that the ancestors of British Celtic people spread from central Europe during the Iron Age sometime around or after 1200 BC and were then later displaced by the Anglo-Saxons. Another idea is that the Celts were descended from those who at the end of the last glacial first colonised north-western Europe before 10 000 years ago. What Jeremy Searle and colleagues did was to note that common shrews (*Sorex araneus*) and water voles (*Arvicola terrestris*) from specific mitochondrial DNA (mtDNA) lineages have peripheral western/northern European distributions that are strikingly similar to that of the Celtic people. They showed that the mtDNA lineages of three other small mammal species (bank vole *Myodes glareolus*, field vole *Microtus agrestis* and pygmy shrew *Sorex minutus*) also form a 'Celtic fringe'. They looked at animals from northern and western areas of Britain to find out whether they have different mtDNA from their counterparts in other parts of the British Isles. They argue that these small mammals most reasonably colonised Britain in a two-phase process following the LGM, with climatically driven partial replacement of the first colonists by the second colonists, leaving the first colonists with a peripheral (Celtic-fringe-like) geographical distribution. They suggest that these natural Celtic fringes provide insight into the same phenomenon in humans and support its origin in processes following the end of the LGM.

What is happening is that only those individuals with the genetic propensity to survive and migrate with climate change migrated laterally any great distance. They took with them only their traits, leaving behind the genetic diversity of their species community's ancestry. (Those that stayed perhaps migrated only a short distance vertically or into tolerable ecological niches.) Consequently, range expansion by a species in response to climate change reduces genetic variation at the margins of the species range. This leads to an interesting speculation. Species range expansion might well compromise the adaptive potential of its marginal populations. Remarkably, this prediction had not been tested up to 2008. Then, in that year, Benoit Pujol and John Pannell from the University of Oxford showed that populations of the plant annual mercury (*Mercurialis annua*) responded to selection on a key life-history trait less well than populations from the species' glacial refugium. This species expanded its range into Spain and Portugal from North Africa following glacials, and then back again in interglacials. Their results provided direct evidence of a decline in adaptive potential across the geographic range of a species after a shift in its distribution. The implications of this to ecological management with global warming is that predicting evolutionary responses to environmental change will need to account for the genetic variability of species and the spatial dynamics of their geographic distributions[6].

The above examples of migration and refugia are all terrestrial but this could apply to aquatic environments too. Furthermore, the above examples using genetic analysis tend to suggest that refugia, where species ride out times of climatic change, often only have a minimal changing effect on species populations within the refugia. Yet, as with climate change per se, the comings and goings from refugia can result in significant evolution. In 2005 results of genetic, morphological (shape) and geographical analyses by a European and South African team of researchers showed how climate

---

[6] This may not seem important because species have survived past glacial–interglacial transitions and so could survive future global warming. However, current and forthcoming climate change is taking place faster than any previous glacial–interglacial changes, which means that ecosystem managers may want to consider the (controversial) option of species translocation.

change in association with a transitional refuge had a lasting effect on biodiversity on a near-continental scale. In this case, which we look at shortly, the example is aquatic (Joyce et al., 2005).

Of course, in one sense all refugia are transitional in that at a certain latitude an area of land mass may at one time favour a species, enabling it to survive, while at another be inhospitable to it. In both hospitable and inhospitable times the land area (be it a mountains or whatever) physically remains, but during inhospitable times the refuge disappears. One example is the palaeo-Lake Makgadikgadi that existed in the lowest part of the Kalahari Plain (present-day northern Botswana) during the last glacial, which was for part of the mid-southern African watershed a time of rain, or pluvial. During this time the lake was large and comparable in size to present-day Lake Victoria further north in eastern Africa. Today little remains other than a mix of salt pan and swamp: primarily the Makarikari Salt Pan and Okavango Swamp.

The importance of palaeo-Lake Makgadikgadi was discovered because of the rich biodiversity of cichlid fish species in sub-watersheds 800 km or so apart that would have once fed into this lake. Conversely, the rivers associated with Lake Victoria, and the rivers flowing into other large African lakes, have only two widespread riverine lineages of haplochromine cichlids. This contrasts with the lakes themselves, which have high cichlid biodiversity. The reason for this, it is supposed, is that these large lakes provide many opportunities for speciation whereas the rivers do not. So why are the rivers that flow into the Kalahari Plain species-rich?

The analyses carried out suggest that palaeo-Lake Makgadikgadi, when it existed, did provide many opportunities for speciation. This species assemblage was physically forced to move into the various surrounding river systems as the lake gradually dried out over the period from when the Holocene began to about 2000 years ago. Had the lake dried out extremely quickly instead, then this movement of populations may not have happened. The same would have been true if the lake were very small. The palaeo-lake's large size also enabled there to be a variety of ecological niches that facilitated speciation. Finally, it is likely that this speciation happened over more than one glacial–interglacial cycle.

Although important, not all species rely on refugia and with some species the entire population migrates as the climate changes. So the biological response to climate change is complex and we will return to this later when considering the response to possible future climatic change.

Things are rarely simple in biology. Unlike in theoretical physics or chemistry, where much work includes laboratory experiments where researchers just look at, and just work with, aspects they wish to study, biological systems are complex. The above studies look at how individual species respond to climate change, but natural systems are more complex. For example, not only may a species respond to climate change but its range migration may well bring it into contact with other species. These other species may otherwise have been able to ride out the climate change had it not been for the incoming migratory species with which they then have to compete. Such an example was provided by a small team of Scandinavian, British and German ecologists led by Heikki Seppä, Teija Alenius and Richard Bradshaw in 2009.

At the end of the last glacial, Norway spruce (*Picea abies*), one of the dominant tree species in Eurasia, spread slowly westward in northern Europe, invading eastern Finland about 6500 years ago, eastern central Sweden about 2700 years ago and southern Norway about 1000 years ago. Its range migration is one of the most recent and best constrained invasions of a main tree species in northern Europe and it allowed Seppä and colleagues to assess using pollen analysis (see section 2.1.5) colonisation patterns and the associated competitive replacement processes with other species. There was an average spreading rate of 0.2 km year^{-1} from eastern to western Finland. They discovered that the *P. abies* population in northern and western Scandinavia (Norway, Sweden and Finland) increased in size from the time of the initial expansion to levels comparable with the modern situation in just 100–550 years. At each site *P. abies* invaded a dense, intact Scots pine–birch–alder–lime–hazel (*Pinus sylvestri–Betula–Alnus–Tilia–Corylus*) forest. The resident mixed forest provided no or weak resistance to the colonisation of Norway spruce. The tree species that most clearly declined was small-leaved lime (*Tilia cordata*), the ecological niche of which considerably overlaps with that of *P. abies*. *T. cordata* in Sweden was also responding to climate change and was already in decline before the arrival of *P. abies* (and probably less likely due to human interference 1000 years ago as suggested by some hypotheses, although as we shall see human interference does often affect a species' response to climate change). Other species coped better with the climate change and the invasion of *P. abies*. Scots pine (*P. sylvestris*) showed no or little change during the *P. abies* colonisation and population growth, probably because *P. sylvestris* predominantly grows in drier and more nutrient-poor soils than *P. abies*. *Alnus* also seemed resilient. Birch (*Betula*), hazel (*Corylus avellana*) and elm (*Ulmus*) did decline in central Sweden as a result of *P. abies* invading.

This work demonstrates that ecosystems are transient, and subject to change when new species attain dominance and replace one (or several) key components of the previous ecosystem. It shows how the former species-rich mixed ecosystem was invaded by *P. abies* and gradually changed to a modern boreal forest, dominated by Norway spruce and Scots pine (*P. abies* and *P. sylvestris*). This ecosystem change happened gradually from east to west, as the spread of *P. abies* forest caused a decrease in the distribution of the mixed conifer/deciduous forest.

Today, it is likely that future warming will promote a major northward spread of temperate deciduous tree species including small-leaved lime (*T. cordata*). By the year 2100 *T. cordata* will be a major component of – although Scots pine is likely to dominate – the forest in central Sweden, its range reaching along the Swedish east coast up to 68°N (north of the Baltic Sea coast where Sweden meets Finland).

Biological change did not just occur at the time of glacial–interglacial transition, it also took place with climatic change within glacials themselves. For example, the Italian palaeorecords mentioned in section 4.5.1 show that the Lago Grande di Monticchio region saw switches between three broad biome types. Oak (*Quercus* spp.) and elm (*Ulmus* spp.) temperate woodland dominated both the current and previous Eemian (or Sangamon as it is sometimes known) interglacial. Conversely, pine (*Pinus* spp.) and juniper (*Juniperis* spp.) dominated the LGM and other cooler parts of the last glacial, whereas warmer interstadial climatic events during the glacial saw wooded steppe dominate.

The genetic consequences of all these biome transitions both within and between glacial and interglacial conditions were profound for many plant species. For tropical forests, they were not only reduced during glacials but their present distribution areas were dissected. Aridity increased in glacials and so there was the expansion of desert areas by the time of the LGM to about five times their present extent. Also, as mentioned, in present-day Europe and North America high-latitude boreal species moved south ahead of the European (Fennoscandian) and North American (Laurentide) ice caps as they expanded. There was a considerable dynamic as ecological communities re-organised in terms of composition, latitude and altitude.

Typically, when a plant species migrates it does so through seed dispersal. The distance that seeds travel from a source plant is largely determined by probability with a curved distribution: the majority of seeds travel a short distance and a few travel longer distances. But largely is not the same as always. Circumstances, such as a storm or a seed getting stuck in an animal's fur, can see seeds transported considerable distances beyond the norm. So, if one were to plot a graph of the number of seeds against distance one would not only see a probability curve but also a small leptokurtic kink representing a small distribution further afield. (This leptokurtic phenomenon is the same as those bedevilling concerns over the separation of genetically modified crops from their non-modified counterparts, and concerns over invading species. For while one may plant a genetically modified crop so far apart from its non-genetically modified counterpart as to be assured that 99.9% of pollen from one crop will not reach the other, there is still a small chance of contamination.)

That a small proportion of seeds will travel exceptionally long distances has profound genetic consequences during climate change events. Plants migrating out of refugia due to changing conditions will be moving into lands that up to the change had been occupied by other species, many of which will now be ill-suited to the new conditions. So those few seeds travelling the farthest first grow up with little competition and subsequent generations can grow with comparative impunity. Northern hemisphere interglacial populations at the edge of their range would, with glacials, migrate south into large areas of suitable territory with leaders establishing new colonies long before the rest arrive. Indeed, those behind now have to compete to migrate. Since only a small number of plants make up this leading wave, as mentioned above, the genetic diversity of the new colonies covering a large area is reduced compared with those found in the refugia, which cover smaller areas. Indeed, this also happens with some animal species, for example, as illustrated by an analysis of 41 species of North American fish (Bernatchez and Wilson, 1998).

In addition to expanding populations having restricted genetic diversity, so a retreating population will suffer dissection and extinctions so that the last survivors at the edge of the population's range will also tend to have restricted genetic diversity. And it is these individuals that will lead the charge with the next climatic cycle (Hewitt, 1996).

With migration due to climate change, animals and plants will inevitably encounter some environmental parameters different to those found in their refuge. For example, changing latitude, so as to broadly maintain a constant thermal regimen, also sees a change in day length with season. This is not entirely insignificant. The reproduction of many plants is determined by temperature and day length, as are the migration dates of some birds, and day length even affects the reproductive cycle of some insects. So while genetic diversity is restricted, there is some genetic variation, which is necessary for survival.

It is also important to note that whereas one place may be a climatic refuge for some species, it may not be for another. A consequence of this is that a single region in an interglacial may receive different species from different refugia. For example, at the end of the last glacial Britain received oaks (*Quercus* spp.), shrews (*Sorex* spp.), hedgehogs (*Erinaceus* spp.) and bears (*Ursus* spp.) from Spain, but grasshoppers (*Chorthippus* spp.), alder (*Alnus* spp.), beech (*Fagus* spp.) and newts (*Triturus* spp.) from the Balkans. There are also co-evolutionary implications if a population is to spend some of its time with one community and other times with a quite different community in terms of genetic make-up, if not species make-up.

Matters in Europe and North America differed. In Europe the Alp, Pyrenee and Carpathian mountains ranges provided latitudinal barriers, as indeed did the Mediterranean. In North America the mountain ranges run more north–south and the area of uninterrupted land is more extensive than in Europe. The species and the genetic diversity of the south-eastern USA suggest that the area was a refuge for many species during glacials (Hewitt, 2000).

However, Europe and North America were not totally isolated. As previously mentioned (Chapter 3), during glacials the sea level lowered and there was a Beringial land bridge between present-day Russia and Alaska. Circumpolar species repeatedly expanded and fragmented with the glacial–interglacial cycles. This repeated fragmentation led to the formation of subspecies and sometimes new species. Animals so affected included guillemots (*Uria* spp.), dunlins (*Calidris* spp.), reindeer/caribou (*Rangifer* spp.) and bears (*Ursus* spp.; Hewitt, 2000).

Another instance of such population change and fragmentation was provided late in 2004 by Beth Shapiro from the University of Oxford, who was the lead author of a paper arising out of an international collaboration of more than 15 institutions. The team looked at bison (*Bison* spp.) and sequenced mtDNA from some 450 individuals from the present to more than 60 000 years ago. By ascertaining the geographical distribution of genes, as well as which genes survived to the present and when others died out, they were able to deduce how the bison population changed. They showed that the bison population across Beringia declined catastrophically from 42 000 to 35 000 years ago. This was the time in the run-up to the LGM. Unlike with mammoths (see section 4.3), this was before humans had established a significant presence in the area. Given that this was a time of declining sea levels (hence an expanding Beringian land area) the most likely explanation is that the collapse was climatic. The research also revealed the fragmentation that the bison population underwent during the LGM. They found that today's North American bison (*Bison bison*) was separated from the rest of the bison population (including those that ultimately became the European bison, *Bison bonasus*). The North American bison arose from a subpopulation separated from the rest by the Laurentide ice sheet. These subpopulations survived in ecological refugia to the south of the major northern hemisphere ice sheets.

As mentioned, mountains can also provide refugia vertically away from areas that have become inhospitable. Some tropical mountains especially have been ecologically important in providing stable refugia where there is vertical migration with climatic change but, importantly, comparatively little horizontal migration. Species can move up and down the mountain as the vertical ecological zonation changes with climate

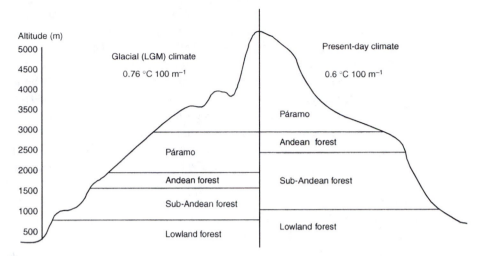

Altitude (m)

Glacial (LGM) climate

0.76 °C 100 m⁻¹

Present-day climate

0.6 °C 100 m⁻¹

Páramo

Andean forest

Páramo

Sub-Andean forest

Andean forest

Lowland forest

Sub-Andean forest

Lowland forest

Fig. 4.10 Altitudinal distribution of main vegetation belts across a Colombian Andean mountain. After Hooghiemstra and Van der Hammen (2004). See also Figure 6.2a.

change (see Figure 4.10). Having said this, do not assume that the vegetation zones all change uniformly. Often there are fundamental differences in altitudinal ecological zonation between glacial and interglacial times. For example, a study of a Colombian Andean mountain (Hooghiemstra and Van der Hammen, 2004) has shown that at the LGM the rate of temperature decrease with altitude was 0.76°C 100 m⁻¹ compared with 0.6°C 100 m⁻¹ today. This is due to the drier conditions at the LGM (cool air holds less water than warm air, which is a different, but related, phenomenon to cooler seas evaporating less; the moisture content of the air affects its heat-carrying properties). Furthermore, the sub-Andean forest occupied a narrower band that may be due to the lower altitudinal limit of glacial night frost, again due to the drier conditions. The message here is that proportional and linear relationships of climate and ecological parameters between different climate modes cannot be assumed.

Nonetheless, some tropical mountains, according to DNA divergence, have provided a home for populations for over 6 million years. This is a similar duration to that associated with some regions of African and Amazonian forest that were refugia for many species. These tropical mountain and lowland refugia are so stable that they also exhibit great genetic diversity that in some instances facilitates speciation itself (Hewitt, 2000). Because of the short migration distances (i.e. short vertical rather than long horizontal) and because mountain ranges sport so many environmental niches, tropical mountains may *partially* explain greater species richness in the tropics.

Finally, some species in some areas could not cope with Quaternary glacial–interglacial transitions. As Martina Pacher and Anthony Stuart (2008) point out, the cave bear (*Ursus spelaeus*) was one of several spectacular megafaunal (large mammal) species that became extinct in northern Eurasia during the late Quaternary. Vast numbers of their remains have been recovered from many cave sites, almost certainly representing animals that died during winter hibernation. *U. spelaeus* probably disappeared from the Alps and adjacent areas – currently the only region for which there is fairly good evidence – around 27 800 years ago, at the time of the LGM. Climatic cooling and inferred decreased vegetation productivity were probably responsible for its disappearance from this region. The cave bear survived significantly later elsewhere, for example in southern or eastern Europe. This point is important for two reasons. First, climate change extinctions are often local and not global. Secondly, the Quaternary ice age has seen several glacial cool periods interspersed with briefer warmer interglacials (such as the Holocene one we are now in). This in turn raises the question as to why some species died out at the LGM and not previous ones. As mentioned when discussing mammoths (see section 4.3), the new factor in some parts of the world in the LGM was that of the spread of early humans.

The sea-level rise and flooding of the Beringian Russian–Alaskan land bridge took place between the LGM and the early parts of our current (Holocene) interglacial. It is known that some now-extinct Pleistocene fauna survived well into our current Holocene interglacial. For example, mammoths (*Mammuthus* spp.) survived on St. Paul Island as late as 7900 years ago and Wrangel Island 3700 years ago, which is well after the end of the last glacial. Conversely, mammoths that were less isolated on the Californian Channel Islands became extinct far earlier, around the time of human arrival *c*.11 000 years ago (Guthrie, 2004). The relationship between this mammoth extinction and human arrival is thought to be causal. Had the Beringian islands been much larger, so as to enable larger populations (and hence reduced inbreeding), then it is possible that mammoths might have survived for longer in the absence of humans, perhaps even to the present day. (Beringia should not be considered as just a land bridge but, for reasons alluded to above, also a land area where many characteristic Pleistocene species and communities evolved.)

This prehistoric extinction demonstrates how stress from both climate change and from humans affects species: a negative synergy of much relevance to current and future climate ecology. The story of this megafaunal extinction spans the LGM, and so is part of this section's discussion on the last glacial, but it also continued on through the glacial–interglacial transition and into the Holocene. So we will return to this topic in sections 4.6.3 and 4.6.4 on the Holocene.

4.6 Interglacials and the present climate

4.6.1 Previous interglacials

We are currently in an interglacial, the Holocene interglacial, but there have been several other Quaternary interglacials. The last one before ours was the Eemian (sometimes known as the Sangamon) interglacial some 125 000–115 000 years ago. It began just like our current interglacial with a geologically short period (around a century or two) of rapid warming. Indeed, the warming may well have been faster due to changes in ocean circulation but tempered due to the massive ice sheets that took time to melt following the previous glacial's maximum. What is known is that the temperature early in the Eemian interglacial was a degree or so warmer than at present. This lasted for a few centuries before cooling to present-day temperatures. Also during the last interglacial period, 'North Atlantic Deep Water' was $0.4 \pm 0.2°C$ warmer than today, whereas 'Antarctic Bottom Water' temperatures were unchanged.

The warming gave a boost to sea levels during the Eemian, which were at least 3 m and possibly 5–8 m higher than today. Regional crust movement in northern North America and northern Europe, due to the weight of the glacial Laurentide and Fennoscandian ice sheets, lowered the land in those areas and these only slowly recovered once the ice melted: indeed, following the LGM these regions are still slowly rising today. This vertical crust movement on top of the mantle is known as isostasy, and the resulting apparent change in sea level is known as 'isostatic' change. It complicates attempts to identify past sea levels. (As we discussed earlier and will develop in Chapter 6, sea levels can also change due to terrestrial ice formation and melting as well as thermal expansion.)

One important question is where did this extra water early in the last interglacial come from? The reason why this question is so important is because if we are currently warming the planet, then we might see sea levels similar to those back in the Eemian. If we knew where the extra sea water came from then we would know where to monitor matters today. At first the preferred theory was that the melting of the West Antarctic Ice Sheet (or WAIS) contributed much of this rise. However, in 2000 computer modelling tied in with the ice-core records suggesting that Greenland may have made a significant, if not a major, contribution (Cuffey and Marshall, 2000).

Isostasy and isostatic change aside, it now seems it was the warmth that contributed the most to sea-level rise. (Remember that in an interglacial there is less terrestrial ice so isostasy serves at the height of an interglacial for the land to rise and hence relative sea levels to fall.) In 2009 US researchers led by Robert Kopp and Frederik Simons compiled a database of sea-level proxies during the warmest part of the last interglacial. They concluded that global sea levels at this time were even higher than the year 2000 estimates, being at least 6.6 m higher than today, and gave a two-thirds probability that they exceeded 8.0 m, and a one-third probability that they exceeded 9.4 m. If this level was 4 m or above it is likely that Antarctica melt makes a significant contribution to that of Greenland. So the implication of this research is that west Antarctic and Greenland ice sheets were significantly smaller at the height of the last interglacial. (As we shall see, more recent research on today's sea-level rise suggests that ice on islands off Greenland and North Canada's mainland may also have made a significant contribution.)

The researchers also looked at the rate of sea-level rise during the last interglacial when sea level was similar to today's. They concluded that the sea level then was rising at over 0.56 m each century but probably less than 0.92 m each century. Their results highlight the vulnerability of ice sheets to even relatively low levels of sustained global warming.

Previous interglacials were often considered in the 1990s to be a palaeo-analogue of our current interglacial. However, there are notable differences between the various glacial and interglacial cycles, both as predicted from Milankovitch theory and seen practically from the geological record. Fortunately the latter tend to confirm the former. Over the years this understanding has grown as analyses of the geological record have extended further back from one interglacial cycle in the 1970s through to eight in 2004 from the European Project for Ice Coring in Antarctica (EPICA, 2004) and associated analyses (Siegenthaler et al., 2005; Spahni et al., 2005). Glacial–interglacial cycles, we now know, are not exactly the same, and other evidence corroborates this.

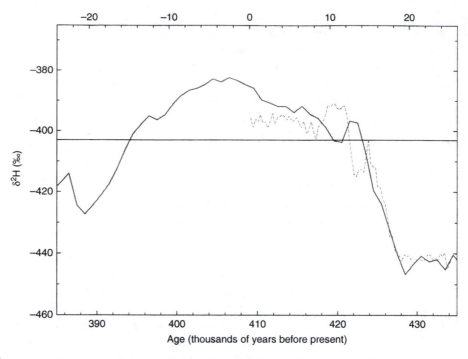

The regional temperature of the present Holocene interglacial (broken line) superimposed on the one following Termination V (solid line) as represented by the change in Antarctic ice-core ^{2}H. The top horizontal timescale represents that of the current interglacial to the present (0) and the bottom timescale relates to the interglacial following Termination V. The positioning of the superimposition is arbitrary but an alignment has been made between the respective glacial terminations. If our interglacial and that of the one following Termination V can be considered similar (in Milankovitch terms) then the superimposition suggests that we are roughly half way through our current interglacial from EPICA (2004). Reproduced with permission.

Even before the 2004 EPICA cores we knew that not all interglacials were the same, especially the current Holocene and previous Eemian interglacials. First, the present Holocene interglacial appears to be more climatically stable; climatic stability has been found during the previous Eemian interglacial in France (Rioul et al., 2001) and Greece (Frogley et al., 1999) according to a number of biological palaeoclimatic indicators, but less so elsewhere. Second, ocean circulation during the Eemian appears to be have been different (Hillaire-Marcel et al., 2001), which may explain the stability contradiction. Third, the early Eemian saw a period of warmth that was significantly warmer than the early Holocene. Finally, Milankovitch theory suggests that there was greater seasonality in the Eemian, with warmer summers and cooler winters. There is evidence for this, such as from coral oxygen-isotope climatic indicators from the Middle East (Fells et al., 2004). One of the main reasons for the differences between the Eemian and Holocene interglacials is that although the Milankovitch curve insolation climatically forces interglacials, its various contributing orbital factors can combine in a number of ways.

Therefore, early in the 21st century, when the EPICA ice-core data from previous cycles became available, the fourth interglacial before ours – beginning 420 000 years ago – was preferred for comparison with the Holocene (EPICA, 2004; Figure 4.11).

This is because of the aforementioned complexity of the Milankovitch curve, which incorporates three orbital variables (see Chapter 1). At times (allowing for climate-system inertia) of low Milankovitch forcing we see a glacial, and conversely with high forcing we see an interglacial. There are a number of ways in which the three orbital components to Milankovitch can come together to form these peaks and troughs. Unlike the previous Eemian interglacial, the fourth past interglacial (following what geologists call Termination V) was borne out of Milankovitch circumstances similar to those of the current Holocene interglacial. That the Milankovitch circumstances were different for the more recent Eemian compared to our Holocene is the likely explanation for the differences in ocean circulation (the continental tectonic arrangement in the last interglacial being barely different from today's). It also affected natural systems. Pollen records from marine sediment off south-west Greenland indicate important changes of the vegetation in Greenland over the past million years. Abundant spruce pollen (*Picea* spp.) indicates that boreal coniferous forest developed some 400 000 years ago during the height of that interglacial (de Vernal and Hillaire-Marcel, 2008). Indeed, at that time far more of Greenland was ice-free than today. There are two reasons for this. First, four glacials ago there were fewer previous glacials during which ice built up, and also the further back we go with Quaternary glacials (past 2 million years) the general trend (with the odd exception) is that the glacials were less severe. Secondly, as mentioned, the interglacial 400 000 years ago was far longer than the intervening three up to our present Holocene interglacial.

The aforementioned Milankovitch complexities in turn possibly explain why high-latitude palaeorecords indicate a different temperature profile for the Eemian interglacial compared with the present Holocene, while lower-latitude European palaeo-records suggest a similar climatic stability.

Importantly, the previous (Eemian) interglacial was short, of the order of 10 000 years or so. Were the Holocene interglacial truly similar, then this would suggest (in the absence of present-day anthropogenic global warming) that we are due to enter another glacial in the geologically very near future, as the Holocene is already 11 700 years old. (Indeed, this worried climatologists in the 1970s.) However, it transpires that the fourth interglacial ago was longer, lasting about 25 000 years. It also differed in its structure compared to the Eemian. The Eemian began with an extremely warm period of a thousand years, a time far warmer than our Holocene has been to date. Our Holocene's initial millennium was also warmer, but only marginally, compared to the rest of the interglacial. The fourth (Termination V) interglacial ago similarly lacked the Eemian's notably warmer initial spell and in that sense was similar to the Holocene. However, whereas our Holocene is comparatively stable to within a degree or so, the temperature of the fourth interglacial ago seems to have risen to a peak and then fallen more gently. Superimposing Antarctic deuterium data from both interglacials, one over the other, shows that the deuterium figures between the Holocene and the fourth interglacial ago (see Figure 4.11) are closer than that between the Eemian and the Holocene, but still are not the same. The way temperature changes within the current interglacial is different from how it changed in the fourth interglacial ago. Could this be due to other factors such as the gradual evolution of carbon reservoirs over successive glacial–interglacial cycles? Or perhaps solar forcing, changes in the Sun's output, affected the climate differently between the two interglacials? After all, approximately 10% of changes in positive radiative forcing of our current climate since 1850 are estimated as being due to changes in the Sun's energy output, although there is considerable uncertainty associated with this (IPCC, 2001).

If our Holocene interglacial is, in Milankovitch terms, similar to the one 420 000 years ago following Termination V, then the Holocene could last 25 000 years. This would mean that, some 10 000 years in, we are not quite half way through our present interglacial. This contrasts with palaeoclimatologists' views back in 1972, when comparisons were made with the Eemian and they thought that all interglacials were as short, suggesting that the next glacial was imminent (Kukla et al., 1972). However, more recent analyses (Berger and Loutre, 2002) of future Milankovitch curves and climate models suggest that we are indeed at least only half way through our current interglacial, just as the latest ice-core evidence suggests. Furthermore, continuing current carbon dioxide emissions throughout this century could conceivably mean extending our current interglacial, possibly as much as another 50 000 years, but almost certainly another 30 000 years, assuming depletion over the coming centuries of all economically recoverable fossil fuels without any carbon abatement. Indeed, if we trigger something analogous to the Initial Eocene Thermal Maximum/Palaeocene–Eocene Thermal Maximum (IETM/PETM) then the warming might extend 100 000 years or more (see section 3.3.9). Either way, the next glacial is not now thought to be imminent.

Why the shapes of the temperature curves of the Milankovitch-similar Holocene and Termination V interglacials are different is a good question. The aforementioned reasons are all likely in that the cause lies in the other variables in the system. But there are other possibilities too. For example, the Earth's plate tectonic arrangement, while only slightly different in the last interglacial, was more different still four glacial–interglacial cycles ago. For instance, the Himalayas were lower 420 000 years ago and this would have affected albedo and atmospheric circulation. Even so, these differences over hundreds of thousands of years are paltry compared to the difference in plate tectonics over tens of millions of years mentioned in Chapter 3. Another difference between the Termination V interglacial and ours is that our interglacial is four more along the series of Quaternary interglacials that have seen a broad trend of deepening intervening glacials. So, the Termination V interglacial may not be analogous in all ways to the Holocene, but both Milankovitch theory and ice-core evidence suggest that their durations may be similar.

4.6.2 The Allerød, Bølling and Younger Dryas (14 600–11 600 years ago)

The end of the last glacial in the northern hemisphere was marked by a series of major events of sufficient strength to be reflected in the southern hemisphere, albeit more weakly. This is reflected in Figure 4.12, which portrays ice-core records of the past 15 000 years. Figure 4.12a shows detail of the deuterium record from Vostok, Antarctica, which reflects regional temperature change. Figure 4.12b shows detail of the northern hemisphere's Greenland ice core showing its electrical conductivity measurements (ECMs). Here conductivity of ice is improved with acid ions and decreases when these acids are neutralised by dust. Ammonia – such as from nearby biomass burning (water-soluble ammonia is easily absorbed by water vapour and then washed out of the air) – will also neutralise acid, but the ammonia record in Greenland's

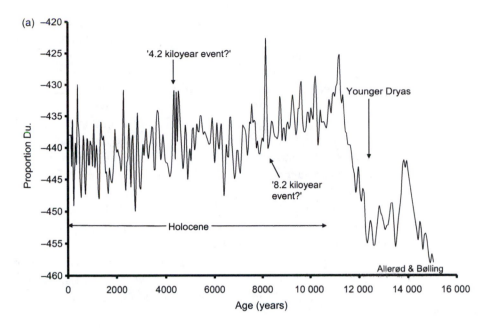

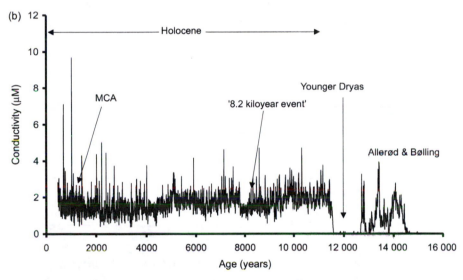

Fig. 4.12   (a) Vostok deuterium record over the past 15 000 years, reflecting Antarctic regional temperature change (data are shown as deuterium [Du] in parts per thousand compared to Standard Mean Ocean Water [SMOW] deuterium). Data from Pétit et al. (2001), *Vostok Ice Core Data for 420,000 Years*, IGBP PAGES/World Data Center for Paleoclimatology Data Contribution Series #2001-076. NOAA/NGDC Paleoclimatology Program, Boulder, CO, USA. Regarding major glacial and interglacial heat change and exchange between hemispheres via the Broecker ocean circulation there is a time delay (see section 4.5.3). But there is also more immediate, albeit weak, atmospheric heat exchange. For this reason could it be that some northern hemisphere climate events are reflected in the southern hemisphere Antarctic record? (Possibly not?) (b) Greenland ice-core electrical conductivity over the past 15 000 years reflecting North Atlantic Arctic precipitation, and hence – indirectly – temperature. MCA, medieval climatic anomaly. Data source: IGBP PAGES/World Data Center for Palaeoclimatology; as used in Taylor et al. (1993). See section 4.6.3 for a discussion about the 8.2 kiloyear event.

cores (fairly distant from biomass burning) is reasonably constant from 11 000 to 15 000 years ago so the ECM record reflects acid-neutralising calcium dust, which comes from the erosion of calcium carbonate rocks such as limestone or chalk. However, there is another factor: the presence of dust reflects the amount of precipitation. The more rain, the more acid-neutralising dust is washed out of the atmosphere. Dust is less sensitive than ammonia to washing out of the air and so, in the absence of nearby extensive vegetation, dust is a good indicator of regional precipitation as water vapour will tend not to take it out of the air, although rain will. Put the aforementioned data all together and the Greenland ice-core ECM record reflects regional precipitation. Furthermore, this is one more example of the climatic relevance of dust.

As we have already frequently noted, the warmer the planet the more ocean evaporation, hence rain: increased temperature and rain go together. Consequently dust, which would only be present when it is dry, is in turn a secondary indicator of hemispheric temperature (Taylor et al., 1993). More exactly, as temperature is not being directly reflected, the level of dust indicates whether the Earth is in a glacial (low rainfall/high dust) as opposed to an interglacial (high rainfall/low dust) mode. Finally, from Chapter 2 we know that deuterium is an indicator of regional climate. Given this, it possible to see how the panels of Figure 4.12 relate to climate.

Some 15 000 years ago the Earth was warming after the LGM. However, this warming was not permanent and cooler conditions returned, mainly in the northern hemisphere. The temporary warm period itself was comprised of two shorter warm periods, or interstadials, called the Allerød and Bølling, which in turn were followed by a cool period, the Younger Dryas, from 12 850 to 11 650 years ago before there was a warming again to the present Holocene interglacial. The results of a high-resolution deuterium (see section 2.3.1) Greenland ice-core analysis by an international team led by Jørgen Peder Steffensen concluded in 2008 that the northern hemisphere climate mode initially switched mode. The warming transition 14 700 years ago is the most rapid and occurs within a remarkable 3 years while the warming transition at 11 700 years ago lasted 60 years, as revealed by Greenland cores. This is fast! (And of relevance to the 'climate surprise' discussion at the end of Chapter 6.)

Returning to Figure 4.12a, it is an enlargement of part of Figure 4.2 from an Antarctic Vostok ice core and shows that the Younger Dryas was not so strongly reflected in the southern hemisphere: in other words, the Younger Dryas was predominantly a northern hemisphere event. Looking closely at Figures 4.12a and b it can be seen that the Antarctic cooling in fact began roughly 1000 years before the Younger Dryas in the northern hemisphere. So, although the two events are linked, they are different and the cooling in Antarctica is known as the Antarctic Cold Reversal. There is other evidence that the northern hemisphere's Younger Dryas cooling had less impact south of the equator, such as in the pollen-grain record from lake sediments in southern Chile (Bennett et al., 2000). However, not all the evidence is so minimal, and little impact compared to the northern hemisphere is not the same as no impact. Other pollen evidence from Chilean lakes (Moreno et al., 2001) demonstrates that there was enough alteration of the climate for some significant ecosystem change in the southern hemisphere. Coral evidence from the south-western Pacific suggests a sea-surface cooling somewhere between 3.2 and 5.8°C. Furthermore, the Younger

Dryas cool period was of sufficient strength for both Swiss and New Zealand alpine snowlines to drop to about 300 m below the lowest levels reached in the subsequent Holocene. In short, there was linkage of *effects* to the southern hemisphere from the northern, but was there hemispheric linkage regarding the *cause* of the Younger Dryas?

What did trigger the Younger Dryas? It has been suggested that the early melting of the North American Laurentide ice sheet added fresh water to the Gulf Stream in the Gulf of Mexico as the presence of the remaining ice sheet to the north prevented the meltwater discharging into the North Atlantic. The latter would, as we shall see, most likely disrupt the Broecker thermohaline circulation. Conversely, the addition of fresh water further south in the Gulf of Mexico could have enhanced the Broecker circulation and so helped the transportation of heat from the tropics via the Gulf Stream. This could perhaps have been a significant additional factor in pulling the planet so rapidly out of the last glacial. The consensus today is that the thermohaline circulation was disrupted, so causing the cooling of the Younger Dryas. Although there is still some debate as to exactly where the fresh water came from that did the disrupting, geological evidence clearly shows that there was flooding south of the Laurentide ice sheet. Indeed, there is some computer-model evidence to suggest that about half of the northern hemisphere meltwater came from the Laurentide sheet and that here the largest pulse (and even if the models are wrong, certainly a significant one) came, not via the Mississippi or Hudson routes, but via the Fram Straits off the east of northern Greenland (Tarasov and Peltier, 2005). (We will return to this shortly.)

But are changes in the North Atlantic circulation the only possible causes for the end of the short Younger Dryas cool period? Could a marine methane release, such as happened in the depth of the last glacial, be a cause? The answer is probably not. Terrestrial and marine biological sources of methane both have the same amount of hydrogen (four atoms per methane molecule) but some of these hydrogen atoms are in the form of the deuterium isotope. The amount of deuterium from marine and terrestrial sources of methane differs markedly: marine methane has more deuterium. Analysis of the $^2H/^1H$ ratio (deuterium to normal hydrogen) of air trapped in the Greenland ice sheet suggests that the marine methane clathrates were stable during these particular warming periods and did not contribute to this episode of climate change (Sowers, 2006). (This may perhaps be because methane releases from clathrates earlier in the glacial took place when the sea level – hence stabilising pressure on the clathrates – was lower. At the time of the Younger Dryas sea levels were already rising and so even though the planet was warming clathrate stability would have benefited from the extra pressure of water.)

If, as it seems, the Younger Dryas was caused by some North Atlantic disruption of the thermohaline circulation then the part of the Earth most significantly affected would be north-west Europe and the British Isles, which would no longer be warmed by the Gulf Stream. Here we do find that in England's Lake District the glaciers returned (or started growing again) so quickly that living trees were *caught* in their advance.

Conversely, the Antarctic Cold Reversal, due to it starting earlier, cannot be entirely explained by any disruption in the North Atlantic of the thermohaline circulation that led to the Younger Dryas. The Antarctic Cold Reversal could be explained by Antarctic melt as the Earth left the LGM behind. This could disrupt part of the Broecker thermohaline circulation in the southern hemisphere. If this happened it would still weaken the overall global thermohaline circulation. Because the northern hemisphere had large land-based ice sheets outside of the Arctic Circle (unlike the Antarctic Circle) these would have taken longer to react to coming out of the LGM. If the global thermohaline circulation had already been part-disrupted by the Antarctic Cold Reversal then it would be more susceptible to subsequent northern hemisphere disruption. Again the lesson to learn is that changes in ocean circulation (and atmospheric circulation for that matter) have a profound effect on the climate.

As discussed in the previous section, while the pacemaker of the glacial–interglacial cycles is the change in Milankovitch insolation, the vagaries of warming and cooling within glacials were driven either by changes in greenhouse gases and or changes in ocean or atmospheric circulation and in particular the Broecker thermohaline circulation. Indeed, Milankovitch insolation in July at 60°N had already peaked following the LGM by the time of the Younger Dryas and remained at a high value throughout it, and for a time into the early Holocene. We have already noted how greenhouse gases parallel each other in both northern and southern hemispheres, which means that the likely candidate *primarily* causing the northern hemisphere's Younger Dryas return to glacial conditions was connected with atmospheric and/or ocean circulation and not greenhouse climate forcing. Having said this, there was a noticeable dip in both carbon dioxide and methane concentrations at the time of the Younger Dryas, but it is likely that this was the result of a biotic response to the cooling rather than a cause of it.

Yet if it was the melting of the northern hemisphere ice caps that caused the Younger Dryas temporary cooling, why did this happen suddenly and not gradually as the ice caps melted? Furthermore, why was there not cooling of a similar degree in the southern hemisphere? Here, the geography of the northern and southern hemispheres is markedly different. A key factor is that in the northern hemisphere the landmasses of North America, Europe and Siberia enabled large ice sheets to form during the LGM. Conversely the southern hemisphere's ice sheet was centred on Antarctica and the Antarctic continental shelf and not on land at lower latitudes as in the north. Unlike their marine counterparts, terrestrial ice sheets push (terminal) moraine in front of them and as the ice retreats these can act as dams. Local terrestrial geography can also serve to constrain meltwater. So it is likely that as the Earth warmed, leaving behind it the depth of the last glacial, the melting Laurentide ice sheet created a volume of inland fresh water. It is this water that ultimately discharged into the salty northern Atlantic, probably in a pulse, thus switching off the Broecker thermohaline circulation.

This hypothesis is worth further consideration because, as we shall see, there was at least one other pulse of meltwater likely to have disrupted the Broecker circulation. So is there evidence for such a reservoir of fresh water? Indeed there is. The geological evidence for large lakes off of the Laurentide ice sheet at the end of the last glacial is clear and there were also numerous smaller bodies of water. Two lakes in particular were bigger than Canada's present Great Lakes and so they represented a considerable volume. Indeed, it is likely that the majority of freshwater fish in Canada and the northern USA today have ancestors that lived for some time in these two lakes. Both lakes were bounded on their north-eastern side by the Laurentide ice sheet. On their western side were the upland plains and hills that lie in the shadow of the Rocky Mountains. The largest meltwater lake, when at its greatest, was Lake Agassiz, at some 350 000 km^2, and it is named after the naturalist who coined the term ice age. These lakes were only temporary and just existed at the end of the glacial because of the ice dams. They were also dynamic. When the Laurentide ice sheet began to melt sufficiently, more water was fed into the lakes and part of the Laurentide ice retreated. The lakes therefore changed shape and size. Many of the lakes were from time to time connected to each other. The variability of ice aided this. There was not only melting but also re-freezing. In addition, not only would there be terminal moraines and local geography but ice dams were also possible. Geologists think that occasionally an ice dam that had held back water would, once there was sufficient water behind it, float and so allow a pulse of water to escape. If the ice dam was at a key location then the volume of water released would be considerable. In North America sometimes the drainage would be to the north through the Mackenzie River and sometimes south through the Missouri and Mississippi rivers: throughout these regions there is ample geological evidence of brief torrential flooding dating from the end of the last glacial. It could well be that it was one of the larger of these pulses that precipitated the Younger Dryas with the suddenness we see on the Greenland ECM record. Indeed, this is the preferred theory, although it should be noted that there are others.

The other thing to note about the Greenland ice-core ECM is that it appears that the regional climatic changes associated with the brief interstadials after the LGM, and also the beginning and end of the Younger Dryas itself, took place extremely quickly: perhaps as fast as 5–50 years. Furthermore, a detailed examination of the ECM record reveals climatic flickers as short as 1 year: these could be changes in atmospheric circulation (Taylor et al., 1993) and possibly represent flickers between two semi-steady states. Such rapid change serves to remind us of the IPCC's warning about being wary of surprises with regard to future warming. We shall return to this in Chapter 6.

Before leaving the topic of Greenland's palaeodust record, it is worth remembering the effect of iron on plankton blooms (see section 1.3). The Greenland ice-core ECM markedly reflects the transition from a dusty glacial to a less-dusty interglacial environment. There is no reason to suppose that other types of dust than the calcium acid-neutralising dust were not similarly affected. The transition from glacial to interglacial would have also affected iron-containing dust. In a glacial mode more iron dust would be blown on to the oceans so encouraging algal blooms that draw down atmospheric carbon dioxide from the atmosphere and so helping to keep the planet cool (see Chapter 1). Similarly, under wetter, warmer interglacial conditions some iron-containing dust would be washed out of the atmosphere before reaching the oceans. This would reduce algal blooms, reduce carbon dioxide drawdown and so help maintain the atmospheric concentration and keep the planet warmer. These are other examples of feedback systems (again, see Chapter 1) that regulate the biosphere much in the way engineers describe a cybernetic system (Figure 4.13).

Although the climate warmed quickly after the last glacial, some other environmental parameters took longer. Sea-level rise took thousands of years. The Baltic Sea

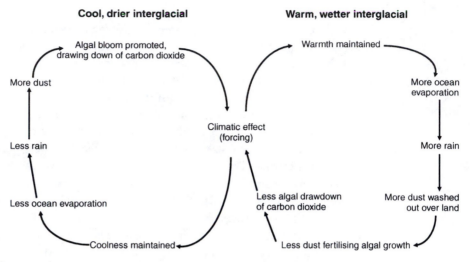

Cool, drier interglacial   **Warm, wetter interglacial**

Algal bloom promoted,   Warmth maintained
drawing down of carbon dioxide

More dust   More ocean
                                                         evaporation

                      Climatic effect
Less rain   (forcing)   More rain

Less ocean evaporation   Less algal drawdown   More dust washed
                              of carbon dioxide   out over land

        Coolness maintained   Less dust fertilising algal growth

Fig. 4.13 Climate–dust interactions combine to tend to maintain either a cool (glacial) or a warm (interglacial) mode. This is just one of the processes tending to keep the Earth in one of a number of climate states.

filled and Britain became an island only as recently as about 7000 years ago, some 4000 years after the end of the last glacial. From this we can see that the approximate average rate of sea-level rise over this time was a little more than half a metre a century. Having said this it is important to note that this is an average rate and there were times between the LGM and 7000 years ago when sea-level rise was faster (see Chapter 6), and others when it was slower.

4.6.3 The Holocene (11 700 years ago–the Industrial Revolution)

About 11 600 years ago the Earth finally emerged from the last glacial into a warmer climate mode that has lasted until the present. (Although the official date for the commencement of the Holocene is 11 700 years ago, based on key – Global Standard Section and Point – geological strata.) Average snowlines around the planet (where latitudinally climate allowed this) rose around 900 m after the last glacial maximum (LGM). On the basis of today's temperatures this suggests an atmospheric warming of about 5°C and of the tropical sea surface of some 3°C. Of course, the huge amount of ice left over from the last glacial did not disappear as quickly as the Milankovitch-forced climate had changed. It took some 3000 years for the Fennoscandian sheet over northern Europe and Russia to largely disappear and over 5000 years for the North American Laurentide ice sheet to collapse. This slower collapse may have been in part because of the vagaries of the climate due to atmospheric circulation but was probably more connected with the proximity of the higher latitude Greenland ice sheet, which is still with us today, and of course which provides us with valuable palaeoclimatic records from its ice cores.

The melting of the ice sheets themselves continued to have an effect on the northern hemisphere's climate around 8200 years ago. This is sometimes known as the 8.2 kiloyear event. It is thought (Barber et al., 1999) that a pulse of meltwater from around 8470 years ago (Rohling and Pälike, 2005) from the North American glacial lakes of Agassiz and Ojibway, until then held back by the ice, disrupted the Broecker thermohaline circulation by freshening the cold, salty water in the North Atlantic. Again, as with the Younger Dryas (discussed in the previous section), this prevented the water there from sinking, one of the processes driving the global conveyor. It subsequently cooled the temperature in central Greenland by 4–8°C and regionally in the surrounding North Atlantic sea and land by 1.5–3°C for very roughly a century around 8200 years ago, again in a similar (albeit lesser) climate response as with the Younger Dryas. The graph in Figure 4.12b shows Greenland ice core conductivity, an indicator of dust trapped in the ice, which in turn reflects precipitation in that part of the northern hemisphere, and in turn very loosely reflects temperature (the cooler the air the less moisture it can hold). Because of the length of this chain of interpretation, some caution is needed: the usefulness of dust in ice cores is more that it illustrates how fast the regional climate can flip (see section 4.6.2). What is known is that this early Holocene climate blip does show up in a number of northern hemisphere climate proxies around 8200 years ago. While not as great a return to more cooler glacial times as was the Younger Dryas around 12 000 years ago, the 8200- and 4200-year events do highlight the need for concern as to what might happen if the Broecker thermohaline conveyor became disrupted, as some hypothesise it might with Greenland melt due to current and near future global warming. Yet this northern hemisphere event is comparatively trivial to the changes in the Earth system that took place between its glacial and interglacial modes.

The warming since the LGM, although triggered by change driven by Milankovitch orbital variation in the Sun's energy reaching the northern hemisphere, was amplified by changes in the carbon cycle. Among these was considerable growth in peri-Arctic peatland. Glen MacDonald, David Beilman and colleagues reported in 2006 their ^{14}C radioactive isotope dating of these peatlands. This revealed that the development of the current circumarctic peatlands began 16 500 years ago and expanded explosively between 12 000 and 8000 years ago in concert with high summer insolation and increasing temperatures. Their rapid development contributed to the sustained peak in methane and modest decline of carbon dioxide during the early Holocene, and likely contributed to methane and carbon dioxide fluctuations during earlier interglacial. So, as the Earth left the last glacial and entered the current interglacial the peri-Arctic peatlands grew. This would have sequestered (drawn down) carbon in the form of carbon dioxide from the atmosphere. This sequestering of atmospheric carbon would have lowered the global temperature or dampened the warming taking place, but some of this carbon returned to the atmosphere as methane. Methane is a stronger greenhouse gas than carbon dioxide (see Table 1.2) and this made up for the cooling effect of the carbon dioxide drawdown.

Some 10 000 years ago Milankovitch forcing peaked (as determined by the 60°N July insolation). However, inertia within the planet's climatic system, and other forcing factors combined, delayed the peak in the planet's climate until 6000 years ago: the peak of the Holocene climatic maximum. Whereas the climate below latitudes of about 40° was only a little warmer (mostly about 0.5°C or less) than today (the early 21st century), the climate at higher latitudes was roughly 1.5°C above the Holocene average. Further, at extremely high latitudes over much of the North Pole and

Greenland, as well as a pocket in Russia north of the Caspian Sea, the climate was as much as 3°C warmer than at the end of the 20th century. At lower latitudes this warm period manifested itself not so much with a warmer climate but in other ways as more energy entered parts of the climatic system, such as the monsoon's summer circulation. So, there was more precipitation in many parts of the subtropics. Consequently, the tropics during the maximum were warmer and wetter than today. However, due to current global warming, from the various IPCC scenarios (see section 5.3.1), it is likely that we will reach temperatures equivalent to the Holocene climatic maximum well before the mid-21st century and possibly as soon the end of its first quarter.

It is worth noting that in the past the Holocene climatic maximum (9000–5000 years ago) has been described as the Holocene climatic optimum. Many palaeoclimatologists have ceased to use this term as the use of the word optimum does not mean that the climate was in any sense climatologically or biologically optimal, but that the time represented a thermal maximum for the pre-21st-century Holocene. Having said this, we are still in the Holocene, and present-day global warming is taking us beyond what was the Holocene climatic maximum, and so it may be that terminology will again change in a decade or two.

In addition to a warmer climate, and resulting sea-level rise, the Holocene saw humans contribute their own ecological impact. The degree these various factors of climate, sea-level rise and humans together impacted on species varied with location and, of course, the species in question. The resulting picture was complex. Yet, just one of these factors was unique to the end of the last glacial and the current interglacial and that was the spread of modern humans (*H. sapiens*). There is no single mechanism responsible for the extinctions of every species in every location in the last glacial and the current interglacial. As noted earlier in section 4.3, some Pliocene large animals (megafauna) survived to quite recently in geological terms. Mammoths (*Mammuthus* spp.) survived up to as recently as 4000 years ago due to isolation on Wrangel Island. Another species, the Irish elk (*Megaloceros giganteus*), survived to about 7000 years ago. Before 20000 years ago the Irish elk was widespread from Ireland to Russia (west–east) and Scandinavia to the Mediterranean (north–south) but at the time of the LGM (18000–22000 years ago) they appear to have become more restricted, probably to refugia in the shrub steppes of central Asia. From there the Irish elk recolonised much of their lost ground as the Alpine and Scandinavian ice sheets retreated. It is possible that the productive shrub and open grass community necessary for the Irish elk persisted in the eastern Urals (Stuart et al., 2004). The picture is further complicated in that some large herbivore species (possibly including the Irish elk) not only responded to changes in their food-plant communities but themselves affected those plant communities and plant productivity, through nutrient redistribution and grazing. Further, by trampling tundra mosses and stimulating grasses through grazing the woolly mammoth might have helped maintain and even expand its own habitat (Pastor and Moen, 2004). Just as a prolonged period of climate change (albeit vulcanism-induced) put pressure on the dinosaurs (the extinction of which was then sealed with an asteroid impact), so climate change appears to have put pressure on Pliocene megafauna which were finally made extinct by human hunting; after all, the megafauna had survived a number of previous glacial–interglacial climatic shifts.

Human contribution to species extinction actually began before the Holocene, so we must briefly return to the beginning of our current interglacial, and indeed earlier, to look at this story.

Evidence from Alaska and the Yukon territory (Guthrie, 2006) suggests that there woolly mammoths (*Mammuthus primigenius*) and a regional species of horse (*Equus ferus*) died out around 12 000 years ago, around the time when humans arrived, with very little overlap. Again the picture is more complex than first appears, for along with the arrival of humans there is also the appearance of moose (*Alces alces*). Pollen analysis shows that the area saw a transition through three principal types of biome. The first was a cold, dry steppe up to around 13 500 years ago in which mammoths thrived. The second was a transitional moister biome that existed from 12 000 to 13 500 years ago. Finally there was the rise after 12 000 years ago of mesic taiga and tundra with dwarf birch (*Betula*). The suggestion is that plant species in this ecosystem did not support many large herbivores. That humans arrived around that time may have added critical pressure on top of the radical environmental change taking place, both at the time of their arrival and the episode of ecological change 1500 years earlier. This environmental change was driven by changes in the global climate as the planet moved out of the glacial. In this instance, humans themselves were not (solely) responsible for the extinction.

In 2011 a large international team led by Eline Lorenzen, David Nogués-Bravo and Ludovic Orlando reported their formidable study on late Quaternary megafaunal extinctions. Beginning around 50 000 years ago, Eurasia and North America respectively lost approximately 36 and 72% of their megafauna. In general, the proportion of species that went extinct was greatest on continents that experienced the most dramatic climatic changes, which implies the major role of climate change. However, the pattern of megafaunal extinctions in North America and Australia approximately coincides with the appearance of humans and this suggests a possible anthropogenic contribution to these species' extinctions. To disentangle the processes underlying population dynamics and extinction we investigate the demographic histories of six megafauna herbivores of the Late Quaternary: woolly rhinoceros (*Coelodonta antiquitatis*), woolly mammoth (*M. primigenius*), horse (wild *E. ferus* and domestic *Equus caballus*), reindeer/caribou (*Rangifer tarandus*), bison (*Bison priscus* and *B. bison*) and musk ox (*Ovibos moschatus*). These taxa were characteristic of Late Quaternary Eurasia and/or North America and represent both extinct and extant species. Their analysis was based on 846 radiocarbon-dated mtDNA sequences, 1439 directly dated megafauna remains, 1557 indirectly dated megafauna remains and 6291 radiocarbon determinations associated with Upper Palaeolithic human occupations in Eurasia. They then reconstructed the demographic histories of the megafaunal herbivores from the ancient DNA data, modelled past species distributions and determined the geographical overlap between humans and megafauna over the past 50 000 years. They found that whereas climate change alone largely explains the extinction of Eurasian musk ox and woolly rhinoceros (both extinct just before the Holocene), for many other now-extinct late megafauna Quaternary species changes in abundance clearly depended on more than one factor. By themselves, neither climate change nor humans alone can explain most megafaunal extinctions of this time. A combination of the two is required. The balance of importance of these factors in such extinctions varies but both these species went extinct around the start of the Holocene. Of course, not all megafauna became extinct: reindeer currently remain relatively unaffected by either of these factors. This study could not elucidate the causes underlying the extinction of woolly mammoth but it is hard not to conclude that humans did play a part, given that mammoths survived previous interglacials and their demise in some places does seem suspiciously concurrent with the arrival of humans.

It is interesting to note that the late Pleistocene and early Holocene extinctions were largely driven by a combination of both climate change and human pressure, for today's current extinctions are similarly driven. The only difference between these two extinction periods is that the late Pleistocene and early Holocene extinction was partly driven by *natural* climate change whereas the current extinction involves *anthropogenic* climate change. It is also interesting to note that the degree of extinctions on each continent varied depending on the length of human existence (see Table 4.1). Because humans evolved in Africa, and then spread to Eurasia, animals here had a chance to co-evolve. (We shall shortly return to this in the next section on biological response.)

Taking all the evidence together, what seems to be happening with regards to human and climate impact on megafauna? First, we need to note that megafauna had survived the comparable climate change that took place at the ends of previous glacials. The only substantially different factor at the end of the last glacial and beginning of this (the Holocene) interglacial was the presence of humans. This suggests that humans were involved in the megafauna extinctions that took place. That the considerable global climate change of the LGM–Holocene has been a factor in species demise is not surprising. As Lorenzen et al. (2011) point out, a species' response, or sensitivity, to this climate change varies with the species so it seems likely that those more sensitive to climate factors were more prone to impact from humans. The size of megafaunal populations do seem to have been in decline during the LGM–Holocene transition ($\approx 20\,000$–$10\,000$ years ago). A number of the megafaunal extinctions appear to have taken place around the start of the Holocene, $12\,000$–$10\,000$ years ago. But remember that there are problems with dating accuracy. It could be that the false end of the last glacial at $14\,000$ years ago, with the brief (Younger Dryas) abrupt return to part-way more glacial conditions (Figure 4.12), could have provided additional climate stress that made species more vulnerable to human impact. (Remember that the Younger Dryas was more pronounced in the northern hemisphere where the Palaeolithic megafauna extinctions took place.) The conclusion by Lorenzen et al. that 'no evidence that Palaeolithic humans greatly impacted musk ox populations' is at first curious as the musk ox survived well into the Holocene until just a few hundred years ago and during all this Holocene time ($\approx 10\,000$ years) the climate was stable. Their conclusion seems to be based on 'Musk ox remains [being] found in only 1% of European archaeological sites and 6% of Siberian sites, and do not overlap noticeably in range with Palaeolithic humans in either Europe or Siberia'. True, musk ox populations seem to have declined in the Palaeolithic and this may well be because their habitat is one of a very cold climate. However, the musk ox would have certainly encountered humans *after* the Palaeolithic and been particularly vulnerable to human impact having a low population and reduced area of habitat at that time. All in all, the megafauna extinctions we know about seem to have been a combination of both climate change and human impact, even if some species were more affected by one factor than the other and, indeed, that the strength of these factors changed with time up and into the Holocene, beginning 10 000 years ago. Since then, as previously mentioned, the climate was comparably stable within a degree or so of present-day temperatures. Nonetheless, during this time there were periods when the climate was warmer and some when it was cooler than the Holocene mean.

The time from 8000 to 5000 years ago marks the aforementioned period known as the Holocene climatic maximum (see also Figure 5.3). In Britain trees grew on land 180–300 m higher than the present-day tree line, and species such as lime (*Tilia* spp.) and elm (*Ulmus* spp.) were more common than today. Trees also grew in the Orkney Islands, Faroe Islands and Iceland. Winters then were probably far milder than today (Lamb, 1965).

However, there is some ambiguity in the climate record as some parts of the Earth appear to have cooled for part of the climate maximum period, especially in high latitudes (Antarctica, Greenland and eastern Canada). One suggestion is that this high-latitude cooling decreased precipitation and reduced terrestrial biomass. The carbon from this biomass then entered the atmosphere and indeed a small increase of 10 ppm in atmospheric carbon dioxide is detected between 7000 and 5000 years ago. (However, this is small by late-20th-century increases of more than 60 ppm over the century.) The Holocene climatic maximum increase in greenhouse forcing may in part account for the climate maximum and this period of time may represent a redistribution of carbon between various pools (Steig, 1999). Such problems are the stuff of some current research.

Palaeoclimatic records for the last 4500 years generally indicate that temperatures were lower than the Holocene climatic maximum. There was a general cooling of 1–2°C, known as the Iron Age neoglaciation, which took place between 4500 and 2500 years ago.

There was then a return to warmer conditions, but not nearly as warm as the Holocene climatic maximum, around the beginnings of the Roman Empire. Changes of just a degree, or at most around two, mark the nature of much of the climatic changes seen within the Holocene interglacial. (Remember that regional warming and cooling can be greater than the global average change. Furthermore, a comparatively small global change in temperature might become manifest in greater or less seasonality.)

A return to a cooler climate took place 1500–1000 years ago (AD 500–1000), which coincided with the historical period in Europe known as the Dark Ages. This was followed by the medieval climatic anomaly (MCA) or medieval warm period (MWP; broadly around AD 900–1300). The MCA is also sometimes called the medieval climatic optimum but human ecologists and environmental scientists are wary of this term as it implies that this time was somehow optimal. During the MCA European temperatures reached some of the warmest levels for the last 4000 years, since the Holocene climatic maximum. Although not as warm as the Holocene maximum was for Europe, the MCA saw Europe roughly a degree warmer than in the last quarter of the 20th century. Estimates for continental Europe's climate are mainly 1.0–1.4°C, and for Greenland 2–4°C, warmer during the MCA compared to the end of the 20th century.

There is evidence that the MCA was not regional (confined to Europe) and was to some degree at least hemispheric and, if not in terms of temperature but precipitation, more global. Preserved tree stumps have been found in marshes and lakes in California's Sierra Nevada that have been radiocarbon dated and analysed by dendrochronology (see section 2.1.1) to determine the climate. This has revealed that the Sierra experienced severe drought for over two centuries before AD 1112 and for more than 140 years before AD 1350 (Steine, 1994). Diatom (common members of algal flora) analysis of sediment cores from Moon Lake, in the Upper Colorado River, infer some level of salinity. This suggests that during the MCA, further to the north west, the northern great plains that cover much of the US states of Montana, Wyoming and North and South Dakota saw increased drought frequency compared to more recent, hemispherically if not globally, cooler times. Today the area is one of considerable agricultural productivity, but which was famously undermined by droughts in the 1890s and 1930s (Laird et al., 1996). The implications for today are that should 21st-century emissions of greenhouse gases continue then that area could again suffer from severe and prolonged drought. (It is interesting to note that since the late 1980s part of western North America has experienced a marked decline in rainfall.)

There is, however, an absence of evidence for the MCA from some parts of the northern hemisphere (such as dendrochronological evidence from the Urals). Further, it may be that the warming was confined to certain seasons. The picture is unclear and readers interested in the MCA are advised to read the references and do their own literature searches. It *might* be that there was a general warming but accompanied with some circulation changes that counteracted the warming in some areas. This is not too unsurprising as climate change is never completely uniform all over the globe. Given this, it *is* somewhat surprising that the MCA northern hemisphere temperatures were in part reflected in the southern hemisphere. Using silver pine (*Lagarostrobos colensoi*) in 2002 Edward Cook, Jonathan Palmer and Rosanne D'Arrigo reported the first dendrochronological temperature analysis of past temperatures in New Zealand's summer (January–March) that extended back through the proposed MCA AD 1000–1300 interval. They found that its expression in terms of the New Zealand summer temperatures is composed of two periods of generally above-average warmth – AD 1137–1177 and 1210–1260 – that are punctuated by years of below-average temperatures and a middle period that is near average. Overall, this translates to a MWP that was probably 0.3–0.5°C warmer than the overall 20th-century average at Hokitika and, for the AD 1210–1260 period, comparable to the warming that has occurred since 1950. However, to date (2012) the emerging picture of the MCA is not yet as coherent as climate researchers would like. In 2009 US researchers, led by Michael Mann and one British team member, reporting in *Science*, used a global climate proxy network of tree-ring samples, ice cores, coral and sediments to reconstruct surface temperature patterns over this interval. The medieval period was found to display warmth that matches or exceeds that of the early first decade of the 21st century in some regions, but which falls well below recent levels globally. This period was marked by a tendency for conditions like the La Niña (part of the ENSO) in the tropical Pacific. The authors suggest that the North Atlantic Oscillation may be involved in the MCA as real-life proxies combined with computer modelling show a large temperature anomaly over the North Atlantic just below Greenland and Iceland[7].

One of the more significant of the 'minor' ($\pm0.5°C$) climate fluctuations in comparatively recent historic times was the Little Ice Age (or the LIA). This occurred between 1550 and 1850 (and principally 1400–1700 from global proxies) but from more a European perspective it was divided into two parts, the latter half of the 16th to close to the end of the 18th century (about 1550–1770) and then a return in the 19th century (about 1800–50). It appears that during the Little Ice Age a tongue of Arctic ice reached down as far as the Faroe Islands while glaciers advanced in Europe, Asia and North America. The Little Ice Age even affected ocean circulation, and the sea-surface temperatures off west Africa were markedly reduced, by 3–4°C (deMenocal et al., 2000). The Little Ice Age brought extremely cold winters to many parts of the Earth, but is most thoroughly documented in Europe and North America. In the mid-17th century, glaciers in the Swiss Alps advanced, gradually engulfing farms and crushing entire villages. The River Thames in England, and the canals of the Netherlands, froze over during the winter, and people skated and even held fairs on the ice. In the winter of 1780 New York Harbour froze, allowing people to walk from Manhattan to Staten Island. Sea ice surrounding Iceland extended for miles in every direction, so closing the nation's harbours.

That the Little Ice Age is not so well documented outside of Europe and North America suggests that it may not have been a completely global event, or if it was global it may have had a focus in one hemisphere. Although the temperature evidence is less clear, there are data (Verschuren et al., 2000) showing climatic resonance with the Little Ice Age derived from lake-sediment strata and fossil diatom and midge assemblages in equatorial East Africa. These suggest a drier climate there during the MCA (AD 1000–1300) and wetter conditions broadly corresponding with the Little Ice Age (AD 1270–1850). In South Africa the time of the Little Ice Age was more arid compared to the earlier MCA when it was wet, and in Kenya it was dry (Tyson et al., 2002). It had been thought that if the Little Ice Age were global but less pronounced outside Northern Europe we may not have picked up such a shallow climate change from natural proxies and it may simply be that it was the rise of meteorological instrumentation in Europe and North America, which naturally led to much of the meteorological recording taking place there and so better documentation of the Little Ice Age. Interestingly, ice cores reveal that atmospheric carbon dioxide did not show a marked decline during this time. (There was a dip of around 10 ppm in concentration from 284 ppm in the middle of the 16th century, which is minor compared to the subsequent increase from the 18th century onwards of around 80 ppm.) So, at the time of the Little Ice Age there was no significant decline in the forcing of climate from one of the major greenhouse gases. Yet Michael Mann and colleagues in 2009 did detect the Little Ice Age (in addition to the MCA) in their meta-analysis

---

[7] If I may make a personal speculation, I note that this area is a key part of the Broecker thermohaline circulation and so wonder whether changes in this circulation may be involved. Furthermore, given that ocean basin flushing time is the order of 10^2–10^3 years could it be that there is a long-term oscillation in the strength of the Broecker ocean conveyor?

of global climate proxies. So if the Little Ice Age was a real global climate phe-nomenon, and carbon dioxide change was not involved, what other factors might have caused it?

Of the non-greenhouse contenders that could have generated the Little Ice Age there are two more likely: volcanic eruptions (see Chapter 2 and Figure 2.1) and variations in solar output, hence insolation on the Earth. Indeed, there were a number of major volcanic eruptions during the Little Ice Age, including Billy Mitchell in the south-west Pacific (approximately 1580), Huaynaputina in Peru (1600), Parker in the Philippines (1641) and Long Island in New Guinea (approximately 1660). Then in the second part of the Little Ice Age there was Laki in Iceland (1783), Tambora in the Lesser Sunda Islands (1815) and Krakatau west of Java (1883). These can only have served to cool, or further cool, the planet away from the late Holocene average. Of these, Huaynaputina was one of the worst and the acid spike in the Greenland ice cap is higher for 1600 than for 1883 (Krakatau). After Huaynaputina the following few years saw very low temperatures. In Europe the summer of 1601 was cold with freezing weather in northern Italy extending into July and overcast skies for much of the year. In parts of England there were frosts every morning throughout June. Less violent and more recent was the eruption of Laki in Iceland in 1783. It is estimated that 80 Mt of sulphuric acid aerosol was released by that eruption. To put it into a modern context, this is four times more than El Chichon (Mexico, 1982) and 80 times more than Mount St. Helens (USA, 1980). In terms of a vulcanologists' perception of climate interactions, the Laki eruption illustrates that low-energy, large-volume (14 km^3 basaltic lava and over 100 million t of sulphur dioxide for Laki) and long-duration (8 months) basaltic eruptions can have climatic impacts greater than quick, large-volume, explosive, silica-rich eruptions. The sulphur contents of basaltic magmas are 10–100 times higher than silica-rich magmas. Immediately following the Laki eruption the winter in the eastern USA was about 4.8°C below the 225-year average, while the very broad estimate across the northern hemisphere was of a 1°C drop (to be treated with caution). What is clearer is that it took less than 2 years to recover from half the temperature drop and somewhere between 5 and 6 years for a complete recovery to occur. Another reason for viewing the effects of the Laki eruption with a little caution is that the same year saw the eruption of a smaller volcano, Asama, in Japan. Even so, whereas high-sulphur volcanic eruptions may cool the planet for a few years, they by themselves cannot explain all of the Little Ice Age's climate.

In 2012 Gifford Miller, Áslaug Geirsdóttir, Yafang Zhong and colleagues suggested that a series of possibly four volcanic eruptions over a period of 50 years could explain the sudden expansion of Canadian and Iceland ice in 1275. They had ^{14}C-dated the remains of vegetation that had been buried under an apparently sudden onset of ice in the Canadian Arctic and, using climate models, the explanation that seemed to make the most sense was that of sulphur-rich volcanic eruptions. Further, the temporary cooling effect of this series of eruptions was sustained due to a critical transition of ocean circulation causing a small (compared to glacial–interglacial transitions) climate threshold being crossed (see section 6.6.8). However, the locations and exact dates of these eruptions remain unknown and so have not been included in Figure 2.1.

With regards to solar output, during the period 1645–1715 solar activity – as inferred by sunspot activity – was low, with some years having no sunspots at all. This period is known as the Maunder Minimum after the astronomer E. W. Maunder who noticed, from records, that there were few sunspots during that period. For example, during one 30-year period within the Maunder Minimum astronomers observed only about 50 sunspots, as opposed to a more typical 40 000–50 000. What the precise link is between low sunspot activity and cooling temperatures has not been firmly established, but the coincidence of the Maunder Minimum with the deepest trough of the Little Ice Age is highly suggestive.

One might suspect that if solar output, hence Earth's insolation, during the Little Ice Age was less, this would have a uniformly global effect. This is not necessarily so, given (as noted in Chapter 1) that the northern hemisphere is predominantly land and the southern ocean and that these react differently to incoming energy. Second, in terms of biology, reduced insolation throughout the year may only affect part (and/or the length) of the thermal growing season. This may explain why Himalayan tree-ring series show little evidence of the Little Ice Age as these dendrochronological series are mostly affected by pre-monsoon temperatures and rainfall (Boragaonkar et al., 2002).

In addition, there may have been changes in atmospheric and/or ocean circulation that may have exacerbated any global change in some regions and ameliorated them in others. In 2006 David Lund, Jean Lynch-Steiglitz and William Curry reported on their use of Foraminifera in the Florida Straits. Their work suggests (there are some reasonable working assumptions) that during the Little Ice Age (from around 1200 to 1850) the Gulf Stream's normal warm current flow of 31 Sverdrups (Sv)[8] was reduced by 10%. This would have reduced heat flow to the Atlantic off western Europe and Greenland, so is a likely explanation for why the Little Ice Age was so pronounced there.

Of course, in terms of nomenclature, the Little Ice Age was neither an ice age nor a glacial. That such a palaeoclimatologically powerful term has come to describe what was a short and modest shift in climate is simply due to it being the major climatic event taking place within modern recorded history. In terms of psychology and the social science dimensions of climate change, the use of the term Little Ice Age demonstrates that hemispheric and global climate change is hugely relevant to human society. We will return to the biological and human ecological effects of Holocene climate change shortly, but for now it might be interesting to note that our cultural heritage of a Dickensian white Christmas owes itself to the Little Ice Age.

The recovery from the end of the Little Ice Age is well documented by over a century of direct observation in Europe and North America. Globally, snowline retreat since 1860 to the end of the 20th century suggests a global warming of 0.6–0.7°C. And between 1880 and 1930 records at Oxford, England, show that there was a 10% increase in the thermal growing season.

Notwithstanding shorter-term events such as the warm MCA and the Little Ice Age, similar to the long-term study of Mann et al. (2009) using climate proxies, Darrell

---

[8] 1 Sv $= 10^6$ m^3 s^{-1}. The Sverdrup is a unit of measure of volume of water transport used exclusively in oceanography. It is not an SI unit, and its symbol is the same as that for the sievert, Sv (a unit referring to radiological dose of ionising radiation).

Kaufman, David Schneider and Nicholas McKay led a North American and European team to bring together a synthesis of decadally resolved proxy temperature records, published in *Science* in 2009. Unlike the Mann team's meta-analysis of global climate proxies this study looked purely at northern hemisphere proxies from poleward of 60°N and covering the past 2000 years. Although not global, like Greenland and Antarctica ice-core derived temperature analyses, this analysis is of relevance for much of the northern hemisphere even if caution is needed interpreting it at the whole-hemisphere, or even global, scale. The conclusion they drew from their analysis was that the long-term trend (i.e. excluding things like the MCA and Little Ice Age) of northern hemisphere temperatures was one of gradual cooling through to the 20th century. Indeed, the northern hemisphere cooling seems to have ended at the end of the Little Ice Age. What then happens in the 20th century is warming that takes temperatures above anything this proxy analysis reveals in the previous nineteen centuries.

In terms of present-day questions of global warming, this means that this Little Ice Age recovery period is critical. It was during this time (after inception of the Industrial Revolution) that significant amounts of greenhouse gases (though small at first) were added to the atmosphere from human activity. Whereas before the 20th century the global climate was largely (but not solely) determined by non-human factors, it was during the 20th century that anthropogenic warming increasingly became manifest.

This point is an important one because it is possible to mistake (and indeed some vocal people have mistaken) anthropogenic global warming for some other factor. That the early 20th century saw a warming out of the Little Ice Age does not mean that there was just one factor at work. Some (such as the popular thriller author Michael Crichton) put this climate change down to 'natural' climatic vagaries, whereas others (such as some astronomers) put it down to changes in solar output. However, the IPCC, the considered scientific authority on climate change, is clear that many factors affect climate, including changes in the Sun's output, but that this does not prevent human-generated greenhouse gases becoming an ever increasingly powerful factor, from small beginnings with the Industrial Revolution to a dominant (if not the dominant) climate change factor by the end of the 20th century.

Before leaving our climate review of the pre-industrial Holocene, it is worth noting that there has been some discussion with regards to human influences on the global climate system over this time (see Anthropocene, section 4.4). Analysis of methane concentrations from Antarctic ice cores covering the past 2000 years reveal an inter-esting ^{13}C isotope pattern embedded within the record of change in overall methane concentrations, containing both ^{13}C and ^{12}C.

As noted previously, plant photosynthesis sequesters more ^{12}C from the atmosphere than ^{13}C but this preference not only differs between C_3 and C_4 plants. The proportions of ^{13}C in methane given off by microbial decomposition of organic material in wetlands and of ^{13}C in fossil fuels are also different (this last itself has only been a significant factor for the past 200 years and is reasonably well documented). Therefore, in theory it should be possible to begin to disentangle the way various methane sources contributed to atmospheric concentrations. Analysis of methane trapped in ice-core bubbles has showed that changes in $^{13}CH_4$ (represented as δ $^{13}CH_4$) were such that the $\delta^{13}CH_4$ was about 2‰ above expected values between the years AD 0 and 1000. Conversely, $\delta^{13}CH_4$ values were about 2‰ below expected values in the subsequent 700 years, up to AD 1700. The MCA also shows up as expected in the methane pattern. However, interpreting the various isotopic $\delta^{13}CH_4$ values is difficult and so, as yet (without other corroborating evidence), conclusions should be viewed with caution. Yet the suggestion is that both human activities (such as land clearance and use of wood as a fuel) and natural climate change have influenced the pre-industrial carbon cycle and that, up to the early 21st century, researchers have understated the human influence on the late pre-industrial Holocene methane budget (Ferretti et al., 2005).

Having said this – that we may perhaps have underestimated human influence on the global carbon cycle the past few centuries (or even a score of centuries) – this is a long way off from saying that humans were discernibly influencing the global carbon cycle 5000 years ago or earlier (see early in section 4.4). Looking at the agricultural methane argument, let's assume that 10% of humans on the planet at the time had a 1 ha paddy field. This is surely an exaggeration because agriculture was then only in its infancy and productivity (yield per unit area) was low, but rice and farm ruminants have emissions in the same order of magnitude and so this might be considered an average. A back-of-the-envelope calculation (even using present, high paddy field fluxes) reveals that we are talking about less than a 1% change in the pre-industrial natural methane flux (both of sources and of sinks) and which is equivalent to even less of current anthropogenic (let alone combined with the natural) methane flux. So, this is within the natural variability of the system. (Conversely, for loose comparison, modern fossil fuel carbon and land-use emissions are several per cent of the natural terrestrial flux.) Consider also that very early human rice cultivation would more likely have taken place in existing wetlands (and not artificially created ones) and so not significantly add to overall methane fluxes, and similarly early ruminant grazing simply replaced existing wild ruminant grazing. Would there be a significant additional early human anthropogenic methane flux? It is therefore unlikely that pre-Iron Age society had much of an impact on the global carbon cycle as has been proposed.

Of course, back-of-the-envelope calculations are not good enough for anything other than as an indication of a likelihood, and the debate as to whether humans were discernibly affecting the global cycle 8000–5000 years ago continued for a few years. The computer model analysis by Joy Singarayer, Paul Valdes and colleagues in 2011 was particularly welcome (as an addition to other previous counters to the early-Holocene start of the Anthropocene, interested readers may see the Quaternary review by Wanner et al., 2008). The simulations by Singarayer and colleagues captured the declining trend in real methane concentrations ascertained from ice cores that dated from the end of the last interglacial period (115 000–130 000 years ago) to recent times, which had been used to diagnose the Holocene methane rise as unique, and hence arguably signalling the start of the Anthropocene. Their findings suggested that no early agricultural sources are required to account for the increase in methane concentrations in the 5000 years before the industrial era and additionally that the methane rise was most likely natural, from wetlands, with a probable significant contribution from South American ecosystems.

Having looked at the climate forcing factors and change of the Holocene with just broad reference to biosystems in this section, we now look in detail at the biological and ecological response to these changes in climate.

4.6.4 Biological response to the last glacial, LGM and Holocene transition

The biological response to glacial and Holocene interglacial climate change is, in the main, manifested in the form of species migration (see also section 4.5.4). However, fixing one's self in the landscape, an observer would see any one location undergo considerable ecological change as species come and go. For example, the Beringian land bridge, the aforementioned area that was ecologically important before the Quaternary, re-appeared during glacials with their lower sea levels. Due to marine geological cores containing pollen, plant macrofossils and insect remains, it is possible to determine how the Beringian environment changed over a glacial–interglacial cycle. Close to the LGM Beringia's ecology was one of predominately heathland with some shrubs and the occasional birch tree (*Betula* spp.). During the Younger Dryas birch increased, as did grasses, and there was a patchwork of ponds filled with weeds. By the time of the early Holocene, Beringia's open ground – which was being reduced by sea-level rise – was that of mesic tundra, half way between polar tundra and a wetter temperate biome, where summer temperatures were warmer than present-day Alaskan northern slopes (Elias et al., 1996).

The transformation between glacial and interglacial conditions was arguably the greatest in North America as, at the LGM, its glacial ice sheets were at least 50% greater than those in Europe and Asia combined. As such, the ice sheet was as large as that of Antarctica, covering all of Canada and the north of the USA (see Figures 4.3 and 4.4). The Canadian ecologist Chris Pielou (1991) has reviewed many of the biological changes and some of these are summarised here along with other research observations.

A key problem facing species riding out climatic change is where can they live? The concept of 'refugia' has been introduced in previous sections and so, just as there were refugia for present-day temperate species during the last glacial, so there must have been refugia for peri-Arctic species when the ice sheets were at their greatest extent. It is likely that during the LGM exposed mountains peaks rising above the glacial ice sheet (nunataks) would have provided a possible refuge for Arctic stony species such as the woolly louse-wort (*Pedicularis lanata*) and Ross's sandwort (*Minuartia rossii*), although Beringia would have undoubtedly been an important refuge for such species (and this would explain their presence in both present-day Siberia as well as Canada and Greenland). Nunataks (that is, mountain tops appearing through an ice sheet) nearer the coast would have seen a comparatively milder climate (although still very cold) and so possibly be refugia for species such as the green spleenwort (*Asplenium viride*) and the mountain holly fern (*Polystichum lonchitis*), the present-day ranges of which cover much of the Rockies at or near the tree line, as well as parts of Greenland.

Long-distance horizontal, as opposed to short-distance vertical, migration was an important mechanism for many species to survive the glacial–interglacial climatic changes. However, the migrations southwards (at the beginning of glacials) were different from those northwards (at a glacial's end) as the environment around expanding ice sheets is different to that around contracting ones. At a time of a cooling climate, the cooling takes place before ice sheets expand. This would lead to freezing ground (permafrost) before the ice sheet, or an arctic tundra-like environment, and if there was such an environment already there then this zone would have widened. Either way, plants cannot live in this zone and any that had previously lived there would die. As noted earlier in this chapter, plants themselves do not migrate, as such; rather, the descendents from the seeds of those furthest away from the ice distribute even further still, and so it is the respective plants' ranges that shift. The reverse process tends to take place with contracting ice sheets, except that contracting sheets do not have soil underneath them and so colonisation is delayed until soil forms. Conversely, contracting sheets expose clear ground devoid of trees and this void can facilitate seed dispersal.

Horizontal species-range expansion in response to major glacial–interglacial climate change often results in species migrating over considerable distances. As has been mentioned, in Europe latitudinal migration was restricted by the Mediterranean, Pyrenees and Swiss Alps. However, in both North and South America it was possible for many species to migrate 1700–2400 km, as did the lodgepole pine (*Pinus contorta*) from Washington State to the Yukon, or the maple (*Acer* spp.) and chestnut (*Castanea* spp.) from Louisiana to Ohio, Pennsylvania and New York State. Migration began at the end of the Younger Dryas, although there is evidence of some earlier migration during the Allerød/Bølling interstadial, especially along north–south mountains ranges that also allowed for vertical migration. (The interstadial prompted both vertical and horizontal migration. Vertical migration, where possible, especially enabled species to outride the cool Younger Dryas.) However, some species reacted more quickly than others. For example, the eastern white pine (*Pinus strobes*) began its migration about 1000 years before the eastern hemlock (*Tsuga canadensis*).

While this migration broadly began with the onset of climate change, species ranges do not all shift at the same rate. In North America following the last glacial, the range of pine trees (*Pinus* spp.) shifted at an average speed of just fewer than 200 m/year. Travelling at such speeds it can take millennia to cover the 3000 km that many species did. Indeed, in North America the lodgepole pine reached its northern limit probably only less than a century ago, and of course with current anthropogenic warming its range can move still further north. Conversely, the white spruce (*Picea glauca*) migrated some 50% faster, at an average of around 300 m/year.

One of the key factors determining the speed with which a species' range can change is the nature of the plant's seeds. Maple (*Acer* spp.), with its winged seeds, changes its range faster on average than the chestnut (*Castanea* spp.), with its heavy seeds: 200 and 100 m a year, respectively. It should be noted that not all winged seeds enable their parent species to move faster than their non-winged competitors. Seeds of the oak (*Quercus* spp.) were sufficiently large to be attractive to animals but small enough for them to be carried more easily and faster than a number of species with winged seeds. In North America, oak ranges moved by an average of 350 m/year. Indeed, climate change exerts an evolutionary pressure favouring migration. It has been noted that the lodgepole pine's seeds at the northern end of their range had smaller wing loadings (lower weight/wing area ratio) than those at the southern end (Cwynar and Macdonald, 1987).

From intra-species genetic variation it can be seen that the LGM–Holocene transition also facilitated the speciation of animals. Indeed, many species would have to change and sometimes in ways fundamental to their life cycle. For example, in a warmer world species' migration patterns would change. Three North American bird species – the common flicker (*Colaptes auratus*), the yellow-rumped warbler (*Dendroica coronata*) and the dark-eyed junco (*Junco hyemalis*) – each have their own eastern and western North American race or subspecies. These races have their present-day geographical distribution either side of a line that runs from Canada's north-western British Columbia to New Orleans, Louisiana, in the USA. That each of these species has two subspecies with the same geographical range is not accidental but reflects that during the glacial the two populations migrated south to two different parts of the USA separated by a large dry desert, so enabling allopatric speciation, before migrating back at the glacial's end.

Other pairs of bird species are more distinct (but still owe their evolution to the last glacial–interglacial cycle). These include the northern shrike (*Lanius excubitor*) and loggerhead shrike (*Lanius ludovicianus*); the Bohemian wax-wing (*Bombycilla garrulous*) and the cedar wax-wing (*Bombycilla cedrorum*); and the northern three-toed woodpecker (*Picoides tridactylus*) and the black-backed three-toed woodpecker (*Picoides arcticus*). It is thought that for each of these pairs there was a common ancestor but that during the LGM one population rode out the worst of the glacial in Eurasia and the other in America. Then after the glacial the two species moved north to overlap their distribution in North America.

One adaptation that plants can make is to adjust to the different day lengths throughout the year at different latitudes. Photoperiodism (the phenomenon by which plants flower at a specific time of the year and day) needs to change when species migrate latitudinally. Many high-latitude plants are long-day plants: they cannot flower until spring has become summer. Conversely, many temperate-zone plants are short-day plants. Change in photoperiodism manifests itself in species at the level of different plant races. So, the glacial–interglacial climate change was the impetus for the creation of numerous new races of plants. But it was not all new races and speciation. The end-glacial climate change brought extinctions too.

We briefly looked at the Earth's major extinction events in the previous chapter and saw that many of these were climate-related: even if the immediate cause was not climatic (such as an asteroidal or bolide impact), climate change often followed. So, not surprisingly the Quaternary series of glacials and interglacials have been associated with extinctions. In the main these extinctions happened at the end of the glacials. One of the biggest of these took place at the end of the last glacial.

True, North America had the largest non-polar ice sheet during the last glacial, but the extinction at the end of the last glacial was not confined to that continent: there were extinction events in South America, Africa, Asia, Australia and Europe. Not all species extinctions take place at the glacial's end, but there are more species threatened during this time if only because that is the part of the glacial–interglacial cycle that sees the greatest rapid climatic change. However, there are other times during glacials where sudden, albeit smaller, climatic change takes place and extinctions (especially locally) can result.

One key question is why was the end of the last glacial associated with a particularly major extinction event compared to those in other glacial cycles? One answer, mentioned previously, concerns humans. As noted (see Chapter 3 and also section 4.3) the late Cenozoic spurred hominid evolution. Indeed, the repeated series of glacials and interglacials would favour hominids that were generalists and could adapt to different environments, and especially those that could artificially modify their own personal environment, such as through using fire. Although there is evidence of an earlier minor human presence in North America before 13 000 years ago (which we will discuss shortly) – there was the hunting of megafauna in the south west of the present USA (perhaps as early as 18 000 years ago) – the first *major* incursion of humans towards the glacial's end 13 000 years ago had crossed the Beringia land bridge from Asia into North America. To establish themselves in that continent they probably kept close to the coast and away from the Laurentide ice sheet that still dominated inland North America. Prominent among these were people belonging to what is termed the Clovis culture. The Clovis were big-game hunters, which is typical of nomadic people as opposed to agrarian cultures that necessitate fixed abodes.

Note that there are also clear signs of a pre-Clovis human presence. Among the pre-Clovis signs are those at Monte Verde in Chile that date from around 12 500 years ago. Further, human footprints have been found fossilised in volcanic ash in the Valsequillo Basin, Mexico, which have been dated to before 40 000 years ago. So, while the Clovis migration was an important event, they were clearly not the first on that continent.

The timing of the Clovis migration itself into North America is confined by the melting of the Laurentide ice cap and around the time of the Younger Dryas (section 4.6.2). The only likely terrestrial route they could have taken would have been a narrow gap between the ice sheet and the glaciers of the coastal Pacific mountains in the Yukon. From carbon dating we know that this passage was not available much before 13 000 years ago as the Laurentide ice sheet would have been too big. Conversely, the problem with considering the pre-Clovis cultures' route on to the continent is how did they overcome the glacial obstacles? One answer may be because the sea was lower and they could have skirted along the coast. The coastal termination of glaciers would have made this difficult but, especially with primitive boats, not impossible. Unfortunately, evidence for such a passage is now under water (Marshall, 2001).

Earlier, in the Old World, humans had moved through Africa and into Eurasia. There are signs of tool making going back as far as 125 000 years and earlier, back to the interglacial before the last glacial. In short, humans had the means and opportunity to provide added pressure on species already stressed by climate change.

Yet this situation is at first perplexing. If humans were in Africa well before the American continents, why has the New World seen a greater extinction of large animals compared to Africa? (see Table 4.1). The answer is likely to be co-evolution.

Table 4.1 The numbers of mammalian megafaunal genera (groups of closely related species) in the Pleistocene extinction (from Barnosky et al., 2004)

Continent	No. of genera extinct	Percentage extinct on each continent
Africa	5	18
Australia[a]	14	88
Eurasia[b]	5	36
North America	28	72
South America	49	83

[a] Australia also has seven extinct (and no surviving) genera of megafaunal bird.
[b] This includes only northern Asia due to insufficient data on southern Asia.

African large animals evolved alongside hominids. Those animals that learned to fear and avoid humans survived, while those that did not perished. Conversely in the New World, where human co-evolution had not taken place, large species were less able to cope with the arrival of humans.

The Quaternary (from 2.6 mya) and its series of glacial–interglacial cycles was the period of time of considerable evolution of the human line. The earliest direct evidence of hominid stone-age technology dates from 2.6 mya in the Ethiopian Rift Valley. Conversely, the bones and tools of *Homo erectus* are found both in Africa and Eurasia from about 1.5 mya (1.9 mya, *H. erectus* was the first hominin to leave Africa).

Human Evolution and Climate

As this chapter's discussion of the Oligocene to the Quaternary is beginning to include more and more mentions of humans, it is worth briefly pulling together the key salient facts as we now know them regarding human evolution and climate.

The taxonomic superfamily Hominoidea, of the very first apes, arose around 75 mya when the world was far warmer than today. These were much smaller than and unlike humans. There is evidence of Hominoidea outside of Africa a little over 17 mya (in Germany). They subsequently (over many millions of years) diversified into chimpanzees, bonobos, gorillas, orang-utans, lemurs and of course humans. The family of Hominidae (the great apes) arose sometime around 15 mya at a time when there had already been ice for several million years in East Antarctica but the Earth 15 mya was a period of further cooling with the appearance of ice in West Antarctica

The subtribe Hominina (bipedal apes) first evolved some time around 5 mya. By then ice was beginning to appear in Greenland and elsewhere in the northern hemisphere. Remains of *Ardipithecus kadabba* dating from 5.6 mya and *Ardipithecus ramidus* some 4.4 mya were found in Africa (section 4.3). *Australopithecus afarensis* arose around then (possibly overlapping with *Ardipithecus*) but there is evidence of *A. afarensis* from 3.9 mya. (See Wood and Harrison, 2011, for details of the relevant biological evolutionary timeline.) *Australopithecus* was but one of a number of *Homo* species precursors. At that time the Quaternary glacials and interglacials had not begun but the Earth had long been cooling since the Eocene maximum over 50 mya. Ice was present the year round in both hemispheres and seasonality was more pronounced. Throughout this time Milankovitch orbital forcing of the climate had taken place but with the onset of the

Quaternary (2 mya) the effects of this forcing were going to get more pronounced and there started to be the first glacials. By then a number of *Homo* species had evolved in Africa, including *Homo habilis* and *Homo erectus*, with a cranial size roughly three-quarters (or less) that of modern humans. One of the closest *Homo* species to ourselves (*H. sapiens sapiens*; or *H. sapiens*) was *Homo sapiens neanderthalensis* (or *Homo neanderthalensis* for short). It evolved from earlier *Homo* species but outside of Africa some 350 000 years ago. The diminutive *Homo floresiensis* was somewhat of an evolutionary offshoot oddity that probably died out 18 000 years ago, well before the onset of the current interglacial. Both archaeological and genetic analysis suggests that modern humans, *H. sapiens*, evolved in (possibly sub-Saharan) Africa with some remains found in Ethiopia dating from 195 000 years ago. Modern humans subsequently colonised the world. (For details of human genomic analysis see Li et al., 2008, and for a more complete list of early dates and sites of hominin remains see Carrión et al., 2011.) For a while *H. sapiens* shared southern Europe with *H. neanderthalensis* before the latter went extinct around the time of the LGM. The current interglacial (the brief return to cooler times in the Younger Dryas aside) began properly around 10 000 years ago with *H. sapiens* then being the only human species on the planet.

Dates regarding *H. sapiens'* African evolution and diaspora need to be treated with caution due to likely problems, hence error, in dating such field samples. It also should be noted that much evidence is not clear. Stone tool remains, unless accompanied by fossil early human remains, cannot necessarily be attributed to a specific hominin with certainty. Indeed, the phrase anatomically modern human (or AMH) lacks the precision needed to elucidate a clear picture: anatomically modern humans can sometimes relate to more than one hominin species. For example, there is some indication that an anatomically modern human (possibly modern human, possibly not) population migrated out of Africa before the Toba super-eruption 74 000 years ago (Armitage et al., 2011). The study of early human evolution and migration has seen a number of recent groundbreaking discoveries, which means that our understanding of what happened has changed and probably will continue to do so.

One evolutionary offshoot (quite probably from the *H. erectus* line) is that of *Homo floresiensis* (named after the island of Flores, east of Java, where the remains of the first specimens were found in September 2003). Analysis (^{14}C, luminescence, uranium series and electron-spin resonance) suggests that *H. floresiensis* existed from before 38 000 years ago (probably well before and perhaps as much as 800 000 years ago) to as recently as 18 000 years ago (Morwood et al., 2004), although there has been some speculation of continued survival to more recent times (now thought unlikely). Either way, they therefore evolved either in the last glacial or over several glacials, but whichever it is, it is likely that a glacial maximum, with its lower sea levels and extensive land bridges, facilitated the island of Flores' initial habitation, even if some sea crossing was required.

H. floresiensis were short hominids of about 1 m in height with a chimpanzee-sized brain (approximately 380 cm^3), but which did use crude stone tools and fire. Whereas climate change may very well have facilitated their getting to Flores, it was the limited size of Flores itself, together with its semi-isolation, which facilitated the evolution of *H. floresiensis'* short stature.

So, from Africa early human (*Homo*) species then migrated to other lands. Note: modern humans (*H. sapiens*) evolved 195 000 years ago and then later left Africa, possibly sometime between 85 000 and 55 000 years ago (most likely earlier in this window than later) and probably entered Europe around or before 45 000 years ago.

In Europe Neanderthals (*H.*) evolved from 300 000 years ago (or about three glacials ago), although DNA analysis suggests their genetic divergence from a common ancestral stock some 600 000 years ago. Archaic Neanderthal remains have been found in Spain dating from somewhere between 530 000 and 250 000 years ago and there is some suggestion that this could have been a core Neanderthal population for western Europe; western European Neanderthals could certainly have outridden glacials there. Although competition with modern humans undoubtedly pressurised the Neanderthals, there is radiocarbon dating evidence from southern Spain that climate change contributed to their final demise around 24 000 years ago, as that was the time of a Heinrich event (see section 4.5.3) and its associated northern hemisphere climate fluctuation (Tzedakis et al., 2007).

Neanderthals had some DNA sequences not found in *H. sapiens*. Having said that, there is genetic evidence that *H. neanderthalensis* did breed with *H. sapiens* in the Middle East and Europe sometime around 90 000–65 000 years ago (the more recent date fits other evidence), as well as in the Far East around 50 000–40 000 years ago (see a review by Callaway, 2011, along with other references in this section).

There was a *H. sapiens* population bottleneck with a reduced population of about 10 000 individuals in Africa during the Pleistocene. There is some debate as to what caused this bottleneck and speculation includes global environmental change due to a super volcano, possibly the eruption of Toba 74 000 years ago, which is thought to have lowered temperatures by 3–5°C for up to 5 years (see section 6.6.5). It may be that it was due to the combination of a glacial and these factors, or even another factor entirely. In 2011, genetic comparison of 21st-century humans (a Chinese, a Korean, three Europeans and two Yorbas) was conducted by Heng Li and Richard Durbin of the Wellcome Trust Sanger Institute in Cambridge, UK. Put briefly and simply, the larger a population is the more genetic diversity it can have. Due to mutation and other processes, genetic sequences change over time, from generation to generation. Put these two together and (admittedly making assumptions about the rate of change) it is possible to get a very rough and ready history of a population's size over time. This can be even more rough and ready if you are working with a small number of people (genomes), which Heng Li and Richard Durbin were. Yet they were able to loosely detect a bottleneck in the human population somewhere around 60 000 years ago. Furthermore, they estimated the populations of European and Asian humans to have been of a similar size before 20 000–10 000 years ago, with a far greater constricted population than that of the African humans. The African population bottleneck seems to have been smaller and to have recovered quicker. Could this be the genetic echo of the Toba eruption 74 000 years ago? Toba is in Indonesia and would have had a bigger impact on any human population in Asia than Africa. In addition, the higher-latitude European population would have been more susceptible to a sudden cooling of the climate for a few years following such an eruption than those in warmer Africa.

Anyway, subsequently this African population expanded some 100 000–50 000 years ago, during which some left Africa and then, somewhere around 95 000–65 000 years ago (in the last glacial) there was an expansion in Europe, where *H. sapiens* met

H. neanderthalensis. However, *H. sapiens* began replacing *H. neanderthalensis* more recently, from about 41 000 years ago (which coincides with a colder Europe due to a short-term Heinrich event that made the North Atlantic cooler than it was at that time in the already cold glacial). Neanderthals subsequently markedly declined about 30 000 years ago. At the height of the LGM, 24 000 years ago, humans in Europe would have been forced south. mtDNA analyses of modern humans today show six principal lineages from this Upper Palaeolithic age. (It is important to stress, though, that these dates are not exact. We will return to this shortly.)

Consequently much of the Pleistocene (2.6 mya–11 700 years ago) and its series of deepening glacial–interglacial cycles saw considerable human evolution and speciation as well as migration. *H. neanderthalensis* and *H. sapiens* were the most similar to each other, whereas *H. floresiensis* was the most different. The shorter *H. floresiensis* appears to have evolved from a more encephalised (larger-brained) ancestor so its discovery markedly added to the biological spectrum of humanity. At one time, about 40 000 years ago, all three species of hominin must have flourished simultaneously, albeit separately. *H. floresiensis'* small size is possibly due to their island location, which also was the habitat for dwarf elephants (a *Stegodon*) and the large Komodo dragon (*Varanus komodensis*). (The island of Flores was home to both the smallest elephant species and human but a large predator. Small prey tends to escape notice in a confined environment compared to large prey but large predators outcompete rivals where resources are clearly finite.)

Genetic analysis has yet to reveal the full picture of modern human expansion across the globe. This is partly because the genetic techniques only began to become practically available to researchers around the turn of the millennium. It is also partly because some nations, especially in the Middle East, are resistant to Western researchers conducting genetic analysis of their populations. As said, modern humans arose before 150 000 years ago in sub-Saharan or East Africa during the glacial before last. We can now see that East African human genetic diversity is to this day particularly high and early colonization within Africa can be seen by branches in mtDNA – which is transferred down the female line – and Y-chromosome branches, carried down the male line. These branches are known as haplogroups and can be found as Kalihari Bushmen (Khoisan) and certain pygmy populations. Haplogroups also contribute to our understanding of the aforementioned modern-human expansion into Europe. Haplogroups therefore help illuminate understanding of modern human exodus from Africa and migration to Australia. Peter Forster and Shuichi Matsumura of Cambridge University usefully summarised this research in *Science* (2005) by way of introducing two quite independent papers in the same issue by Macaulay et al. and Thangaraj et al.

Up to 2005 there was some debate as to how modern humans (or indeed their precursors) originally left Africa for Asia and the initial colonization of India and the Malay peninsula and beyond. Perhaps the most obvious route would be up the Nile and across into Sinai (the route likely used for the aforementioned European expansion) or, alternatively, some 2400 km south, across the narrow Bab-el-Mandeb Strait at the southern end of the Red Sea and into what is present-day south Yemen.

Both research teams looked at small, long-established, indigenous populations (that are today dwindling fast and becoming diluted genetically). They both found evidence that corroborates a common picture. A population containing some 600 women (possibly slightly fewer and maybe as many as 2000 or more) left East Africa somewhere very roughly around 60 000 years ago or perhaps a little earlier. This was in the middle of the last glacial but a good 30 000 years before the glacial maximum. However, it was a cooler part of the glacial and although sea levels were not as low as during the later glacial maximum it is possible (if not pragmatically likely) that conditions were then favourable for migration across the Middle East. This early colonising population appears to have taken a coastal 'express' route arriving in southern Australia by 55 000–46 000 years ago (see below). This suggests a migration speed of somewhere between 0.7 and 4 km/year and is of the same order of magnitude as other genetically dated inland journeys of migrant species populations during the last glacial.

It should be stressed, as discussed in Chapter 2, that (as with all dating techniques) ^{14}C dating is prone to error. In the case of ^{14}C dating some error comes from the rate of ^{14}C production in the atmosphere not being constant (it varies with solar activity) and variations in the Earth's magnetic field. Other error stems from sample contamination by more recent carbon (say through carbon carried by percolating water), which affects potential sample ^{12}C:^{14}C ratios. This particularly affects the dating of older samples (with the least ^{14}C). For example, if a 40 000-year-old sample was contaminated by just 1% with modern carbon then the dating error would be around 7000 years. With regards to the dating of human remains there have been two recent innovations. In 2004 a technique was developed for extracting collagen (a family of fibrous proteins found throughout vertebrates) from *within* bone, so helping to eliminate contamination. Second, since 1988 there has been a continually improved understanding of variations in ^{14}C production over the past 50 000 years. Although this does not substantially affect the picture portrayed here, it has been argued that modern humans migrated through Europe faster than had been thought, and that their period of co-existence with Neanderthals was shorter (Mellars, 2006).

Early humans, and allied hominids, were hunter–gatherers. The remains of plant grinders or pounders (such as grinding slabs, mortars and pestles) make their first appearance in the Upper Palaeolithic from 45 000 years ago and suggest simple food processing. The beginning of baking was an important step in human nutrition, facilitating carbohydrate entering the bloodstream and being metabolised to glucose. While such evidence is compelling, there are only a few examples where food remains have survived to permit identification in association with such preparatory artefacts. (After all, pestles and mortars could be used to make dyes, not food.) The earliest direct evidence of simply processed food remains comes from Galilee about 23 000 years ago with remains of wild barley (*Hordeum spontaneum*) and emmer wheat (*Triticum dicoccoides*). This was close to the LGM and the plant species were wild. Domestication of wheat and barely subsequently took place around 10 000–9000 years ago, during the beginning of our Holocene interglacial. It therefore appears that a series of key developments in human food security took place under different climatic conditions and across glacial–interglacial change. (We will return to food security when addressing current and future food-security concerns in Chapter 7.)

The Holocene also saw human migration across the Pacific Ocean. The nature of this migration has been hotly debated. However, as new tools for genetic analysis become available it has been possible to obtain a clearer picture. mtDNA analysis indicates that the ancestors of the Polynesians originated from somewhere in eastern Indonesia and then migrated east and south, through and around Australasia approximately 11 000 years ago, before moving on to Polynesia, reaching Fiji 3000 years ago and then on to Samoa and Tonga before 2500 years ago. There is a genetic cohesiveness among the Polynesians, from Hawaii in the north to New Zealand in the south and including Melanesia, the peoples of which Captain Cook thought were different (Gibbons, 2001).

This wave of migration was subsequent to the earlier colonising 'express' route down to Australia by 55 000 and 46 000 years. The Polynesians clearly had to be accustomed to substantial sea voyages even if they did as much island-hopping as possible. However, the earlier coastal express route would have benefited from the lower sea levels, and the resulting land bridges, in the latter third of the last glacial, which would have greatly reduced the need for extensive sea voyages.

The arrival of humans in Australia marked the onset of a megafaunal extinction event. There has been much debate as to whether this was due to climate change or the presence of humans (although the coincidental timing makes the latter difficult to ignore). It is true that there were climate fluctuations during the last glacial. However, there were similar fluctuations earlier in the glacial and indeed previous glacial and interglacial climatic cycles. It is therefore unlikely that climate alone was the cause of this extinction.

Some 60 Australian taxa are known to have become extinct, including all large browsers, whereas large grazing forms, such as the red (*Macropus rufus*) and great grey (*Macropus giganteus*) kangaroos, survived. The selective loss of browser-dependent species is suggestive of environmental change but, because of the afore-mentioned argument, climate change alone cannot account for this. So what may have happened?

As discussed in the previous chapter when looking at the late-Miocene expansion of C_4 grasses, herbivorous animals' food can be determined through ^{13}C analysis. C_3 plants (including some grasses) have ^{13}C values of $-26‰$ while C_4 plants (which are mostly grasses) have ^{13}C values of $-12‰$; consequently, things like fossil teeth or eggshells provide a record of the proportion of C_4 and C_3 plants in past herbivore diets. Research by Gifford Miller and colleagues (2005) looked at the proportion of ^{13}C in the eggshells of the now-extinct flightless bird *Genyornis newtoni* and the Australian emu *Dromaius novaehollandiae* to provide a record covering the past 140 000 years in South Australia's semi-arid zone. They found that a marked shift in the proportion of ^{13}C took place 50 000–45 000 years ago. They also tested the ^{13}C record against other glacial–interglacial times (in 15 000-year groupings) to see if climate change was affecting the ^{13}C record. It was not. The suggestion therefore is that the human newcomers were changing the environment through fire. The drought-adapted mosaic of trees, shrubs and grasslands became more like the modern fire-adapted desert scrub. Animals that could adapt survived; those that could not became extinct (Miller et al., 2005). A similar story emerges from the work of Gavin Prideaux, John Long and colleagues (2007).

Whereas this affirms the role of human activity in causing environmental change and so being an extinction driver, it does not, as it may at first seem, discount the impact synergistically of climate change together with that of humans on the extinction event. The above testing of climate change against the ^{13}C record only demonstrates that previous climate did not affect the ecology sufficiently to affect the animals' diets. It could be that human-induced (anthropogenic) ecological change did a certain amount of environmental modification and that this could have been furthered by climate change at the time. An alternative variation could be that anthropogenic ecological change put a certain level of pressure on the species and that climate change raised this to extinction level. Both these scenarios are relevant to aspects of 21st-century anthropogenic climate change in that human activity has modified a significant part of the modern global terrestrial landscape and that this, coupled with 21st-century climate change, is synergistically detrimental to many species as well as local ecosystems and regional biomes.

Meanwhile, the human migration down North America to South America continued. One of the earliest major South American agricultural civilizations that depended on a range of domesticated plants arose in Peru 5000–3800 years ago. The plants used included cotton (*Gossypium barbadense*), squash (*Cucurbita* spp.), chilli (*Capsium* spp.), beans (*Phaseolus vulgaris* and *Phaseolus lunatus*), lucuma (*Pouteria lucuma*; formerly known as *Lucuma obovata*), guava (*Psidium guajava*), pacay (*Inga feuillei*), camote (*Ipomoea batatas*), avocado (*Persea americana*), maize (*Zea mays*) and achira (*Canna edulis*). This civilisation provided a foundation for many of the features of subsequent Andean civilisations through to early historic times (Haas et al., 2004; Perry et al., 2006). Other evidence suggests earlier American use of some species.

Returning to Eurasia, once the climate had settled in the current Holocene interglacial, the Neolithic agricultural revolution began to spread from the Near East around 9000–5000 years ago (Hewitt, 2000). Agricultural practice grew and became an increasing part of the landscape and helped the human population to expand. Domestication of agricultural species did not always happen as a single event for each species. For example, genetic evidence from Peter L. Morrell and Michael T. Clegg from California University in 2007 suggests that barley (*Hordeum vulgare*) was domesticated more than once, once within the Fertile Crescent and a second time 1500–3000 km further east. At Jeitun in southern Turkmenistan domesticated forms of barley and einkorn wheat (*Triticum monococcum*) were being cultivated, and domesticated goats and sheep were being herded, by 6000 BC, indicating a well-developed agrarian society. Another example, from a team led by Jason P. Londo and Yu-Chung Chiang in 2007, is that of rice (*Oryza* spp.). Here, India and Indochina may represent the ancestral centres of diversity for *Oryza rufipogon*. Additionally, the data suggest that cultivated rice was domesticated at least twice from different *O. rufipogon* populations and that the products of these two independent domestication events are the two major rice varieties, *Oryza sativa indica* and *Oryza sativa japonica*. Based on this geographical analysis, *O. sativa indica* was domesticated within a region south of the Himalaya mountains range, likely eastern India, Myanmar and Thailand, whereas *O. sativa japonica* was domesticated from wild rice in southern China. Turning to archaeological evidence, the first use of rice took place in eastern China in lowland swamps. Pollen, algal, fungal spore and micro-charcoal data from sediments suggest that Neolithic communities used fire to clear alder (*Alnus* spp.)-dominated wetland scrub to prepare sites for rice cultivation 7700 years ago, although other evidence indicates earlier use of rice in the region before 9000 years ago (Zong et al., 2007).

The Neolithic agricultural revolution was not confined to plant species. The earliest evidence (dated fatty deposits in pottery remains) of milk in the Near East and south-eastern Europe was provided by an international team of researchers; led by Richard Evershed and Sebastian Payne in 2008. They showed that milk was in use by 6500 years ago (some 2000 years earlier than was previously thought).

The Neolithic agricultural revolution was for humans a successful event conferring stability to human populations and allowing them to increase in size. But a subsequent population-increasing event was to come.

Since the Industrial Revolution in the 18th-century human population has shown a further marked increase. The global population has been estimated to have been around 425 million in the year 1500 and 600 million in 1700. But from then it grew to 4430 million by 1980, an increase of nearly 640% (see Chapter 7). Commensurately, the 1700–1980 increases in cropland were over 466%, or 1.2 billion ha, whereas the larger area of forest and woodland worldwide declined by 18.7%. As we shall see, this will impede natural ecosystem adaptation to climate change. During this time agricultural efficiency increased, which is why the growth in agricultural lands was somewhat less than the growth in population. (As we shall see in Chapter 7, increasing direct and indirect energy inputs largely drove the improved agricultural productivity.) But it was the Industrial Revolution that was to spark the major human impacts on climate change that concern us today. The Industrial Revolution began in the latter half of the Little Ice Age and both these events begin our examination of the present climate and biological change in the next chapter.

4.7 Summary

The past glacial and the current interglacial have seen much biological change, which again clearly demonstrates that biology and climate are interrelated. Of relevance to readers, the past glacial–interglacial cycle also influenced the evolution that saw the spread and (with the comparatively stable Holocene interglacial) the expansion in numbers of modern humans. That humans have had a profound effect on ecology in various parts of the planet is most germane to present climate change, which has an increasingly anthropogenic component to it. That the climate has been warmer – marginally at times during this and previous interglacials and markedly before the Quaternary – provides palaeo-analogues from which lessons can be learned as the

Earth warms again. To do this we need to see how the climate is changing today and this is the focus of the next chapter.

Past glacials and interglacials have also had a genetic impact on species as they have expanded from, and retreated to, their respective refugia. In a world without humans this broad pattern of genetic forcing and glacial–interglacial ecological change would have continued with future glacials and interglacials. However, our planet does have humans (and hence has a changed and more ecologically fragmented landscape) and we do have global warming. Once this warming takes us beyond past interglacial analogues (likely in the 21st century), then species' respective genetic mix (hence characteristics) may not always enable them to flourish as they have during the past 2 million years of the Quaternary or even earlier, in the Pliocene. Further, given that component species may not flourish, there is the possibility in some circumstances that global temperatures greater than past interglacials, let alone human land-use impacts, could result in ecosystem dysfunction.

Finally, the growth of human civilisations during the current (Holocene) interglacial has largely been dependent on domesticating species for food. This domestication has taken place numerous times over the past several thousand years during which the global climate has been comparatively stable compared to the glacial–interglacial range. As we shall see, 21st-century warming is likely to take us beyond this range and so is likely to impact on global food security. The future of food security and climate change will be discussed in Chapter 7.

4.8 References

Abbott, R. J., Smith, L. C., Milne, R. J., Crawford, R. M. M., Wolff, K. and Balfour, J. (2000) Molecular analysis of plant migration and refugia in the Arctic. *Science*, 289, 1343–6.

Allen, J. R. M., Bradnt, U., Brauer, A. et al. (1999) Rapid environmental changes in southern Europe during the last glacial period. *Nature*, 400, 740–3.

Armitage, S. J., Jasim, S. A., Marks, A. E. et al. (2011) The southern route "Out of Africa": Evidence for an early expansion of modern humans into Arabia. *Science*, 331, 453–6.

Barber, D. C., Dyke, A., Hillaire-Marcel, C. et al. (1999) Forcing of the cold event of 8,200 years ago by catastrophic drainage of Laurentide lake. *Nature*, 400, 344–8.

Barker, S., Diz, P., Vautravers, M. J., Pike, J. et al. (2009) Interhemispheric seesaw response during the last deglaciation. *Nature*, 457, 1097–1102.

Barnosky, A. D., Koch, P. L., Fernac, R. S., Wing S. L. and Shabel, A. B. (2004) Assessing the causes of late Pleistocene extinctions on continents. *Science*, 306, 70–4.

Bennett, K. D., Haberle, S. G. and Lumley S. H. (2000) The last glacial-Holocene transition in southern Chile. *Science*, 290, 325–8.

Berger, A. and Loutre, M. F. (2002) An exceptionally long interglacial ahead. *Science*, 297, 1287–8.

Bernatchez, L. and Wilson, C. C. (1998) Comparative phylogeography of nearctic and palearctic fishs. *Molecular Ecology*, 7, 431–52.

Bond, G., Broecker, W., Johnsen, S. et al. (1993) Correlations between climate records from North Atlantic sediments and Greenland ice. *Nature*, 365, 143–7.

Boragaonkar, H. P., Kolli, R. K. and Pant, G. B. (2002) Tree-ring variations over the Western Himalaya: little evidence of the Little Ice Age. *Pages News*, 10(1), 5–6.

Broecker, W. S., Bond, G. C., Clark, K. G. and McManus, J. F. (1992) Origin of the northern Atlantic Heinrich events. *Climate Dynamics*, 6, 265–73.

Callaway, E. (2011) Ancient DNA reveals secrets of human history. *Nature*, 476, 136–7.

Carrión, J. S., Rose J. and Stringer, C. (2011) Early human evolution in the Western Palaearctic: ecological scenarios. *Quaternary Science Reviews* 30, 1281–95.

Cerling, T. E., Wang, Y. and Quade, J. (1993) Expansion of C4 ecosystems as an indicator of global ecological change and palaeoecologic indicators. *Nature*, 361, 344–5.

Cerling, T. E., Harris, J. M., MacFadden, B. J. et al. (1997) Global vegetation change through the Miocene/Pliocene boundary. *Nature*, 389, 153–8.

Cook, E. R., Palmer J. G. and D'Arrigo, R. D. (2002) Evidence for a 'Medieval Warm Period' in a 1,100 year tree-ring reconstruction of past austral summer temperatures in New Zealand. *Geophysical Research Letters*, 29, 12–16.

Cowie, J. (1998) *Climate of Human Change: Disaster or Opportunity?* London Parthenon Publishing.

Cuffey, K. M. and Marshall, S. J. (2000) Substantial contribution to sea-level rise during the last interglacial from the Greenland ice sheet. *Nature*, 404, 591–4.

Cwynar, L. C. and MacDonald, G.M. (1987) Geographical variations of lodgepo-lepine in relation to population history. *American Naturalist*, 129, 463–9.

DeConto, R. M. and Pollard, D. (2003) Rapid Cenozoic glaciation of Antarctica induced by declining atmospheric CO2. *Nature*, 421, 245–9.

de Garidel-Thoron, T., Rosenthal, Y., Bassinet, F. and Beaufort, L. (2005) Stable sea surface temperatures in the western Pacific warm pool over the past 1.75 million years. *Nature*, 433, 294–8.

deMenocal, P., Ortiz, J., Guilderson, T. and Sarnthein, M. (2000) Coherent high- and low-latitude climate during the Holocene. *Science*, 288, 2198–2202.

de Vernal, A. and Hillaire-Marcel, C. (2008) Natural variability of Greenland climate, vegetation, and ice volume during the past million years. *Science*, 320, 1622–5.

Elias, S. A., Short, S. K., Nelson, C. H. and Birks, H. H. (1996) Life and times of the Bering land bridge. *Nature*, 382, 60–3.

EPICA (2004) Eight glacial cycles form an Antarctic ice core. *Nature*, 429, 623–7.

Evershed, R. P., Payne, S., Sherratt, A. G. et al. (2008) Earliest date for milk use in the Near East and southeastern Europe linked to cattle herding. *Nature*, 455, 528–31.

Fells, T., Lohmann, G., Kuhnert, H. et al. (2004) Increased seasonality in Middle East temperatures during the last interglacial. *Nature*, 429, 164–8.

Ferretti, D. F., Miller, J. B., White, J. W. C. et al. (2005) Unexpected changes to the global methane budget over the past 2000 years. *Science*, 309, 1714–16.

Forster, P. and Matsumura, S. (2005) Did early humans go north or south? *Science*, 308, 965–6.

Frogley, M. R., Tzedakis, P. C. and Heaton, T. H. E. (1999) Climate variability in northwest Greece during the last interglacial. *Science*, 285, 1886–8.

Gibbons, A. (2001) The peopling of the Pacific. *Science*, 291, 1735–7.

Guthrie, R. D. (2004) Radiocarbon evidence of mid-Holocene mammoths stranded on an Alaskan Bering Sea island. *Nature*, 429, 746–9.

Guthrie, R. D. (2006) New carbon dates link climatic change with human colonization and Pleistocene extinctions. *Nature*, 441, 207–9.

Haas, J., Creamer, W. and Ruiz, A. (2004) Dating the late Archaic occupation of the Norte Chico region in Peru. *Nature*, 432, 1020–3.

Haug, G. H., Ganopolski, A., Sigman, D. M. et al. (2005) North Pacific seasonality and the glaciation of North America 2.7 million years ago. *Nature*, 433, 821–5.

Heinrich, H. (1988) Origin and consequences of cyclic ice rafting in the Northeast Atlantic Ocean during the past 130000 years. *Quaternary Research*, 29, 142–52.

Henn, B. M., Gignoux, C. R., Jobin, M., Granka, J. M. et al. (2011) Hunter-gatherer genomic diversity suggests a southern African origin for modern humans. *Proceedings of the National Academy of Sciences USA*, 108(13), 5154–62.

Herbert, T. D., Peterson, L. C., Lawrence, K. T. and Liu, Z. (2010) Tropical ocean temperatures over the past 3.5 million years. *Science*, 328, 1530–4.

Hewitt, G. (1996) Some genetic consequences of ice ages, and their role in divergenceand speciation. *Biological Journal of the Linnean Society*, 58, 247–76.

Hewitt, G. (2000) The genetic legacy of the Quaternary ice ages. *Nature*, 405, 907–13.

Hillaire-Marcel, C., de Vernal, A., Bilodeau, G. and Weaver, A. J. (2001) Absence of deep-water formation in the Labrador Sea during the last interglacial period. *Nature*, 410, 1073–7.

Hooghiemstra, H. and Van der Hammen, T. (2004) Quaternary ice-age dynamics in the Colombian Andes: developing an understanding of our legacy. *Philosophical Transactions of the Royal Society B*, 359, 173–81.

Hornbach, M. J., Saffer, D. M. and Holbrook, W. S. (2004) Critically pressured free-gas reservoirs below gas hydrate provinces. *Nature*, 427, 142–4.

Imhoff, M. L., Bounaoua, L., Ricketts, T., Loucks, C., Harriss, R. and Lawrence, W. T. (2004) Global patterns in human consumption of net primary production. *Nature*, 429, 870–3.

Intergovernmental Panel on Climate Change (2001) *Climate Change 2001: Technical Summary of the Working Group II Report*. Cambridge: Cambridge University Press.

Ivany, L. C., Brey, T., Huber, M., Buick, D. and Schöne, B. R. (2011) El Niño in the Eocene greenhouse recorded by fossil bivalves and wood from Antarctica. *Geophysical Research Letters*, 38, L16709.

Joyce, D. A., Lunt, D. H., Bills, R. et al. (2005) An extant cichlid fish radiation in an extinct Pleistocene lake. *Nature*, 435, 90–5.

Kaufman, D., Schneider, D. P., McKay, N. P. et al. (2009) Recent warming reverses long-term Arctic cooling. *Science*, 325, 1236–9.

Kennett, J. P., Cannariato, K. G., Hendy, I. L. and Behl, R. J. (2000) Carbon isotopic evidence for methane hydrate instability during Quaternary interstadials. *Science*, 288, 128–33.

Knutti, R., Fluckiger, J., Stocker, T. F. and Timmermann, A. (2004) Strong hemispheric coupling of glacial climate through freshwater discharge and ocean circulation. *Nature*, 430, 851–6.

Kopp, R. E., Simons, F. J., Mitrovica, J. X., Maloof, A. C. and Oppenheimer, M. (2009) Probablistic assessment of sea level during the last interglacial stage. *Nature*, 462, 863–7.

Kukla, G. J., Matthews, J. M. and Mitchell, Jr, J. M. (1972) The end of the present interglacial. *Quaternary Research*, 2, 261.

Laird, K. R., Fritz, S. C., Maasch, K. A. and Cumming, B. F. (1996) Greater drought intensity and frequency before AD 1200 in the Northern Great Plains, USA. *Nature*, 384, 552–4.

Lamb, H. H. (1965) Outline of Britain's climatic history. In Johnson, C. G. and Smith, L. P., eds., *The Biological Significance of Climate Changes in Britain*, pp. 3–31. London: Academic Press and The Institute of Biology.

Li, H. and Durbin, R. (2011) Inference of human population history from individual whole-genome analysis. *Nature*, 475, 493–6.

Li, J. Z., Absher, D. M., Tang, H. et al. (2008) Worldwide human relationships inferred from genome-wide patterns of variation. *Science*, 319, 1100–4.

Londo, J. P., Chiang, Y.-C., Kuo-Hsiang Hung, K.-H., Chiang, T.-Y. and Schaal, B. A. (2007) Phylogeography of Asian wild rice, *Oryza rufipogon*, reveals multiple independent domestications of cultivated rice, *Oryza sativa*. *Proceedings of the National Academy of Sciences USA*, 103(25), 9578–83.

Lorenzen, E. D., Nogués-Bravo, D., Orlando, L. et al. (2011) Species-specific responses of Late Quaternary megafauna to climate and humans. *Nature*, 479, 359–64.

Lund, D. C., Lynch-Stieglitz, J. and Curry, W. B. (2006) Gulf Stream density structure and transport during the past millennium. *Nature*, 444, 601–4.

Macaulay, V., Hill, C., Achilli, A. et al. (2005) Single, rapid coastal settlement of Asia revealed by analysis of complete mitochondrial genomes. *Science*, 308, 1034–6.

MacDonald, G. M., Beilman, D. W., Kremenetski, K. V. et al. (2006) Rapid early development of circumarctic peatlands and atmospheric CH_4 and CO_2 variations. *Science*, 314, 285–8.

Mann, M. E., Zhang, Z., Rutherford, S. et al. (2009) Global signatures and dynamical origins of the little ice age and medieval climate anomaly. *Science*, 326, 1256–60.

Marincovich, Jr, L. and Gladenkov, A. Y. (1999) Evidence for an early opening of the Bering Strait. *Nature*, 397, 149–51.

Marshall, E. (2001) Pre-Clovis sites fight for acceptance. *Science*, 291, 1730–2.

Mellars, P. (2006) A new radiocarbon revolution and the dispersal of modern humansin Eurasia. *Nature*, 439, 931–5.

Miller, G. H., Fogel, M. L., Magee, J. W., Gagan, M. K., Clarke, S. J. and Johnson, B. J. (2005) Ecosystem collapse in Pleistocene Australia and a human role in megafaunal extinction. *Science*, 309, 287–90.

Miller, G. H., Geirsdóttir, Á., Zhong, Y. et al. (2012) Abrupt onset of the Little Ice Age triggered by volcanism and sustained by sea-ice/ocean feedbacks. *Geophysical Research Letters*, 39, L02708.

Moreno, P. I., Jacobson, Jr, G. L., Lowell, T. V. and Denton. G. H. (2001) Interhemispheric climate links revealed by a late-glacial cooling episode in southern Chile. *Nature*, 409, 804–8.

Morrell, P. L. and Clegg, M. T. (2007) Genetic evidence for a second domestication of barley (*Hordeum vulgare*) east of the Fertile Crescent. *Proceedings of the National Academy of Sciences USA*, 104(9), 3289–94.

Morwood, M. J., Soejono, R. P., Roberts, R. G. et al. (2004) Archaeology and age of a new hominin from Flores in eastern Indonesia. *Nature*, 431, 1087–91.

Negre, C., Zahn, R., Thomas, A. L., Pere Masqué, P. et al. (2010) Reversed flow of Atlantic deepwater during the Last Glacial Maximum. *Nature*, 468, 84–8.

Pacher, M. and Stuart, A. J. (2008) Extinction chronology and palaeobiology of the cave bear (*Ursus spelaeus*). *Boreas*, 38(2), 189–206.

Pagani, M., Freeman, K. H. and Arthur, M. A. (1999) Late Miocene atmospheric CO_2 concentrations and the expansion of C4 grasses. *Science*, 285, 876–9.

Pastor, J. and Moen, R. A. (2004) Ecology of ice-age extinctions. *Nature*, 431, 639–40.

Pearson, P. N., Foster, G. L. and Wade, B. S. (2009) Atmospheric carbon dioxide through the Eocene-Oligocene transition. *Nature*, 461, 1110–13.

Perry, L., Sandweiss, D. H., Piperno, D. R. et al. (2006) Early maize agriculture and interzonal interaction in southern Peru. *Nature*, 440, 76–9.

Peterson, L. C., Haug, G. H., Hughen, K. A. and Röhl, U. (2000) Rapid changes in the hydrological cycle of the tropical Atlantic during the last glacial. *Science*, 290, 1947–51.

Pétit, J. R., Jouzel, J., Raynaud, D. et al. (1999) Climate and atmospheric history of the past 420 000 years from the Vostok ice core, Antarctica. *Nature*, 399, 429–36.

Pétit, R. J., Aguinagalde, I., de Beaulieu, J.-L. et al. (2003) Glacial refugia: hotspots but not melting pots of genetic diversity. *Science*, 300, 1563–5.

Pielou, E. C. (1991) *After the Ice: the Return of Life to Glaciated North America*. Chicago: University of Chicago Press.

Premoli, A. C., Kitzberger, T. and Veblen, T. T. (2000) Isozyme variation and recent biogeographical history of the long-lived conifer *Fitzroya cupressoides*. *Journal of Biogeography*, 27, 251–60.

Prideaux, G. J., Long, J. A., Ayliffe, L. K., Hellstrom, J. C. et al. (2007) An arid-adapted middle Pleistocene vertebrate fauna from south-central Australia. *Nature*, 455, 422–5.

Pujol, B. and Pannell, J. R. (2008) Reduced responses to selection after species range expansion. *Science*, 321, 96.

Ravelo, A. C., Andreasen, D. H., Lyle, M., Lyle, A. O. and Wara, M. W. (2004) Regional climatic shifts caused by gradual global cooling in the Pliocene epoch. *Nature*, 429, 263–7.

Rioul, P., Andrieu-Ponel, V., Rietti-Shati, M. et al. (2001) High-resolution record of climate stability in France during the last interglacial period. *Nature*, 413, 293–6.

Rohling, E. J. and Pälike, H. (2005) Centennial-scale climate cooling with a sudden cold event around 8,200 years ago. *Nature*, 434, 975–9.

Rothwell, R. G., Thomson, J. and Kähler, G. (1998) Low-sea-level emplacement of a very large late Pleistocene megaturbide in the western Mediterranean Sea. *Nature*, 392, 377–80.

Ruddiman, W. F. (2003) The anthropogenic greenhouse era began thousands of years ago. *Climate Change*, 61(3), 261–93.

Ruddiman, W. F. and Thomson, J. S. (2001) The case for human causes of increased atmospheric CH_4 over the last 5000 years. *Quaternary Science Reviews*, 20, 1769–77.

Searle, J. B., Kotlík, P. Rambau, R. V. et al. (2009) The Celtic fringe of Britain: insights from small mammal phylogeography. *Proceedings of the Royal Society of London Series B*, 276, 4287–94.

Seppä, H., Alenius, T. and Bradshaw, R. H. W. et al. (2009) Invasion of Norway spruce (*Picea abies*) and the rise of the boreal ecosystem in Fennoscandia. *Journal of Ecology*, 97, 629–40.

Severinghaus, J. P. (2009) Southern see-saw seen. *Nature*, 457, 1093–4.

Shackleton, N. J. (2000) The 100,000 year ice-age cycle identified and found to lag temperature, carbon dioxide and orbital eccentricity. *Science*, 289, 1897–1901.

Shackleton, N. J., Cave, I. N. and Weedon, G. P. (eds) (1999) Astronomical (Milankovitch) calibration of the geological timescale. *Philosophical Transactions of the Royal Society B*, 357, 1731–2007.

Shapiro, B., Drummond, A. J., Rambaut, A. et al. (2004) Rise and fall of the Beringian steppe bison. *Science*, 306, 1561–5.

Siegenthaler, U., Stocker, T. F., Monnin, E. et al. (2005) Stable carbon cycle-climate relationship during the late Pleistocene. *Science*, 310, 1313–17.

Sigman, D. M., Jaccard, S. L. and Haug, G. H. (2004) Polar ocean stratification in a cold climate. *Nature*, 428, 59–63.

Sigman, D. M., Hain, M. P. and Haug, G. H. (2010) The polar ocean and glacial cycles in atmospheric CO_2 concentration. *Nature*, 466, 47–55.

Singarayer, J. S., Valdes, P. J., Friedlingstein, P., Nelson, S. and Beerling, D. J. (2011) Late Holocene methane rise caused by orbitally controlled increase in tropical sources. *Nature*, 470, 82–5.

Smith, L. C., MacDonald, G. M., Velichko, A. A. et al. (2004) Siberian peatlands a net carbon sink and a global methane source since the early Holocene. *Science*, 303, 353–7.

Sowers, T. (2006) Late Quaternary atmospheric CH_4 isotope record suggests marine clathrates are stable. *Science*, 311, 838–40.

Spahni, R., Chappellaz, J., Stocker, T. F. et al. (2005) Atmospheric methane and nitrous oxide of the late Pleistocene from Antarctic ice cores. *Science*, 310, 1317–21.

Steffensen, J. P., Andersen, K. K., Bigler, M., Henrik B. Clausen, H. B., Dahl-Jensen, D. et al. (2008) High-resolution Greenland ice core data show abrupt climate change happens in few years. *Science*, 321, 680–4.

Steig, E. J. (1999) Mid-Holocene climate change. *Science*, 286, 1485–6.

Steine, S. (1994) Extreme and persistent drought in California and Patagonia during the Mediaeval time. *Nature*, 396, 546–9.

Stokes, S., Thomas, D. S. G. and Washington R. (1997) Multiple episodes of aridity in southern Africa since the last interglacial period. *Nature*, 388, 154–8.

Stuart, A. J., Kosintsev, P. A., Higham, T. F. G. and Lister, A. M. (2004) Pleistocene to Holocene extinction dynamics in giant deer and woolly mammoth. *Nature*, 431, 684–9.

Stute, M., Forster, M., Frischorn, H. et al. (1995) Cooling of tropical Brazil (5°C) during the last glacial maximum. *Science*, 269, 379–83.

Tarasov, L. and Peltier, W. R. (2005) Arctic freshwater forcing of the Younger Dryas cold reversal. *Nature*, 435, 662–5.

Taylor, K. C., Lamorey, G. W., Doyle, G. A. et al. (1993) The 'flickering switch' of late Pleistocene climate change. *Nature*, 361, 432–5.

Thangaraj, K., Chaubey, G., Kivisild, T. et al. (2005) Reconstructing the origin of Andaman islanders. *Science*, 308, 996.

Trauth, M. H., Maslin, M. A., Deino, A. and Strecker, M. R. (2005) Late Cenzoic moisture history of East Africa. *Science*, 309, 2051–3.

Tripati, A., Backman, J., Elderfield, H. and Ferretti, P. (2005) Eocene bipolar glaciation associated with carbon cycle changes. *Nature*, 436, 341–6.

Tzedakis, P. C., Hughen, K. A., Cacho, I. and Harvati, K. (2007) Placing late Neanderthals in a climatic context. *Nature*, 449, 206–8.

Tyson, P., Odada, E., Schulze, R. and Vogel, C. (2002) Regional-global change linkages: Southern Africa. In Tyson, P., Fuchs, R., Fu, C. et al., eds., *Global Regional Linkages in the Earth System*, pp. 3–73. Berlin: Springer.

Vandergoes, M. J., Newham, R. M., Preusser, F. et al. (2005) Regional insolation forcing of late Quaternary climate change in the Southern Hemisphere. *Nature*, 436, 242–5.

Verschuren, D., Laird, K. R. and Cumming, B. F. (2000) Rainfall and drought in equatorial east Africa during the past 1,100 years. *Nature*, 403, 410–13.

Wang, X., Auler, A. S., Edwards, R. L. et al. (2004) Wet periods in northeastern Brazil over the past 210 kyr linked to distant climate anomalies. *Nature*, 432, 740–3.

Wanner, H., Beer, J., Bütikofer, J., Crowley, T. J. et al. (2008) Mid- to Late Holocene climate change: an overview. *Quaternary Science Reviews*, 27, 1791–1828.

Wara, M. W., Ravelo, A. C. and Delaney, M. L. (2005) Permanent El Niño-like conditions during the Pliocene warm period. *Science*, 309, 758–61.

Watanabe, T., Suzuki, A., Minobe, S. and Kawashima, T. et al. (2011) Permanent El Niño during the Pliocene warm period not supported by coral evidence. *Nature*, 471, 209–11.

Weaver, A. J., Saenko, O. A., Clark, P. U. and Mitrovica, J. X. (2003) Meltwater Pulse 1 A from Antarctica as a trigger for the Bølling-Allerød warm interval. *Science*, 299, 1709–12.

White, T. D., Asfaw, B., Beyene, Y., Haile-Selassie, Y. et al. (2009) *Ardipithecus ramidus* and the paleobiology of early hominids. *Science*, 326, 75–86.

Wolff, C., Haug, G. H., Timmermann, A. et al. (2011) Reduced interannual rainfall variability in East Africa during the last ice age. *Science*, 333, 743–7.

Wolff, E. W., Fischer, H., Fundel, F. et al. (2006) Southern ocean sea-ice extent, productivity and iron flux over the past eight glacial cycles. *Nature*, 440, 491–6.

Wood, B. and Harrison, T. (2011) The evolutionary context of the first hominins. *Nature*, 470, 347–52.

Zachos, J., Pagani, M., Sloan, L., Thomas, E. and Billups, K. (2001) Trends, rhythms and aberrations in global climate 65 MA to present. *Science*, 292, 686–93.

Zazula, G. D., Froese, D. G., Schweger, C. E. et al. (2003) Ice-age steppe vegetation in east Beringia. *Nature*, 423, 603.

Zong, Y., Innes, J. B., Chen, C., Wang, Z. and Wang, H. (2007) Fire and flood management of coastal swamp enabled first rice paddy cultivation in east China. *Nature*, 449, 459–62.

Present climate and biological change

5.1 Recent climate change

5.1.1 The latter half of the Little Ice Age

The 17th century was not just the time of the Little Ice Age, it is also noted (and for some better noted) for the Renaissance, which saw the gathering of scientific understanding that in turn was to drive the Industrial Revolution of the 18th and 19th centuries. In Britain in the 1640s and 1650s scientists sought what they termed 'a great insaturation', which drew on the philosophies exposed by the likes of Francis Bacon. Among these, Bacon's principles of exact observation, measurement and inductive reasoning provided the intellectual tools for scientific advance. These advances had yet to percolate through to day-to-day application in technology, so life, society and its economy were still largely powered by humans and animals together with the burning of wood. Major global impacts from human activity were not yet manifest (although, of course, trace global signatures such as metals in Greenland ice cores can be found dating to thousands of years earlier).

In terms of climate and weather, 1659 – within the Little Ice Age – is an important date. Before that date we rely solely on the proxy indicators (see Chapter 2) for climatic information. After 1659 there began a source of new information: direct meteorological measurement. The first significant series of measurements began in 1659 and (much later) were compiled into a monthly series of temperature readings for rural sites in central England by Gordon Manley (1974). This is the longest homogeneous record and is still kept up to date by the UK Meteorological Office. The earlier measurements were varied but increasingly included, and were soon dominated by, instrumental measurements. The Central England Temperature (CET) records were soon accompanied by others to ultimately be built into a series such as those for De Bilt in The Netherlands from 1705. Such records are fundamentally important. As we have seen, although we can use a variety of proxies to build up, piece by piece, quite a good picture of past climate, deep-time climatic proxy indicators simply are either not sensitive or representative enough to tell us what was going on. This is especially true with regard to finer changes. For example, ice-core isotopic records are very fine for charting regional glacial and interglacial transitions of a few degrees but are less useful for discerning trends in changes of fractions of a degree within our current interglacial. For instance, Figure 5.1 is a portrayal of seven centuries of Greenland deuterium up to the late 20th century. That it only represents the deuterium evapo-fractionation from the ocean surrounding Greenland restricts its usefulness.

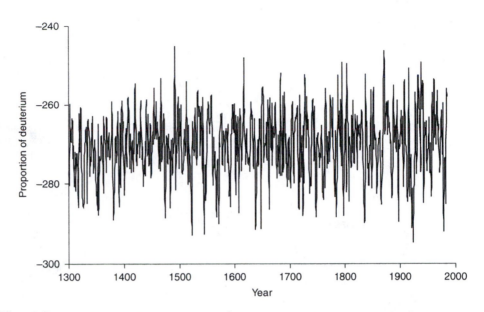

Fig. 5.1 Seven centuries of Holocene Greenland deuterium (^{2}H) measurements. The proportion of deuterium in the ice, compared to the common isotope of hydrogen (^{1}H) in water, is proxy for regional temperature (see Chapter 2). It demonstrates the high rapid variability of this one dimension. Smoothing data helps, as does including more than one dimension. (By contrast, Figure 5.3 is based on Greenland isotopes and snow accumulation as well as being smoothed.) Data from the National Oceanic and Atmospheric Administration (see Acknowledgements).

This is not just because of regionality but because different times of the year see different amounts of water transportation ending up as Greenland ice, so whereas some parts of the year may be getting warmer, others may be getting cooler, and what ends up in the ice may not be as representative as researchers would like. Also, the climate is not one-dimensional. Temperature alone does not determine climate but a range of factors, such as humidity, precipitation, sunshine, seasonality and so forth. Instrumentation provides standard, more sensitive, forms of measurements independent of the constraints of many individual physicochemical, and biological, palaeoclimatic proxies. There is a somewhat ironic symmetry that we have relied on proxies, especially biological ones, to determine past climates before the Renaissance, and then instrumentation afterwards. The Renaissance enabled there to be an Industrial Revolution that in turn saw the beginnings of extensive fossil fuel use, which is now beginning to perturb biological systems. The post-Renaissance period enabled calibration of the bioproxies used in the pre-Renaissance period. Since the Renaissance, instrumentation has enabled the discipline of meteorology. Its forecasts have been useful primarily because the weather affects our biologically based lives and, more recently, because our anthropogenic (i.e. human-generated) perturbations of the climate are affecting many biological systems globally. Nonetheless, at first (in the Little Ice Age) instrumental records were neither as widespread nor as rigorously compiled as they became later. Combining early records with a variety of palaeoclimatic indicators does provide a useful picture, such as that in Figure 5.2.

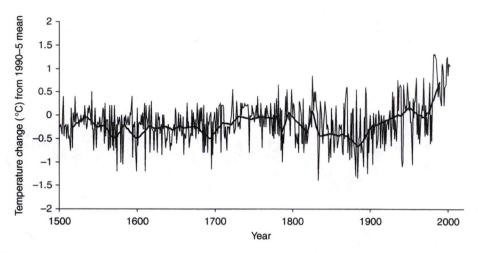

Fig. 5.2 Northern hemisphere climate trends over 500 years as discerned from instrumental, dendrochronological and Greenland ice-core records. Based on Luterbacher et al. (2004).

Against the context of the Quaternary glacial – interglacial cycles that helped shape human evolution, the Little Ice Age is a comparatively minor blip in our current Holocene interglacial. There are, of course, even smaller blips and regionally there are extreme meteorological events of even shorter duration. Yet these can have an acute impact on human activities, as we shall see in the next section. For now, one 20th-century example will serve to illustrate the point: the dust-bowl years of the 1930s, especially 1934 and 1936, in the midwest USA. This extreme arose because temperate continental interiors are subject to great seasonality with extremely cold winters and hot summers. For farmers in the midwest this last was not too much of a problem because the summer heat enabled vast quantities of water to be transported, which fell as rain. However, the years 1934 and 1936 were both extremely hot *and* extremely dry. This, combined with poor agricultural practice, resulted in not only crop failure but also severe soil erosion. The thing to note here is that just as in more recent times there have been extreme regional events, so there were also in the Little Ice Age.

Notwithstanding isolated extreme events, the Little Ice Age itself was not a period of sustained cold. This is why perhaps there is not as much clear evidence from across the planet as one might like to define such a period as one of great global significance. As with the dust-bowl years, the Little Ice Age mainly shows up in specific climatic dimensions. The 1590s were cold years but frequent subsequent years were more notable specifically for their colder winters. However, one of the first climatic events (albeit it a mini one) picked up by the CET record is that after the exceptionally cold 1690s there was a time of warming, around the turn of the century. The 1720s and 1730s were as warm as the late 20th century before suddenly returning to cooler conditions in 1740.

The climate, as far as Europe was concerned, showed no discernable trend over the next 180 years: there was nothing warmer than the 1730s and nothing cooler than the 1690s. This shows up not only in the CET and the De Bilt series but also in the records from other European cities. The summers were particularly consistent. What signals the emergence from the Little Ice Age is the less-harsh winters. So, just as the 20th-century dust-bowl years in the midwestern USA are defined climatologically by their hot dry summers, so the Little Ice Age is defined by harsher winters and shorter summers that may have been a little wetter and only marginally cooler than those in the 20th century. This resulted in a shorter thermal growing season with atypical conditions at its beginning (sowing time) and end (harvest time). It is estimated that in England during the coldest decades of the Little Ice Age the thermal growing season was shortened by 3–4 weeks. As we shall see later in the chapter this is the same order of magnitude of biological effect we are seeing early in the 21st century compared with the mid-20th century due to global warming. The contrast is that the effect is obviously the other way around, with an extension of the growing season (see Chapter 6). As we shall also see in the next section, this shortening of the thermal growing season in the Little Ice Age resulted in considerable human suffering.

Added to this change, in the late 18th and early 19th centuries there were also a few volcanic eruptions that seem to have had a global effect, with the years following the eruptions being even colder than the other years of the Little Ice Age. These included Laki (1783), Tambora (1815), Cosiguina (1835), Cotopaxi-Awu (1856) and Krakatau (1883; see Figure 2.1). A team of chemists from the USA and France, led by Jihong Cole-Dai, in 2009 found compelling evidence of a previously little documented large volcanic eruption that occurred in 1809. The following years were very cold; the entire decade from 1810 to 1819 was one of the coldest of the past 500 years. Now, we have long been aware that the eruption in 1815 of an Indonesian volcano Tambora, which killed more than 88 000 people, had caused the global cold weather in 1816 and subsequent years. Indeed, at a symposium in 2001, Marc Prohom, with co-authors Pere Esteban, and Javier Martin-Vide from the University of Barcelona, presented the case for a strong volcanic influence on the climate around this time. They noted that cooling in the early part of the decade, before the Mount Tambora eruption, suggested that Tambora alone could not have caused the climatic changes of the decade and even cited an unknown eruption in 1809. Cole-Dai et al. in 2009 found a large amount of volcanic sulphuric acid in the snow layers of 1809 and 1810 in both Greenland and Antarctica, indicating that there had been a major volcanic incident at that time, confirming Marc Phom's reference to an 'unknown' volcanic 1809 eruption.

Interestingly, in eastern Asia the 17th century was also a cold period, as it was in Europe, especially around 1800. However, 19th-century eastern Asia did not see years that were as cold as those in Western Europe or North America. The Little Ice Age should therefore not be seen as a period of consistently uniform cold across the planet. There were, though, a few short cold episodes, either across the northern hemisphere or globally (which brings us to the ocean circulation changes discussed in the previous chapter). These were the decades 1590s to 1610s, 1690s to 1710s and 1800s to 1810s. There were also a few slightly warm periods, including the 1730s and 1820s, before reaching the 20th-century warming, which commenced in the 1930s (Burroughs, 1997).

The third quarter of the 19th century saw a brief easing of conditions, as signalled by glacier retreat in Europe. This did not last and the century's final quarter saw glacial expansions in Europe and Canada. In terms of high-latitude northern hemisphere climate this quarter represented the harshest period of the Little Ice Age. The Little Ice Age clearly came to an end in the early 20th century.

Although we do not know the exact nature of the Little Ice Age (whether it was truly global and not just restricted mainly to the northern hemisphere) it may still have been a period of the harshest climate on Earth since the beginning of the Holocene. Certainly, European glaciers were at their most advanced during the Little Ice Age compared with their decline early on in the Holocene. *If* the Earth were slowly sinking back into a glacial (and it would be slow, given that Milankovitch forcing will tend to promote an interglacial over the next 10 000 years; see section 5.1.4), then the Little Ice Age might have been representative of a cooling episode as part of a longer-term trend. If this is really the case then recent greenhouse climatic forcing due to human emissions of greenhouse gases has had a tremendous impact, for recent greenhouse warming would not only have to be compared against, say, an early 20th-century standard, but also against a theoretical, cooler early 20th century had the Earth's slow slide into a future glacial continued unaffected. This perspective was supported in 2009 by work of Darrell Kaufman, David Schneider and colleagues, who compiled a synthesis of decadally resolved proxy temperature records from poleward of 60°N covering the past 2000 years. This indicates a pervasive Arctic cooling in progress from 2000 years ago that continued through the Middle Ages and into the Little Ice Age. They then used those data coupled with a research community climate model to discern what may have been happening. Their inference was that this long-term trend was caused by the steady Milankovitch orbitally driven reduction in summer insolation. The cooling trend was reversed during the 20th century, with four of the five warmest decades of our 2000-year-long reconstruction occurring between 1950 and 2000. The implication of this research is that the generation of greenhouse gases by humans caused this Arctic warming.

Looking at the Little Ice Age from a greenhouse perspective raises a question. Carbon dioxide levels during that time showed no particular deviation from the norm, not nearly sufficiently to account for the cooling that was seen. So, what caused it? There are arguments that other factors played a part, such as the solar Maunder Minimum (see Chapter 4) and the number of significant volcanic eruptions. What is known (even if the exact extent beyond somewhere between half-a-degree to a degree is still unclear) is that climate forcing from anthropogenic greenhouse emissions has increased markedly since the end of the 19th century.

5.1.2 20th-century climate

Greenhouse gas emissions from the burning of fossil fuels and land-use change (mainly deforestation) increased dramatically in the 20th century. More carbon dioxide was emitted in the second half of the 20th century compared to the 100 years leading up to 1950. Further, emissions from burning of fossil fuels were four times higher by the end of the century (an increase of 300%) compared to 1950 (Cowie, 1998; we will return to the pattern of emissions later). As a result, the atmospheric concentration of carbon dioxide rose from its pre-industrial (mid-18th-century) level of 279 ppmv to 369.4 ppmv in the year 2000 (Worldwatch Institute, 2003) and 389.8 ppmv in 2010.[1]

Apart from some wavering (which is mainly the result of natural variability), the 20th century was notably a century of warming. This was to such an extent that the 1990s saw a concentration of record-breaking hot years.

5.1.3 21st-century climate

Such has been the anthropogenic injection of greenhouse gases into the atmosphere since the Industrial Revolution, and especially during the 20th century, that there is little doubt that global warming will continue throughout the 21st century. The international scientific consensus on 21st-century warming has been determined by the UN Intergovernmental Panel on Climate Change (IPCC). This will be covered in the section on the IPCC Business-as-Usual scenario (section 5.3.1).

5.1.4 The Holocene interglacial beyond the 21st century

In looking at the likely future climate of the current interglacial (the Holocene) and beyond, we need to consider how our perceptions have changed over the past few decades. Since the 1970s, when ice-core analysis revealed the broad nature of the glacial–interglacial cycle, there has been a general assumption that the length that glacials last is the order of 100 000 years; conversely, interglacials are an order of magnitude smaller and are about 10 000 years long. Furthermore, as noted in Chapter 4, ice cores have shown that in the past the transition between glacial and interglacial states has been sudden and not gradual. In the 1970s there was a concern that as the Holocene was already more than 10 000 years old the Earth might be about to enter a glacial period (see section 5.1.1). As noted in Chapter 1, global warming and human-driven greenhouse concerns had long abounded and so the question for the late 1970s and early 1980s focused on the balance between these two warming and cooling forcing agents. The IPCC's first report in 1990 (see below) provided an international scientific consensus, concluding clearly that anthropogenic global warming factors superimposed on the natural climate forcing agents would dominate through the 21st century and, by implication, decades beyond, if not more.

So, in light of recent record-breaking years, this leads us to the question of whether these years are part of a longer-term change in the global climate system involving climate and biosphere parameters other than temperature. It is a question to which we shall return later when looking at the IPCC's conclusions (section 5.3) and climate change's impact on biological systems (section 6.1). However, in the summer of 2007 the first signs of human influence on global precipitation patterns were announced by a small group of researchers led by Xuebin Zhang and Francis Zwiers. They looked at meteorological records for precipitation by latitude band throughout the 20th century and then compared them with 14 climate models. They estimated that anthropogenic forcing of the climate by the end of the century had contributed significantly to an increase in precipitation in northern hemisphere mid-latitudes, drying in the northern

---

[1] See the National Ocean and Atmosphere Administration website: www.esrl.noaa.gov/gmd/ccgg/trends/.

hemisphere's subtropics and tropics, and moistening in the southern hemisphere's subtropics and deep tropics. Finally, the observed meteorological changes were larger than those suggested by the models, which says something about the nature of the change together with the capability of models to simulate the future. This also builds on preliminary work in 2007 by Frank Wentz and colleagues (see section 1.7).

The year 2007 saw other work corroborating the notion that by the beginning of the 20th century we were influencing the global hydrological cycle. The autumn of 2007 saw corroboration of the Wentz team's conclusion that precipitation and total atmospheric water have increased with warming at about the same rate over the past two decades, of around 6% $°C^{-1}$. A British group led by Katherine Willett also looked at meteorological data and compared it with a climate model. They identified a significant global increase in surface humidity that is mainly attributable to human influence as distinct from natural forcing. Finally, and importantly as this text concerns climate change biology, another British team, this time led by Betts, in 2007 used an ensemble of 244 runs of a climate model that incorporated a terrestrial vegetation dimension. They investigated whether terrestrial vegetation would affect the hydrological cycle of a warmer world: would a warmer world see more continental run-off? (By the early 21st century climate modellers wer² increasingly using models with biological components in addition to the physical at.nosphere, land and ocean components that had by the end of the 20th century become commonplace.) There are two trends in plant physiology that are likely in a world warmed by more carbon dioxide. First, being warmer (and with enough moisture) plants will transpire more water. Second, with an increase in atmospheric carbon dioxide, stomata (the openings on the surface of leaves through which water and gas can pass) will tend to close and/or become fewer in number: either way, transpiration will be reduced. The question is, which trend – more or less transpiration – will dominate and how will this affect continental run-off? The conclusions of Betts and colleagues were that in some areas plants will fail to get enough water because of greater transpiration, and that there will be more evaporation from land, leading to some regions being prone to drought. Overall, this reduction in transpiration together with the increase in the hydrological cycle (more water being cycled) will result in increased continental run-off. This means that there will be more water in streams and rivers, and hence some areas may see greater flooding.

Together, all the above work illustrates that the global climate system is changing profoundly and this has implications for the hydrological cycle in the 21st century and beyond.

As ever, I must sound a word of caution. The end of the 21st century, and a world that is 4°C warmer, are a long way off. A lot could happen to global evapotranspiration between now and then. In 2010 a Western European and US team led by Martin Jung, Markus Reichstein and Philippe Ciais provided a data-driven estimate of global land evapotranspiration from 1982 to 2008, compiled using a global monitoring network, meteorological and remote-sensing observations and the use of a computer model. In addition, they assessed evapotranspiration variations over the same time period using an ensemble of process-based land-surface models. Their results suggest that global annual evapotranspiration increased on average by 7.1 ± 1.0 mm per year per decade from 1982 to 1997. After that, coincident with the last major El Niño event in 1998, the increase in global evapotranspiration seemed to cease until 2008. This change was driven primarily by moisture limitation in the southern hemisphere, particularly in Africa and Australia. In these regions, microwave satellite observations indicate that soil moisture decreased from 1998 to 2008. So, increasing soil-moisture limitations on evapotranspiration largely explain the recent decline of the global land evapotranspiration trend. Whether the changing behaviour of evapotranspiration is representative of natural climate variability or reflects a more permanent re-organisation of the land water cycle is a key question for Earth system science.

Another question regarding the longer-term prospects for the global climate is that of where we are in the natural glacial–interglacial cycle. Has the anthropogenic release of greenhouse gas affected this cycle? By the 1990s two things had become clear. First, in the northern hemisphere at least, there was a marginal cooling trend of 0.2°C between AD 1000 (especially since AD 1350) and AD 1900. This cooling trend was both slight and only just above the background noise of climatic variability. Nonetheless, it might suggest that, as hinted above, in the absence of 20th-century global warming, the beginning of the end of the current Holocene interglacial would be upon us soon. (The fear in the 1970s had been that there would be global cooling into a glacial, and not global warming.) However, as noted above, this long-term cooling trend ended with the 20th-century warming. The second thing that became clear in the 1990s was that both glacials and interglacials are complex, with vagaries of their own (such as Bond cycles and associated Heinrich events within a glacial). Furthermore, as noted in section 4.6.1, glacials and interglacials have their own individual characteristics. Just because the Eemian interglacial (which in some texts is referred to as the Ipswichian or Sangamon) 130 000 years ago was roughly similar in length to its predecessor interglacial 220 000 years ago, this does not necessarily mean that our current Holocene interglacial will be a similar length.

As noted in Chapter 1, the pacemaker of glacials and interglacials is the Milankovitch solar radiation curve. Again (see Chapter 1) this curve is made up of three cycles of varying lengths and so it is likely that, unless the Milankovitch circumstances are similar, each glacial and interglacial will have its own characteristics. Indeed, as discussed in the last chapter, the closest of the interglacials similar to our own in a Milankovitch sense was the one following Termination V (five glacials ago), called the Hoxnian interglacial, which began 425 000 years ago (see Figure 4.11). It is this interglacial, and not the Eemian, that should be used as an analogue for our current Holocene interglacial (Augustin et al., 2004).

Projections into the future using Milankovitch curves (Berger and Loutre, 2002) suggest that, unlike the Eemian, the Holocene interglacial will be exceptionally long. Looking back at the Waalian interglacial (sometimes called the Pastonian) following Termination V, we can also see from the Antarctic ice-core record that that interglacial lasted some 30 000 years, from 425 000 to 395 000 years ago (Augustin et al., 2004). In other words, today, some 10 000 years into our current Holocene interglacial, *without* global warming it is *unlikely* that we would soon see a return to full glacial conditions. Without global warming it is thought that we might have an overall cooling over the next 25 000 years, only after which would there then follow a sharper return to glacial conditions. However, assuming that we will continue to add substantial amounts of carbon dioxide to the atmosphere over the next couple of centuries, we will be forcing the climate well above what would have been expected for our current interglacial.

One question that those with some knowledge of climate change, and Earth systems (or biosphere) science, often ask is whether there will be changes to the ocean and/or atmospheric circulation system. Here the short answer is very likely to be yes. Yet specifically the question arises as to whether there will be changes in the Broecker thermohaline circulation (Figure 4.7) and the meridional overturning circulation (MOC) in the North Atlantic; that is, the drawdown of surface water in the North Atlantic into the abyssal depths. The fear is that if the MOC shuts down then warm water in the Gulf Stream (more accurately the North Atlantic Drift) will cease to travel so far north. (It will continue to operate because the Earth will carry on rotating, so providing a Coriolis force, but it would not be drawn to the current North Atlantic down-welling region on the latitude of Greenland, Iceland and Norway.) If this were to happen then north-west Europe would begin to experience harsher winters as does Canada or other parts of continental Eurasia on the same latitude.

This question is a difficult one. The IPCC's 2007 report said that:

> It is *very likely* that the Atlantic MOC will slow down over the course of the 21st century. The multimodel average reduction by 2100 is 25% (range from zero to about 50%) for SRES emission scenario A1B.

Scenario A1B is an IPCC Business-as-Usual scenario with some switching away from fossil fuel use in the 21st century. The term 'very likely' is precise IPCC parlance for 'more than 90% probability'. This does seem to confer some sort of certainty. However, note the range the IPCC also cite: 'range from zero to about 50%'. In short, there is a chance that the decline in the MOC could be close to zero. In Britain the Natural Environment Research Council (NERC; the UK independent government-funded agency for distribution of environmental research funds), together with other national stakeholders (from the USA), started in 2004 to run a monitoring programme called RAPID-WATCH, which monitored the Atlantic MOC on a daily basis. It used an array of moored instruments deployed along latitude 26.5°N in the N. Atlantic. The programme is currently (2012) scheduled to run until 2014, but early results were published after just a couple of years in 2007 with Stuart Cunningham as the lead author. Their preliminary conclusion was that the Atlantic MOC was currently highly variable. The problem with this result is that by 2004–6 the Earth was already warming and moving away from its late-Holocene Little Ice Age state. The issue that consequently arises is whether this variability is 'natural' for the MOC across the whole Holocene (past 11 700 years of our current interglacial) or is it a sign that the Earth system is close to a 'critical transition' (a so-called tipping-point) and that the global climate is about to cross a 'climate threshold'? (As we shall see in section 6.6.8, system variability rises when close to critical transitions.) If the system is close to a critical transition, then the planet may not just 'gradually' warm as the IPCC's assessment reports (1990, 1995, 2001b, and 2007) consensually conclude;[2] it may

---

[2] This is not to say that the IPCC up to and including its 2007 assessment was wrong: far from it. the science of critical transitions as applied to the Earth system has only been developed in recent years and in any case the IPCC does warn us to be wary of climate 'surprises'.

flip to a new state. It is important to note that there are other parts of the Earth system that may harbour a critical transition point: the MOC is just one (see also section 6.6.8).

In short, irrespective of whether or not critical transitions are involved, the future sees us entering a climate mode that bucks the natural trend and which has not been seen during the past 2 million years of glacials and interglacials. Either way, the Holocene interglacial is not likely to end (geologically) soon (within a few thousand years). Conversely, not only is it likely for the interglacial to continue for another 10 000 or even 25 000 years (as it would without anthropogenic warming), but with human-induced warming it may *possibly* (in the context of the last chapter's discussion of the early-Eocene carbon event) last for well over 100 000 years: here at the very least the next Milankovitch-paced glacial would be skipped. Meanwhile, the next few centuries are likely to see global warming with temperatures unprecedented during previous interglacials, let alone our current one, which saw the rise of early human civilisation.

5.1.5 Holocene summary

Whereas the Holocene represents a period of warmth as well as of climatic stability, the past 10 000 years have not seen a static climate. For example, the Holocene climatic maximum saw considerable shifts in biomes, both before and after that time. More recently there was a second medieval (warm) anomaly and then the Little Ice Age cooling. This last in particular serves to demonstrate that there are many factors involved in the forcing of the global climate and not just greenhouse gases, be they natural or anthropogenic. Indeed other factors, such as volcanic and solar activity, play their parts.

So far we have only considered climate change in the broadest of contexts and with little detail. There is not the space in a book that introduces the interface between climatology, biology and human ecology to dwell at length on the intricacies of climate. For instance, one could spend much time debating where weather variability ends and climate change begins. Nonetheless, such intricacies do exist and the reader needs to be aware of this. For example, there are several oscillations taking place in the oceans and atmosphere that determine dry or wet, warm or cool, or combinations thereof in the seasons in any year.

For instance, in the South Pacific there is an oscillation, the El Niño Southern Oscillation (ENSO), that results in El Niño years. The El Niño is a current that flows along the coasts of Ecuador and Peru between January and March and which terminates the local fishing season. In some years it is warmer and lasts longer, so preventing a nutrient-rich upwelling. It is part of the bigger Southern Oscillation that sees the Pacific waters build up due to the prevailing winds in the western South Pacific. This build-up takes time and is released both when the build-up is sufficient and meteorology allows. There are a number of other key oscillations. One affecting North America and Europe is the Arctic Oscillation. Another, affecting the Antarctic, is the Antarctic Oscillation, which is sometimes known as the Southern Annular Mode.

A question researchers have posed in recent years is whether these oscillations are themselves modulated over a longer time frame. For example, one recent analysis of layers in cores of 13 lakes in north-eastern USA has determined how the degree of storminess has varied over the Holocene. The results show a characteristic period of about 3000 years and that this pattern appears consistent with long-term changes in the Arctic Oscillation (Noren et al., 2002). Such oscillations are but one grouping of many climatic intricacies that shape the complex detail of climate change.

Notwithstanding this complexity, or the natural vagaries and climate-related oscillations, the Holocene's most recent hundred years (the 20th century) have largely been a period of warming. Be it based on palaeoclimatic proxies (such as the high-latitude dendrochronological records in Figure 2.1), or instrumental (satellite- or ground-based), or a combination of both (Figure 5.2), late-20th-century warming was unprecedented in the time since instrumentation began. Further, using palaeoproxies, the late 20th century saw the greatest warming for 1000 years. Indeed multi-proxy reconstructions of monthly seasonal surface temperatures for Europe back to 1500 show that the late 20th and early 21st centuries were warmer than any other time during the past 500 years. This agrees with findings over the entire northern hemisphere. European winters were 0.5°C warmer in the 20th century compared to the previous four centuries. Conversely, summers have not been that much warmer, although individual ones have had warmer spells. Finally, in the half millennium up to 2000 the coldest European winter was that of 1708–9, which was not in contemporary times but in the Little Ice Age (Luterbacher et al., 2004).

The 20th century's warming trend is almost certainly going to continue, which is of tremendous concern. The Earth may be entering a climatic mode that is unprecedented in human history. Indeed, it is likely that in the 21st century the Earth will achieve temperatures not seen since the height of the last interglacial (140 000 years ago) and just possibly temperatures not seen during the Quaternary (the past 2 million years). Beyond this, further warming in the next century may result in temperatures not seen since before the Quaternary, more than 2 million years ago. If so – and it is a real possibility – it will transform our planet in ways that many have not yet contemplated. The shift to this warmer world will not necessarily be smooth: it could involve 'critical transitions', of which a profound change to the Broecker thermohaline circulation is but one. It is because we are biological creatures that are involved in, and dependent on, many biological systems, which in turn are affected by climate, that we need to take note. Such climate change will certainly affect virtually everyone's lives, both individually and on a societal level. To recognise the serious nature of this possibility one needs only see how the more modest climatic changes have affected human activities from the medieval climatic anomaly (MCA) to the present.

5.2 Human change arising from the Holocene climate

5.2.1 Climatic impacts on early human civilisations

Human civilisation (settlements and cities) arose early during the Holocene interglacial. These settlements were small; minute by modern standards. Importantly, sea-level rise following the last glacial continued for a few thousand years after the

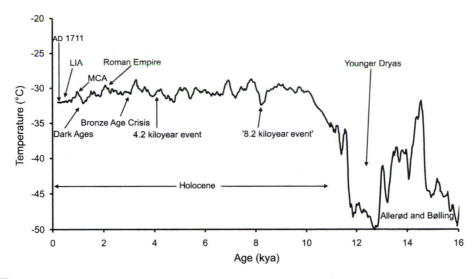

Fig. 5.3 Central Greenland temperature from the beginning of the 18th century (end of the pre-industrial era) to 16 000 years ago. As early civilisations were based in the northern hemisphere and as the Greenland ice core provides an approximate proxy for principal changes in that hemisphere's climate, so this record contains elements of hemispheric climate change that affected early civilisations. *Note:* regional variation within the hemisphere takes place as circulation patterns change with global temperature. So look for change in this temperature (irrespective of direction) as a marker of societal impact. Also note that central Greenland temperature is not the same as global temperature. Year 0 on the *x* axis is 1950. Data from Alley, 2004, and obtained from GISP2 Ice Core Temperature and Accumulation Data. IGBP PAGES/World Data Center for Paleoclimatology Data Contribution Series #2004–013, downloaded from ftp://ftp.ncdc.noaa.gov. The original journal publication was Alley (2000). LIA, Little Ice Age; MCA, medieval climatic anomaly.

Holocene's commencement and continued through to about 7000 years ago. Consequently some of the early coastal civilisations became inundated, as is evident from the submarine remains of settlements in the Mediterranean and off the coast of India. This was probably the earliest major impact of global climate change on human civilisation. That is to say, an impact as a direct result of climate on the most advanced type of human activity at the time (unless one were to argue that, prior to settlements of buildings, the development of the use of fire during some of the preceding Quaternary glacials/interglacials amounted to civilisation).

Getting an idea of the principal times of Holocene climate change in various parts of the world is difficult. However, an indication of the timings of principal change in the northern hemisphere can be gleaned from Greenland ice cores (see Figure 5.3). These times of climate change saw changes in the pattern of atmospheric circulation, and hence the distribution of precipitation and seasonal patterns.

Even without a complex civilisation, just small settlements, humans have been subject to the vagaries of climate change from the beginning of the Holocene (and presumably earlier in prehistory). The greening and ultimate desiccation of the Sahara is probably among the single most severe climatic fluctuations of the Holocene (but not the single most severe abrupt event). Ecosystem succession in the Sahara is well known from many lines of evidence such as pollen spectra, palaeolake levels and high-resolution palaeolake sediment cores. It was driven by variation in orbital insolation and magnified by feedback between monsoonal rainfall and vegetation. Human adaptation during this climate fluctuation is best known in the Eastern Sahara to the west of the Nile valley. This region saw continuous occupation from 10 500 years ago, when hunter – gatherers expanded across open grass savannah habitats, to about 5500 years ago, when drought drove pastoralists from most areas of the Sahara. Regions of occupation elsewhere in the Sahara most closely resemble those of the Eastern Sahara during the early Holocene ($\approx$10 000–9000 years ago), when pottery-producing hunter – fisher – gatherers resided beside palaeolakes. In other words this area can be considered as fairly representative of the broader region. By the mid-Holocene, occupational patterns diverged in the Central and Western Sahara due to wet – arid cycles. During periods of little rain small communities moved to transient bodies of water, rivers and upland refugia. Despite increasing knowledge regarding occupational succession in the Sahara from early to late Holocene, this record is based on individual sites that typically preserve short intervals of occupation, include few if any intact burials and rely largely on indirect dating of human remains and artefacts.

In 2008 it was reported that a new site complex called Gobero, located at the western tip of the hyperarid Ténéré Desert in the southern Sahara in Niger, on the north-western rim of the Chad Basin, provided a clearly variable habitation record (Sereno et al., 2008). It showed evidence of periods of human occupation and in-between times when the land was devoid of any settlement. Approximately 200 burials were found, ranging in age over five millennia. As such, Gobero preserves the earliest and largest Holocene prehistorical cemetery in the Sahara, so its archaeological exploration opened a new window on the funerary practices, distinctive skeletal anatomy, health and diet of early Holocene hunter – fisher – gatherers, who expanded into the Sahara when climatic conditions were favourable. Associated middens (rubbish dumps) and an exceptional faunal and pollen record help illuminate the area's history of episodic human occupation under conditions of severe climatic change. The time 16 000–9700 years ago, at the end of the last glacial (the Devensian or Weichsel as it is sometimes known in Europe, or Wisconsin in North America) and beginning of the current Holocene interglacial, was mainly arid, and dune sands accumulated. At the time 9700–8200 years ago there were wet climatic conditions that sustained a population of hunter – fisher – gatherers who were largely sedentary with lakeside burial sites that include the earliest recorded cemetery in the Sahara, dating to approximately 8500 years ago. Then, around approximately 8500–8300 years ago, the level of palaeo-Lake Gobero rose, submerging the dunes and forcing humans out. Well-aerated permanent water at depths of 5 m or more is suggested by the remains of Nile perch (*Lates niloticus*). This was followed 8200–7200 years ago by a millennial drought; a time that correlates well with the arid interruption in the central Sahara, which was a somewhat shorter interval ($\approx$8400–8000 years ago, with a duration of around 400 years) of severe climatic deterioration across the Chad Basin that was linked to cooling events in the North Atlantic. (It was the time that a pulse of fresh water from Lakes Agassiz and Ojibway entered the North Atlantic disrupting the Broecker thermohaline circulation. This period of climate change is sometimes known as the 8.2 kiloyear event; see section 4.6.3.) This was immediately followed by the return of humid conditions 7550–6750 years ago and the start of a lengthy period of human habitation to 4780 years ago: human settlement ended sometime around 4910–4760 years ago. Interestingly, the human remains found show a morphological, presumably evolutionary, change across the periods of habitation (Sereno et al., 2008).

The disruption by Lakes Agassiz and Ojibway of the Broecker thermohaline circulation 8200 years ago, which significantly cooled the northern hemisphere for around a century (and less so for around three centuries), is also thought to have affected Neolithic people in Europe and around the Mediterranean. In 2006 Bernhard Weninger, Eva Alram-Stern, Eva Bauer and colleagues compared the climate record with archaeological sites before and after this event using radiocarbon dating. They found that there were major disruptions of Neolithic cultures in the Levant, northern Syria, south-east Anatolia, Central Anatolia and Cyprus, all at the same time. The influence of the event 8200 calibrated-years ago is best recognised in Central Anatolia. There, the large and long-flourishing settlement at Catalhöyük-East, which had been inhabited for around a thousand years, was deserted quite abruptly. Of course, as rainfall and climate belts shift there were winners as well as losers, and many other major archaeological sites in the Eastern Mediterranean were first occupied around 8200 years ago. For example, in north-west Anatolia there was Hoca Çeşme; in Greece there was Nea Nikomedeia, Achilleion and Sesklo; and in Bulgaria there was Ovcarovo-Platoto. Others became deserted around this time, such as in Cyprus: Khirokitia and possibly Kalavassos-Tenta. Interestingly, the researchers found no site where there was continuous settlement across (before and after) the event 8200 years ago. The researchers propose that the event triggered the spread of Neolithic farmers out of Anatolia, into Greek Macedonia as well as into the fertile floodplains of Thessaly, and simultaneously into Bulgaria and probably other regions as well (but these were not yet researched at the time of Bernhard Weninger and colleagues' work). Their results explain the transition from what archaeologists and prehistorians call the Late Neolithic to the Early Chalcolithic and so the event could have caused some very significant, irreversible and unexpectedly rapid changes to the contemporary social, economic and religious lifestyle in large parts of western Asia and south-east Europe. Yet small settlements (be they many, spread over a wide area) are one thing, what of larger, more complex societies with some coordinating urban habitation?

Early civilisations (with cities as opposed to small settlements) were also vulnerable to climate change. For example, it is likely that the demise of the Akkadian empire in Mesopotamia was climate-induced. This civilisation was arguably the Earth's first empire. It was established sometime in the century beginning 4300 years ago by Sargon of Akkad and ultimately covered Mesopotamia from the headwaters of the Tigris and Euphrates Rivers to the Persian Gulf during the late third millennium BC. Archaeological evidence has shown that this highly developed civilisation collapsed abruptly 4170 ± 150 years ago, and quite probably this was related to a shift to more arid conditions as part of what is sometimes called the 4.2 kiloyear event. At the time there were widespread abandonments of the agricultural plains of northern Mesopotamia and dramatic influxes of refugees into southern Mesopotamia. There populations grew so much that a 180 km-long wall, the 'Repeller of the Amorites', was built across central Mesopotamia to stem nomadic incursions to the south (Weiss,

2000). While detailed palaeoclimate records to test the climate-induced decline asser-tion from Mesopotamia are rare, changes in regional aridity are preserved in adjacent ocean basins. For example, in 2000 Holocene changes in regional aridity were determ-ined using geochemical analyses of a marine sediment core from the Gulf of Oman. (This is directly downwind of Mesopotamian dust-source areas and archaeological sites.) These data suggest a very abrupt increase in dust, and hence Mesopotamian aridity, around 4025 ± 125 years ago that persisted for roughly 300 years. Further-more, radiogenic (Nd and Sr) isotope analyses confirm that the observed increase in mineral dust was derived from Mesopotamian source areas. This provides a direct link between Mesopotamian drought and social collapse; hence drought was a key, if not critical, factor contributing to the collapse of the Akkadian empire (Cullen et al., 2000). The onset of the sudden drought in Mesopotamia around 4100 years ago coincides with a widespread cooling in the North Atlantic. During this event, sometimes called the Holocene Event 3, the Atlantic subpolar and subtropical surface waters cooled by 1°C to 2°C.

Similarly, 4500–3500 years ago the Indus valley civilisation, and the principal cities of Harappa and Mohrnjo-daro, ended due to drought. There is no evidence of war but we do know that there was a major decline in rainfall across the Middle East and northern India 3800–2500 years ago. The same drought may also have affected civilisations around the eastern Mediterranean such as the Hittite empire, the Mycenae and the Egyptians (whose longer-term fate may also have been climate related, see below). I say 'may have' because their decline is sometimes attributed to a raiding, aggressive, sea-going people.

Regarding climate's socio-economic importance to the success and failure of early civilisations, it is interesting to note that the Roman Empire thrived over a period that was coincident with a benign climate. Indeed it was a Roman, the writer Saserna, who was one of the first to suggest that a biologically related human activity, viticu-lture, was a proxy indicator of climate change. More recently dendrochronological evidence (Büntgen et al., 2011) points to Central European summer precipitation and temperature variability with wet and warm summers taking place during periods of Roman and medieval prosperity. Conversely, increased climate variability from approximately AD 250 to 600 coincided with the demise of the Western Roman Empire and the subsequent European turmoil.

Meanwhile, in Western Europe the Argaric culture emerged in south-eastern Spain around 4300 years ago. This civilisation, which inhabited small fortified towns, was one of the first in Western Europe to adopt bronze working. But about 3600 years ago the culture died out. The area in which the Argaric people lived is one of Europe's driest areas even today (indeed, especially today with recent warming) and so the Argarics were living in a marginal or near-marginal environment. Pollen analysis suggests that ecological change took place around that time. Part of this is thought to be due to humans clearing forests, but there is evidence conditions were becoming progressively arid from about 5500 years ago onwards as indicated by a broad reduction in forest cover, the appearance of plants adapted to dry conditions and a drop in lake levels. What it means is that the Argaric arose somewhere that was already climatically stressed and this together with the pressure they themselves exerted on the environment meant that their society was no longer sustainable.

Between around 1200 and 1150 BC, just as the Bronze Age was giving way to the Iron Age, there was a period sometimes known as the Bronze Age Crisis (see Figure 5.3). This time saw the decline of the Mycenaean kingdoms, the Hittite Empire and the much of the Egyptian Empire. During this time some major cities, such as Hattusa, Mycenae and Ugarit, were destroyed. This crisis has been attributed to many things including volcanic activity, migrating populations and, of course, the technological revolution of iron itself. Now, while there were some small and medium volcanic events, these would have had a far more local impact than the Eurasia-wide decline. And while there were migrating and warring peoples, this begs the question as to why there was migration and war? Climate change seems to be a key factor (Weiss, 1982). There are the alluvial deposits near Gibala-Tell Tweini that provide a unique record of environmental history and food availability estimates covering the Late Bronze Age and the Early Iron Age (Kaniewski et al., 2010). Indeed, a refined pollen climatic proxy suggests that drier climatic conditions occurred in the Mediterranean belt of Syria from the late 13th/early 12th centuries BC to the 9th century BC. This period corresponds with the time frame of the Late Bronze Age collapse and the subsequent Dark Age. The abrupt climate change at the end of the Late Bronze Age caused region-wide crop failures, leading towards socio-economic crises and sustainability failure, forcing regional habitat-tracking whereby people moved as the locales where favourable conditions shifted. Archaeological data show that the first conflagration of Gibala occurred simultaneously with the destruction of the capital city Ugarit currently dated between 1194 and 1175 BC. Gibala redeveloped shortly after this destruction, so the environmental conditions in that area could not have been that bad even if its destruction might be attributed to displaced populations arriving from places where the climate was not clement.

The hemispheric if not global climate did seem to change and while there is no clear evidence of global warming or cooling there does appear to have been a period of European cooling and drought. At the very least there was some temporary change in oceanic and atmospheric circulation patterns. The impact of the Bronze Age Crisis on early civilisation was sufficiently profound as to be considered a distinct episode in prehistory.

Again, the decline of many civilisations after AD 536 could also have been climate-related, due to a major volcanic eruption (perhaps Rabaul in New Guinea, suggested by Stothers, 1984; see also Figure 2.1). Chroniclers of the day from Rome to China report that the Sun dimmed for up to 18 months, enough to have a marked effect on some temperate species, and there were widespread reports of poor harvests. This was early in the post-Roman European Dark Ages.

An ocean away, AD 536 was also the time of the fall of Teotihuacan city in Mexico, whose influence extended to Veracruz and the Maya region. At its height, in the first half millennium AD, Teotihuacan was the largest city in the pre-Columbian Americas. Its population may have reached more than 200 000, placing it among the largest cities of the world at that time. Human remains from the time of the collapse show signs of malnutrition, suggesting famine.

A couple of centuries later in Central America, ^{18}O sediment analysis from Lake Chichancanab reveals that over the past 8000 years it was driest in AD 750–900 (Hodell et al., 1995). This may explain the collapse of the Mayan civilisation that had previously thrived for more than 2000 years. The Mayan culture culminated in a coherent society covering much of Guatemala, Belize, Honduras and Mexico, sustaining a population at a level higher than today's. Although the evidence is not absolutely conclusive, the pattern of Mayan decline reveals that the cities next to rivers survived the longest, which is what you would expect in a drought-driven decline. It has also been suggested that in South Africa the medieval climatic anomaly (MCA; or medieval warm period) around AD 900–1300 may have been responsible for changes in settlement patterns (Tyson et al., 2002). There is a considerable body of evidence to support this, but to cite just one from Gerald Haug and colleagues (2003), river flow evidence in northern tropical South America shows that the collapse of Maya civilisation in the Terminal Classic Period probably occurred during an extended regional dry period, punctuated by more intense multi-year droughts respectively centred at approximately AD 810, 860 and 910. The data suggest that a century-scale decline in rainfall put a general strain on resources in the region, which was then exacerbated by abrupt drought events, contributing to the social stresses that led to the Maya demise.

There is debate in the academic literature concerning the decline of the Maya because factors other than climate change can cause societal collapse (even if climate change played a part). And the evidence is not always clear (Aimers, 2011). Climate proxies do not always give as accurate dating as one would like, and there are regional variations in climate which mean that if the proxy is not located in the immediate locality then accuracy is compromised. Our current understanding of the relationship between climate and the decline of the Maya therefore remains fuzzy. Sharpening the evidence will require new high-resolution, accurately dated proxy records, preferably on a local scale, from both archaeologists and palaeoclimatologists, necessitating mutual cooperation when interpreting the results (Hodell, 2011).

Subsequently, and elsewhere in South America, the rapid expansion of the Inca civilisation from the Cuzcoarea region of highland Peru approximately AD 1400–1532 produced the largest empire in the New World. Although this rise may in part have been due to the adoption of innovative strategies, supported by a large labour force and a standing army, it would not have been possible without increased crop productivity, and this in turn was linked to more favourable climatic conditions. In 2009, a multi-proxy, high-resolution 1200-year lake sediment record from Marcacocha, located 12 km north of Ollantaytambo in the heartland of the former Inca Empire, revealed a period of sustained drought that began from AD 880, followed by increased warming from AD 1100 that lasted beyond the arrival of the Spanish in AD 1532. After AD 1150 these increasingly warmer conditions allowed the Inca and their immediate predecessors the opportunity to exploit higher areas through the construction of agricultural terraces with glacial-fed irrigation, in combination with agroforestry techniques. The irrigated terraces themselves may have been increasingly necessary in the regions to overcome conditions of seasonal water stress, so allowing efficient agricultural production at higher altitudes. The outcome of these strategies was greater long-term food security and the ability to feed large populations. The climate of the region did change after 1540 but it was the arrival of the Spanish (and diseases with them) that sealed the Incas fate (Chepstow-Lusty et al., 2009).

Moving north, the Anasazi, Ancient Pueblo People or Ancestral Puebloans were an old native American culture, in the south west of what is today the USA, that arose around 1200 BC and who are best known for their stone and adobe dwellings built along cliff walls. The period AD 700–1130 saw a rapid increase in population due to consistent and regular rainfall patterns. Yet the cliff settlement was abandoned by AD 1300. What is known is that after approximately 1150 North America experienced climatic change in the form of a 300-year drought, which also led to the collapse of the Tiwanaku civilisation around Lake Titicaca. The contemporary Mississippian culture also declined around this time. This was just one of the 'mega-droughts' the south-west USA has seen in the past 1000 years. A number of these and their social impacts were outlined by Peter deMonocal from Columbia University in a review paper in *Science* in 2001. In this he notes that that a comprehensive dendrochronological analysis (see section 2.1.1) of hundreds of tree-ring chronologies from across the USA has established that there was a series of summer droughts extending back to AD 1200. Furthermore, this chronology points to there being much more persistent droughts before the 1600s than subsequently. These so-called mega-droughts were intense and persisted over many decades, recurring across the USA south west roughly once or twice every 500 years. This dendrochronological evidence has also been corroborated by lake sediment records. Peter deMonocal further points out that palaeo-oceanographic data indicate that these events were associated with changes in subpolar and subtropical surface as well as deep-ocean circulation.

This picture became clearer in 2011 when the first dendrochronological record going back over a thousand years was constructed. Through the deduced drought timings this showed that the south-western USA drought of 1150 extended down to Mexico (Stahle et al., 2011). (This dendrochronological record also shows a mega-drought in AD 897–922 and readers will note that this too is coincident with the drought in northern South America that affected the Mayans in AD 750–900: American mega-droughts may have covered a wider area than previously thought.)

Turning to the eastern hemisphere, work by Pingzhong Zhang, Hai Cheng, R. Lawrence Edwards and colleagues in 2008 used an isotope record from Wanxiang Cave, China, to correlate the Asian monsoon history over the past 1810 years. The summer monsoon correlates with solar variability, northern hemisphere and Chinese temperature, alpine glacial retreat and Chinese cultural changes. The Asian monsoon was generally strong during Europe's medieval warm period (MWP). Conversely, it was weak during Europe's Little Ice Age, as well as during the final decades of the Tang, Yuan and Ming Dynasties, all times that were characterised by popular unrest. It was strong during the first several decades of the Northern Song Dynasty, a period of increased rice cultivation and dramatic population increase.

This 1810-year timeline from China sees changes that chime with the timing of (sometimes different) changes in climate represented in the central European dendro-chronological data of Ulf Büntgen et al. (2011) and other proxy records. Exceptional climate variability is reconstructed for circa AD 250–550, and coincides with some of the most severe challenges in Europe's political, social and economic history in the early medieval period. Distinct drying in the 3rd century paralleled an aforementioned period of serious crisis in the Western Roman Empire marked by barbarian invasion, political turmoil and economic dislocation in several provinces of Gaul, including Belgica, Germania superior and Rhaetia. Reduced climate variability from circa AD 700–1000 coincides with new and sustained demographic growth in north-west Europe, and even (as we shall shortly see) the establishment of Norse colonies in the then not-so-cold environments of Iceland and Greenland. Humid and mild summers paralleled the rapid cultural and political growth of medieval Europe under the Merovingian and Carolingian dynasties and their successors. Average precipitation and temperature with fewer fluctuations during the subsequent circa AD 1000–1200 period was also one of medieval socio-economic growth. However, this was not sustained. Wetter summers during the 13th and 14th centuries and a first cold spell circa AD 1300 marked the onset of the Little Ice Age.

From all of this we can see that climate change has been a key factor in the success (and/or failure) of early human societies. In his 2001 review paper Peter deMonocal observes that what makes these ancient events so relevant to modern times is that they simultaneously document both the resilience and vulnerability of large, complex civilisations to environmental variability. Early complex societies are neither powerless pawns nor infinitely adaptive to climate variability. As with modern cultures, the ancients adapted to and thrived in marginal environments with large interannual climate variability. Again, as with ancient cultures, modern civilisations (regrettably) gauge their ability to adapt to future climate variations on the basis of what is known from historical (oral or instrumental) records. What differentiates these ancient cultures from our own is that they alone have witnessed the onset and persistence of unprecedented drought that lasted from many decades to a couple of centuries.

Added to this germane view, another point needs to be made. In the past, those belonging to societies that became debilitated by climate events had a number of choices. If they had the wisdom and means they might adapt. However, sometimes this was not an option, but they still had the choice to stay where they were or migrate (with sometimes the halfway option of raiding societies elsewhere for the resources they needed to survive). However, today, with more than 7 billion people on the planet, the option of wholescale population migration is not a viable one: at least not viable without considerable social conflict.

5.2.2 The Little Ice Age's human impact

The Little Ice Age (see sections 4.6.3 and 4.6.4 on the Holocene) and the period that led up to it has probably one of the best-recorded historical impacts of climate on human activity. Prior to the Little Ice Age the warm MCA had enabled viticulture in the English midlands. Since then and up to the late 20th century viticulture's main northern boundary (other than a few pockets) was in Normandy, some 480 km to the south. Because in Europe the human impact of this transition from medieval warmth to the Little Ice Age is so well recorded we know what took place. It is less well documented outside of Europe, although it is thought that possibly the decline of the Mapungubwe state (in the current semi-arid Shashi-Limpopo basin, in the area where the Botswanan, Zimbabwean and South African boundaries meet) may have been a Little Ice Age impact. As noted in the previous chapter the palaeo-evidence for the severity of the Little Ice Age outside of the northern hemisphere is not so clear, although there is evidence of other climatic change, other than temperature, such as changes in aridity/precipitation. Similarly, in the Kuiseb River Delta of Nambia settlements were abandoned between 1460 and 1640 (Tyson et al., 2002). Was this due to the Little Ice Age? There is oxygen isotopic evidence from stalagmites in Makapansgat that the termination of what are assumed to be Little Ice Age effects in South Africa was abrupt. Also, as discussed in the last chapter, irrespective of its extent, even in the northern hemisphere the reasons for the Little Ice Age are not entirely clear. This should alert policy-makers considering future climate change that distinctive climatic-forcing events are not necessary for climate change associated with considerable human impact: combinations of lesser forcing factors may result in marked human impact.

In Western Europe during the Little Ice Age there was a general cooling of the climate between 1150 and 1460 and a very cold climate between 1560 and 1850 that had dire consequences for its people. The colder weather impacted agriculture, health, economics, social strife, emigration and even art and literature. Increased glaciation and storms also had a devastating effect on those who lived near glaciers and the sea. In the northern hemisphere the more severe climatic changes were not restricted to Europe, however. In China the population fell from around 100 million in the mid-13th century by about 40% over the next century or so.

What happened in terms of human impact has been pieced together from a variety of sources. Records of English, French and German harvests and/or prices survive to this day and these correlate well with the weather in good and bad years. So, it is easy to see that each of the peaks in prices corresponds to a particularly poor harvest, mostly due to unfavourable climates, with the most notable price peak in the year 1816 – 'the year without a summer' – following the Tambora eruption of 1815 (see Figure 2.1).

The Little Ice Age harvest losses not only caused price increases but death too. Famines became more frequent and in 1315 one killed an estimated 1.5 million in Europe. This was probably the first big famine of the Little Ice Age and may have been aggravated because agricultural systems were still partly embedded in MCA mode that lasted through much of the 13th century. This happened elsewhere in the northern hemisphere. In the east, as mentioned, there was famine in China. In the west of the northern hemisphere, at the century's end the Anasazi abandoned their long-established communities in the now desert of the south-western USA. At the time, North American climatic belts shifted, including those suitable for economically important crops: in what is now Wisconsin the northern limit of maize cultivation moved southwards by up to 320 km.

At the beginning of the 14th century Northern Europe saw two exceptionally severe winters, in 1303 and 1306. However, it was the run of cool and wet summers between 1314 and 1317 that particularly decimated harvests, hence precipitating the aforementioned great famine of 1315. Records show that this reduced the food supply from Scotland to the Pyrenees and as far east as Russia. In some places, even when plant growth was good, the wet, as is clearly recorded in England, made the harvest difficult. In London in the spring of 1316 the price of wheat rose 8-fold over that of late 1313: these price levels were not to be reached again until later in the Little Ice

Age in the late 15th century. Meanwhile, across the English Channel in France in just 6 months grain prices rose by more than 3-fold up to May 1316.

The climate also had military impacts. In France King Louis X's attempt to combat Count Robert of Flanders came to a halt when Louis's army became bogged down in mud: it was so bad that horses could not pull wagons. Meanwhile, for the average person matters were dire. People ate dogs and reports of cannibalism were not uncommon; graves were robbed and criminals cut down from gibbets for food. Similar reports exist from other Little Ice Age famines but those from the 1314–17 period were particularly bad and this famine undoubtedly exceeded the others in terms of its severity (Burroughs, 1997).

It is worth emphasising that the aforementioned human impacts experienced early in the 14th century were not simplistically proportional to the poor climate. Their severity was augmented by communities having grown during the preceding warmer climatic anomaly and having expanded into lands that were ideal for the warm anomaly but which became marginal during the Little Ice Age. Such demographic pressures can exacerbate the impacts of climate change. This is a theme we will return to again when looking at future climate change (Chapters 6 and 7) as currently the early-21st-century global population is at a record-breaking high, at more than 7 billion and counting.

It is not known exactly how many people died in the 1314–17 famine but in Ypres in Belgium (an advanced city of the time) records suggest a mortality of more than 10%. Because the Black Death (or the Great Mortality as it was known then) ravaged England in 1348 and Scotland in 1349, 14th-century mortality from both plague and various climate impacts reduced England's population by around a third. Looking at the 14th century as a whole, much of Europe and Eurasia was similarly affected.

Climate (or in this case perhaps the weather) also affected the Black Death's impact. While the Little Ice Age saw many cool years, there were also occasional warm ones. One of the warmest was that of 1348 and this may account for the rapid spread of plague throughout England in that year. But more typical of the Little Ice Age was cold and wet, and this too, under certain circumstances, helped spread plague: pathogens tend to thrive more in wet than arid conditions. Great floods in China in 1332 not only killed several million people but disrupted large areas of the country, causing a substantial movement of both people and wildlife. Regarding emergent diseases, this was a recipe for disaster and so it is not unreasonable to conclude that climate played its part in the way that the plague spread (Burroughs, 1997).

Although the 1314–17 famine was probably the worst in terms of human impact, it was not the only one during the Little Ice Age. There were vagaries of the weather but harvests were not usually so bad as to provoke sharp and major price rises in grain, although there were some, such as in 1439, 1482, 1608, 1673 and 1678. Another severe famine, particularly according to French records, took place in the 17th century due to the failed harvest of 1693. Millions of people in mid-Western Europe – France and surrounding countries – died. This time was again one of poor climate (see Figure 5.2) and the 1690s saw a number of very cold winters and cool, wet summers affecting much of Western Europe. Indeed, 1693 saw one of the worst famines since 1314–17, although the impact was worse in continental Europe than in

England. There was also a bad year in 1697 and it is estimated that in Finland roughly a third of the population died.

In 2011 a team of Chinese researchers, led by David Zhang and Harry Lee, correlated 16 European variables between the 13th and 17th centuries that themselves are affected by European climate and with northern hemisphere proxy temperatures. These included tree rings and grain yield as indicators of bioproductivity, agriculture production from the historic agriculture production index, food supply as suggested by grain price and the historic wage index, social disturbance, migration, famine mortality, nutritional status as revealed by average population height, epidemics, war as quantified by a fatality index, and human population. One hundred and twenty cross-correlations of these variables were statistically significant ($\geq$95%) and of these 116 were highly significant ($\geq$99%). The workers used a European temperature series as another indicator of conditions of harmony or crisis to simulate the 'golden' and 'dark' ages in Europe over the past millennium. They set a temperature-variation limit equal to -0.1 σ (σ means standard deviation) according to the 100-year smoothed European temperature series as the general crisis threshold. The periods in which the temperature was lower than -0.1 σ or greater than -0.1 σ of the long-term mean represented dark ages and golden ages, respectively. With that threshold, the dark ages calculated were AD 1212–1381, the Crisis of the Late Middle Ages, and AD 1568–1665, the General Crisis of the 17th Century. Conversely, the golden ages were the 10th–12th centuries (the High Middle Ages), the late-14th to early 16th centuries (the Renaissance) and the late-17th to 18th centuries (the Enlightenment). This is largely in agreement with time intervals delimited by historians. The mild cooling in Europe in the late 18th and 19th centuries brought about a rise in consumer prices, social disturbance, war and migration, but not demographic crisis, because of social buffers such as cross-continental migration, trade and industrialisation. In the study, Zhang et al. (2011) all criteria for confirming the causal mechanisms between climate change and human crisis were met. The alternation of historical golden and dark ages in Europe and the northern hemisphere, which often has been attributable to sociopolitical factors was indeed rooted in climate change, even if they were exacerbated by sociopolitical circumstances. Climate change determined the fate of agrarian societies via the economy (the ratio between resources and population). The findings of the Zhang and Lee team have important implications for industrial and post-industrial societies. They concluded that any natural or social factor that causes large resource (supply) depletion, such as climate and environmental change, overpopulation, overconsumption or non-equitable distribution of resources, may lead to a general crisis.

Let us put the quantitative analysis of Zhang et al. (2011) into some sort of qualitative context with what we already knew. As noted, the effect of the harsher Little Ice Age climate on those already on marginal land was far worse than those on formerly good lands of the MCA that the Little Ice Age in turn made marginal. Those already living at high altitudes in mountain pastures were very badly affected. Consequently, Scandinavian, Scottish and Swiss farmers suffered particularly. Problems were not solely confined to direct climatic impacts. The new Little Ice Age environmental conditions favoured some crop pathogens. Since snow covered the ground deep into spring in the mountainous parts of Europe a fungus known as pink snow mould (*Fusarium nivale*) devastated crops. It thrives under melting snow cover or prolonged, cold, drizzly weather. Additionally, due to the increased number of days of snow cover, the stocks of hay for the animals ran out in some pre-Little Ice Age benign areas, so livestock were fed on less-nutritious straw and pine branches. Many cattle had to be slaughtered. Others still died in the heavy snow. Marginal lands at high latitudes were similarly pressured. In northern Norway many farms were abandoned for better land in the bottoms of valleys. By 1387, Norwegian agricultural production and tax yields were far less compared with what they had been around 1300. By the 1460s it was being recognised that this change was long-term. As late as 1665 the annual Norwegian grain harvest is reported to have been only 67–70% of what it had been in the year 1300 (Lamb, 1995).

The Little Ice Age did not just affect marginal lands but also sensitive crops in less harsh areas; crops that required specific climatic conditions at specific times in the thermal growing season. Illustrative, as previously noted, is the Roman Saserna's suggestion of viticulture as a climate proxy. Grapes require a narrow range of conditions for ripening and so are particularly sensitive to climate change. The Little Ice Age saw many countries with a strong viticulture industry, such as France, have bad years for wine coincident with climate swings. Naturally those areas, such as southern England, that had benefited from the MCA and which had become the new northern margin for viticulture, were not able to grow vines much after 1400. German wine production between 1400 and 1700 was never above 53%, and at times was as low as 20%, of the production before 1300.

The sociological impacts of the Little Ice Age have also been suggested to have resulted in one particularly gruesome aspect of European life of the time and particularly around the 1590s: persecution. It is human nature to want to point the finger of blame when detrimental events take place. Local food, being a key biological driver, and food security, were more important locally than today, when developed nations have access to harvests from different parts of the planet. A poor harvest for a farmer in the 16th century caused personal hardship and poor harvests for a village or town would undermine the economy. It was therefore easy to blame someone for casting a spell that ruined crops, and so the 1590s were particularly notorious for witch hunts and executions. Indeed, as would be expected from a climate perspective, southern Europe saw fewer witch executions than more northern countries such as Germany, where there were many. Indeed, warmer Spain, which is famed for the Inquisition, had comparatively few executions. Here crop damage was far less of a factor than the desire to convert people to the 'true faith'. Crop damage was worse in the north and again catastrophic on the northern marginal lands. So much was this so that the descendents of those who previously had taken advantage of the northern migration of marginal lands at the beginning of the MCA found their fortunes reverse with the Little Ice Age. This is particularly highlighted by the pattern over the centuries of international migrations. During the climatic anomaly years of AD 800–1200 Iceland and Greenland were settled by the Vikings from Scandinavia. By around AD 1100 the population of Iceland peaked at some 70 000 and this is clearly climate-related. Human remains from Norse burial grounds in Greenland have been found that in the 20th century were in permanently frozen soil that was virtually impossible to dig by hand. This suggests, for the bodies to have been buried, that the average local temperature at the time of Norse occupation was some 2–4°C higher than at the end of the 20th century. In reality, the coastlands of the pleasantly named Greenland were not that much different from Iceland at the time and, if anything, worse, but it was felt that a country with a nice-sounding name would be more likely to attract new settlers.

Archaeological evidence has revealed that Icelandic Vikings had large farmsteads with cattle, pigs, sheep and goats, and there were ample pastures to sustain these as well as crops such as barley. However, the growing season in Greenland, even in the MCA, was very short. Frost typically occurred in August and the fjords froze in October. Nonetheless, by the 1300s more than 3000 colonists lived on 300 farms scattered along the west coast of Greenland.

This was all to change with the Little Ice Age. Indeed, well into this cool period, in 1492, the Pope complained that no bishop had been able to visit Greenland for 80 years on account of the sea ice. In fact, by that time his Greenland congregation was in all likelihood either already dead or had migrated. Hermann (1954) notes that during the mid 1300s many Greenlanders had moved on to Markland (today's Newfoundland) in search of a more benign home. Formal communication between Greenland and Europe had ceased in 1410 and was not re-established until the 1720s. Europeans did not recolonise that area until the 1800s. Meanwhile, Icelandic grain production was given up in the 14th century and in 1695 sea ice completely surrounded Iceland except for one port. This ties in with the aforementioned reports that in Western Europe in the 1690s crop failures were common. It was the coldest decade of the 17th century.

As noted at the start of this chapter, systemised meteorological recording began in the 18th century and has continued to the present day. Not surprisingly, the coldest year on this record occurred during the Little Ice Age. It was 1816 and followed the eruption in 1815 of Mount Tambora in Indonesia. The eruption distributed aerosols into the stratosphere which cooled the Little Ice Age climate further. In addition, 1816 saw the latest grape harvests ever recorded: November. Meanwhile, in eastern Canada and New England three cold waves in June, July and the end of August (with frosts) devastated harvests and damaged crops as far south as Boston, MA, USA. Harvests were down across the northern hemisphere and there is even a record of famine in Bengal that triggered an outbreak of cholera. The 1816 cholera epidemic even became global (the world's first pandemic of the disease), reaching Western Europe and North America by 1832.

In terms of human mortality, correlation of the CET record between 1665 and 1834 with mortality reveals that mortality increased in both cold winters and hot summers. Quantitatively, warming of 1°C in winter and cooling of 1°C in summer reduced seasonal mortality by 2 and 4%, respectively. The combination of such mild winters and cool summers was to raise life expectancy by about 2 years. We will revisit this when discussing health impacts of future climate change in Chapter 7, and it will be important to remember that the age profile of early 21st-century developed nations is far different to their 18th-century counterparts, with a higher proportion of elderly people. Nonetheless, climate-related health effects, albeit to an increasingly different degree, were still relevant through the 19th and 20th centuries. As recently as 1940, UK mortality from bronchitis and pneumonia was, at more than 850 000, double the remainder of the decade's average, and correlated with that year's cold winter. The year 1941 also saw a harsh winter, as did 1942, but there were successively fewer pneumonia and bronchitis cases as the most vulnerable had already died (Burroughs, 1997). This last point is also relevant to longer-term climatic change over a number of years compared to the impact of an isolated atypical year.

Returning to the question of food and the security of supply, one famine is particularly notorious in the 19th century: the Irish potato famine. This famine of 1845 was caused by the potato blight fungus, *Phytophthora infestans*. The climate of the autumns of 1845 and 1846 was particularly conducive to its spread. However, that a million Irish peasants died (and another million and a half emigrated as a result) cannot solely be laid at climate's door. Little then was known about such fungi, let alone their life cycle and habitat preferences: these were to be later elucidated by the German biologist Anton de Bary. The use of monocultures, the few potato cultivars (strains) available, the high population density as supported by near-subsistence farming (living on the edge) and a virulent new strain of the pathogen, together with the weather, all combined synergistically to aggravate the famine. As with climate and seasonal mortality, there are parallels with the position in which we find ourselves today. True, today the developed nations are not dependent on subsistence farming, but this just means that the ante has been upped: we have today a far higher population (more than 7 billion in 2012 and more than 6 billion at the end of the 20th century as opposed to fewer than 2 billion at its beginning) that is supported by globally connected, intensive agricultural systems. Today we are reliant on a slightly broader base of crop cultivars, but the conservation of genetic strains and wildlife biodiversity (the source of new natural genetic material) are still problematic. We are still close to the edge; possibly too close? As we shall see, climate change and weather can still serve to aggravate agricultural problems in the 21st century.

The Little Ice Age ended towards the end of the 19th century and the global climate began to warm by around half a degree Celsius towards roughly the millennial average (AD 1000–2000). This was followed by a 50-year period of stability. Computer models, both including and excluding the human addition of greenhouse gases, suggest that the Little Ice Age would have ended as it did through to the mid-20th century, but that the longer-term trend from the middle of the 20th century would have been one of gradual cooling to broadly somewhere around $-0.2°C$ of what was in reality the AD 1961–90 mean had not anthropogenic greenhouse gases been added (see IPCC, 2007, chapter 9). As it was, not only did the Little Ice Age end, but additional warming took place in the latter third of the 20th century. Figure 5.4 shows the actual 3-year-smoothed global temperature change since 1850 relative to the 1961–90 average as determined by systemised meteorological recording.

Looking at the recent year-by-year global climate as is directly measured (as opposed to the earlier longer-term climate changes inferred by all the various climate proxies), is what causes the shorter-term drunkard's walk of climate change. Since the beginning of chapter 3 we have explained the major shifts in climate as the biosphere has developed, switched between modes, and been subject to sudden vagaries such as the carbon isotope excursion (CIE) or volcanic events. But, for example, given that there were no major eruptions or the like in the period 2000–9, and given we were still pumping greenhouse gases into the atmosphere, why did global warming halt and

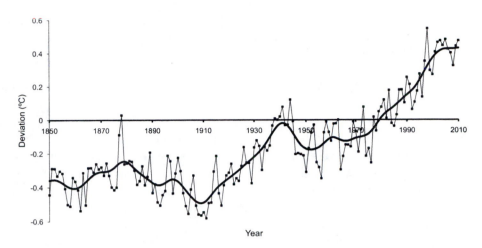

Fig. 5.4 Global temperature by year and also using a 3-year-smoothed filter, relative to the 1961–90 average and covering the period 1850–2010. The data for this graph was originally compiled jointly by the Climatic Research Unit and the UK Meteorological Office, and downloaded from www.cru.uea.ac.uk/cru/info/warming/ in July 2011.

even some cooling occur at this time? The answer is that not only are there chaotic meteorological vagaries but also the cycles of change in ocean circulation such as the El Niño are still continuing, as are changes in other forcing factors. Indeed in 2011 Robert Kaufmann, Heikki Kauppi, Michael Mann, and James Stock found, that the 1998–2008 hiatus in warming coincides with a period of little increase in the sum of anthropogenic and natural climate forcings. There was a declining in solar insolation as part of a normal 11-year cycle, and a cyclical change from an El Niño to a La Niña dominate our measure of anthropogenic and rapid growth in reflective sulphur emissions (driven by large increases in coal consumption in Asia in general, and China in particular). Individually, each of these cooling factors would not have halted the previous decade's warming trend, but together they did. Altogether these vagaries can sometimes conspire to make current global warming seem to go away for a few years. Actually this is only part of the story: the biosphere has probably continued to warm, despite possibly receiving less heat and reflecting more. We need to remember also that meteorological measurements that make up what we consider to be the average temperature of the planet do not actually do this: such measurements are an estimate of surface temperature. You will recall that heat entering the biosphere is not confined to its surface. Some of it, on a very short-term (diurnal) basis, is distributed throughout the atmosphere, and some of it absorbed by the oceans where in the longer term it is transported both laterally and vertically. Computer models of the biosphere by Gerald Meehl, Julie Arblaster and colleagues in 2011 show that, in the years between 2005 and 2010, heat was drawn from the globe's surface into the abyssal depths through downwelling currents and processes such as the Broecker thermohaline circulation. Even so, the anthropogenic warming following the Little Ice Age has been such that even if all the 'normal' vagaries in the various climate forcing factors simultaneously happened to coincide in a cooling way, the Earth would not return to the climate of the Little Ice Age. I say 'normal' because abnormal events (such as super-volcanic eruptions) can spring 'climate surprises', and we will look at some of these in Chapter 6.

5.2.3 Increasing 20th-century human climatic insulation

The climate's impact is not confined to more organised societies, although these societies may better document such episodes. Less-developed nations have always been subject to the vagaries of the weather and vulnerable to climate change. This is as true in the high-tech present as it was a century ago and earlier. However, those living in developed nations, while not completely escaping the ravages of weather and climate extremes, have become increasingly insulated from the weather and seasons. At the beginning of the 20th century in Europe, North America and Japan (the centres of industrialisation at the time), even those living in urban areas would need to note the weather forecast and dress accordingly: wrap up well in winter and wear waterproof clothing when appropriate. Travel at the start of the 20th century for most in these nations was still by foot or animal power. Steam was beginning to provide occasional long-distance travel for some, but most travel miles were by those exposed to the elements. Venturing outside meant preparedness dictated by the weather, seasons and climate. Similarly, the diet in these nations was also largely climate-determined. The fresh produce available was largely local and very seasonal. Indoors, most dwellings had just one or two primary sources of heat, usually a fire in the main living room, sometimes with a back-boiler to provide hot water. Gas cookers in urban areas were common, as denoted by the term 'town gas', which actually was largely a mix of methane and carbon monoxide derived from heating coal.

Jump forward to the end of the 20th century and things were markedly different. The developed world consisted of more nations, not to mention that the Earth's population had roughly quadrupled. Venturing out commonly involved using the car, the global passenger fleet having grown from virtually nothing in 1900 to more than 500 million vehicles by 2000 and proportionally with a nation's wealth per capita (Worldwatch Institute, 2003). By this time cars came with heaters as standard and many included air conditioning, which is fast becoming a standard fitting, especially in wealthier countries. Nor was diet in these nations any longer restricted by season or indeed locality. Visit any supermarket in a wealthy country today and the range of out-of-season and foreign foods is readily apparent. Meanwhile, homes themselves were increasingly cocooned by central heating and, where summers are regularly warm, air conditioning. Many families in north-western Europe celebrated the turn of the millennium at home in a shirt-sleeve environment. In short, the mundane lives of those in developed nations have become apparently increasingly insulated from the environment. 'Apparently' because people in developed nations are still vulnerable to extreme weather events.

Yet the very means that enabled this independence – energy use – was itself responsible for the generation of greenhouse gases that were, and are, altering the climate. Notwithstanding, this insulation from the environment has its limits. If there is ice on railway lines or on the road system, even those railway carriages or cars with the best heating systems are impeded. As for the future, as we shall see, developed

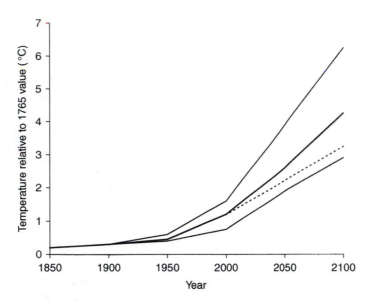

Fig. 5.5   The 1990 IPCC Business-as-Usual (B-a-U) scenario simulating the anticipated increase in global mean temperature with best estimate (thick line) and upper and lower estimates (thin lines) showing the temperature rise above the 1765 value. The IPCC subsequently revised the way it presented its scenarios and its forecasts in 1992 and adhered to these for its 1995 review. The nearest equivalents to the IPCC 1990 B-a-U scenario are the IPCC IS92 scenarios a and b and they are portrayed approximately by the dashed line. As these scenarios are within the upper and lower estimates of the 1990 B-a-U forecast, and as all the IPCC forecasts are decided by committee (hence are based on scientists' opinions rather than scientific fact) there was, at the time of the second report, little advantage in considering other forecasts than the 1990 B-a-U. The take-home message over the intervening half decade was essentially the same, be it regarding energy-policy questions raised or the possible environmental impacts to consider. In part after IPCC (1990).

nations will not be immune to impacts arising from likely climate change. Meanwhile, those in the less-developed nations have been, and are, less fortunate and vulnerable to direct impacts from climate change.

5.3   Climate and business as usual in the 21st century

5.3.1   The IPCC Business-as-Usual scenario

No one can see into the future, but we can attempt to forecast what might happen given certain circumstances. One set of circumstances is known as the Business-as-Usual scenario or B-a-U scenario (Figure 5.5). The next chapter looks at our greenhouse gas contributions, especially from energy, at the moment and for the rest of the 21st century. For now, while looking specifically at climate, it is worth appraising the scientific consensus for global warming forecast under the B-a-U scenario. In the 1980s there was such concern about the possibilities of global warming, as well as confusion, with doubt that anthropogenic climate change was even likely, that politicians called for international scientific guidance through the United Nations (UN). In 1988 two UN agencies – the World Meteorological Organization (WMO) and the UN Environment Programme (UNEP) – established the IPCC under the chairmanship of Professor Bert Bolin.

The panel was charged with:

1. assessing the scientific information related to the various components of the climate change issue, such as the emissions of greenhouse gases and the modification of the Earth's radiation balance, as well as the science needed to assess environmental and socio-economic consequences;
2. formulating realistic response strategies for the management of climate change issues.

It needs to be emphasised that there were some (especially in what came to be known unofficially as the fossil fuel lobby) who did not believe – or perhaps did not wish to believe – that human-induced climate change was real. This belief still exists among some energy-related and political quarters today, whereas many in the public are still uncertain, or unaware, of the science underpinning current climate issues. Some countries too (again especially those consuming prodigious quantities of fossil fuels) felt that the issue was overblown. Nonetheless, politicians needed clear unambiguous guidance. The IPCC aimed to provide this by ascertaining the scientific consensus. For its first report (1990) the IPCC brought together 170 scientists from 25 countries through 12 international workshops. A further 200 scientists were involved in the peer review of the report. Although there was a minority of opinions that could not be reconciled with the panel's conclusions, the extensive peer review helped ensure a high degree of consensus among the authors and reviewers. Even so, reaching this consensus was difficult. One of the ways the panel did this was to present both a best estimate as well as high and low estimates of the likely changes in global temperature. Further, they looked at a number of scenarios assuming that politicians introduced three levels of controls, as well as the B-a-U scenario which supposed that nothing would be done and that the late-20th-century trend in emissions would continue to grow. It has to be acknowledged that this methodology was as much to bring the disparate views of the scientists underpinned by various degrees of climatological understanding on board as it was to provide margins for error. Consequently, as we shall see, the difference between the high and low estimates under the B-a-U scenario for 21st-century warming was greater than the difference between the low estimate and nothing happening at all (zero climate change). Furthermore, the best estimates for the various scenarios all fell within the range between the B-a-U scenario's high and low estimates.

For these reasons the IPCC's 1990 B-a-U estimate (Figure 5.5) was of central importance to international policy-makers, whose initial problem was whether or not to acknowledge the problem and then, if so, to establish the environment in which subsequent detailed climate change policy might be formulated and enacted. A degree more or less warmer on top of the IPCC's 1990 anticipated 21st-century warming simply did not figure in policy discussions in the political arena: the high – low range was just stated and the policy debate in the early 1990s focused on whether doing anything about such warming actually mattered. Even today policy-makers are

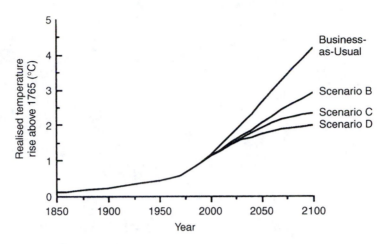

Fig. 5.6 Graphs of the IPCC 1990 scenarios. Redrawn with permission.

not seemingly concerned with the degree of climate change, but what they should do about managing or controlling change per se. As we shall see in Chapter 8, even when leading politicians met in December 1997 in Kyoto, Japan, the policy questions largely boiled down to whether countries would sign up to such a modest international agreement that (according to the IPCC scientific conclusions) it would not halt the increase in global warming but might slightly slow down the rate of increase in warming.

Meanwhile, the IPCC's 1990 B-a-U envelope has not yet (in 2012) been *effectively* superseded. Even in each of the IPCC's subsequent reports – 1992, 1995, 2001, and 2007 – the range of scenarios had the 'best estimates' for warming still within the 1990 high – low envelope of the B-a-U scenario. Only the high – low envelopes of just a few of the special 2001 policy scenarios went outside the IPCC 1990 high – low B-a-U envelope and that was just at the low end (1.15°C less warming for 2100 since 1750 than the low B-a-U estimate presented in 1990). So, while Figures 5.6 and 5.7 provide best estimates for various IPCC 1990 scenarios, the 2001 IPCC all-scenario and all-model envelopes respectively, and the 2007 IPCC selected scenarios, it is important to note that Figure 5.5, showing the original 1990 B-a-U scenario, is so similar to the 2001 all-model best-estimates envelope (Figure 5.7a) and the IPCC 2007 principal scenarios (Figure 5.7b) that we will use the 2007 best-estimates envelope against which to set the policy discussion for much of the rest of this book. Specifically, the discussion will be compared with what we know actually happened from palaeoclimatic indicators of the real Earth (and not an expensive computer model) at times of past warming and from when it was actually warmer than today. The former is analogous as to how biomes react to warming (climate change), whereas the latter is analogous to where we are heading with global warming.

These various IPCC scenario forecasts all portray a similar picture but because there were four IPCC assessments up to 2007 (and a fifth is due in 2013), when looking at the various assessment scenario forecasts together it can be a little confusing. So, it might be helpful to bring all the IPCC assessment Business-as-Usual scenarios (1990–2007) together.

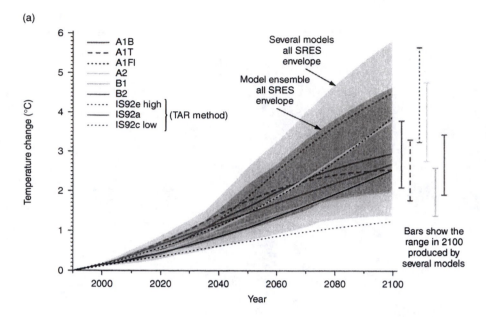

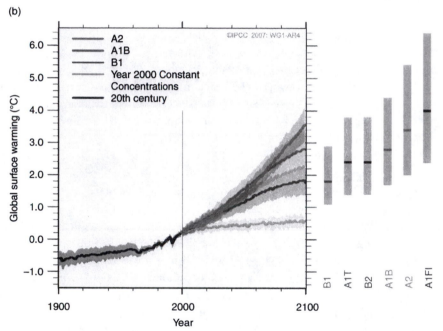

Fig. 5.7 (a) The IPCC (2001b) all Special Report on Emission Scenario (SRES) temperature-change forecasts, which provide an envelope of predicted 21st-century global climate change relative to the *1990 temperature*. Reproduced with permission of the IPCC. TAR, IPCC 2001 third assessment report. (b) The IPCC (2007) principal scenarios for the 20th and 21st centuries based on temperatures relative to the *1980–99 average*. Note that the line curving upwards to end at the highest forecast temperature for 2100 is the A2 B-a-U analogous scenario (if we do not significantly attempt to curb greenhouse gas emissions). The line leading to the lowest forecast rise for the year 2100 the IPCC include for purely comparison purposes as it is the scenario in which the Earth magically has atmospheric greenhouse gas concentrations held at what they were in the year 2000; this scenario is of course impossible fantasy but shows that whatever we do we will not escape at least 0.6°C warming by 2100. This reproduction is of figure SPM.5 from *The Physical Science Basis. Working Group I Contribution to the Fourth Assessment Report of the Intergovernmental Panel on Climate Change*, Cambridge University Press, is with permission.

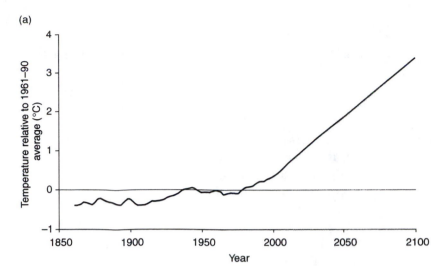

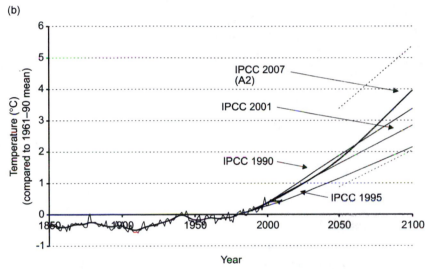

Fig. 5.8 (a) Global temperature (3-year smoothed) for 1861–2002 and IPCC 1990 B-a-U scenario for the 21st century, relative to the 1961–90 average. (b) Global temperature (3-year smoothed) for 1850–2010 actual measurements and the various 'best-estimate' IPCC B-a-U scenarios up to 2100 from the 1990, 1995, 2001 (very slightly simplified) and 2007 (bold line 2000–2100 under the A2 scenario) assessment reports, but all standardised to a zero temperature equivalent to the 1961–90 mean (the reference temperature that the University of East Anglia's Climate Research Unit uses). The dotted lines between 2050 and 2100 are the IPCC 1990 B-a-U 'high' and 'low' estimates. Note that the subsequent IPCC B-a-U analogous best-estimate assessments are all between the 1990 B-a-U high and low estimates (a testimony to the robustness of the IPCC's 1990 B-a-U scenario).

Figure 5.8a is a 3-year-smoothed graph (as is the bold line in Figure 5.4) of the global temperature since 1861 to 2002, with Figure 5.8a the IPCC 1990 B-a-U forecast added from 1990 to the end of the 21st century. (The upper and lower limits [that were in Figure 5.5] prior to 2050 have been removed leaving just the 'best estimate' for the IPCC's 1990 Business-as-Usual scenario which remains a useful rough-and-ready, loose approximation of the long-term climatic trend for the next few decades at least as far as policy-makers are concerned.) From this it can be seen that should we continue to increasingly add greenhouse gases to the atmosphere as we have been, then the planet will end up warmer than it has been since before the beginning of the Industrial Revolution. Even if we included the lowest of the best estimates in the various IPCC 2001 scenarios, the forecast still would have been far warmer at the end of the 21st century than at any other industrial time. (This can be seen in Figure 5.5a.)

It is important to note that the temperature scale varies between IPCC assessments: this is a feature that most people overlook. The 1990 assessment refers to temperatures above pre-industrial times (Figure 5.6), taken as the temperature in 1765. Conversely, the 1995 and 2001 IPCC assessment forecasts for the 21st century refer to temperatures above the 1990 global temperature. Different again, the 2007 assessment's forecasts were based on temperatures relative to the 1980–99 average. Meanwhile the Climate Research Unit at East Anglia University uses the 1961–90 average as its zero temperature against which anomalies are presented. We need to note that the warming base line is not the same for all the IPCC assessment. So what I have done with the Figure 5.8b is to adjust all the IPCC assessments estimates under B-a-U (or the analogous scenario depending on the respective IPCC report) relative to the Climate Research Unit 1961–90 baseline standard. The IPCC scenario forecasts presented in Figure 5.8b are: the IPCC 1990 Business-as-Usual forecast; the IPCC IS92a scenario from the 1995 assessment; the IS92 equivalent to 1990 in the 2001 assessment[3]; and the A2 scenario from the IPCC's 2007 assessment. From Figure 5.8b it can be seen that all the IPCC's various B-a-U (or to be strict the B-a-U analogous) scenarios forecast a temperature rise of between 2 and 4°C above the 1961–90 mean global temperature for the year 2100. Indeed, remove the IPCC's 1995 B-a-U estimate and all remaining estimates are between 2.85 and 3.96°C above the 1961–90 mean for the year 2100. The reason why the 1995 IPCC estimate is lower is largely because they assumed a greater cooling forcing from anthropogenic aerosols.

The take-home message from this confusing variation in the different IPCC assessments is that there is actually reasonable agreement between the future greenhouse gas scenarios. That they differ slightly is due to (1) continually improving scientific understanding of the various climate forcing factors and (2) the (hopefully) better appreciation of the likely additions of greenhouse gases to the atmosphere as we progress into the 21st century, especially as, if nothing else, the future is closer.

Taking all the IPCC assessment reports together (irrespective of whether or not we place greater emphasis on the latest), the greatest likelihood is that in the 21st century the Earth will become warmer than at any other time in the Holocene interglacial. In fact, using IPCC 2007 warming high estimates (as opposed to best estimates) the end of the 21st century could easily see temperatures higher than Earth has had since the last interglacial some 140 000 years ago, or indeed the previous interglacials during which humans evolved. However, the Earth is a very different place today compared to 140 000 years ago (or earlier). To begin with, back then there were not more than

---

[3] Note that that some of 2001 assessment's other B-a-U-type scenarios were slightly higher than the one meant to be the most analogous to the 1990 assessment and so are even more in line with the 1990 assessment.

7 billion humans, and rising. Nor was the planetary landscape so managed or the global commons, both atmosphere and oceans, perturbed by human action. Therefore, the biological and human ecological consequences of such climate change will be more marked than the climate impact during the warmest of these earlier warm times.

If we carry on this century much as we did throughout the last, and so end up with marked climatic change, the question arises as to what would happen if greenhouse gas emissions were curbed. The IPCC looked at a range of such scenarios in each of their assessments (1990, 1995, 2001 and 2007), which assumed various controls of gas emissions, improvements in energy use and so forth. These scenarios were realistic in that they took into account that the starting point was the present (1990 or 2000 depending on which IPCC report), so humanity was stuck with the then actual current pattern of energy production and use, and could only modify matters as old power-station plants came to the end of their economical (and even carbon-economical) life. (By 'carbon economical' I mean if we scrapped a young coal-fired power station and built a wind farm then the energy investment in building the coal-fired station would not have been recouped.) The IPCC knew that the present time also includes a growing global population and they used demographers' quite reliable ranges of forecasts through to the middle of the 21st century and beyond. They took into account that it would take time and investment to introduce alternatives. Finally, their scenarios varied in that some assumed that there would be only modest emission reductions compared to the B-a-U scenario, while others assumed that more aggressive policies would mean greater savings. Among the strongest climate-combating IPCC 1990 scenarios, carbon dioxide emissions would be reduced to 50% of their 1985 levels by 2050 and even this unlikely scenario still resulted in a climbing global temperature through to 2100 and beyond. So, every IPCC scenario shows that global warming will continue through the 21st century. Obviously, if the cuts in greenhouse gas emissions were sufficiently great there must be a point at which the global climate would stabilise without further warming. The IPCC did look at such climate stabilisation but did not include this as one of their detailed scenarios. The reason was that the magnitude by which we would need to change the global economy to realise these stabilising cuts is so great; to stabilise the climate we would need to reduce greenhouse gas emissions planet-wide by 60–80% of 1990 levels! To do this, against a backdrop of global economic growth and growing population (both multiplicative and not additive factors; see discussion on population and Figure 5.10, below) such cuts would have an even greater socio-economic impact. This cut to 60 – 80% of 1990 levels is therefore not even remotely likely. It would appear virtually certain that we are committed to seeing the Earth warm up.

For its 2001 assessment, the IPCC not only reviewed the literature as to the latest science and looked at improved computer models, they developed a range of more complex greenhouse gas emissions scenarios arising from different policy assumptions. These they developed in a Special Report on Emission Scenarios (or SRESs), of which six were highlighted but a full set of 35 were used in the IPCC's 2001 report to provide a high–low envelope inside which all the SRES best-estimate forecasts lay. Further, an outer envelope was ascertained, representing the scenarios used by several different climatic models. This outer envelope provided the IPCC with new high and low estimates for anticipated climate change in the 21st century (see Figure 5.7a).

Table 5.1 Summary conclusions of the IPCC's 1990, 2001 and 2007 assessments for forecast temperature and sea-level changes 1990–2100. Subsequent research suggests that these may be low, although this is an area of considerable debate

Change 1990–2100	1990 B-a-U above pre-industrial levels	2001 Several models, all SRES envelope above 1990	2007 A2 (B-a-U) scenario above 1980–99 mean
Warming			
High estimate	5.25°C	5.8°C	5.4°C (A1F1 scenario[b] 6.4°C)
Best estimate	3.25°C	2.0–4.5°C (model ensemble)	3.4°C
Low estimate	1.8°C	1.4°C	2.0°C (B1 scenario[c] 1.1°C)
Sea level			
High estimate[a]	110 cm	88 cm	51 cm (A1F1 scenario[b] 59 cm)
Best or average estimate[a]	65 cm (best estimate)	22–70 cm (all models, all SRES averages)	37 cm (A2 median value)
Low estimate[a]	22 cm	9 cm	23 cm (B1 scenario[c] 18 cm)

[a] Not including land-ice uncertainty and above 1990 level.

[b] The A2 scenario is the 2007 scenario the IPCC used in their summary global warming graph. This scenario is broadly analogous to the IPCC's 1990 B-a-U and does factor in some very modest climate change policies. However, the 2007 assessment also includes the A1F1 scenario which is the scenario that includes fast economic growth and more intense fossil fuel use. This is included in the table not just for interest but because some reports and summaries by independent academics and science writers of the 2007 assessment (including the stop press appendix in this book's first edition, published in 2007) quote the 2007 IPCC report for 2100 warming of 1.1–6.4°C and sea-level rise estimate as 18–59 cm.

[c] The B1 scenario is one of low economic growth with global cooperation and is not considered business as usual.

The IPCC applied a similar methodology in 2001 to sea-level rise. The seas are expected to rise due to the melting of glaciers, the Greenland and Antarctic ice caps and the thermal expansion of the oceans. Of these, thermal expansion is the dominating factor and here much depends on the degree of heat transferred to the oceans. Equally, there is uncertainty as to the rate of glacier and ice-cap melt. As melting affects cold terrestrial systems, much of the subsequent sea-level discussion in this book on biological and human aspects will focus on this dimension, and also total sea-level rise, as this affects terrestrial ecosystems lost due to marine encroachment. (Readers will have to look elsewhere for detailed discussion on aspects of thermal expansion.)

Antarctica *may*, however, offset part of this rise by trapping more ice from the increased snow precipitation expected in a warmer world. However, while its 2001 best-estimate range for 1990–2100 warming essentially captured its 1990 best estimate, the 2001 estimates for sea-level rise were lower (see Table 5.1). This was because of improved computer models. The 2001 IPCC assessment refers to average SRES forecasts and a high–low range for the same. Yet they also wisely included in their graphs a higher–lower estimate due to uncertainties in land-ice changes, permafrost and sediment changes. Importantly, even with this extension of range, they did not allow for uncertainty relating to the stability of the West Antarctic Ice Sheet (WAIS). This point is often overlooked. This is because the IPCC's 2001 summary for policy-makers and the technical summary, published together in *Climate Change 2001: The Scientific Basis* (IPCC, 2001b) only touched briefly upon this uncertainty, even if this brief mention is perfectly clear. The WAIS 'dynamics are still inadequately under-stood', and a part of a lengthy IPCC graph caption states that its 'range does not allow for uncertainty relating to ice dynamic changes in the WAIS' (IPCC, 2001b). These uncertainties and caveats are worth noting for just half a decade later we are seeing unexpected degrees of melting in both Antarctica and Greenland. These suggest that the IPCC's sea-level rise estimates in 2001 may have been too. In 2007 the IPCC did indeed raise their lower estimate for overall sea-level rise, but also lowered their high estimate.

The IPCC's 1990, 2001 and 2007 conclusions for the anticipated change forecast between 1990 and 2100 are summarised in Table 5.1. We will return to sea-level rise shortly and then again when discussing possible 'surprises' in the next chapter.

The IPCC 1990 report summarised that it was 'certain' that the forecast emissions would result 'on average in both an additional warming of the Earth's surface' and sea-level rise. The 1992 IPCC supplement concluded that the subsequent scientific understanding 'either confirm[ed] or [did] not justify alteration of the major conclu-sions of the first IPCC Scientific Assessment'. The IPCC's 1995 second assessment was if anything more cautious in its conclusion, in part due to the fierce political debate surrounding the first assessment's conclusions. In 1995 the IPCC concluded that 'the balance of evidence suggests a discernable human influence on global cli-mate'. By 2001, the IPCC had regained some confidence and concluded that 'an increasing body of observations gives a collective picture of a warming world and other changes in the climate system'. Further, that 'there is now new and stronger evidence that most of the warming observed over the past 50 years is attributable to human activities'. Then in 2007 the IPCC said that 'the understanding of anthropo-genic warming and cooling influence on climate has improved since the TAR [2001 third assessment report], leading to *very high confidence* that the global average net effect of human activities since 1750 has been one of warming'. Here the italics is the IPCC's own emphasis and 'very high confidence' in IPCC parlance has the very specific meaning of being a probability of 9 out of 10.

The 2007 report also said that that the 'warming of the climate system is unequi-vocal' and that 'most of the observed increase in global average temperatures since the mid-20th century is *very likely* due to the observed increase in anthropogenic greenhouse gas concentrations'. Again the use of italics is the IPCC's. This was a development since the IPCC 2001 assessment, which only put it as being 'likely'. Again here, the use of terms likely and very likely by the IPCC have a specific meaning: of more than 66 and more than 90% probability, respectively.

The reasons for the IPCC's careful wording and caution (and also note for example the prudent use of the word 'most') were 4-fold. First, the IPCC were presenting a 'scientific consensus' of a range of views and this range's breadth had to be signalled.

(Contrary to the views of climate sceptics and deniers, nearly all climate researchers go along with the IPCC view [Oreskes, 2004], but read the small print of the IPCC's full text and there are many caveats and views do vary, but not in the way climate sceptics say.) Second, the IPCC have only assessed work published in peer-reviewed journals. Preliminary work, work underway and even work that has been accepted by eminent peer-reviewed journals but not yet published was not considered[4]. Third, as a consequence of this personal assessment of the emerging picture, discussions over lunch at symposia and scientists' other unpublished insights were simply not admissible for consideration by the IPCC. Finally, IPCC researchers and leading climate change programmes in a number of countries (such as in the UK and USA) spent much of their research investment on constructing incomplete – but ever more sophisticated and hence more useful – climate computer programmes.

In financial terms, 1990s investment in research on whole-organism biological palaeo-responders to change, and even monitoring of current change, came very much second to investment in computer models. In the UK at least, in the early 21st century this is beginning to change, and even regarding computer use the biological components are increasingly being incorporated into models (in contrast, the climate models of the 1980s, on which the first IPCC assessment was based, essentially modelled the planet's physicochemical dimensions as if it were lifeless). The 2001 report did include model outputs that incorporated some biologically determined factors. However, there is still much biology to include in climate models. Indeed, although policy-makers continue to place great faith in computer models, that models still have a long way to go is emphasised occasionally in the scientific literature. Considerable development is required before policy-makers should consider them a reliable tool for the detailed regional analysis that is required for a nation's climate change policy.

One recent example (from the time of this book's first edition) of such a warning in a high-impact journal was a letter in *Nature* from three US researchers (MacCracken et al., 2004) from the Climate Institute and the H. John Heinz Center for Science, Economics and the Environment. They concluded by saying 'We strongly agree that much more reliable regional climate simulations and analyses are needed. However, at present, as [the *Nature* news report] makes clear, such simulations are more aspiration than reality.'

Model estimates of future warming are also only as good as the assumptions that go into the model. As our understanding improves, so models improve. This does not mean they are now perfect, but with greater understanding comes model improvement. Present models will be better than older ones at forecasting the future as atmospheric greenhouse gas concentrations change. One such (of many) changes in our appreciation of the way the global climate system works came in 2008 and in this instance concerned the mass of warm air circulated (Pauluis et al., 2008). Global

---

[4] The IPCC is meant to draw on peer-reviewed academic literature and nearly always does so. Its Working Group I reports on the science in successive assessments are the most rigorous. Working Group II covers impacts and adaptation and Working Group III covers the mitigation of climate change. Away from Working Group I there have been some complaints regarding some of the detail in the other Working Group reports and these seem to stem from the use of non-peer-reviewed (so-called grey) literature. This has fuelled climate sceptic debate (especially in 2008–2010).

circulation transports warm air from the tropics polewards to higher latitudes. What Oliver Pauluis and colleagues showed was that when averaged on moist isentropes (an entropy measure), as opposed to dry isentropes, the total mass transported by the circulation is twice as large. This is of importance in climate models as it could account for perhaps up to half of the air in the upper troposphere in high latitude, polar regions. Such new understanding needs to be considered by climatologists and incorporated into their climate models. Although the above is not a biological dimension to climate change (the principal focus of this book) it is worth mentioning because climate models are not so good at reflecting high-latitude (polar) climate change: the poles (especially the north pole) are warming faster than elsewhere and computer models do not fully reflect this. This faster warming of the poles, especially in the northern hemisphere, is sometimes called Arctic amplification or polar amplification, and models in the past have had difficulty reflecting this; even today they do not fully capture the effect even though they include elements such as albedo feedback effects due to declining reflective surface ice (Figure 1.8b). Other evidence that it is atmospheric energy transport from lower latitudes into the Arctic that may explain (at least part of) the Arctic amplification mystery came in 2008 from Rune Graversen, Thorsten Mauritsen and colleagues from Stockholm University. This phenomenon may be explained by heat transported by air above the lowest part of the atmosphere (the middle troposphere).

As said, the climate system is made up of many dimensions and interacting factors of various, and often changing, strength. And so, as our understanding of the climate system improves, changes are made to models so that the next generation is improved. In addition to the new, purely climatological considerations there is continually improved understanding of Earth system biology and the latest computer models include biological dimensions. One example of an area of biological concern is how much carbon dioxide is drawn through photosynthesis each year and how much is released through respiration. The amount of atmospheric carbon drawn down through photosynthesis is clearly related to global gross primary production (the fixation of inorganic carbon into organic forms by autotrophs). This is a key part of the carbon cycle the understanding of which is continually improving through research (see section 1.3 and Figure 1.3). Here the annual biological addition and removal of carbon dioxide to the Earth's atmosphere is part of the biological components included in recent climate models. In 2011 those developing models could not fail to take notice in the work of Lisa Welp and Ralph Keeling who led an international team that presented 30 years' worth of data on $^{18}O/^{16}O$ in carbon dioxide from the Scripps Institution of Oceanography global flask network (Welp et al., 2011). They used this isotopic ratio together with what happens in an El Niño year (part of the El Niño Southern Oscillation, or ENSO) in a simple model to infer global primary production. Their analysis suggests that current estimates of global gross primary production, of 120 petagrams (Pg; or Gt) of carbon per year, may be too low, and that a better estimate might be of 150–175 Pg of carbon per year; that is, a 25–46% increase. Of course, if there is greater carbon draw dawn by plant and algal photosynthesisers in a year then equally there must be greater respiration releasing carbon (otherwise atmospheric carbon dioxide would decline), and this in turn means that the fast carbon cycle (as opposed to the deep carbon cycle; see section 1.3) must be cycling even faster than thought. (See also Cuntz, 2011, for an overview of this work and also other earlier evidence for increased global primary production discussed in section 6.1.3.)

That we need to treat climate computer models with caution was highlighted in 2007 by Stefan Rahmstorf in Germany and Anny Cazenave in France together with colleagues in Britain and the USA. They looked at (then) recently observed climate trends for carbon dioxide concentration, global mean air temperature and global sea level and compared them to previous model projections as summarised in the IPCC 2001 assessment report (IPCC, 2001b). They reported that data for the period since 1990 to 2005 raised concerns that the climate system, in particular sea level, may be responding more quickly to climate change than our current generation of models indicates. (This research just missed getting included in the IPCC's 2007 assessment but will no doubt help inform its 2013 report.)

In 2010 the Royal Society (Britain's academy promoting scientific excellence) also urged caution when considering climate models. In its *Climate Change: A Summary of the Science* report, it said

> The ability of the current generation of models to simulate some aspects of regional climate change is limited, judging from the spread of results from different models; there is little confidence in specific projections of future regional climate change, except at continental scales.

As noted, one of the biggest problems with global computer models is that to date (2012) they do not successfully reflect climate change at high latitudes, both when modelling the present Earth and in the past with a different continental distribution. Their failure to capture the mid-Cretaceous warm Arctic climate (see Chapter 3) is a case in point, although we know from palaeoclimate proxies (fossil remains, biomolecular climate proxies and oxygen isotope analysis) that the Arctic was warm at that time. It was even warmer later, during the Initial Eocene Thermal Maximum (section 3.3.9), and global models also have difficulty in capturing this. So, even being aware of polar amplification from actual measurements, the implication from palaeodata is that high latitudes will warm far more than models currently suggest. Indeed, Antarctica is currently the fastest-warming place on Earth, having warmed some 3°C in the 50 years up to 1990 (which compares to a global average warming of about 0.6°C for the entire 20th century). Whereas model failures may not so markedly affect the IPCC's climate-scenario forecasts for tropical and temperate zones, implications for the IPCC's projections of sea-level rise are considerable, as the major ice caps (Antarctica and Greenland) capable of contributing to sea-level rise are at high latitudes. (We will return to this in the next section as well as next chapter.)

Part (although not all) of this high-latitude, polar amplification problem – especially with regard to the northern hemisphere, which has a greater land area just outside the Arctic Circle compared to the Antarctic Circle – is the effect of biology on albedo. Again, it must be emphasised that computer climate models have been, and are, getting ever more sophisticated (and including more biological dimensions). Up to the end of the 20th century models did include surface cover in a simplistic sense and surface-cover albedos so that land without snow (and sea without ice) absorbed more, and reflected less, solar energy than land with snow (or sea with ice). Forests too, in real life (as opposed to a computer model), see areas prone to snow have a lower albedo (are darker) than snowfields without trees. Conifers have evolved to allow snow to fall off of them comparatively easily. All well and good, and most of the more sophisticated models at the beginning of the 21st century accounted for such factors and included a (coarse) model map of land cover. These models naturally took into account one of the positive-feedback cycles of warming reducing the area of snow cover, so reducing the solar energy reflected, so in turn causing warming, which then further reduces snow cover (see Figure 1.8b). However, only a handful of models in the first half of the 2000s decade included the effects of a changing regional ecology and hence a changing and dynamic landscape. For, in addition to the above feedbacks (instead of having a static biological landscape), there is the biological dynamic. As the Arctic region warms then so it is open to colonisation from species that could not have survived the previously cooler climate. Even if the ground is snow-covered, if it has shrubs and trees that have migrated into the area due to warming, then the albedo will change. No doubt, as we come to the end of the decade in 2010, such biological dynamic factors have (at least on a coarse level) been included in models. But in 2005 it was becoming clear that such factors were important in the field, in real life (as yet not so much with model makers). At that time a wide range of innovative Alaska data sets were assembled covering the decades from the 1960s (and in some cases from the 1930s) to 2004. They included surface-temperature records, satellite-based estimates of cloud cover, ground-based estimates of ground cover and albedo, field observations in changes of snow cover and, importantly, changes in vegetation cover. Alaska was a worthy study area for it has seen a number of thresholds crossed with regional warming that relate to abrupt physical and ecological change near the freezing point of water. And the region has warmed. Palaeoclimatic evidence shows that Arctic Alaska was in 2005 warmer than at any time in at least the past 400 years and so there should be a large impact on water-dependent processes. The warming, and especially summer warming, cannot be attributed to climatic cycles such as the North Atlantic Oscillation or the Arctic Oscillation, while cycles such as the El Niño mainly affect Alaskan winter temperatures and not summer ones. Changes in sea ice around Alaska would have an effect in the spring and autumn, when the ice melts and freezes, and so, whereas this time may have changed by days or weeks with regional warming, it would not affect matters in the depth of winter or in the middle of summer. The study team (Chapin et al., 2005) state that the degree of change in Arctic Alaskan summer warming, above that of the global average, is best explained not by landscape drivers but by the lengthening of the snow-free season.

Indeed, snow melt has advanced by 1.3 days per decade at Barrow (a coastal site) and an average of 2.3 days per decade over several other coastal sites. Inland, in the northern foothills of the Brooks Range, warming had taken place at a rate of 3.6 days per decade, and 9.1 days per decade for the entire Alaskan North Slope. Since 1950 the cover of tall shrubs within the North Slope tundra has increased by 1.2% each decade from 14 to 20%. The study team pooled, in a meta-analysis, field-warming experiments showing that increasing the summer temperature by 1–2°C generally triggers increased shrub growth within a decade (which is consistent with both recent field observations and the palaeo record of Holocene shrub expansion). However, it is thought that the increase in shrubs only accounts for 2% of recent warming. A further 3% may be attributable to increased tree cover, especially that of the white spruce (*Picea glauca*) whose cover has increased, decreasing the treeless area by 2.3%. Although this makes only 5.3% of Arctic Alaskan summer warming accountable to changes in vegetation cover (and the increased snow-free season accountable for much of the rest) the researchers warn that the region is on the cusp with a number of thresholds about to be crossed. The conversion of Arctic Alaskan tundra to spruce forest (very much a possibility by the early 22nd century) has never before occurred during previous Holocene warm intervals; that is, the slightly warmer episodes over the past 11 700 years since the end of the last glacial. Given that Alaska would have been covered in ice during the glacial, it means that it could not have been covered by forest for more than 100 000 years since the brief height of the last interglacial. In addition to ice-water-related thresholds, as shrubs become established so the nitrogen cycle is enhanced, which in turn encourages further vegetation. This is another positive feedback serving to accelerate matters. In short, the research suggests that the already anomalous increased warming of Arctic Alaska, as part of polar amplification, is likely to become even more pronounced in the future (Chapin et al., 2005). Environmental change thresholds (critical transitions) are currently a priority area of climate research and, as the abovementioned research suggests, are central to high-latitude warming over and above those that computer models have predicted to date.

Including biology and landscape change factors in computer models is now begin-ning to show that there are greater differences between the various of the IPCC's 2000 SRESs for the 21st century (IPCC, 2000). Even between scenarios with sim-ilar carbon dioxide at a given time there can be differences in temperature due to the different biological landscape in different scenarios. The IPCC's A2 storyline of these SRESs is one of more regional resource use and self-sufficiency with a slightly slower economic growth. This is in contrast to the B1 storyline with the same population up to the mid-21st century (but which declines thereafter), and again increased sustainability, but through more global solutions and without extra climate initiatives. In 2005 Johannes Feddema and colleagues showed that incor-porating differences in the biological landscape from these two scenarios resulted in somewhat different temperature profiles compared to that portrayed in the IPCC 2001 assessment. The agricultural expansion in the A2 scenario appeared to result in significant additional warming over the Amazon and cooling of the upper air column as well as nearby oceans. These effects in turn influenced the Hadley and monsoon circulations and so modified some climate regimes outside of the tropics. Meanwhile the model of Feddema and colleagues showed that agricultural expansion in mid latitudes produced cooling and decreases in the mean daily temperature over many areas (Feddema et al., 2005). Although it is early days for such biological land-scape inclusions in global models, they will undoubtedly soon become an increas-ingly routine, if not fundamentally intrinsic, addition to global climate computer simulations.

So, not only do models need to accommodate more biological components and in considerable detail, modellers need to keep abreast of biological research outputs. For example, a 2004 paper, from a team led by Colin O'Dowd from Ireland, showed that the contribution of organic matter to marine aerosols over the North Atlantic during plankton blooms made up 63% of the sub-micrometre aerosol mass (aerosols being a climate forcing agent). However, in winter, when biological activity is at its lowest, the organic fraction decreases to just 15%. Further, their simulations suggest that organic matter can enhance cloud-droplet concentration by between 15% and more than 100% and so is an important component of the aerosol-cloud climate-feedback system involving marine biota. Such cross-disciplinary research is common in climate change science.

It must not be forgotten (if the earlier non-biotic example of isentropes was not enough) that computer modellers also need to continually reappraise the physical processes they are modelling, especially when new evidence and ideas come along. For example, at the end of 2004 a team of US geologists (Ufnar et al., 2004) drew attention to field evidence that mid- and high-latitude rainfall in the mid-Cretaceous was far higher than today. They suggested that this might have warmed high latitudes, as ocean evaporation causes cooling and precipitation results in latent heat release, hence regional warming. In the warmer Cretaceous there would have been far more evaporation from the tropical oceans and this could provide another mechanism for transporting heat to high latitudes (in addition to atmospheric and ocean circulation mechanisms).

Having said all this, while computer models still have a long way to go before detailed regional forecasts can be made with sufficient confidence for local policy-makers and planners, a rough-and-ready near-future or even past (historic) global approximation is another matter. The latter is particularly revealing (and important in the climate change debate) because running models of the historic climate, during which meteorological measurements of the real world have been made, means that the model can be tested against reality. If this reality test works then it is possible to run the model again but without the factors of anthropogenic climate change. As we have already noted (see the end of section 5.2.2) this has been done. By 2005 the Hadley team in the UK had a model that successfully captured the vagaries of the climate from the middle of the 19th century to the early 21st century. Its output seemingly tracks with reasonable accuracy actual meteorological measurements of global temperature. Both the Hadley model and real measurements show that the planet from the middle of the 19th century through to the early 20th century had a varying climate but the overall trend was more or less steady (an output not too dissimilar from the relevant years portrayed in Figure 5.2). But after the early 20th century *both* the Hadley computer model and real meteorological measurements show a rising temperature. Clearly, the model works broadly. However, take out the anthropogenic greenhouse component from the model and its output continues the 19th-century stable trend (albeit fluctuating a bit year to year) through to the present. As noted, this is good corroborating evidence that human factors are behind the 20th- and 21st-century warming of climate.

In short, while computer models are a work in progress (as is all science), they still have considerable value and their outputs can be checked to a certain degree by looking at the current situation as well as what happened in the past. If they get both these right then we can have some confidence in their modelling of the future. Indeed, even when they get some past of present detail wrong, we can use this to get an idea of the potential types of error in their future forecasts.

5.3.2 Uncertainties and the IPCC's conclusions

The above discussion demonstrates that we need to be aware of uncertainties and view the IPCC's conclusions in each of their successive assessment reports with these in mind. Related to the mid-Cretaceous warm-Arctic problem above, one of the major concerns is why the greatest warming actually measured during the 20th century took place at high latitudes. This is a concern doubly worrying because, as noted, the global climate models do not yet properly reflect such warming (although they are improving). This is likely to be reconciled once we have a better idea of how heat is distributed about the planet: the previous subsection's suggestion of low-latitude air movement is but one idea.

Another concern is that of a phenomenon nicknamed global dimming. This at first may seem paradoxical in a world that is supposed to be warming up. Yet those concerned with agricultural biology have known for some years that the amount of sunlight reaching parts of the Earth's surface has been in marked decline. Normally weather instruments for temperature and humidity are kept in the shade, but water evaporation is not just a matter of temperature and humidity: it is also a matter of sunshine. The relevance of this has been demonstrated by a series of long-term (multi-decade) measurements of what is called 'pan evaporation' taking place in several countries. Pan evaporation refers to the daily evaporation of water from a pan exposed to the elements including sunlight and is of importance to biologists studying agricultural systems, as pan evaporation is relevant to irrigation methods. It has therefore been monitored for a number of decades in various countries. What has been found is that pan evaporation has been decreasing markedly in many countries for more than 50 years. Sunlight, separate to temperature, excites water molecules and so enhances evaporation. In 2002 two Australian researchers, Michael Roderick and Graham Farquar, discovered that reduced pan-evaporation rates appeared to be linked to a decrease in sunshine: hence the expression global dimming.

What appeared to be happening is that fine-aerosol pollution from urban areas and industry is providing seeds for the formation of clouds that consist of finer-than-usual droplets (the indirect aerosol effect as opposed to the direct aerosol effect of the aerosols without the atmospheric water). These clouds are more reflective than usual. Also, the finer droplets are less likely to result in rain.

The consequences of dimming were considerable. First, there is the human implication: it is now considered by some that global dimming caused less ocean evaporation in parts of the northern hemisphere, such that in the 1980s and 1990s it resulted in the failing of the monsoon and mass famine in east Africa. But it also means that if there is a greater cooling force than we thought over much of the temperate northern hemisphere, then there must be an even greater warming force than we thought for there to have been the net result of warming in the 20th century. In 2004 a small team of US researchers led by Joyce Penner reported that they had a measure of this indirect aerosol effect by comparing conditions in two separate areas of that country, one of which had clean air and the other of which was polluted with aerosols. They showed that this effect was markedly greater than estimates used up to that time by IPCC climatologists. Aeroplane condensation trails also cause significant global dimming and this was demonstrated in a rare experiment following the attack on the World Trade Center in New York on 11 September 2001, when for 3 days virtually all of the US civilian jet air fleet was grounded (Travis et al., 2002). Furthermore, the IPCC estimates for aerosol forcing for their first three assessments (1990, 1995 and 2001) were based on satellite measurements only over the ocean. Subsequently, with improved orbital remote sensing, it was possible to make estimates of aerosol forcing over land. This indicated that aerosol forcing was in fact at the high end of the earlier range the IPCC used. These results suggest that present-day direct radiative cooling forcing due to aerosols is stronger than that represented in the IPCC assessments up to and including their third report. The implication is therefore that future atmospheric warming due to an increase in anthropogenic greenhouse gases will be greater than the IPCC had predicted (Bellouin et al., 2005).

As for the longer-term trends in global dimming, 2005 saw the publication of the summaries of two sets of data. One was from a set of hundreds of terrestrial sunlight measurements gathered by an international team led by Martin Wild, an atmospheric scientist at the Swiss Federal Institute of Technology in Zurich. The other study looked at satellite data and was led by Rachel Pinker from the University of Maryland. Both found a similar trend. It would appear that since the 1950s up to the early 1990s global dimming had been taking place. However, since the early 1990s there had been some recovery, with 'global brightening'. This is thought to be due to the decline in the Soviet Block's dirty industries as well as the increased use of improved cleaner technology by Western developed nations. Currently work is underway to estimate how much global dimming has offset late-20th-century global warming. The IPCC's 2007 assessment notes that global dimming does not seem to have continued after 1990.

If global dimming is something the IPCC have (prior to 2007) not fully taken into account then this could, at least in part, explain why high-latitude warming has actually been greater than their models predicted. It could be that the cleaner air of the poles (protected by circumpolar air currents) was not as globally dimmed. This is but one possibility. In this instance much of the academic research relating to global dimming and brightening was published after 2000. The IPCC only considered science research published prior to its 2001 report, so the IPCC's first three assessments could not take this dimension into account. The IPCC's report for 2007 began to include such considerations. It noted that 'global dimming' is not global in extent and it has not continued as a significant effect after 1990. Decreases in solar radiation at the Earth's surface from 1970 to 1990 had an urban bias. Further, there have been increases in solar radiation received at the Earth's surface since about 1990. An increasing aerosol load due to human activities decreases regional air quality and the amount of solar radiation reaching the Earth's surface. In some areas, such as Eastern Europe, recent observations of a reversal in global dimming to that of brightening link changes in solar radiation to air quality improvements.

A fresh twist to the aerosol climate forcing story came in 2007 when Veerabhadran Ramanathan, Muvva Raman, Gregory Roberts and a few colleagues mainly from the Scripps Institution of Oceanography in the USA used three lightweight, remote-controlled aircraft to monitor the energy flux in aerosol brown clouds over the Indian Ocean caused by biomass burning and fossil fuel consumption. By flying the craft in formation, they were able to get a mass of data on aerosol concentration, amounts of soot and solar fluxes. They flew 18 missions in March during the dry season when air flow brings pollution from Asia across the Indian ocean. The IPCC, up to and including its 2007 report, largely considered aerosols to have a cooling effect: sulphate aerosols reflect, there is a global dimming component and there is indirect cooling forcing (aerosols can seed moist air and so facilitate the forming of clouds, which are reflective). But some components of aerosol clouds (black carbon – soot – from combustion for example) can absorb solar radiation and so heat the atmospheric layer in which the aerosol resides (1–3 km high). What they found was that there was a much stronger warming factor than previously thought and that this would warm the climate of the region as much as the recent increases in greenhouse gases. Indeed, they propose that this has further spurred the observed retreat of Himalayan glaciers. (See also a review article by Peter Pilewskie, 2007.)

But the Indian Ocean aerosols may have other environmental consequences. In 2011 Amato Evan, James Kossin and colleagues reported on the increase in the intensity of pre-monsoon Arabian Sea tropical cyclones (what would be called hurricanes in the Atlantic) in the northern Indian Ocean during the period 1979–2010. They suggested that these could be a consequence of a simultaneous upward trend in anthropogenic black carbon and sulphate emissions. They used a combination of observation over more than 30 years, and model data to demonstrate that the anomalous circulation, which is radiatively forced by these anthropogenic aerosols, reduces the basin-wide vertical wind shear, so creating an environment that allows tropical cyclones to become more intense. This is not a trivial concern. In 1998, a major cyclone resulted in more than 1100 deaths in western India, and Cyclone Gonu in 2007 caused more than US$4 billion in collective damage to Oman, the United Arab Emirates and Iran (Sriver, 2011). The problem is that a longer time series than 30 years is required and so, as often is common with research, this work will need to continue.

In 2011 United Nations Environment Programme (UNEP) and the World Meteorological Organization published the report *Integrated Assessment of Black Carbon and Tropospheric Ozone*. It highlighted that soot (black carbon) aerosols act to enhance short-term warming forcing of the climate. It estimated that controlling emissions of black carbon aerosols could reduce short-term climate forcing by around 0.2°C in addition to reducing human health impacts and damage to crops.

Overall, various phenomena and factors, including those we may not even yet appreciate, should make us pause and realise that there are limits to certainty in the IPCC assessments. This is something the IPCC themselves appreciate.

Successive IPCC reports have provided improved appraisals of our scientific understanding of the way global and regional climates are changing. Nonetheless, there is much we do not know and, for instance with regards to both remote sensing and computer modelling, we are constantly pushing at the advancing edge of technical capability. Yet we are still left with the overall question of how robust are the IPCC's conclusions? As noted, the IPCC arrive at their decisions by committee, which in turn bases its conclusions on its members' own interpretations of a variety of evidence, both palaeoclimatic and palaeoclimatic combined with computer models. Even so, these interpretations are subject to human psychological factors that make it additionally difficult to assess how reliable the IPCC's conclusions may be. For instance, to take our computer model case further, it is possible to begin to estimate how robust the IPCC's conclusions are against the types of computer model the IPCC uses. Following the publication of the IPCC's 2001 assessment, researchers from the University of Bern, Switzerland, attempted to do just this (Knutti et al., 2002). They took a simple climate model (therefore, one with an output subject to relatively large error margins) but calibrated it to the same three-dimensional ocean/atmosphere models used by the IPCC in their third assessment. Other than ensuring that the model's responses reflected known past climate change (i.e. that the model's responses to established historic circumstances were 'realistic'), they varied the model's parameters by the estimated uncertainties of its various climate forcing factors. In this way they obtained a number of climate forecasts for the 21st century. The conclusions were clear. Just 5% suggested that the IPCC's 2001 high – low estimates were too high. In other words, there was a 5% chance that the IPCC was being too alarmist in predicting greater warming than is likely. Conversely, the researchers concluded that there is a 40% probability that the increase in global mean surface temperatures will exceed the range predicted by the IPCC. In short, they estimated that the balance of probability would be that global warming in the 21st century will be greater than the IPCC forecasted in 2001. If the researchers' conclusions were valid, it means that the consequences of the warming discussed in subsequent chapters will be greater. One literature review in 2005 on aerosol cooling, by Meinrat Andreae, Chris Jones and Peter Cox, suggests that warming forcing since the Industrial Revolution up to the end of the 21st century might be between 2 and 8°C, although a mid-estimate of 6°C is unlikely because there will probably be some natural background aerosol cooling in addition to that from human activity.

With research like this, it was not surprising that the 2007 IPCC analogous scenario estimates for 21st-century warming were a little greater than in 2001. The best models (including the ones the IPCC used) may not be accurate on a sub-regional basis but they do *broadly* paint similar pictures to each other on a global scale. Getting similar results by different computer methods lends a certain credibility to the results, hence the models. This is worth noting, as it is model development and the increase in their sophistication that will one day surely prove most useful for determining change at increasingly local levels in numerous ways.

A second instance of concern that the IPCC's earlier 2001 forecast may have underestimated possible future climatic change was presented by a European research team led by Hans von Storch in 2004. They examined the validity of previous interpretations of the palaeoclimatic proxy record for the northern hemisphere that were in turn used by the IPCC. Their conclusions were that such past analyses of noisy (highly variable) proxy data may have underestimated past climate change. If the 2001 IPCC had been relying on underestimates of past change to make their future forecasts, then their future forecasts would also be underestimates. Past variations may have been at least a factor of two larger (see section 2.5).

The above gives just two examples of studies that have examined the reliability of the 2001 IPCC scenarios. These examples are scientific attempts to quantify confidence in scenario conclusions that, in addition to the previous subsection, provide further reason to treat the climate models on which the IPCC rely with caution. New information and discussion is ongoing in the scientific literature and it is likely that the forthcoming 2013 IPCC assessment will further tweak the 2007 forecasts.

In the greater scheme of things, does such tweaking of warming forecasts matter? As noted, policy-makers are concerned with a bigger socio-economic and political picture than the exactness of warming magnitudes. It is of negligible importance to a government that warming 100 years from now may be half a degree or so more or less than a current estimate. The key conclusion of importance (to policy-makers) is that warming is almost certain to take place within such-and-such a range. This seems sacrilegious to some scientists, who rightly take pride in their work and who in conversation with politicians in Westminster, or on Capitol Hill, correct policy-makers when talking about half a degree, or x centimetres of sea-level rise, or whatever. This point may seem trivial, but it is not. It is increasingly important (and not just regarding climate change) for scientists to have good communications with politicians. Scientists being pedantic as to the accuracy of the current vogue of predictions, however scientifically worthy, can give the impression of fact and importance when in actuality there is greater uncertainty about the detail. In truth there is considerable uncertainty as to whether a specific number of x degrees warming has taken place since the Industrial Revolution, let alone whether there will be specific y degrees of warming in the future. Conversely, there is great certainty as to the bigger picture, of which we can be confident: the Earth has been getting warmer since the Industrial Revolution and will continue to do so in the 21st century. Indeed, the IPCC reports themselves are very good about the various degrees of certainty and confidence they can attribute to their conclusions, but one has to read the full text and not just the executive summaries.

For the full range of its SRES scenarios (not just the A2 scenario presented in Table 5.1) the IPCC 2007 assessment concludes that the Earth will warm by between 1.1 and 6.4°C above the 1980–99 mean by 2100, and it has considerable confidence in this. Improved environmental understanding (much of which will come from the biosciences) and computer modelling (not to mention processing power) will further refine our view of the future, as indeed will the passage of years. However, the IPCC has for the past three assessments given one additional key warning, and that is to be prepared for surprises. We need to be aware that a volcanic eruption, or abrupt changes in ocean circulation and other events, can happen, and that these will have considerable climatic consequences that take us outside the IPCC B-a-U scenario and other scenario forecasts. This may be an unsettling conclusion but in terms of policy it suggests the need for us to plan for the worst-case scenarios and surprises rather than easier options (see section 6.6).

It would be prudent – despite it being early days – to consider options outside the IPCC forecast envelope. So are we moving into an even warmer world than the IPCC forecast or not? A major volcanic eruption, such as of that underneath Yellowstone National Park in the USA, would cool the world for several years. However, currently changes in most of the other probable climate forcing variables are more likely to force the climate the other way, suggesting that if anything the IPCC forecast is conservative and that global warming will be greater rather than lesser. (Indeed, even an out-of-the-ordinary Yellowstone-type event would generate carbon dioxide, the warming effect of which may well outlast its cooling impacts.) Since 1990, global temperatures, atmospheric concentrations of carbon dioxide and carbon emissions have all been marginally higher than the IPCC's SRESs. Indeed, impacts on human well-being by extreme weather events are already becoming manifest, as the IPCC have predicted (see Chapter 7).

Finally, notwithstanding the real possibilities of an even warmer world than the IPCC predict, there is additional concern over sea-level rise. The IPCC have always been cautious about sea-level estimates in each of its reports and the main IPCC texts repeatedly refer to the considerable uncertainties in the science. This is for a variety of reasons, including that the component factors (such as ice-melt and thermal expansion) of existing estimates from the past two centuries of sea-level rise (measured at various sites in Europe) have large estimated errors associated with them. In short, we do not know for certain what has caused the annual slight 1–2 mm changes over this time. With regards to future rise, there is considerable uncertainty about how Antarctica and Greenland will behave with warming. Indeed, the early 21st century has seen much academic debate as to the anticipated sea-level rise by the year 2100, and some might even say that the IPCC's 2007 estimates were too conservative, even if they did raise the lower estimate from what it was in 2001.

By monitoring the Greenland and Antarctic ice caps in the early 21st century (a body of work not available for consideration by those compiling the 2001 IPCC report) we began to see far greater discharge from both these ice caps than anticipated. Thomas et al. (2004) suggest that Antarctica, far from having a negative to $+2$ cm change in sea level between 1990 and 2100 (as suggested by the IPCC in 2001), had a discharge in 2002–3 from West Antarctica that was much greater than that observed during the 1990s. This discharge rate is of some 250 km^3 year^{-1}, so by itself (without other countering factors) it provides a rising factor to sea level of more than 0.2 mm year^{-1}. Even if the rate of discharge stops increasing and remains the same for the rest of the 21st century, then West Antarctica's contribution to sea-level rise could – and note that this is a speculation – be a factor of ten higher than the IPCC's 2001 forecast for all of Antarctica up to 2100.

However, even these high rates, should they come to pass, are unlikely to remain the same in the future, as research has now revealed that most of the West Antarctica's continental ice shelves (which are off shore but grounded) are hundreds of metres deeper than thought. Without the near-surface underpinning rock that was until recently assumed to be there, and with the shelves collapsing, ice will flow more freely from further inside Antarctica (Thomas et al., 2004). For this reason alone Antarctica's possible contribution to sea-level rise may even be found to become greater than the 10-fold excess above the 2001 IPCC rate now being observed. This is unless, and this caveat is important, as mentioned above, increased precipitation (snowfall) in a warmer world freezes on East Antarctica, so having the net effect of removing water from the Earth's oceans.

There is a second reason for the concern that Antarctica may contribute more to sea levels in the future, and that is, if global warming is itself greater, then Antarctica's margins (although not so much its interior, which retains considerable thermal inertia) could become even more unstable and experience greater ablation. Research on Antarctic ice using satellite data gathered between 1992 and 2003 reveals a mixed picture (Davis et al., 2005). It shows that West Antarctica is losing ice but that East Antarctica is gaining it, with its ice sheets mainly becoming thicker (Figure 5.9a). This is because of increased snowfall, which itself is an expected feature of global

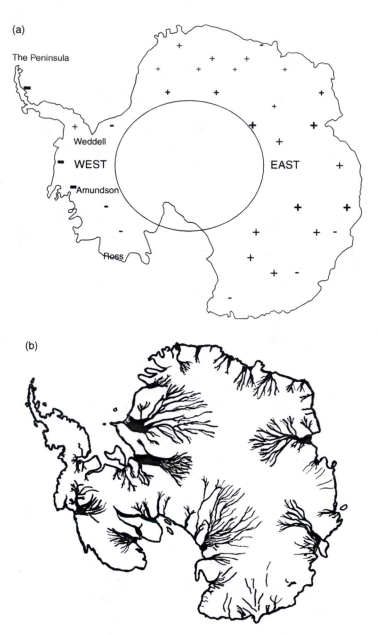

Fig. 5.9 (a) In parts of Antarctica (mainly the east) the ice is thickening (+) and in other parts the ice is thinning (−). The sketch map is based loosely on Vaughan (2005), which is in turn based on satellite data from Davis et al. (2005). The circle in the centre represents the high-latitude zone not covered by the satellite's orbit. The IPCC 2007 report portrays a similar but slightly different picture for change between 1992 and 2003, with greater mass loss around the Amundson area and little to no gain over much of East Antarctica. (b) Sketch after an animation from European Space Agency, Canadian Space Agency and Japanese Space Agency data and processed using NASA funding at California University. It reveals narrow river-like (very slow) flows of ice from the continental interior. Such irregularities largely govern how the ice sheet reacts to changes imposed at its edges. Whereas panel (a) shows where ice is accumulating and lost, this image (b) shows how it is being transported a considerable way from Antarctica's interior. *Note*: this illustration only shows the areas with the fastest of these very slow flows. The data presentation was also summarised in Rignot et al. (2011). You can getter a better idea by looking at the animation at www.bbc.co.uk/news/science-environment-14592547#panel1.

warming. The gains in East Antarctic ice between 1992 and 2003 north of 81.6°S (below which the satellite never orbited) was about 45 ± 7 billion t. This is enough to offset Antarctica's net contribution to sea-level rise by 0.12 ± 0.02 mm year^{-1}. The IPCC's 2007 report put Antarctica's net average annual contribution to sea level rise for the period 1993–2003 as being 0.21 ± 0.35 mm year^{-1}. However, whereas a far more detailed assessment of Antarctica's contribution to sea-level rise is required, the picture that emerges is that the continent, being the fastest-warming continent on the planet, may well be on the cusp. If so then ongoing detailed monitoring will be required. We will return to this and the related research in the next chapter's subsection on sea-level rise (section 6.6.3).

With regards to Greenland, in 2009 work by Michiel van den Broeke, Jonathan Bamber, Janneke Ettema and colleagues used satellite gravity observations from the Gravity Recovery and Climate Experiment (known as GRACE). They found that the total 2000–8 mass loss from Greenland was approximately 1500 Gt, which is equivalent to 0.46 mm year^{-1} of global sea-level rise (and note that this is excluding thermal expansion). Now, Greenland (as well as Antarctica) also accrues ice from snowfall. Taking this into account (that is, without the moderating effects of increased snowfall and refreezing) post-1996 Greenland ice-sheet mass losses would have been 100% higher than this approximate 1500 Gt. However, since 2006, high summer melt rates have increased Greenland ice-sheet mass loss to 269 Gt year^{-1} (0.75 mm year^{-1} of equivalent sea-level rise). This compares to the 2007 IPCC assessment (which noted that Greenland melt has risen very considerably since the 1960s) reporting that Greenland's contribution to sea-level rise rose from 0.05 ± 0.12 mm year^{-1} during 1961–2003 to 0.21 ± 0.35 mm year^{-1} for the period 1993 to 2003. This is broadly in line with the IPCC's A2 scenario estimate for Greenland's contribution to sea-level rise from its 1980–99 mean to the mean for the final decade of the 21st century, which is about a centimetre. This may seem trivial, and it is compared to other factors that are thought to dominate 21st-century sea-level-rise estimates (especially thermal expansion), but the 2007 IPCC assessment also notes that 'models do not yet exist that address key processes that could contribute to large rapid dynamical changes in the Antarctic and Greenland Ice Sheets that could increase the discharge of ice into the ocean'. What the GRACE results show is that the melting that is occurring even now is already pushing at the upper end of IPCC estimates.

In 2012 new estimates from the GRACE satellite were published (Jacob et al., 2012). This gave a global estimate of glacier and ice sheet melt for the period 2003–10 of 536 ± 93 Gt year^{-1}, which contributed 1.48 ± 0.26 mm year^{-1} to global sea-level rise (again note that this excludes thermal expansion factors). The estimate for annual mass loss of just the Greenland/Antarctic and surrounding glaciers and ice sheets between 2003 and 2010 of 384 ± 71 Gt year^{-1} contributed 1.06 ± 0.19 mm year^{-1} to sea-level rise.

To get a good estimate of likely sea-level rise by the end of the 21st century we essentially need three things: good corroboration of mass-loss estimates (this is beginning to come through); a good run of mass-loss data covering two decades to see how things change (this is a work in progress and will last some years); and a good estimate of the likely sea-level rise from thermal expansion (we do not yet have this). In short, we may have to wait for a decade or so before the IPCC can reduce their uncertainty over sea-level rise: of course, there will be significant discoveries and developments along the way.

There will be further discussion of sea-level rise when looking at possible surprises in section 6.6.3.

To summarise, the uncertainties over the conclusions from the 2001 and 2007 IPCC assessments do not centre on whether the Earth is getting warmer per se: the IPCC are confident that global temperatures will rise throughout the 21st century and are 'very likely [90–99% chance] to be without precedent during at least the past 10 000 years' (IPCC, 2001b). Instead, uncertainties centre on just how much warming and sea-level rise will take place. But, because warming and sea-level rise are projected to continue well beyond the end of the 21st century the question really is not by how much will the Earth warm and its seas rise, but when?

One thing we do know is that during the last interglacial temperatures briefly peaked to within the window that the IPCC forecasts for the 21st century. We also know that maximum sea levels during the last interglacial were about 5–6 m higher than they are today (we will briefly return to this in section 6.6.3). However, how long after the peak in the last interglacial's temperature the peak in sea level took place is not known with the certainty we need to plan for the future. The computer models (which are continually being developed) currently suggest that we will not see a 5–6 m rise in sea level during this century. Conversely, although the geological evidence does not have the temporal resolution one might like, the evidence does suggest a rapid rise (geologically speaking). Resolving this question (be it through improved computer models and/or geological evidence) is a research priority. Meanwhile, policy-makers should not be complacent and should take due heed of the IPCC's caveats and warnings of climate-related surprises. We will return to surprises at the end of the next chapter.

This brings us on to what will happen beyond the year 2100. In theory, in three centuries or so, should we, as may reasonably be expected, have exploited virtually all of our present-day economically recoverable fossil fuels, but assuming no adoption of climate-abatement measures, the temperature rise could rise a similar number of degrees as with the Initial Eocene Thermal Maximum (IETM) described in section 3.3.9. (And this does not take into account any warming from marine methane hydrate [clathrate] carbon release or re-organisation of biosphere carbon pools [such as in boreal soils], as was likely in the early Eocene.) Whereas sea-level rise over a timescale of two or three centuries could, in a worst-case scenario, be 10 m or even more, it would be far greater still over a timescale of many centuries, as occurred during the Eocene event. The IPCC in 2001 only briefly considered events beyond 2100, which then was based on science from the 1990s (which is how the IPCC works, quite rightly). This is not to decry the IPCC: science has its limits and research itself functions within constraints. Indeed, the first IPCC report (1990) did much to spur investment in climate-related research. The 2007 IPCC assessment did devote a chapter subsection to the next 1000 years under the A1B scenario (which assumes less longer-term fossil fuel use than the A2 scenario). It suggests that temperatures might stabilise a degree or so above that which it forecasts for the end of the 21st century (Figure 5.7b) by the early 22nd century, and then will remain that high through to the end of the millennium. With regards to sea-level rise, the 2007 IPCC report considers that thermal expansion is likely to add between one and two metres on top of whatever ice-cap melt there is (but they do not give an estimate for this last).

Finally, looking even more centuries ahead it is thought that once the sea level has risen due to the Greenland and Western Antarctica ice caps melting, there would be little scope for levels to decline completely to present values without the onset of a proper glacial. For example, melting of the Greenland ice sheet alone would raise sea levels by 7 m (ignoring thermal expansion or any other contribution, such as from Antarctica). Furthermore, both geological evidence and computer models (run in 2002) suggest that even if the Earth returned to 20th-century temperatures the Greenland ice sheet would not reform (Toniazzo et al., 2004).

Given the latest science, the palaeo-analogues and separately that much of the global economy is based on settlements that have been located for many centuries – and so which are ill-suited to address some future climate change consequences – the impact of climate change in purely human terms cannot be ignored. The IPCC attempt to provide a comparatively near-future picture; the uncertainty relates to how near a future, while planning arguably needs to be undertaken with a longer-term perspective. We have considerable confidence that both warming and sea-level rise will continue beyond the 21st century. So, while today we may build sea defences to cope with the anticipated 21st-century rise plus a safety margin, we should perhaps construct them in anticipation of their needing to be enhanced to cope with further rise in the future. After all, many of today's rail networks are grounded in centuries-old construction. Such planning will be far more cost-effective in the long term, and allow us to better cope with a surprise rise in levels, as warned of by the IPCC.

The above multi-century scenario may sound sensationalist but it is not one of doom and disaster. Many species on Earth survived the IETM, even if the planet changed and there was a (primarily) marine extinction event. Indeed, modern humans (*H. sapiens sapiens*; or *H. sapiens*) survived the considerable environmental changes that have taken place since the last glacial maximum, even if some civilisations have not survived subsequent regional climate change. Of course, earlier in history, our planet did not have to sustain humans in the numbers it does do today (or is likely to in the coming decades). Even so, to maximise the longevity of something similar to modern global human society and some types of robust, biodiverse, biological support system (but not necessarily their location) *all* that is required is understanding (research), anticipation (planning) and application (implementation). Application is beyond the scope of this text, but planning and policy concerns will be addressed in Chapter 8, while climate-related biological and human impacts will be summarised in Chapters 6 and 7, respectively. Meanwhile, there are further matters arising out of the 2007 IPCC report, and subsequent science, to note.

5.4 Current human influences on the carbon cycle

Human influences on the carbon cycle will, the IPCC concluded in 2001, continue to change atmospheric composition throughout the 21st century. As the IPCC provide

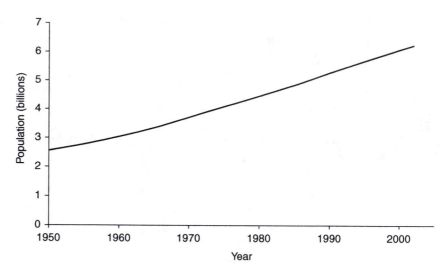

Fig. 5.10   The world population 1950–2002.

the official scientific consensus, it is worth summarising their conclusions with regards to the principal greenhouse gases in the current century.

5.4.1 Carbon dioxide

As noted in Chapter 1, the human contribution to atmospheric carbon dioxide is increasing. The breakdown of the principal sources of this carbon dioxide is given in Table 1.3.

The IPCC's 2001 assessment concludes that carbon dioxide emissions from fossil fuel burning are 'virtually certain' to be the dominant influence on changes in atmospheric concentrations for the 21st century. Further to the greenhouse gas information given in Chapter 1, the reason for this dominance is 2-fold. First, our lifestyles are becoming more energy-intense: people are consuming more energy on a per-capita basis. Second, the global population is growing (see Figure 5.10). Both these factors are multiplicative rather than additive. We will return to global population when looking at the likely future population as part of human ecology in the next chapter.

With reference to the carbon cycle (Figure 1.3), the IPCC (2001b and affirmed in 2007) concludes that as atmospheric carbon dioxide increases the ocean and land will absorb a smaller fraction of subsequent releases. This means that burning a litre of petrol in 10 years time will result in more carbon dioxide residing in the atmosphere for a century or so than a litre burnt 10 years ago. This in turn means that the warming effect from burning a single litre will be greater. (In other words the release of a unit of carbon dioxide will have a greater warming ability, even if its global warming potential [GWP] remains 1 by definition.) With regards to Table 1.1 and the 2010 concentration of carbon dioxide, the IPCC (2001b) conclude that whereas pre-industrial levels of carbon dioxide were around 280 ppm, and the 2006 concentration was around 382 ppm, or a 36% rise (and it was 392 ppm, a 40% increase, in 2011), the projected 2100 concentration is likely to be 540–970 ppm (or

90–240% over pre-industrial levels). The anthropogenic carbon dioxide forcing of climate is therefore going to be some three times greater (possibly more) for the 21st century alone than it was two and a half centuries ago!

In terms of palaeoclimatology, today's carbon dioxide concentration has not been exceeded during the past 420 000 years and probably not during the past several million years. The rate of increase since 1900 is unprecedented in terms of human history and, indeed, evolution. This means that current climate change and that in the near future (one or two human generations) due to anthropogenic greenhouse forcing will be of a nature and degree that humanity has yet to experience.

The IPCC (2001b) also note that 'hypothetically' if all the carbon released due to historic land-use change (forest clearance and such) could be reversed then projected 2100 atmospheric concentrations could be reduced by 40–70 ppm. However, this would not only be a proportionally minor reduction compared to the aforementioned anticipated rises (less than four decades' worth of 1990s-rate atmospheric rise), it would also be a temporary one. For once forests are grown, or soils managed so as to absorb the maximum amount of carbon they can, then the forests and soils (under climatically stable and managed conditions) will at best remain greenhouse-neutral. 'At best' because the necessary continued proper management of such forests and soils cannot be assured (notwithstanding that we still have much to learn about soil carbon). Indeed, soils managed at the beginning of this century to absorb carbon might well see the carbon released again as the climate warms. Given that atmospheric carbon dioxide stabilisation (as discussed above) is unlikely, so that the global climate will continue to warm, and that management cannot be assured (in some circumstances a single ploughing can release many years of a soil's carbon sequestration), it would be most unwise to rely on land management as a principal tool of carbon sequestration. This is just one reason why the IPCC use the word 'hypothetically'. Having said this, whereas the prospects for land management to counter atmospheric increases are very limited, greenhouse gas-sensitive land use can at least slow potentially significant further additions of carbon to the atmosphere. (Deliberate, human-managed sequestration will be dealt with in Chapter 7.)

Aside from climate change concerns, as noted in Chapter 1 the oceans absorb some of the increased atmospheric carbon dioxide from human activities. There are impacts arising from changes in the oceans' chemical environment. The carbon dioxide absorbed by the sea reacts with water to form carbonic acid, so making the oceans more acidic. In 2004 a research team for an international 5-year programme, led by US researcher Chris Sabine, reported on a compiled data set of 9618 hydrographic stations and their dissolved inorganic carbon (DIC) measurements. The global data set gathered was based on ten times more observations than the last time such a survey was conducted, in the 1970s. The researchers estimated that the cumulative oceanic sink from anthropogenic carbon dioxide from about 1800 to 1994 was 118 ± 19 Pg (or Gt) of carbon and that this is possibly about a third of the oceans' long-term sink potential. However, this does not mean that we are a third of the way in time to this ocean-saturation point; we are closer, as anthropogenic emissions per year are greater now than during the 19th and early 20th centuries. Further, as we shall see when discussing the human ecology of climate change, human emissions of carbon have been steadily increasing since 1800, so that instead of four centuries (had the 1800–1994 atmospheric concentration of carbon dioxide been a constant average) we have far less time before saturation is reached. Emissions for the last 50 years of the 20th century were very roughly equal to total emissions for the 200 years up to 1950. Furthermore, as we continue to approach saturation so the oceans become *ever more* acidic: the reduction in pH of a more acidic ocean has a greater biological impact than the same lowering of pH in a less acidic ocean. Alongside the paper by Chris Sabine et al. was another from a team led by Sabine's co-worker, Richard Feely (2004). They concluded that, under the IPCC's 2001 forecasts for the end of the 21st century, atmospheric carbon dioxide levels could be around 800 ppm. At that level ocean pH would further decline, by about 0.4, to a level not seen for more than 20 million years. At lower pH values calcium carbonate shells of forams will begin to dissolve faster and growth will be compromised. Calcification rates can drop by as much as 25–45% at carbon dioxide levels equivalent to atmospheric concentrations of 700–800 ppm. These levels will be reached by the end of this century if fossil fuel consumption continues at projected levels.

Effects will be worse in those parts of the ocean where carbonic acid concentrations are highest. They would have a profound effect on ocean acidity and the oceans' biological community that relies structurally on carbonate. Already one of the researchers on the team (Victoria Fabry) has observed partial shell dissolution in one of the pteropods (a group of marine gastropod molluscs), a live pyramid clio (*Clio pyramidata*) in the sub-Arctic North Pacific. Groups and phyla at possible risk include calcareous green algae, echinoderms (the phylum of starfish and sea urchins), bryozoans (those of a phylum of colonial animals also known as the ectoproctans) and the deep-sea benthic forams.

About half of the anthropogenic carbon dioxide taken up over the last 200 years – the 118 ± 19 Pg of carbon – can be found in the upper 10% of the ocean. In turn, this 118 Pg very approximately represents less than 40% of the carbon released as carbon dioxide by human action (both fossil fuel burning and land-use change). If this carbon had not been absorbed by the oceans, atmospheric concentrations would be higher. Recalling that pre-industrial atmospheric carbon dioxide was 280 ppm (Table 1.1), with the 2003 concentration being 376 ppm, this last would be some 55 ppm higher still were it not for oceanic uptake. So, in one sense what is bad for the ocean is good for the atmosphere (although do note that this use of 'good' and 'bad' is unscientific).

Of the ocean DIC from anthropogenic carbon dioxide, more than 23% is found in the North Atlantic, which might be surprising considering its small size (15% of the global ocean surface) compared to other oceans such as the Indian Ocean and North and South Pacific. However, the North Atlantic is bounded by the north-western European and the North American countries, which together have historically (up to 2000) been responsible for very roughly half the planet's anthropogenic carbon dioxide emissions. This in turn suggests that a good proportion of carbon dioxide enters the ocean by being washed out by rain, as opposed to ocean-surface diffusion and wave mixing. However, because so much of the southern hemisphere is ocean, some 60% of ocean DIC from anthropogenic carbon dioxide is stored in that hemisphere's seas. Taking all this together we might expect signs of a major impact on carbonate species to first begin to manifest themselves in the cool North Atlantic and in waters arriving from it, such as in the Arctic Ocean or the sub-Arctic North Pacific.

Since the Industrial Revolution sea-surface pH has dropped by 0.1 and this is over a 200-year period. A further fall in pH of 0.4 over the next 100 years represents a far greater rate of change. The impact will not just be on calcareous species but, because of ecological relationships, reverberate throughout much of the food web. In August 2004 Britain's Royal Society declared ocean acidification to be a research priority. Ocean acidification could well be one of the surprises of which the IPCC warn as a caveat to their 21st-century forecasts (see the end of Chapter 6). In 2005 the Royal Society published a report of their working party, assembled to look at the matter. It noted that there appears to be no practical way of removing additional carbon dioxide once it has been absorbed by the oceans and that such absorption, should it continue, would impact on the marine ecosystems with possible substantial socio-economic impacts. It concluded that action needed 'to be taken now to reduce global emissions of carbon dioxide to avoid the risk of large and irreversible damage' (Royal Society, 2005).

5.4.2 Methane

Atmospheric methane concentrations have increased by 158% since 1750 (see Table 1.1). It is thought that more than half the current methane flux to the atmosphere is anthropogenic (agriculture, natural gas and landfill). The remainder is from natural sources, such as wetlands. The atmospheric concentration of methane has continued to increase but in the 1990s the rate of this increase declined and in the 2000s concentrations hovered around 1775 parts per billion by volume (ppbv). The reasons for variability in the rate of increase in atmospheric methane in the 1990s are not entirely clear as quantitative measurement of global sources and sinks is far from complete. Methane is emitted from fossil fuels, biomass burning, wetlands, agriculture, landfills and waste water. Disentangling atmospheric methane's origins is not easy. The IPCC's 2007 assessment provides a table that presents different sets of estimates for the various sources in the 1990s to 2001. As a rough guide, Figure 5.11 provides a median estimate[5] from the IPCC (2007) report of the proportional sources of methane in the mid-1990s. Natural sources accounted for less than half of estimated emissions in all sets of estimated sources in the IPCC (2007) table, as they do in Figure 5.11, with rice and ruminants being the dominant anthropogenic sources for most of the estimates.

Methane from energy industries declined in the 21st century's first decade. Exploring how this is estimated provides an insight into how attempts are made to ascertain the annual methane budget. In 2011 two teams of researchers publishing in *Nature* attempted to explain the post-millennial decline in methane's rise. First, Murat Aydin,

---

[5] Table 7.6 from the IPCC (2007) assessment report, 'Sources, sinks and atmospheric budgets of CH_4' from the chapter 'Couplings Between Changes in the Climate System and Biogeochemistry'. Of course, a median estimate from this table does not necessarily mean it correct, but it does serve to broadly illustrate the various methane sources. This is why percentages are given and not quantitative amounts, as presented in the IPCC table. Furthermore, matters change over the years.

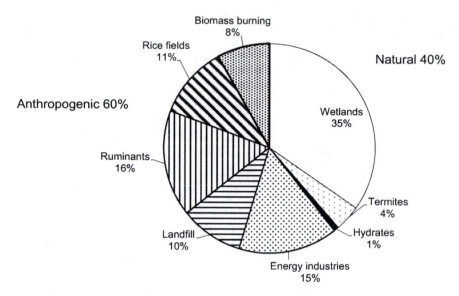

Fig. 5.11 Sources of atmospheric methane from the 1990s to 2001. See text for explanation and source.

Kristal Verhulst, Eric Saltzman and colleagues looked at ethane. Ethane (C_2H_6) and methane are the most abundant hydrocarbons in the atmosphere. However, ethane is only associated with fossil fuels and biomass burning, not the other sources of methane. What the Aydin team did was to look at ethane in firn air (air trapped in perennial snowpack) from Greenland and Antarctica to reconstruct the atmospheric variability of ethane during the 20th century. They found that ethane levels rose from early in the century until the 1980s, when the trend reversed, with a period of decline over the next 20 years. They found that this variability was driven primarily by changes in ethane emissions from fossil fuels; these emissions peaked in the 1960s and 1970s at 14–16 teragrams (Tg) year^{-1} (1 Tg = 10^{12} g = 1 megatonne [Mt]) and dropped to 8–10 Tg year^{-1} by the turn of the century. They concluded that reduction in fossil fuel sources is probably related to changes in light hydrocarbon emissions associated with petroleum production and use. The Aydin team suggest (in the main body of their paper's text) that this decline in ethane might not relate so much to the co-release of methane from 'petroleum production and use' (as stated in their paper's abstract) but more to the venting of gas by petroleum refineries. Putting this quantitative inference into a two-box atmospheric model, they estimated that the total decline in fossil fuel emissions of ethane was 5–6 Tg year^{-1} (or Gt year^{-1}) during 1980–2000. Attributing this decline entirely to decreases in fossil fuel emission sources implies a 15–30 Tg year^{-1} drop in fossil fuel emissions of methane. This decline in fossil fuel methane emissions during the 1990s accounts for less than 60% of the total reduction of atmospheric methane observed during 1980–2000. This in turn raises the question of what other methane emissions were declining.

The second 2011 paper was by Fuu Ming Kai, Stanley Tyler, James Randerson and Donald Blake. They used synchronous time series of atmospheric methane mixing and $^{13}C/^{12}C$ ratios in the northern and southern hemispheres and a two-box atmospheric model. They also used deuterium (isotopic ^{2}H, sometimes represented by D) analysis of the hydrogen in methane to further disentangle some of the methane source options. They concluded that the ^{13}C observations are consistent with long-term reductions in agricultural emissions or another microbial source in the northern hemisphere. Approximately half (51 $\pm$ 18%) of the decrease in northern-hemisphere methane emissions can be explained by reduced emissions from rice agriculture in Asia over the past three decades associated with increases in fertilizer application and reductions in water use.

Taken together, these papers suggest that roughly half (or perhaps a little more) of the decline in methane emissions came from better prevention of methane escape from fossil fuel refining and storage (notably Russia improved its energy infrastructure over this period), and half (or perhaps a little less) from improved rice agriculture that reduced the area of paddy field wetland needed. If this work does interpret what is going on reasonably accurately, then unless further improvements in fossil fuel management occur, and as fossil fuel use continues to grow, methane emissions will rise in the future. Similarly, unless there are continued improvements in rice agricultural production techniques then as the world population continues to grow, and demand more rice, and rice production increases, then so methane emissions will rise. In short, we should expect the atmospheric concentration of methane to rise once more in coming decades.

At the end of 2011, MethaneNet (an affiliation of British researchers) sponsored a symposium on methane research, held at the headquarters of the Geological Society. Various strategies for reducing anthropogenic methane were proposed. First, with regard to methane from agricultural ruminants (principally cattle), studies show that kangaroos (*Macropus* spp.) emit around 2% of the CO_2 equivalent for methane of sheep and even less than that for cattle. Could it be, by altering cattle gut microflora (bacteria) to be more like that of kangaroos, that emissions can be reduced? Progress is difficult because cultivating pure strains of species in ruminant gut flora is not easy. However, attempts are being made to create a meta-genome of the whole gut microbial community. The bad news for prospects of future ruminant methane emissions is that as countries develop, so their protein consumption per capita increases. The good news is that the growth in protein consumption in Asia is largely related to chicken and pork, and consequently less beef.

Second, with regard to rice cultivation, research shows that methane emissions can be greatly reduced by allowing paddy fields to periodically dry out. In areas where the water supply is gravity-fed this is easy to do and it also, given appropriate water-management infrastructure, lowers water consumption. Of course, it should be noted that the carbon does not just vanish. The methane emissions are partly compensated for by the mineralisation of carbon, hence the resulting increased carbon dioxide emissions. But given methane's higher GWP (see Table 1.2) this is a far more preferable option.

Finally, it should be noted that methane from melting permafrost regions in high latitudes largely comes from wetlands and frozen lakes. Methane released within permafrost soils gets converted to carbon dioxide which, as noted, has a lower GWP: this is still a great concern but a lesser one than would otherwise be.

5.4.3 Halocarbons

Halocarbons are carbon compounds, though not strictly part of the carbon cycle due to their resilience and their being human-made. Following international agreements, principally the 1987 Montreal Protocol (a rare success story), the atmospheric concentrations of many stratospheric ozone-depleting and greenhouse halocarbons are either decreasing (CFC-11, CFC-113, CH_3Cl_3 and CCl_4) or the rate of their increase has markedly slowed. The combined abundance of ozone-depleting gases peaked in 1994 and is slowly declining (although ozone depletion will be with us for decades to come). The atmospheric concentrations of these gases tie in with recorded emissions and so in this regard we can be more confident in the latest computer models of the global climate.

If carbon dioxide and methane concentrations were declining, as is the combined abundance of ozone-depleting gases, then greenhouse concerns would not necessitate nearly so much attention. However, as stated above, this is not the case and future carbon dioxide and methane reductions seem unlikely. In this context halocarbons are of importance as they can further enhance warming. This in turn can affect the carbon cycle through the release of carbon from soils and other carbon pools (see Chapters 1 and 7).

One problem with halocarbons is that some of the replacements for the ozone-depleting ones are themselves greenhouse gases. The abundance of the hydrochlorofluorocarbons and hydrofluorocarbons is rising as they are increasingly used as chlorofluorocarbon substitutes. Although their atmospheric concentrations are very small (hence their mass impact on the carbon cycle is small) compared to methane and carbon dioxide, they are very powerful greenhouse gases.

Fully fluorinated chemicals have no biological function (hence are not a major focus of this book) and cannot be removed by biogenic action. Furthermore, they are chemically stable, so they have a long atmospheric residence time, typically thousands of years. However, they are at very low atmospheric concentrations but generally have high GWPs. As such, they will have a significant role in greenhouse warming, albeit in a lesser way compared to methane or carbon dioxide, for a number of centuries.

5.4.4 Nitrous oxide

We briefly touched on nitrous oxide (N_2O) in Chapter 1. As noted, there are many uncertainties as to quantifying the sources and sinks of this long-lived and significant, albeit comparatively minor, greenhouse gas. It is related to the carbon cycle in that nitrous oxide is part of the nitrogen cycle, which includes all living things, as the amino-acid building blocks of proteins contain nitrogen as well as carbon. Hence plants rely on nitrogen and plant activity affects carbon sequestration and soil storage. Consequently, understanding the components of carbon cycle dynamics helps us relate to nitrous oxide concentrations (and vice versa), as well as greenhouse warming through carbon dioxide, methane and – of course – nitrous oxide itself.

Again, as with carbon dioxide, methane and halocarbons, the elevated post-industrial atmospheric concentrations of nitrous oxide are related to human activity.

Human activity, and climatic impacts thereon, will form the subject of Chapter 7. Nitrous oxide contributes even less to anthropogenic warming than methane (see Figure 1.2). Consequently much of this book focuses principally on carbon dioxide.

As to nitrous oxide's anthropogenic sources, these are principally from the agriculture sector.

5.5 References

Aimers, J. (2011) Drought and the Maya: The story of the artefacts? *Nature*, 479, 44.

Alley, R. B. (2000) The Younger Dryas cold interval as viewed from central Greenland. *Quaternary Science Reviews*, 19, 213–26.

Andreae, M. O., Jones, C. D. and Cox, P. M. (2005) Strong present-day aerosol cooling implies a hot future. *Nature*, 435, 1187–90.

Augustin, L., Barbante, C., Barnes, P. R. F. et al. (2004) Eight glacial cycles from an Antarctic ice core. *Nature*, 429, 623–8.

Aydin, M., Verhulst, K. R. and Saltzman E. S. (2011) Recent decreases in fossil-fuel emissions of ethane and methane derived from firn air. *Nature*, 476, 198–201.

Bellouin, N., Boucher, O., Haywood, J. and Reddy, M. S. (2005) Global estimate of aerosol direct radiative forcing from satellite measurements. *Nature*, 438, 1138–41.

Berger, A. and Loutre, M. F. (2002) An exceptionally long interglacial ahead. *Science*, 297, 1287–9.

Betts, R. A., Boucher, O., Collins, M., Cox, P. M. et al. (2007) Projected increase in continental runoff due to plant responses to increasing carbon dioxide. *Nature*, 448, 1037–41.

Büntgen, U., Tegel, W., Nicolussi, K. et al. (2011) 2500 years of European climate variability and human susceptibility. *Science*, 331, 578–82.

Burroughs, W. J. (1997) *Does The Weather Really Matter? The Social Implications of Climate Change*. Cambridge: Cambridge University Press.

Chapin, III, F. S., Sturm, M., Serreze, M. C. et al. (2005) Role of land-surface changes in Arctic summer warming. *Science*, 310, 657–60.

Chepstow-Lusty, A. J., Frogley, M. R., Bauer, B. S. et al. (2009) Putting the rise of the Inca Empire within a climatic and land management context. *Climate of the Past*, 5, 375–88.

Cole-Dai, J., Ferris, D., Lanciki, A., Savarino, J. et al. (2009) Cold decade (AD 1810–1819) caused by Tambora (1815) and another (1809) stratospheric volcanic eruption, *Geophysical Research Letters*, 36, L22703.

Cowie, J. (1998) *Climate and Human Change: Disaster or Opportunity?* London: Parthenon Publishing.

Cullen, H. M., deMonocal, P. B., Hemming, S. et al. (2000) Climate change and the collapse of the Akkadian empire: Evidence from the deep sea. *Geology*, 28(4), 379–82.

Cunningham, S. A., Kanzow, T. O., Rayner, D. et al. (2007). Temporal variability of the Atlantic Meridional Overturning Circulation at 26.5°N. *Science*, 317, 935–8.

Cuntz, M. (2011) A dent in carbon's gold standard. *Nature*, 477, 547–8.

Davis, C. H., Li, Y., McConnell, J. R., Frey, M. M. and Hanna, H. (2005) Snowfall-driven growth in East Antarctic ice sheet mitigates recent sea-level rise. *Science*, 308, 1898–1901.

deMonocal, P. B. (2001) Cultural responses to climate change during the late Holocene. *Science*, 292, 667–73.

Evan, A. T., Kossin, J. P., Chung, C. and Ramanathan, V. (2011) Arabian Sea tropical cyclones intensified by emissions of black carbon and other aerosols. *Nature*, 479, 94–7.

Feddema, J. J., Oleson, K. W., Bonan, G. B. et al. (2005) The importance of land cover change in simulating future climates. *Science*, 310, 1674–6.

Feely, R. A., Sabine, C. L., Lee, K. et al. (2004) Impact of anthropogenic CO_2 on the $CaCO_3$ system in the oceans. *Science*, 305, 362–6.

Graversen, R. G., Mauritsen, T., Tjernström, M. et al. (2008) Vertical structure of recent warming. *Nature*, 541, 53–6.

Haug, G. H., Günther, D., Peterson, L. C., Sigman, D. M. et al. (2003) Climate and the collapse of Maya civilization. *Science*, 299, 1731–5.

Hermann, P. (1954) *Conquest by Man*. New York: Harper.

Hodell, D. A., Curtis, J. H. and Brenner, M. (1995) The role of climate and the collapse of the Mayan civilization. *Nature*, 375, 391–4.

Hodell, D. (2011) Drought and the Maya: Maya megadrought? *Nature*, 479, 45.

Intergovernmental Panel on Climate Change (1990) *Climate Change: the IPCC Scientific Assessment*. Cambridge: Cambridge University Press.

Intergovernmental Panel on Climate Change (1992) *Climate Change 1992: the Supplementary Report to the IPCC Scientific Assessment*. Cambridge: Cambridge University Press.

Intergovernmental Panel on Climate Change (1995) *Climate Change 1995: the Science of Climate Change*. Cambridge: Cambridge University Press.

Intergovernmental Panel on Climate Change (2000) *Emission Scenarios–IPCC Special Report*. Geneva: World Meteorological Organization.

Intergovernmental Panel on Climate Change (2001a) *Climate Change 2001: Impacts Adaptation and Vulnerability–a Report of Working Group II*. Cambridge: Cambridge University Press.

Intergovernmental Panel on Climate Change (2001b) *Climate Change 2001: the Scientific Basis–Summary for Policymakers and Technical Summary of the Working Group I Report*. Cambridge: Cambridge University Press.

Intergovernmental Panel on Climate Change (2007) *Climate Change 2007: the Physical Scientific Basis–Working Group I Contribution to the Fourth Assessment*. Cambridge: Cambridge University Press.

Jacob, T., Wahr, J., Pfeffer, W. T. and Swenson, S. (2012) Recent contributions of glaciers and ice caps to sea level rise. *Nature*, 482, 514–18.

Jung, M., Reichstein, M. and Ciais, P. (2010) Recent decline in the global land evapotranspiration trend due to limited moisture supply. *Nature*, 467, 951–4.

Kai, F. M., Tyler, S. C., Randerson, J. T. and Blake, D. R. (2011) Reduced methane growth rate explained by decreased Northern Hemisphere microbial sources. *Nature*, 476, 194–7.

Kaniewski, D., Paulissen, E., Van Campo, E. et al. (2010) Late second-early first millennium BC abrupt climate changes in coastal Syria and their possible significance for the history of the Eastern Mediterranean. *Quaternary Research*, 74, 207–15.

Kaufman, D. S., Schneider, D. P., McKay, N. P. et al. (2009) Recent warming reverses long-term Arctic cooling. *Science*, 325, 1236–9.

Kaufmann, R. K., Kauppi, H., Mann, M. L. and Stock, J. H. (2011) Reconciling anthropogenic climate change with observed temperature 1998–2008. *Proceedings of the National Academy of Sciences USA*, 108(29), 11790–3.

Knutti, R., Stocker, T. F., Joos, F. and Plattner, G.-K. (2002) Constraints on radiative forcing and future climate change from observations and climate model ensembles. *Nature*, 416, 719–26.

Lamb, H. H. (1995) *Climate History and the Modern World*, 2nd edn. London: Routledge.

Luterbacher, J., Dietrich, D., Xoplaki, E., Grosjean, T. and Wanner, H. (2004) European seasonal and annual temperature variability, trends and extremes since 1500. *Science*, 303, 1499–1503.

MacCracken, M., Smith, J. and Janetos, A. C. (2004) Reliable regional climate model not yet on the horizon. *Nature*, 429, 699.

Manley, G. (1974) Central England temperatures: monthly means 1659 to 1973. *Quarterly Journal of the Royal Meteorological Society*, 100, 389–405.

Meehl, G. A., Arblaster, J. M., Fasullo, J. T. et al. (2011) Model-based evidence of deep-ocean heat uptake during surface-temperature hiatus periods. *Nature Climate Change*, 1, 360–4.

MethaneNet (2011) *Methane Hack: Strategies for Reducing Growth in Atmospheric Methane*. Symposium held 1 December 2011, Geological Society, London. www. methanenet.org.

Noren, A. J., Blerman, P. R., Steig, E. J., Lini A. and Southon, J. (2002) Millennial scale storminess variability in the northeastern United States during the Holocene epoch. *Nature*, 419, 821–4.

O'Dowd, C., Facchini, M. C., Cavalli, F. et al. (2004) Biogenically driven organic contribution to marine aerosol. *Nature*, 431, 676–80.

Oreskes, N. (2004) The scientific consensus on climate change. *Science*, 306, 1686.

Penner, J. E., Dong, X. and Chen, Y. (2004) Observational evidence of a change in radiative forcing due to the indirect aerosol effect. *Nature*, 427, 231–4.

Pauluis, O., Czaja, A. and Korty, R. (2008) The global atmospheric circulation o moist isentropes. *Science*, 321, 1075–8.

Pilewskie, P. (2007) Aerosols heat up. *Nature*, 448, 541–2.

Pinker, R. T., Zhang, B. and Dutton, E. G. (2005) Do satellites detect trends in surface solar radiation? *Science*, 308, 850–4.

Prohom, M., Esteban, P. and Martin-Vide, J. (2001) Surface atmospheric circulation over Europe following major volcanic eruptions, 1780–1995. Presented at the symposium *PAGES–PEPIII: Past Climate Variability Through Europe and Africa* on 27–31 August 2001, Centre des Congrès Aix-en-Provence, France.

Rahmstorf, S., Cazenave, A., Church, J. A. et al. (2007) Recent climate observations compared to projections. *Science*, 316, 709.

Ramanathan, V., Ramana, M. V., Roberts, G. et al. (2007) Warming trend in Asia amplified by brown cloud solar absorption. *Nature*, 448, 575–8.

Rignot, E., Mouginot, J. and Scheuch, B. (2011) Ice flow of the Antarctic ice sheet. *Science*, 333, 1427–30.

Roderick, M. L. and Farquar, G. D. (2002) The cause of decreased pan evaporation over the past 50 years. *Science*, 298, 1410–12.

Royal Society (2005) *Ocean Acidification due to Increases in Atmospheric Carbon Dioxide*. London: Royal Society.

Royal Society (2010) *Climate Change: A Summary of the Science*. London: Royal Society.

Sabine, C. L., Feely R. A., Gruber, N. et al. (2004) The oceanic sink for anthropogenic CO_2. *Science*, 305, 367–71.

Sereno, P. C., Garcea, E. A. A., Jousse, H. et al. (2008) Lakeside cemeteries in the Sahara: 5000 years of Holocene population and environmental change. *PLoS ONE*, 3(8), e2995.

Sriver, R. (2011) Man-made cyclones. *Nature*, 479, 50–1.

Stahle, D. W., Villanueva-Díaz, J., Burnette, D. J. et al. (2011) New drought record from long-lived Mexican trees may illuminate fates of past civilizations. *Geophysical Research Letters*, 38, L05703.

Stothers, R. B. (1984). Mystery cloud of AD 536. *Nature*, 307, 344–5.

Thomas, R., Rignot, E., Casassa, G. et al. (2004) Accelerated sea-level rise from West Antarctica. *Science*, 306, 255–8.

Toniazzo, T., Gregory, J. M. and Huybrechts, P. (2004) Climatic impact of a Greenland deglaciation and its possible irreversibility. *Journal of Climate*, 17, 21–33.

Travis, D. J., Carleton, A. M. and Lauriston, R. G. (2002) Climatology: contrails reduce daily temperature range. *Nature*, 418, 601.

Tyson, P., Odada, E., Schulze, R. and Vogel, C. (2002) Regional-global change linkages: Southern Africa. In Tyson, P., Fuchs, R., Fu, C. et al., eds. *Global-Regional Linkages in the Earth System*, pp. 3–73. Berlin: Springer.

Ufnar, D. F., González, L. A., Ludvigson, G. A., Brenner, R. L. and Witzke, B. J. (2004) Evidence for increased latent heat transport during the Cretaceous (Albian) greenhouse warming. *Geology*, 32(12), 1049–52.

United Nations Environment Programme and the World Meteorological Organisation (2011) *Integrated Assessment of Black Carbon and Tropospheric Ozone: Summary for Policy Makers*. Nairobi: UNEP.

van den Broeke, M., Bamber, J. and Ettema, J. (2009) Partitioning recent Greenland mass loss. *Science*, 326, 984–6.

Vaughan, D. G. (2005) How does the Antarctic ice sheet affect sea level rise? *Science*, 308, 1877–8.

von Storch, H., Zorita, E., Jones, J. M. et al. (2004) Reconstructing past climate from noisy data. *Science*, 306, 679–82.

Weiss, H. (1982). The decline of Late Bronze Age civilization as a possible response to climatic change. *Climatic Change*, 4(2), 173–98.

Weiss, H. (2000) Beyond the Younger Dryas. Collapse as adaptation to abrupt climate change in Ancient West Asia and the Eastern Mediterranean. In Bawden, G. and

Reycraft, R., eds, *Confronting Natural Disasters: Engaging the Past to Understand the Future*. Albuquerque, NM: University of New Mexico Press.

Welp, L. R., Keeling, R. F., Meijer, H. A. J. et al. (2011) Interannual variability in the oxygen isotopes of atmospheric CO_2 driven by El Niño. *Nature*, 477, 579–82.

Weninger, B., Alram-Stern, E., Bauer, E. et al. (2006) Climate forcing due to the 8200 cal yr BP event observed at Early Neolithic sites in the eastern Mediterranean. *Quaternary Research*, 66, 401–20.

Wild, M., Gilgen, H., Roesch, A. et al. (2005) From dimming to brightening: decadal changes in solar radiation at the Earth's surface. *Science*, 308, 847–50.

Willett, K. M., Gillett, N. P., Jones, P. D. and Thorne P. W. (2007) Attribution of observed surface humidity changes to human influence. *Nature*, 449, 710–12.

Worldwatch Institute (2003) *Vital Signs 2003–2004: the Trends that are Shaping Our Future*. London: Earthscan.

Zhang, D. D., Lee, H. F., Wang, C. et al. (2011) The causality analysis of climate change and large-scale human crisis. *Proceedings of the National Academy of Sciences USA*, 108(42), 17296–301.

Zhang, P., Cheng. H., Edwards, L. et al. (2008) A test of climate, sun, and culture relationships from an 1810-year Chinese cave record. *Science*, 322, 940–2.

Zhang, X., Zwiers, F. W., Hegerl, G. C., Lambert, F. H. et al. (2007) Detection of human influence on 20th-century precipitation trends. *Nature*, 448, 461–5.

6   Current warming and likely future impacts

6.1 Current biological symptoms of warming

As noted in Chapter 2, climate change impacts on living species and so in turn these impacts can be used as climatic proxies. As discussed in Chapter 3, climate change throughout the Earth's history has been complex and features a number of distinct and characteristic episodes as well as periods with defining trends. However, as was observed in Chapters 3 and 4, the biological response of species and natural systems to climate change is also complex, as is the biological response of (and impact on) humans. A good proportion of this last is reflected in recorded history. Further, these impact elements – of species response and climatic-event and natural-systems response – are all discernable when applied to climate at the end of the 20th and early 21st century and all exhibit varying degrees of complexity. Some responses are sufficiently complex that they appear as the opposite to what might initially be expected. Even in such cases, when examined carefully, they all clearly reflect the fact that the planet is warming up.

6.1.1 Current boreal dendrochronological response

One illustrative example was briefly mentioned in Chapter 2 when looking at dendro-chronology. Figure 2.1 portrays the pooling together of several dendrochronological series from the northern hemisphere's high latitudes. This clearly shows that in the 20th century the north of the northern hemisphere warmed above the 1601–1974 average. However, some dendrochronological series (Figure 2.1, the dashed line on the far right) seem to exhibit a late-20th-century return to average temperatures even though others reflect continued warming (in line with instrumental measurements).[1] As indicated, this deviation from what would be expected may be due to increased snowfall with the volume of the extra snow delaying the melting of snow cover and so delaying the onset of spring greening. This is not unexpected, because, as previously noted, a warmer planet would have increased ocean evaporation and hence increased precipitation (be it rain or snow). Another explanation mentioned is that some other

---

[1] It was this problem that required researchers to use what one called a 'trick' in a private e-mail when presenting some dendrochronological records as climate proxies over the past few centuries. The researchers' e-mails were hacked and then released in November 2009 just prior to the Copenhagen climate change conference. Subsequently a number of enquiries into the affair confirmed, in the words of the House of Commons Science and Technology Select Committee report (2011), that the use of 'phrases such as "trick" or "hiding the decline" were colloquial terms used in private e-mails and the balance of evidence is that they were not part of a systematic attempt to mislead'.

environmental factor impedes growth. As noted in Chapter 2, species respond to a range of environmental factors, some of which act synergistically, sometimes in a positive way and others in a negative way.

The response of the Alaskan white spruce (*Picea glauca*) to late-20th-century warming exemplifies these complexities. Our current warming period has at high latitudes extended the thermal growing season (TGS) and this is clear from instruments both on the ground and in satellites. Of course, when there are days in the year in which plants can grow, primary biological productivity (plant biomass increase per unit area) increases. This means that more carbon dioxide is sequestered by photosynthesis and this in turn increases water consumption (above and beyond that from climate warming itself). Increased water demand is dependent on there being the necessary extra water present. True, in a warmer world we would expect more rainfall, but this is only an expectation (although there is certainly more water vapour; see Chapter 1) and an expectation we would expect to be reflected in, on average, more precipitation across the whole planet. Yet, again as previously noted, there are frequently regional and even comparatively local marked deviations from the average expectation. So consider a species that is on the border of its environmental range. Whereas one environmental factor may increase (in this case temperature) it does not necessarily mean that other environmental factors will increase (even if we might expect them to) so as to accommodate a species' healthy growth. Then again, whereas in a warmer world we would expect more precipitation, equally we should expect more evaporation.

In short, in some places and for some species whose environment has changed, water availability could be a problem. In 2000, researchers from the University of Alaska Fairbanks found that, whereas warming had extended *P. glauca*'s TGS, growth had not increased as expected (Barber et al., 2000). The problem was one of temperature-induced drought stress. The interior of Alaska is semi-arid and in many places the potential evapotranspiration is equal to annual precipitation. Increase the potential evapotranspiration (through increased temperature and hence extended TGS) without an appropriate increase in precipitation, and species experience drought stress.

This phenomenon is not just of intellectual interest as far as climate change is concerned, but of practical relevance too. *P. glauca* is one of the most productive and widespread forest tree species in the boreal forest of western North America. Consequently it plays an important role in the terrestrial sink of the northerly part of the northern hemisphere. This sink sequesters (that is, soaks up) some of the excess carbon dioxide released through fossil fuel burning and other human activities (over and above natural fluxes of carbon dioxide). If this changes with warming then the sequestration rate will in theory change, which in turn will affect the rate of future warming from carbon emissions, although in practice the scale of such increased carbon sequestration is highly unlikely to make a meaningful difference at the policy level. Nonetheless, such biological work is timely for computer modellers of climate, who, having begun to incorporate some terrestrial and marine biological factors into their simulations, are now doing so with increasing complexity, which is beginning to include the dynamics of biological response to the climate change being modelled. Such developments are vitally important, for, as discussed previously, although computer models have improved they are still a long way from reflecting climate change in a detailed way in both space and time, or in recreating past major change with anything other than an approximation.

6.1.2 Current tropical rainforest response

Tropical rainforests are among the most species-rich tree communities on Earth and the most productive of terrestrial ecosystems. Consequently tropical rainforests are affected by global warming and concomitant increases in atmospheric carbon dioxide. A recent medium-term study, over two decades long in 2004, of 18 tropical Amazonian rainforest plots that were otherwise undisturbed, has shown a number of changes (Laurance et al., 2004). In 2001 the Amazonian forest covered about 5.4 million km^2, approximately 87% of their original extent, with 62% of that which remains in Brazil. The Amazonian forest is home to perhaps a quarter of the Earth's terrestrial species, and about 15% of global terrestrial photosynthesis occurs there. Evaporation and condensation over Amazonia contribute to the global atmospheric circulation, having downstream effects on precipitation across South America, and further afield across the northern hemisphere. The Amazon's forests contribute a significant part of the Earth system's functioning (Malhi et al., 2007).

All trees in each plot of the Laurance et al. study that were more than 10 cm in diameter at breast height were tagged and mapped. In total nearly 13 700 trees were recorded. Because rainforests are so biodiverse some 88% of tree species were too rare for there to be a robust statistical analysis. Furthermore, there were so many species and subspecies that their analysis was done at the genus level to capture the changes with statistics. There were 244 genera, of which 115 were sufficiently abundant for statistical analysis.

They found that within the plots the rates of tree mortality, recruitment and growth have all increased to some degree or other over the 20 years of the study. Of the 115 relatively abundant tree genera, 27 changed significantly in population density or basal area to a degree that was nearly 14 times greater than might be expected from chance alone. This result is similar to other studies. However, contrary to some predictions there was not an increase in pioneer species. Pioneer species are those that are commonly found colonising new or disturbed ground, and higher mortality associated with ecological change would be expected to disturb ground with tree falls.

The researchers also found an increasing number of faster-growing trees, including many canopy and canopy-emergent species. These increased in dominance or population density. Conversely, a number of genera of slower-growing trees, including many sub-canopy species, were declining.

So, what is causing these changes? The researchers noted that it could be that the plots were recovering from some previous disturbance. However, their plots' location – away from areas of human land use, in the depth of forests (so avoiding edge effects) and over an area of 300 km^2 (so avoiding purely local factors) – meant that previous disturbance was an unlikely cause of the changes seen.

Second, the researchers wondered whether the changes might be due to regional climate cycles, such as El Niño-related droughts. These have increased over the last century and are thought to possibly be related to global warming. However, countering this two separate analyses suggest that El Niño-related droughts were not the cause of these tropical rainforest changes, although the researchers advise further checks, so this matter has yet to be resolved. Third, rainfall changes (as distinct from drought, which sees a decline in rainfall, river flow and ground water) might be the cause, but this too was discounted on checking meteorological records.

The final possible reason was that the changes might be driven by accelerated forest productivity. The researchers believed that this was the most likely reason, due to rising atmospheric carbon dioxide, although increased cloudiness and higher airborne nutrient transfer from increased forest fires (both climate-related factors) could be the cause. Either way, greenhouse changes seem to be the spur.

In recent decades the rate of warming in Amazonia has been about 0.25°C per decade. Under mid-range greenhouse gas emission scenarios, temperatures are projected to rise 3.3°C (range 1.8–5.1°C) this century, slightly more in the interior in the dry season, or by up to 8°C if significant forest dieback affects regional biophysical properties. At the end of the last glacial, Amazonia warmed by only approximately 0.1°C per century. Changes in precipitation, particularly in the dry season, are probably the most critical determinant of the climatic fate of the Amazon. There has been a drying trend in northern Amazonia since the mid-1970s, and no consistent multi-decadal trend in the south, but some global climate models project significant Amazonian drying over the 21st century. The zones of highest drought risk are the south east and east, which are also the areas of most active deforestation. In contrast, the north-western Amazon is least likely to experience major drought (Malhi et al., 2007).

These changes could have both local and global implications. Currently, undisturbed tropical rainforests appear to function as carbon sinks and so they help slow global warming. A 1998 long-term monitoring study of plots in mature humid tropical forests concentrated in South America revealed that biomass gain by tree growth exceeded losses from tree death in 38 of 50 Neotropical sites. These more humid parts of the Amazon forest plots have gained 0.71 ± 0.34 t of carbon (t C) ha^{-1} year^{-1} above ground (i.e. excluding soil carbon) (Phillips et al., 1998). A similar 2009 study in Africa (which included the results of previous studies), where a third of the Earth's tropical forests are found, for the period 1968–2007 gave an average above-ground carbon accumulation rate of 0.63 t C ha^{-1} year^{-1}. The data suggest that undisturbed Neotropical forests may (currently) be a significant carbon sink, reducing the rate of increase in atmospheric carbon dioxide. Scaling up to tropical rainforests of the whole African continent suggests total above-ground accumulation of 0.34 gigatonnes of carbon (GtC) year^{-1}. Including this with the aforementioned and other studies up to 2009 suggest that globally tropical forests above ground might be accumulating carbon at a rate of 1.31 GtC year^{-1} within the range 0.79–1.56 GtC year^{-1} in recent decades (Lewis et al., 2009). This compares with an estimated global land sink (all terrestrial biomes) of very approximately 2.2 GtC year^{-1} between 1980 and 2000 (1.4 GtC year^{-1} for 1990–9; IPCC, 2001b). The tentative suggestion is that at the moment tropical rainforests are increasing as a carbon sink.

Yet ecological changes such as seen in the rainforest study by Laurance et al. (2004), if pervasive, could modify this sink effect. Increases in carbon storage by such tropical trees may be slowed by the tendency of canopy and emergent trees to produce wood of lower density as their size and growth rate increases. Further, such species shade the more densely wooded sub-canopy species that were seen to have lower growth and were declining. Indeed, a sufficiently warmer climate may turn a tropical forest into a carbon source as opposed to a sink. Drought and extended dry seasons with climate change would serve to release carbon, whereas increased carbon dioxide concentrations would increase tropical forests as a carbon sink: the balance between these two factors is unknown. Currently there is a research priority to ascertain whether such carbon changes are taking place in the tropics and to confirm their cause as well as to try to forecast the effects of warmer climates on existing tropical forests.

6.1.3 Some biological dimensions of the climatic change fingerprint

The above dendrochronological responses to current climate change (sections 6.1.1 and 6.1.2) are but one aspect of the complexities of biological response to climate change. We have previously noted others, including with regard to climate-driven species migration (both horizontal and vertical), extinctions and evolution. We have also seen (Chapter 2) how individual species can serve as indicators of past climate. So the question arises as to whether there is any general effect of current warming across natural systems. In other words, is it possible to discern a global warming fingerprint?

In 2003 biologist Camille Parmesan (who had contributed to the Intergovernmental Panel on Climate Change [IPCC] 2001 Working Group II report) and economist Gary Yohe published the results of their attempt to discern such a fingerprint. They noted that the discussion surrounding the 2001 IPCC Working Group II's analysis (IPCC, 2001a) of climate impacts and vulnerability had featured a 'divergence of opinion' from its contributors with disparate academic views from science to the social sciences. This was in no small part because of the differences in the way economists and biologists work. Economists tend to look at aggregate data over a long time. Biologists, conversely, recognise that biological change is multi-factorial and that much is to do with non-climatic human intervention (such as land-use change) and so they tend to examine unperturbed systems. The economists view this as data selection, whereas biologists consider it a quality-control filter. However, Parmesan and Yohe combined biological and economist approaches, adopting IPCC criteria, to look at natural (as opposed to human-managed) systems. They looked at data sets as well as individual cases found in the scientific literature. They then assessed these using three variables: the proportion of observations matching climate change predictions, the numbers of competing explanations for these observations and the confidence of relating each observation to climate change. Finally, to help overcome any literature bias (scientists might prefer to write up only positive correlations), they used only multi-species studies that reported neutral and negative climate correlations as well as positive ones. The brunt of their focus was on phenological effects: those of season changes on the lives of plants and animals.

They found that the range boundaries of 99 species of northern hemisphere temperate bird, butterfly and alpine herb, at the end of the 20th century, had moved on average 6.1 km (±2.4 km with 95% confidence) northwards, or the same number of metres upwards, per decade. Phenologically a total of 172 species of herb, butterfly, shrub, tree and amphibian revealed an earlier spring timing of 2.3 days per decade with a 95% confidence range between 1.7 and 3.2 days' advancement per decade. With regards to phenology (see section 6.1.4), purely on the basis of either advancement or delay of spring events (i.e. no quantification, just direction of change), such as frog breeding, bird nesting, first flowering, tree burst and arrival of migrant birds and butterflies, of 677 species assessed from the literature 27% showed no change, 9% showed delayed spring events (the opposite of what you would expect with global warming) and 62% showed spring advancement. In summary, about 87% of species that showed any change showed the change one might expect with climatic warming.

Interestingly, published in the same issue of *Nature* as this study was another that also looked at the fingerprint of global warming on wild plants and animals (Root et al., 2003). Again there was an association with the IPCC, in that four out of the six researchers were previously members of the IPCC's 2001 Working Group II and the study by Root and colleagues built on some of the literature examined by that group. There is little doubt that their motivation to express a separate analysis arose out of the Working Group II's own internal debate, of which Parmesan and Yohe were aware (see above), that there was a 'divergence of opinion'. It is important to note that Root et al. only included those studies that (1) spanned 10 years or more, (2) showed that at least one species changed over time and (3) found either a phenological change associated with temperature or vice versa. However, in addition they attempted to see whether there were any correlations between climatic cycles (see section 5.1.5), such as the North Atlantic Oscillation (NAO) and the El Niño Southern Oscillation (ENSO), and phenological change. Of more than 2600 species covered in the 143 studies that met the researchers' criteria they found that more than 80% of species of animal (molluscs to mammals) and plant (grasses to trees) exhibited change in the direction that one might expect with global warming. They concluded that the balance of evidence is that there is a discernable fingerprint of recent decadal warming on animal and plant populations. Whereas the authors admit that their study selection is open to possible bias, it is interesting to note that this figure of more than 80% appears to broadly corroborate Parmesan and Yohe's figure of 87%.

Camille Parmesan and Gary Yohe's seminal work was done in 2003 but, as ever with research, science moves on. In 2011 a British-based team headed by I-Ching Chen and Jane Hill reported an even faster terrestrial species shift in response to warming. As with the Parmesan–Yohe study, Chen et al. used a meta-analysis. Their results indicate that distributions of species have recently shifted to higher elevations at a median rate of 11.0 m per decade, and to higher latitudes at a median rate of 16.9 km per decade. These rates are approximately two and three times faster than previously reported (6.1 km latitudinally, or metres vertically, and per decade). The distances moved by species were greatest in studies from the meta-analysis showing the highest levels of warming, with average latitudinal shifts being generally (but not always) sufficient to track, or serve as a proxies for, temperature changes. The authors also note that not all species responded by shifting proportionally to the amount of warming. As noted before (for example, section 4.6.4), some species can migrate faster than others. Furthermore, climate change is multifactoral and not just temperature-related (water availability, for example, being another key dimension for species response).

These studies are illustrative. Of course, discerning the biological fingerprint of climate change all depends on what you are measuring: types of species, location and climate change. For instance, a 2008 study examined 171 forest plant species between 1905 and 1985, and between 1986 and 2005 up to 2600 m above sea level in Western Europe. This work not only showed, as did the above previous studies, that climate warming has resulted in a significant upward shift in species optimum elevations but that the upward shift averaged 29 m per decade. The shift was larger for species restricted to mountains habitats and for grassy species: these are characterized by faster population turnover (Lenoir et al., 2008). So why the larger altitudinal shift compared to the above studies? Well, as noted, determinants of biotic response to climate change are multifactorial. However, there is perhaps more than a little clue in the warming experienced over the study period in the study area (France and Corsica). Climatic change in France has been characterized by increases in average temperature of far greater magnitude than increases in the world mean annual temperature, of about $0.6°C$ over the 20th century, reaching $0.9°C$ and even close to $1°C$ in the alpine region since the early 1980s. The study period was divided into two: 1905–85 and 1986–2005. (Readers may care to note that loosely speaking these two periods saw global – not French regional – temperatures respectively below and above the 1961–90 mean; see Figure 5.4.) In short, all the aforementioned studies represent parts of the same terrestrial species response picture.

Nonetheless, within this overall picture of species shift there is plenty of detail. Included here, and mentioned earlier, are the climatic cycles. To take just one example of how a regular climate oscillation affects species, the North Atlantic Oscillation index has been shown to correlate with UK populations of a number of butterfly species. Specifically, the correlations are between a so-called collated index of the butterfly species and the NAO index.

The collated index of butterflies is calculated by the UK Centre for Ecology and Hydrology (CEH), which has been maintaining a database since 1976. CEH researchers monitor the populations of butterfly species at 130 sites and data from these is collated into an index. Turning to the North Atlantic Oscillation, the NAO index is calculated from the difference in air pressure between Iceland and the Azores, Lisbon or Gibraltar (the choice of the southern grouping of weather-station sites seems to make little difference). The NAO index can be positive or negative. If positive then depression systems tend to take a more northerly route across the Atlantic, so giving the UK a warmer and wetter winter and autumn. A negative index is associated with drier and cooler weather in the UK as moisture goes more towards the Mediterranean basin. A number of UK butterfly populations have been shown to fluctuate with the NAO index (Westgarth-Smith et al., 2005), including the common blue (*Polyommatus icarus*) and small cabbage white (*Pieris rapae*).

It is important to note that these periodic fluctuations in species populations that correlate with climatic oscillations are distinct from species population change due to long-term climate change. This means that those studying the biological impacts of climate change have to disentangle these two related phenomena. However, the biological impact of climatic oscillations is relevant to the impact of longer-term climatic change. First, regionally, part of a climatic cycle may go with the direction of longer-term climatic change, so that part of the cycle can provide insights into the regional manifestations of longer-term climatic change. Second, understanding the biological impact of climatic cycles is intrinsically important to regional biomes. Third, longer-term climatic change may result in changes in ocean and atmosphere circulation (see section 6.6.6) and these circulation changes can themselves change the characteristics (or even existence) of regional climatic oscillations. If we are unaware of the current effects of regional oscillations then it will be difficult to discern what biological changes are due to contemporary regional climate systems and what changes are due to longer-term climatic change.

Climate change also has some surprisingly unexpected effects on biological systems. Just as the IPCC urge us to be aware of 'climate surprises' when considering their gradual warming scenario forecasts, so biologists need to be alert to surprise biological responses to climate change. Climate change can alter things in ways that one might not normally consider important and this includes changing to the wrong type of snow falling. The millions of Britons who regularly travelled into and out of London at the start of the 21st century will be aware of the seemingly joke excuse that rail operators used one winter when they said that the 'wrong type of snow' had prevented their track de-icing equipment from working. This made the national headlines. For some species the right type of snow is of crucial importance. Norwegian lemmings (*Lemmus lemmus*) need the right type of snow, of the kind that ground warmth melts a small layer immediately above it so leaving a gap between the ground and the underside of the snow layer. This subnivean space not only provides warmth, it allows some protection from predators. Climate change can mean that such spaces do not exist for as much of each year as they used to. Worse, with warmer conditions the snow immediately above the ground can melt more than it otherwise would have done during the day and then refreeze at night, producing a thicker layer of ice that prevents the lemmings feeding on the underlying moss. In 2008 a small team of Norwegian and French researchers led by Kyrre Kausrud and Atle Mysterud published the results of a long-term study and an ecological analysis. They collected disparate sets of long-term records to reveal how changes in snow type not only affected the lemming population but had wider ecological implications too: lemmings are noted for their population cycles and these have often (not without good cause) been attributed to predator–prey interactions. They got university students taking winter ecology courses to collect data on snow condition. The researchers also gathered hunter-reported bird catches, drew on meteorological records, and used long-term rodent trap records and bird census data, and other records from the period 1970–94 and subsequent years. They began to see a pattern as to how climate affected the lemmings, and in turn other species, and used these associations to make a predictive model that forecast an absence of rodent peaks after 1994. By 2007 they were able to look back and compare the actual lemming population, as revealed by trap catch rate, with that predicted by their model. It was clear that the model not only reflected lemming numbers before 1994 and importantly their peak years, but also the absence of lemming population booms after that year. This provided them with compelling evidence that changing snow conditions are indeed a major factor changing lemming population dynamics. What appears to be happening is this. The North Atlantic

Oscillation index, humidity, temperature and snowfall interact to affect the hardness of snow. This affects the lemming population which in turn impacts on predators such as stoat (*Mustela erminea*) and weasel (*Mustela nivalis*), which affects other prey such as birds (Kausrud et al., 2008; and see also the review by Coulson and Malo, 2008).

Finally, in a warmer world with elevated levels of carbon dioxide we would expect increased levels of photosynthesis. Although regionally there may be differences, with some areas seeing decreases in photosynthesis (for example, due to water restriction), and regionally there would be differences in fixing and retaining carbon (both from plants and soils), in purely photosynthetic terms the reasonable expectation would be to see an increase in global primary production (see the definition of production in the Glossary, Appendix 1). Satellite observation of global vegetation between 1982 and 1999 suggests that globally terrestrial (excluding oceans and freshwater systems) net primary production has increased by 6% (3.4 GtC) over 18 years. The largest increase was in tropical ecosystems and the study suggested that the Amazon rainforest accounted for 42% of the global increase in net primary production (Nemani et al., 2003). This ties in with the evidence for the increase in global primary production noted in section 5.3.1. Here we should remember that whereas global climate-induced changes in primary production will not meaningfully affect atmospheric concentrations of carbon dioxide as far as policy-makers (who seek large changes over a short time) are concerned, changes in primary production are a regionally important mechanism for carbon transportation, via the atmosphere, between ecosystems and, of course, primary production is of ecological interest as well as of relevance to biosphere carbon cycling which, as already noted, is increasing in our warming world.

In accountancy terms, gross primary productivity might be considered as analogous to the all-important, proverbial bottom line: what was the net productivity of the ecosystem (how much living matter grew in the year minus that which died)? Yet for ecologists (and indeed entrepreneurs, to continue with the accountancy analogy) of equal interest is how we arrive at that bottom line: what types of primary producer species will thrive as the world warms? You will recall from section 3.3.11 that plants use a number of photosynthetic pathways and the two most common categories are C_3 and C_4, and that the C_4 pathway includes a carbon dioxide-concentrating mechanism so that the plant can photosynthesise at lower atmospheric concentrations of the gas. Indeed, as discussed, atmospheric carbon dioxide levels were falling in the Miocene and this gave C_4 plants an increased advantage over C_3 plants. Now, it might reasonably be thought that leaf photosynthesis and water-use efficiency in C_4 grasses benefit less from higher carbon dioxide in a greenhouse world than in C_3 grasses, but is this true in practice? In the wild, where C_3 and C_4 plants grow side by side, there will also be competition for water, especially in a warmer world with higher evaporation from land. In 2011 a team of researchers based in the USA plus one based in Australia, and led by Jack Morgan, Daniel LeCain and Elise Pendall, examined just this problem in what was called the Prairie Heating and Carbon Dioxide Enrichment (PHACE) experiment. The PHACE experiment evaluated the responses of native mixed-grass prairie to 1 year of carbon dioxide enrichment (from the then present ambient 385 parts per million by volume [ppmv]) to elevated levels (600 ppmv), as well as C_3 and C_4 plants by themselves. This was followed by another

3 years (2007–9) of combined carbon dioxide enrichment and warming (comparing present ambient controls with elevated temperatures). So, did the C_3 plants thrive more than the C_4? It turns out that they did not.

What Morgan et al. (2011) found was that, compared with C_3 grasses, C_4 grasses growing under semi-arid climatic conditions will prosper in a world with higher carbon dioxide concentrations *and* (the 'and' is crucial) warmer temperatures. This is because of the difference in the way C_3 and C_4 plants can manage water. The leaf pores, stomata, of C_4 plants open less widely under high carbon dioxide semi-arid warm conditions than those of C_3 plants, so C_4 plants tend to transpire less water vapour. (See also a review of this work by Dennis Baldocchi, 2011.) The conclusions have implications for ecosystem function and ecosystem service (such as grazing provision).

So far, in our all-too-brief survey of current global warming's biological dimensions, we have looked at terrestrial dimensions. But what is happening to global productivity in the oceans? Aside from being of ecological interest, because oceans contribute roughly half of global primary productivity, this is of practical relevance due to our reliance on the ecosystem services the oceans provide. Not least of these are mundane economics including fishery potentials and the fundamental importance of planetary oxygen regulation.

Although there is a volume of research literature on phytoplankton, the key ocean primary producers, only recently have we been able to gain a truly global appraisal of the situation through space remote sensing. (Sampling using point source, a slow-moving ship [or even a small fleet], cannot provide a global snapshot.) In 2006 US researchers Michael Behrenfeld, Robert O'Malley, David Siegel et al. used NASA's Sea Viewing Wide Field-of-view Sensor, known as SeaWiFS, the first ocean colour sensor with a wide view, launched in 1997. Its spectral analysis capabilities allow it to assess the amount of chlorophyll in the sea surface, hence it can be used to estimate stocks and productivity. Primary productivity is small compared to total phytoplankton stock so detecting annual changes requires accurate long-term measurement. Behrenfeld and colleagues did just this over the period from 1997 to the beginning of 2006. From the annual temperature plot of global temperature in Figure 5.4 you may be able to see the (ENSO-enhanced) rise in global temperature between 1996–8 of a little over 0.4°C followed by more stability to 2006 near the high end of this temperature hike. Comparing this with both total global marine primary productivity and its distribution, the Behrenfeld team were able to gather some preliminary results (a longer-term study would be welcome) and from these they drew some initial conclusions. They showed that global marine net primary productivity (roughly 55 GtC year^{-1}) grew with temperature by some 1.93 GtC year^{-1}. They then continued to track it, as it declined a little in the years to 2005. Between 1999 and 2004 they not only found a smaller decline in primary productivity of 0.19 GtC year^{-1}, but also that there was a difference between high latitudes (very roughly above 40° where more of ocean primary productivity takes place) and the tropical and subtropical low latitudes. Specifically, the low latitudes are where the bands of temperature-stratified ocean occur, where the warm surface lies over the cool, abyssal depths. At the high latitudes the ocean is more unstratified, because there is more oceanic vertical mixing due to up- and down-welling. Seventy-four per cent of the Earth's stratified oceans (with low vertical mixing) are in the warmer low latitudes, where the average sea surface temperature is greater than 15°C. Upward vertical mixing brings nutrients from the abyssal depths to the surface, and cooler waters facilitate gaseous exchange.

Phytoplankton concentrations vary by a factor of around 100 in different parts of the Earth's ocean surface. In the stratified, low-latitude, areas of the ocean Behrenfeld et al. found that there was an inverse relationship between sea surface temperature and primary productivity. It is in the high latitudes where net primary productivity seems to best correlate with sea surface temperature and it is this that (from the limited time window this unique study shows) seems to drives the overall loose proportionality of global temperature with ocean net primary productivity. (See also a summary review with some discussion of the Behrenfeld et al. paper provided by Scott Doney [2006] in the same issue of *Nature*.)

The prognosis for the future in a warming world, should these trends continue (and remember that it is early days for this type of global perspective), is that marine biological productivity will decline in the tropics and mid-latitudes, and productivity at high latitudes will increase.

Primary productivity is an ecological bottom line, but the physiology (health-related aspects) of primary producers is also of interest. There have been some climate change physiological studies but no global, let alone global-time-series, analysis. However, one study of freshwater plankton that illustrates an avenue of likely considerable importance was its response to a simulated 4°C warmer world. This saw reduced mean and maximum size of phytoplankton by approximately one order of magnitude. The observed shifts in phytoplankton size structure were reflected in changes in phytoplankton community composition, although zooplankton species composition was unaffected by warming. Furthermore, warming reduced the total community biomass and total phytoplankton biomass, although zooplankton biomass was unaffected. This resulted in an increase in the zooplankton to phytoplankton biomass ratio in the warmed simulations, which could be explained by faster phytoplankton turnover. Overall, warming shifted the distribution of phytoplankton size towards smaller individuals with rapid turnover and low standing biomass, resulting in a re-organization of the biomass structure of the food webs. These results suggest that future warming may have profound effects on the structure and functioning of aquatic communities and ecosystems (Yvon-Durocher et al., 2010).

We have seen in this section that the fingerprint of global climate change can be ecologically discerned. The fingerprint can be discerned in many different ways. The effects of climate change are many. They are both biotic (for example, species' ranges shift) and abiotic (glaciers retreat and so forth). Such phenomena have been studied many times and also (less often) over long time periods. The global climate has also been monitored instrumentally, with greater accuracy in recent years. So how many of these long-term studies of biotic and abiotic systems show shifts in the direction we would expect with recent anthropogenic climate change? Step up Cynthia Rosenzweig, David Karoly, Marta Vicarelli and colleagues, who in 2008 attempted to answer just this question.

Rosenzweig et al. first needed to see what climate change is taking place and so divided the world into North America, Europe, South America, Africa, Asia and Australasia. They then looked at the climate change taking place between 1970 and

2004 (based on real observations coupled with a computer model dividing these world regions into smaller – 5° by 5° latitude/longitude – cells of different climate change response, and greatly dominated by warming cells). Meanwhile they searched the literature for long-term biotic and abiotic studies and selected only those that lasted more than 20 years. Altogether they compiled around 29 500 time series (28 754 biotic and 688 abiotic) of response. More than 28 000 were European phenological series but the remainder were from across the globe: this study was remarkable. (There will be more on phenology in the next section.) Excluding the European phenology studies, the majority of the remaining biotic time series related to terrestrial biology and only around 100 or so to marine and freshwater biology. A handful of agricultural and forestry time series were also included.

What the Rosenzweig et al. study showed was that in all their global regions abiotic systems exhibited a response consistent with warming to at least 94% significance (in other words the response was so great that if analysing the analogous series from 100 parallel Earths only six could be attributable to chance variation and 94 to warming). With biotic systems this was at least 88% significance (North America and Asia). In Europe where there had been a large volume of phenological studies, and a total of 28 117 biotic time series examined, the biotic response was consistent with warming to 90% significance. The global region abiotic response was at least 96% significant. The conclusion was that anthropogenic climate change is having a significant impact on physical and biological systems. (A summary overview of this work by Francis Zwiers and Gabriele Hegerl, 2008, appeared in the same issue of *Nature* as the study by Rosenzweig et al.)

6.1.4 Phenology

Aside from changes in species ranges (spatial distributions), phenological data are, as mentioned in the previous section, the other main indicator of a climatic fingerprint. Phenology is the study of a plant or animal's progression through its life cycle in relation to the seasons. Phenological records have been built up over the past few centuries. In Britain the oldest phenological record known is a diary entry for the first cuckoo of spring in Worcestershire in 1703. But the association between the arrival of spring and certain species is much older. For example, the Romans in Britain knew all about the arrival of spring and the returning of the swallow from its overwintering in Africa. However, it was many hundreds of years before phenological data began to be assembled systematically. Arguably the first, which began with observations from 1736, was the English Norfolk country gentleman Robert Marsham's collection of '27 indications of spring'. Others, such as Thomas Barker of Rutland and Gilbert White of Selborne, Hampshire, were also keen observers of nature and kept phenological diaries. Gilbert White even compared his own records with those of a friend in Sussex. Phenological observation has been a popular activity for professional biologists and amateur natural historians ever since, so much so that between 1875 and 1947 the Royal Meteorological Society managed a voluntary national network of recorders (Sparks and Collinson, 2003).

As for the changes in phenology that these records reveal in Britain, a very loose rule of thumb is that it appears that flowering and leafing occur 6–8 days earlier

Table 6.1 Some phenological evidence of advances in spring events (from Walther et al., 2002)

Taxon	Location	Changes	Period observed
Many plant species	Europe	Flowering and leaf unfolding 1.4–3.1 days per decade earlier	Past 38–40 years
	North America	Flowering and leaf unfolding 1.2–2.0 days per decade earlier	Past 35–63 years
18 Butterfly species	UK	Earlier appearance by 2.8–3.2 days per decade	Past 23 years
Amphibians	UK	Earlier breeding	Past 25 years
Numerous bird species	Europe, North America	Earlier spring migration by 1.3–4.4 days per decade and breeding by 1.9–4.8 days per decade	Past 30–60 years

for every degree Celsius rise in temperature. However, species do react differently. For example, oak (*Quercus* spp.) responds twice as fast as ash (*Fraxinus* spp.) to an increase in temperature.

European and North American phenological sets have predominantly reflected changes that are probably associated with climate change, especially with regards to earlier spring phenology (Table 6.1). But there is also some indication of later autumnal events, which again would be expected in a warmer world; however, these are less marked. Indeed, birds that show an earlier migration arrival do not show as late an autumnal departure. Autumnal leaf colouring is even more ambiguous within regions. On a long timescale, European autumnal leaf colouring has shown to be delayed at a rate of about 0.3–0.6 days/decade, whereas the length of the thermal growing season (TGS) has increased by 3.6 days/decade over the past 50 years. This correlates well with independent satellite data.

Here it is worth emphasising that although a warming world may be a dominating driver of current phenological change, as mentioned in the previous subsection, a number of the major climate-related oscillations have a phenological effect. In 2002 Gian-Reto Walther and a small group of mainly European scientists published a review of ecological responses to climate. They reported that an analysis of 50 years of data on 13 plant species in 137 northern hemisphere locations revealed responses to the North Atlantic Oscillation in 71% of the data. Early-blooming herbaceous species showed greater response to winter warming than late-blooming and woody plants. Knowledge of regional climatic cycles is fundamentally important when studying biological response to climate change.

Then (as mentioned in the previous section with regards to plankton) there are the physiological affects of changing seasons on species. In 2010 Arpat Ozgul, Dylan Childs, Madan Oli and colleagues, based in Britain and the USA, demonstrated that the earlier emergence from hibernation and weaning of young of a North American rodent, the yellow-bellied marmot (*Marmota flaviventris*), led to a longer marmot growing season and a larger body mass before hibernation between the years 1976 and 2008. Fatter marmots having to hibernate for shorter periods means reduced winter mortality and so the marmot population has increased. The population of marmots at a site in the Upper East River Valley, in Colorado, USA, more than doubled its 2000–3 mean by 2006–8.

One of the most detailed assessments so far of long-term changes in the seasonal timing of biological events across marine, freshwater and terrestrial environments in Britain was coordinated by Stephen Thackeray and Sarah Wanless, with second author Tim Sparks of the Centre for Ecology and Hydrology. They looked some 25 000 long-term phenology data series for 726 species of plants and animals covering the period 1976–2005. The range of species covered included plankton, plants, insects, amphibians, fish, birds and mammals. They found that more than 80% showed earlier seasonal events. On average, the seasonal timing of reproduction and population growth became earlier by more than 11 days over the whole period, and more recently this change has become faster. There were large differences between species in the rate at which seasonal events have shifted. Changes were most rapid for many organisms at the bottom of food chains, which means that plants and the animals that feed on them were not effectively tracking them. Predators showed slower overall changes in the seasonal timing of their life cycle events, even though the seasonal timing of reproduction is often matched to the time of year when food supply increases, so that offspring receive enough food to survive. Species relationships can depend on time as much as space (see the next section on species' shift in ranges).

Ecologists studying biological responses to environmental change are all too aware that responses are not always straightforward. Confounding a simple analysis of species response to climate-driven changes in seasons is that some related species may respond in markedly different ways. One instance comes from a study of the phenology of reproductive migrations in 10 amphibian species at a wetland site in South Carolina, USA, over more than 30 years (1978–2008) (Todd et al., 2010). It showed that two autumn-breeding amphibians were breeding increasingly *later* in recent years, coincident with an estimated 1.2°C increase in local overnight air temperatures during the September-to-February prebreeding and breeding periods. Nothing surprising there, you might think. But they also found that two winter-breeding species in the same community were breeding increasingly *earlier*.

The other interesting aspect of the study by Todd et al. (2010) was that four of the 10 species studied shifted their reproductive timing an estimated 15.3–76.4 days in the 30-year study period. This resulted in rates of phenological change ranging from 5.9 to 37.2 days per decade, so providing examples of some of the greatest rates of phenological change in ecological events reported up to that time.

However, it is the two amphibian species responding to seasonal change differently that is a central concern when it comes to long-term future ecological consequences of climate change. If a species occupies a specific ecological niche at a specific time in its life cycle then it needs two things. These are to have that niche track climate change (including that manifested seasonally) in a similar way (so that it has food and the other environmental resources upon which it relies), and also not to have to face increased competition from responses by other species to the same climate change. We will shortly look at an example of the former. However, with regards to the latter Todd et al. note that owing to the opposing directions of the shifts in reproductive timing, their results suggest an alteration in the degree of temporal-niche overlap experienced by amphibian larvae in this community. This puts additional pressure on both species.

Knowledge of a species' primary requirements is also fundamentally important given that these requirements may in turn be driven by a climatic dimension or dimensions. These can be complex. Consider the migratory black-tailed godwit (*Limosa limosa islandica*), a shore bird that winters between Britain and Iberia while breeding in the summer almost exclusively in Iceland. Breeding pairs exhibit a high degree of partner fidelity. Whereas most migratory pairs winter together, and so migrate together, the black-tailed godwit is one of a number of species where males and females winter in different locations many hundreds of miles apart. Yet breeding pairs arrive in Iceland typically within about 3 days of each other. The question then arises as to how this degree of synchrony is maintained when the environmental conditions at the different sexes' wintering sites are dissimilar (Gunnarsson et al., 2004). The answer will be key in beginning to understand how such species will respond to future climatic change. Indeed, respond to climatic change they will, for clearly as little as 15 000 years ago, towards the end of the last glacial, their migratory patterns would have been most different and so a successful response to glacial–interglacial climate change must have taken place on a number of occasions.

Today climate change seems to be affecting another migratory bird species, the pied flycatcher (*Ficedula hypoleuca*). Its migration is timed to the availability of food for its nestlings. However, phenological changes have meant that in some parts of The Netherlands the caterpillars that are its food peak early in the season. Here the flycatcher populations are in decline (Both et al., 2006). Whether such species will adapt with time in the modern world remains to be seen. As with the earlier amphibian example, such phenological disruption of ecological relationships is likely to increase, and for some species probably become critical, as climate change continues.

So what is happening to seasons on the global scale? First of all we need to remember our school-day geography, that seasons – that is, the cycle of spring, summer, autumn and winter – only take place outside of the tropics (tropics have their own seasonal patterns, principally wet and dry). Outside the tropics there is an annual cycle of change in mean daily temperature driven by a combination of the Sun's elevation above the horizon and day length. This facilitates the TGS during which primary producers, mainly algae and plants, grow, drawing down carbon dioxide. In the autumn, plants begin to die and decompose, and carbon dioxide is released through the respiration of the decomposer (detritovore) community (including earthworms, insects, fungi, bacteria and so forth). And so, outside of the tropics atmospheric carbon dioxide has an annual cycle. This annual cycle is what causes the ripple seen in the graph of atmospheric carbon dioxide over recent decades (Figure 1.4). Looking at just one year of carbon dioxide change in the northern hemisphere, there is a sinusoidal-like pattern. Figure 6.1 has two single years half a century apart superimposed on each other: 1959 and 2010. This shows how the annual pattern of northern hemisphere atmospheric carbon dioxide concentration has changed. The first thing to note is that the amplitude of the wave has increased. This means that the year is seeing more drawing down of carbon dioxide. This drawdown is unlikely to be the result of abiotic gaseous absorption by the oceans because gas is less soluble

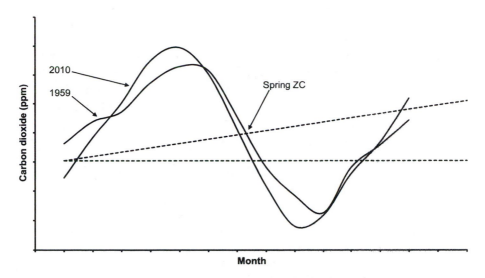

Fig. 6.1 The 1959 cycle of northern hemisphere carbon dioxide superimposed on the 2010 cycle, drawn by the author using data with permission from the Earth Science Research Laboratory of the National Oceanic and Atmospheric Administration. Spring ZC, spring zero crossing point. The horizontal dashed line represents theoretical zero year-on-year growth in average annual carbon dioxide. The upward-sloping dashed line represents the long-term increase in annual carbon dioxide and is calculated as being the average annual increase in CO_2 in 1959 and also 2010.

in the warmer 2010 sea surface. So, the increased drawdown more accurately reflects increased annual global gross primary production. The second thing to note is that the first half of the 2010 curve leads that for 1959. The upward slope (the sloping dashed line on the figure) closely represents the long-term average growth in carbon dioxide (it being the mean increase for the two years 1959 and 2010). Note that in Figure 6.1 the spring zero crossing point (ZC) for 2010 is ahead of that for 1959. It reflects spring arriving earlier and an extension of the TGS. The position for the autumn is less clear. Mathematically, the June/July so-called spring ZC point is easier to establish: from a biological perspective, it is the point at which new growth in April and May takes atmospheric carbon dioxide below the average long-term trend in CO_2 growth. The annual peaks in the northern hemisphere carbon dioxide cycle also are important. The difference between these superimposed peaks (as well as the troughs) is greater than the difference in ZC points.

In 2008, a 20-year data analysis of northern hemisphere carbon dioxide concentrations was carried out by a European and North American team led by Shilong Piao, Philippe Ciais, Pierre Friedlingstein and colleagues, combining satellite chlorophyll observations and a terrestrial biosphere model. Perplexingly, it concluded that the autumn rise in carbon dioxide was happening earlier: intuition suggests with a longer TGS that autumn would come later. This autumnal rise was not so much due to autumn happening earlier per se, but in the warmer autumn there was greater respiration by detritovores (decomposers of biotic material), and this dominated the extended autumnal photosynthesis (see also Miller, 2008, for a summary review of this work).

We have discussed northern hemisphere biological response to changes in the seasons, but how is the northern hemisphere climate changing? Stine et al. (2009) used meteorological records to determine that in terms of temperature the northern hemisphere spring advanced by 1.7 days in the period 1954 and 2007. Of course, this is a hemispheric average and it varies considerably from place to place due to latitude and circulation patterns. It is minimal near the edge of the tropics and considerably greater at higher latitudes. This explains why the phenological response of some species is greater.

6.1.5 Biological communities and species shift

Of course, some species do not change part of their life cycle but will shift their geographical position, or range, in response to climate change. In the previous chapter key characteristics and examples of species' response to glacial–interglacial climatic change were summarised. However, because climate is but one factor (indeed, climate itself is multi-factorial) of many that determine a species' spatial distribution, species rarely move uniformly with each other in response to climate change. For example, a calcareous soil species prefers soils with a high pH and this preference might dominate that of minor climate change. Figure 6.2 illustrates uniform and non-uniform vertical and horizontal migration. (See also Chapter 4 and the discussion on biological and environmental impacts of the last glacial, as well as Figure 4.10.)

Added to the above, and again as noted in the previous chapter, different species migrate at different rates. Consequently, not only is a period of climate change ecologically dynamic but it takes time for ecological communities to stabilise after a period of climate change. Furthermore, again as touched upon in Chapter 4, species at the leading edge of shifts/migrations tend to migrate faster than those already established. As ecologists Walther and colleagues (2002) note, changes in distributions are often asymmetrical, with species invading faster from lower elevations or latitudes than resident species receding upslope or poleward. The result is often an increase in species richness of communities at the leading edge of migrations. However, this biodiversity 'benefit' is transitory. Furthermore, whereas such biodiversity may be seen in natural systems (and so found in the palaeontological record), it is less likely in many of today's systems, which are either managed or bounded by land that is managed by humans. Such managed lands often make for an effective barrier to species migration. This in turn poses problems for the future of species, let alone ecological communities; for even, as noted, if old ecological communities are disrupted, species migration impedes new communities being formed.

An example of all these concerns was described by a largely French team of biologists led by Romain Bertrand and Jonathan Lenoir in 2011, who compared how highland and lowland plant species responded to climate change in France over a 44-year period (1965–2008). They employed a twist on the theory behind species used as climate proxies. Using living species they established the optimal temperature for the species to live but the researchers found that, as a result of climate change, this differed from the temperature of their actual environment. Consequently they could determine how well the French species were migrating to track climate change. They found that forest plant communities had responded by 0.54°C yet the climate had warmed by 1.07°C in highland areas (more than 500 m). In lowland regions the

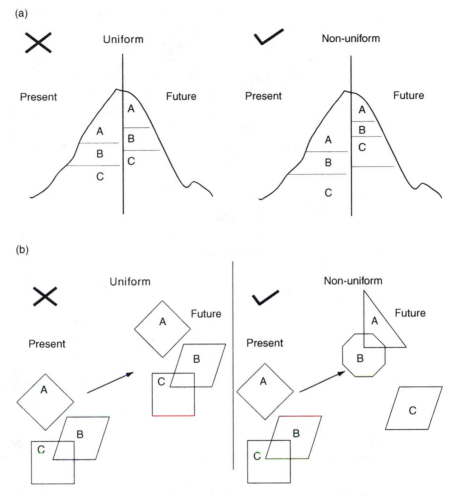

Fig. 6.2 (a) Vertical migration of species and communities due to climatic warming is frequently not uniform. Individual species respond to different environmental factors differently. Consequently there is little reason for uniform migration. An example is given relating to Figure 4.10. (b) Horizontal migration of species is rarely uniform (see Chapter 4 with reference to glacial–interglacial change). As with vertical migration, this is due to different species responding with different sensitivity to different environmental dimensions, of which climate (itself multi-dimensional) is among many other factors, which together will determine a species' spatial distribution. This makes prediction of ecological response to climate change difficult.

disjunction was worse, with plant species responding to only 0.02°C of a 1.11°C warming: there was a larger temperature lag between plant communities in lowland than highland communities.

The explanation for this is multi-faceted. First, highland areas are more mountainous and so species only need to migrate a short distance horizontally to change their altitude considerably, and hence their local climatic conditions. This is not so with lowland species, which have a reduced number of short-distance escapes open to them. Second, highland species, because of short-distance climate vagaries, are naturally adapted to be more able to survive a degree of climate variability, and hence change. Finally, lowland landscapes in developed nations are highly fragmented and this further impedes species range migration.

All of this means that it is virtually certain that in a warmer world we will see new assemblages of species. This provides a challenge for those whose task it is to manage ecosystems, be it for wildlife conservation, amenity or commercial purposes. Ecosystem managers are going to have to deal with new ecological communities involving a new mix, or combination, of species: some call it 'recombinant ecology'.

Given that species shift in response to climate at different rates, could it be that some might even move in the wrong direction with global warming? This is counterintuitive, but remember that not all places on Earth track global changes simultaneously or even in the same way, the vagaries of short-term weather and climate cycles and, finally, that temperature is just one dimension to the climate. In 2011, Shawn Crimmins and colleagues published a paper that subsequently spawned considerable comment[2]. They noted that whereas uphill shifts of species' distributions in response to historical warming were well documented, and that this leads to the expectation of continued uphill shifts under future warming, downhill shifts are considered anomalous and so unrelated to climate change. However, by comparing the altitudinal distributions of 64 plant species between the 1930s and 2010 in California they showed that climate changes have resulted in a significant downward shift in species' optimum elevations. They emphasised that this downhill shift is counter to what would be expected given 20th-century warming but is readily explained by species' niche tracking of regional changes in climatic water balance rather than temperature. So, they said, similar downhill shifts can be expected to occur where future climate change scenarios project increases in water availability that outpace evaporative demand.

Another example comes from a study of treelines. In 2009 Melanie Harsch, from the Bio-Protection Centre in New Zealand, and colleagues conducted a meta-analysis using a global data set of 166 treeline sites with temperature data taken from the closest climate station to each site, to determine treeline change throughout the 20th century. They found that 87 of the 166 site (52%) had changed in the direction as expected, 77 (47%) remained stable, but two (less than 1%) had moved in the other direction. The conclusion they drew was that not all species universally respond in the same way to climate change. They also noted that at the two sites, where treelines had moved the other way, there were signs of disturbance. This study provides a sound lesson that other human-generated factors can confound expectations.

Now, assuming that the methodologies for this and the Crimmins study were sound (some debated the Crimmins results), it is not entirely without precedent to have system change go the other way. In section 6.1.3 we noted Parmesan and Yohe's 2003 result of 87% of species responding to climate change in the direction one would expect with warming, so could mechanisms such as the one suggested by Shawn Crimmins' team explain why the minority of 13% did not? Of course, at the beginning of a period of climate warming, even if a minority of species do not respond in the direction expected, as the climate change continues, even in the presence of

---

[2] See the series of comments from various contributors and the study authors' response in *Science*, 2011, 334, 177.

ameliorating factors, it will be the dominating factor, and species will respond in the direction it determines.

The problem of synchrony of biological community species migration as a response to climatic change is loosely analogous to the synchrony problem for migratory bird pairs overwintering in different locations (see the end of the previous subsection). The big difference, though, is that whereas successful breeding pairs are fundamentally necessary to a species' continued existence – and even a synchronous response across species is necessary for the continued existence of a biological community – the continued existence of a biological community is not always necessary for the continuity of a species. Species may continue, but in new biological communities, as represented in Figure 6.2b. Here the community of species B and C is disrupted by climate change and a new community of species A and B is created.

Let's consider a theoretical case of regional warming of an arbitrary (but fixed) amount over a certain time period. As noted, if species shift at different rates other species in the ecological community on which they rely may not track the climate change in the same way. In 2008 Vincent Devictor, Romain Julliard, Denis Couvet and Frédéric Jiguet expressed concern that the French Breeding Bird Survey results were suggesting that while French March–August climate change over two decades corresponded to a 273 km northward shift in temperature, bird communities had only shifted 91 km northwards. They suggested that bird composition was most likely changing at its maximal possible rate, which is insufficient to catch up with the accumulated delay. This discrepancy, the researchers feel, may have profound consequences on the ability of species to cope with climate change in the long run. However, a contrary view – given that, of all classes of species, birds are among the most mobile – is that the French bird populations surveyed are simply tracking the shifting of their food species in response to climate change and not the climate change itself. Nonetheless, the concern is potentially germane: species track climate at different rates and so established relationships can get disrupted. The important thing, *if* traditional ecological communities are valued, is that the ecological communities as a whole remain intact. Here, providing their coherence is maintained, it may not matter that they do not track climate change exactly, or if their tracking lags behind that of climate change. Alternatively, it may be that it is species survival that is of greater value, in which case it matters less that ecological communities may change with warming.

Another factor to take into account when considering plant species' shift in response to climate is climate buffering: some species at the warmer end of their spatial range lag in their tracking of a warming climate even if at the cooler end of their spatial range there is better tracking. At the warmer end of their range (the lower altitudes or southern end of northern hemisphere ranges), while some aspects of a species' life cycle may be impaired due to warming stress, the individual plants still continue to grow. So it can be that, for example, seed production or flowering may be reduced, but the plant still survives. This potential probably arose due to the evolutionary pressure of those plants being able to outride warmer years arising from the (non-climate change) vagaries of annual weather change (these being driven by climatic cycles). Then, when conditions returned to those more optimal for the species at that location, they were then in place and able to reproduce. What this means in terms of climate change is that there is climate buffering that enables lower-altitude and lower-latitude plants to survive, even though those at the higher-altitude and higher-latitude end of the species' range may migrate. An example of this was provided in 2010 by US bioscientists Daniel Doak and William Morris, of the University of Wyoming and Duke University, respectively. They studied the climate response over 6 years of two tundra plant species, the cushion plant moss campion (*Silene acaulis*) and the geophyte alpine bistort (*Polygonum viviparum*; a geophyte is a herb with perennating buds below the surface). Doak and Morris suggested that an analogous situation may occur with some animal species and so explain why range shifts with climate change are faster at higher latitudes and altitudes than at lower latitudes and altitudes for a given species range. Nonetheless, irrespective of a climate-buffering response, current global warming is a long-term phenomenon (in the biological as opposed to the geological sense) and buffering will only enable species to cope with just so much change. Ultimately, the lower-altitude and lower-latitude ends of species' ranges will shift and the disruption of existing ecological communities could result.

The ecological consequences of variation in species responses (or rate of response) to climate change can be particularly significant. Changes in response to climate change at the lower trophic levels of a community (towards the bottom of the food chain) are likely to have profound effects at higher trophic levels (towards the top). This can happen in time (phenologically as part of the annual cycle) as well as in space (as part of a species' range shift).

This is aptly illustrated, for example, with the results of a long-term (1958–2002) monitoring of marine pelagic species (that is, found on middle levels or surface of the sea) in the North Sea (Edwards and Richardson, 2004). The study looked at the following five functional groups (with their trophic level in parentheses): diatoms and dinoflagellates (the primary producers in this case), copepods (the secondary producers) and non-copepod holozooplankyton (secondary and tertiary producers) and mere plankton, including fish larvae (some of the North Sea secondary and tertiary producers). Because the tertiary producers feed on the secondary producers, which in turn feed on the primary producers, which photosynthesise, species in each trophic level are dependent on the one below. As is fairly common in temperate species, abundance varies throughout the year. Consequently, if the abundance of one species changes at a particular time then the others would have to similarly change if the ecosystem's structure is to remain unaltered, maintaining proportionality between populations. Edwards and Richardson found that species responded differently over the four-decade study period, with some species advancing their abundance peaks and others barely changing. This mismatch in time and between trophic levels as well as functional groups could disrupt energy/nutrient flow to the highest trophic levels. This may explain why (in addition to fishing pressures) North Sea cod (*Gadus morhua*) stocks have declined so markedly in recent years.

Of course, not all marine change can be attributed to climate change. Other things, including anthropogenic factors such as fishing, affect marine ecosystems, which is why the aforementioned North Sea study did not look at the adult commercial species. Also note that the fishing of a few species can affect others in the ecosystem, including those on which they feed and with which they otherwise interact. Again, as with much climate change biology, the picture is complex and so there is often a

Table 6.2 Examples of recent latitudinal and altitudinal shifts in species' ranges (from Walther et al., 2002)

Species/ecotone	Location	Observed changes
Treeline/mountain zonation	Europe, New Zealand, Alaska	Rising in altitude
Arctic shrubs	Alaska	Expansion of range
Alpine plants	European Alps	Altitude rise of 1–4 m per decade
Antarctic plants and invertebrates	Antarctica	Distribution changes
Zooplankton, intertidal invertebrates and fishes	Californian coast, North Atlantic	Increase in abundance of warm-water species
39 Butterfly species	North America, Europe	Northward shifts up to 200 km over 27 years
12 Bird species	Britain	18.9 km average range movement northwards over a 20-year period
Red fox (*Vulpes vulpes*), arctic fox (*Alopex lagopus*)	Canada	Northward expansion of red fox range and retreat of arctic fox ranges

reluctance to draw too firm a conclusion about what is driving change. One instance of debate has been the marine populations in the eastern Pacific associated with the Californian current. Recent US research on planktonic Foraminifera, whose remains accumulate over the years in sediments (the research covered 250 years' worth of sediment), has shown that in the late 20th century there were decreasing abundances of temperate subpolar species such as *Neogloboquadrina pachyderma*. At the same time there was an increase in tropical and subtropical species of forams such as *Globigerina bulloides* and *Neogloboquadrina dutertrei*. Such change seems to reflect 20th-century warming of the current, so lending credence to it being due to climate change and not some other factor (Field et al., 2006). Yet, rather than cite this as hard fact, it is better to view it as one of a number of pieces of corroborative evidence of climate-driven ecosystem change.

As noted previously, altitudinal and latitudinal migration is a common biological response to climate change. There are many recorded examples of such shifts in species' ranges, and a number are summarised in Table 6.2.

The summary review of ecological response by Walther and colleagues (2002) exemplifies climate change impact on ecosystems by noting the change at the extreme ends of an equatorial–polar bioenvironmental gradient. They note that on the one hand tropical oceans have increased in temperature by 1–2°C over the past 100 years, but this average warmth has had superimposed on it short-term warming episodes due to oscillations such as the El Niño Southern Oscillation. Here the trend has been for these episodes to increase in intensity and duration over the past century.

Of the biological impacts due to warming found in tropical seas, the bleaching of corals has been among the most noticeable. Already near their upper thermal limits, mass bleaching events have taken place whenever sea temperatures have exceeded the long-term summer average by more than a degree. Six periods of mass bleaching took place between 1979 and 2002, with their incidence increasing progressively in both frequency and intensity. The most severe took place in 1998, in which it is estimated that some 16% of reef-building corals died worldwide. When the thermal stress is not so extreme only the thin-tissued, branching acroporid and pocilloporid corals have been bleached, leaving larger coral species (such as *Porites* spp.). Thermal stress factors need to be added to the other pressures (such as pollution) on corals. The conservation of 845 reef-building coral species has been assessed using International Union for the Conservation of Nature (IUCN) Red List Criteria. Of the 704 species that could be assigned conservation status, a third (32.8%) are in extinction risk categories (Carpenter et al., 2008).

At the other end of the gradient, Antarctic terrestrial ecosystems have seen visually marked changes in response to warming. These include the colonisation of previously bare or newly exposed ground by mainly mosses, as well as by other plants. Associated with this colonisation, soil invertebrate animals have also spread. At the moment there are few examples of exotic species colonising Antarctica, although those that have are mainly found close to geothermal sources of warmth and the continent's margins. In addition to this climate impact there are a number of human-imported species. Climate change favours increasing the chances of imported species becoming established. As for native species, because many Antarctic species are already at their lower thermal threshold, future warming is likely to see their numbers and range increase, spreading away from the coast. Having said this, imported species will offset part of this range expansion as coastal colonisation takes place and the imported and native species compete.

Meanwhile, in the peri-Arctic environment of Siberia, satellite analysis of some 515 000 km^2 has been carried out since the early 1970s and compared with images taken between 1997 and 2004, in an area containing Arctic lakes (about a seventh of this Siberian biome). The results showed a decline from 10 882 large lakes (those >40 ha) to just 1170; that is, approximately 11% of the original number (Smith et al., 2005). However, most did not disappear altogether but shrank to sizes below 40 ha. Total regional lake surface area decreased by 93 000 ha (approximately 6%). One hundred and twenty-five lakes vanished completely. None of these have refilled since 1997–8. However, where the permafrost has remained continuous (more to the north) lake area has increased by 13 300 ha (+12%). Nonetheless, the more southerly declines in lake area have outpaced the more northerly gains. What appears to be happening is that the more southerly permafrost soils have ceased being permanently frozen, so allowing the lakes to drain. Then again, further to the north the increased precipitation expected with global warming (see section 6.6.6 on ocean and atmosphere circulation) is enabling lakes to become larger. Both the decline in the more southern *and* the increase in the more northern Siberian Arctic lake areas appear to be caused by climate change and warming. Such changes have significant implications for soil carbon storage.

By the end of the 20th century, thaw lakes (as opposed to permanently frozen lakes) comprised 90% of the lakes in the Russian permafrost zone. North Siberian lakes differ from many of those in Europe and North America because they are on top of organic-rich strata. Russian and North American researchers (Walter et al., 2006) reported that thawing permafrost along thaw-lake margins accounted for most of the methane released from the lakes and estimated that the 1974–2000 expansion of thaw lakes, which took place along with regional warming, increased by 58% in the two lakes studied. Furthermore, the methane emitted was dated to 35 260–42 900 years ago. In short, this carbon was not recently (geologically speaking) fixed: the carbon was not fixed in the Holocene interglacial that began about 11 700 years ago, rather in the depth of the last glacial. What this means is that this carbon is not part of current Holocene carbon cycling and it is additional carbon. The release of this carbon due to warming is acting as a positive feedback to current warming (especially as methane has a higher global warming potential [GWP] than carbon dioxide; Table 1.2).

How serious this positive feedback is will depend on other carbon cycle factors: remember, it is the whole climate system that needs to be considered, and this is made up of many interacting feedback factors (see Figure 1.8). For example, another peri-polar counteracting factor is Antarctic ice-shelf melt, which can produce new carbon sinks. In 2009 a report of research by the British Antarctic Survey showed that, in around the past 50 years, in a large body of new open water (at least 24 000 km^2, the area of Wales or, alternatively, Israel) blooms of phytoplankton are flourishing. These new areas of high-latitude open water were left exposed by recent and rapid melting of ice shelves and glaciers around the Antarctic Peninsula. This new natural carbon sink is taking an estimated 3.5 million t of carbon (equivalent to 12.8 million t of carbon dioxide) from the ocean and atmosphere each year (Peck et al., 2009). It is the second largest factor acting against climate change so far discovered (the largest is new forest growth on land in the peri-Arctic). Having said that, the Arctic thaw-lake margins are tentatively estimated to be releasing 3.8 Mt of methane year^{-1} (Walter et al., 2006), which is equivalent to 2.85 Mt of carbon year^{-1}. Yet, given that the Arctic carbon release is in the form of methane (with a larger GWP than carbon dioxide), the Arctic methane release is likely to have a greater warming effect in the coming decades than this new Antarctic carbon sink will act to cool.

It is not just near-polar high-latitude ecosystems such as Arctic lakes that depend on ice. Mountain snowpacks affect the quantity and annual timing of water in streams supplying ecosystems in the surrounding lowlands. Even in a warm interglacial mode, nearly all mountains of sufficient height, save those near the equator, have snow caps. In a warmer world these will be reduced in volume, especially at lower latitudes, although note that smaller mountain snow caps may be seasonally thicker due to extra precipitation. This reduction in snowpack volume is something we are seeing already, not just at lower latitudes but at mid latitudes too, including those in part occupied by China, North America and Europe. Furthermore, in a warmer world snow-cap melt run-off will shift away from summer and autumn, when biological, and indeed human, demand for water is greatest compared to winter and early spring. In short, the annual cycle of water supply for many terrestrial and human systems will see reduced temporal buffering. The human ecology dimension is indicative of the likely biological impact of such a shift, and here this is exemplified by the broad statistic that one-sixth of the human population relies on glaciers and seasonal snowpacks for their water supply. In terms of terrestrial ecosystems, in which more than 50% of river flow is dominated by snow melt, this area includes nearly all of Canada's catchment, the north-western Rocky Mountain states of the USA, nearly all of Scandinavia with the exception of Denmark, alpine Balkan and Carpathian Europe, nearly all the Russian nations and north-eastern China, much of Chile and south-western Argentina and the south of New Zealand's South Island. In these regions ecosystems downstream rely on snow to act as a reservoir to regulate river flow and water supply throughout the year (Barnett et al., 2005a). If in a warmer world there is a smaller volume of snow melt then either other water reservoirs will need to be constructed (or have increased aquifer abstraction) to meet human and ecological need or we must accept the impact. This may be a theoretical and an expensive choice (especially when considering the currently free service that snow provides). It may be that action is taken to help meet human needs, but wildlife outside of semi-natural systems is not likely to see similar investment and will probably be the hardest hit.

In terms of human numbers, perhaps the most critical region in which vanishing snow, ice and glaciers will negatively affect water supply will be China and parts of India, or the Himalaya Hindu Kush region. This area not only supports some 2 billion people but also currently represents the largest volume of ice outside of polar and peri-polar regions. Nearly 70% of the Ganges' summer flow and 50–60% of other of the regions' major rivers comes from melt water (Barnett et al., 2005a).

Once again, as with virtually all areas of the climate–biology relationship, there are complications in the detail. It is undeniable that with global warming we are moving away from a world with many local areas of seasonably variable snow cover (similarly affecting many water catchments) to a world with far fewer areas of snow cover. However, this reduction in cover does act to reduce carbon loss from soils around the margins of snowfields. Such mountains areas of thin snow cover tend also to be ecotones, which are special biological communities in border areas between two habitat types. Because, with climate change, biological communities migrate and change, ecotones are particularly sensitive. As has been mentioned previously (Chapters 3, 4 and 5) carbon in soils can either accumulate or be released, in no small part depending on climate and climate change. Areas of thin snow cover are less insulated from the cold and this in turn affects the rate of carbon release from the soil (Monson et al., 2006). Such detail informs climate computer models.

Of course, all the changes noted in this section relate to change observed to date. Further change is anticipated with further warming. The scientific consensus (from the IPCC) is clear: warming will continue through the rest of this century and beyond. The only questions that remain are how much warming and at what rate. This in turn depends on the future emissions of greenhouse gases. Nonetheless, future species migration and other climate-related biological impacts will continue beyond those already seen.

For example, in the UK biological surveys between 1940 and 1989 have shown that a species of butterfly, the speckled wood (*Pararge aegeria*), occupies much of the UK south of the central Midlands. However, given Business-as-Usual (B-a-U) climate forecasts to the end of the century, the species is expected to be found virtually throughout the UK up to the Shetland Isles, with only the Scottish highlands and the highest North Pennine peaks being excepted. So, in the UK warming is extending the northern limits of *P. aegeria*. Meanwhile, at the other end of a species' thermal range the opposite can happen. For example, in North America, where Edith's checkerspot (*Euphydryas editha*) has a range that extends in places from Mexico to Canada, there is a real risk that with warming its southern limits will move north, and in the south its lower altitudinal limits will rise (Hill and Fox, 2003).

It is important to note once again that such shifts in species' ranges due to climate change are likely to be confounded by human land use and landscape fragmentation, and that this is likely to affect some species more than others. Continuing with the example of butterflies, a 30-year study of all 46 non-migratory British butterfly species with northern range margins in the UK illustrates this climate/landscape fragmentation problem (Warren et al., 2001). As these butterflies are at their northern limit, and it is too cold for them further north, it might be assumed that climate warming would favour an extension in their range. The researchers noted that the butterflies could be described in one of two ways. They are either sedentary specialists or mobile generalists. There is an advantage for specialists to be sedentary so that they are always near the specialist ecological community with which they are associated, whereas generalists are able, being mobile, to realise the advantages of new ecosystems; indeed, 26 out of 28 of the specialist species in the study were sedentary, whereas all 18 of the generalist species were mobile.

The study also noted that, from the 1940s to the end of the century, agricultural intensification in Britain led to a reduction of about 70% of semi-natural habitats that butterflies use. So, on one hand, on the basis of habitat alone, it would be expected that most species would decline. On the other, climatic warming in Britain would be expected to benefit these northern margin species. They found that three-quarters (34/46) of the species declined in their range. Further, habitat specialists fared worse than mobile generalists. The researchers took weekly counts of adults from 120 fixed sites between 1976 and 2000 and demonstrated that mobile generalists tended to increase compared to sedentary specialists. Looking at the broader picture, most species of non-migratory butterflies that reach their northern limits of geographical range in Britain have declined, as they have elsewhere in northern Europe. Conversely, most sedentary species have not expanded because their prospective habitat is too isolated. Consequently the scale of expansions of butterflies that took place following the last glacial maximum through to the 20th century is unlikely to be repeated with further warming in the 21st century. Indeed, the researchers conclude, if sedentary specialists continue to decline then British (and northern European) butterflies will increasingly become dominated by mobile generalists (Warren et al., 2001).

For those interested in the history of climate change biology studies and poleward range shifting in response to climate change, it is worth noting that prior to 1999 evidence had been limited to a single species or to only a portion of a species' range. In 1999 European and US researchers led by Camille Parmesan and Nils Ryrholm provided the first large-scale evidence of poleward shifts in entire species ranges. In a sample of 35 non-migratory European butterflies 63% were found to have ranges that had shifted to the north by 35–240 km during the 20th century, and only 3% had shifted to the south. The magnitudes of these range shifts (35–240 km along a single boundary) are in the order of 5–50 times the colonisation distances achieved by comparable butterflies in single colonisation events. Northward extensions therefore seem to have been a result of sequential establishment of new populations, with each giving rise to further colonisations. Interestingly, nearly all northward shifts involved extensions at the northern boundary with the southern boundary remaining stable (about two-thirds of 22 species). Note from our earlier discussion in this section that subsequently, in 2010, Daniel Doak and William Morris cited the work of Parmesan et al. (1999) as including a buffering response to species range shift due to climate change.

Species' climate-induced range-shift problems are great, and complex, in a modern landscape that is highly fragmented by human management and land use. However, they become critical to the level of extinction if the species' thermally determined spatial range becomes restricted and reduced to nothing. Typically this takes place at the high-latitude margins of continents. For example, a terrestrial species in Eurasia continuing to migrate northwards as the climate warms will ultimately reach the continent's edge and the Arctic Ocean. Faced with nowhere else to go, extinction becomes virtually inevitable. ('Virtually' because translocation to another part of the globe may be an option, but species translocation and introductions to new environments are not taken lightly.) Many of the species at risk are those that evolved during the Pleistocene or earlier, when the Earth's climate was cooling. Of the peri-Arctic species at risk one most often cited is the polar bear (*Ursus maritimus*). It evolved in the mid-Pleistocene some 100 000–250 000 years ago, it is thought, from a group of brown bears that became stranded by glaciers. They speciated rapidly, evolving sharper canine teeth (brown bears have far more vegetation in their diet), a longer neck (more suited for keeping their heads above water while swimming), larger paws (for spreading weight on the ice and for swimming) and of course thicker fur that is lighter in colour (from light brown to white). Polar bears require ice platforms from which to hunt and here is the problem with global warming: Arctic ice is currently reducing at a considerable rate. One early-21st-century estimate has been a 7% reduction in ice cover in 25 years and a 40% loss of thickness. Indeed, measurements of the minimum summer extent of sea ice since 1979 do show a clear trend of year-on-year reduction (see Figure 6.3). This reduction in Arctic ice is unprecedented, some suspect, in the Holocene (the past 11 700 years), and certainly as determined in the past 1450 years as revealed by 69 circum-Arctic climate proxy series (Kinnard et al., 2011). Because polar bears are at the top of the food chain, changes in their numbers are likely to affect the populations of many other species. For example, some seal populations have increased in response to fewer polar bears and this in turn is likely to affect fish predation.

The walrus (*Odobenus* spp.) is also declining, possibly because these animals too must work harder to find food with less sea ice. Walrus mothers nurse their young on sea-ice floes. As ice recedes the walrus do too. Further from the coast the mothers must dive longer and deeper from the ice to the sea floor to find clams. Of the three walrus species the population living in the Russian Arctic (*Odobenus laptevi*) has the smallest population, thought at the turn of the century to be between just 5000 and 10 000.

In the southern hemisphere there are analogous problems. In 2001 French researchers Christophe Barbraud and Henri Weimerskirch reported on a long-term survey (1952–2000) of the emperor penguin (*Aptenodytes forsteri*) colony near the Antarctic station of Dumont d'Urville and linked this to meteorological data gathered just 500 m from the animals. They noted that the breeding population was stable until the mid-1970s but then declined abruptly in the late 1970s to around 50% of its former level, coinciding with a time of high winter temperatures. They contemplate

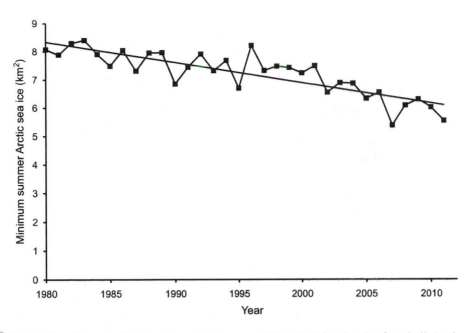

Fig. 6.3 Extent of the minimum Arctic sea ice at the end of the summer (September). Based on data from the National Snow and Ice Data Center (US) at the University of Colorado (http://nsidc.org). Arctic sea ice extent has been measured by satellite microwave remote sensing since 1979. The Earth's surface is measured divided up into cells around 25 km × 25 km. At least 15% of a cell's area has to have ice for it to be defined as ice-covered. The 1979–2000 mean minimum summer Arctic sea-ice extent was 7.7 million km². The 1979–2000 mean therefore gives a long-term trend value (as opposed to actual value) for 1990, the year of the first IPCC report and a baseline year for much climate change policy.

that declines in Antarctic krill (*Euphausia superba*) – which form much of the emperor penguin diet, along with fish and squid – may be linked to this decline in the emperor penguin population. Krill also sustain a number of commercial fisheries and concomitantly are fundamental to much of the Antarctic marine food web and populations. Yet lower krill abundance is associated with less winter ice. That krill populations around the Antarctic are changing with warming waters, and changes in sea ice, as well as commercial fishing (see Schiermeier, 2010), have been corroborated by a body of research. For example, in 2004 a research team, led by Angus Atkinson of the British Antarctic Survey, reported that between 1976 and 2003 the Antarctic south-west Atlantic sector, which contains over half of Southern Ocean krill stocks, saw stock densities decline by more than 50%, while elsewhere those of salps (mainly *Salpa thompsoni*) increased. This represents a profound change in the Southern Ocean food web.

Looking briefly at the marine environment on the global scale, the question of disentangling the impact of commercial fisheries and that of climate change on krill stocks brings us to a key problem for ecologists: is it non-climate-related human activities (such as fishing and pollution) and/or climate change having an impact at the global scale? In 2010 Ove Hoegh-Guldberg and John Bruno from the University of Queensland in Australia, and the University of North Carolina at Chapel Hill,

USA, respectively, had their synthesis on the effects of climate change on the world's oceans published in the journal *Science*. It was arguably the first global assessment of its kind. They concluded that impacts of anthropogenic climate change so far include decreased ocean productivity, altered food-web dynamics, reduced abundance of habitat-forming species, shifting species distributions and a greater incidence of disease. They also noted that although there is considerable uncertainty about the spatial and temporal details, climate change is clearly and fundamentally altering ocean ecosystems. They warned that further change will continue to create considerable challenges and costs for societies worldwide, particularly those in developing countries.

The other place where species' spatial options become constrained with global warming is the tops of mountains. As noted, with global warming thermally determined zonation on mountains changes and rises. If the warming becomes too great then there is nowhere for alpine species to go as species cannot migrate above mountain summits. This is of some ecological and wildlife concern. The alpine biome is just 3% of the vegetated terrestrial surface and such ecological islands are shrinking. For instance, in the Ural Mountains, where temperatures rose by as much as 4°C in the 20th century, tree lines have risen between 20 and 80 m upslope, so reducing the regional alpine zones by 10–20%. Meanwhile, an example of an animal under severe threat comes from the southern hemisphere and is the mountain pygmy possum (*Burramys parvus*). It is a marsupial found in montane heathland from 1400 to 2230 m in south-eastern Australia. It is not a possum in the American sense and is unrelated to true opossums. (The word opossum was taken from the Algonquian Indian word for the American animal.) At the beginning of the 21st century, with just some 2600 adults remaining, the mountain pygmy possum is considered to be under serious threat (and indeed prior to 1970 was thought to be extinct). The species is further pressured as its habitat is favoured by skiers. Human land use fragments the landscape and further aggravates the conservation problems posed by current climate change. Land-use concerns recur in many discussions about current climate change and wildlife conservation.

Another example of threatened montane species, but this time one that includes the changing environmental conditions that accompany climate change, comes from local endemic animal extinctions in the highland forests of Monteverde, Costa Rica. The situation became apparent following a step-like warming of the tropical ocean in 1976; indeed, 20 of the 50 anuran species (frogs and toads) in a 30 km^2 study area went extinct, which suggests pressure on the broader population. In conservation terms, one of the most serious of these local extinctions was that of the locally endemic golden toad (*Bufo periglenes*) in 1987. The population crashes were all associated with a decline in dry-season mist frequency (which had been measured) and it is presumed that this in turn was due to a raising of the cloud-bank base (which unfortunately was not measured). Furthermore, as noted at the beginning of this section, climate change causes modifications in the way that populations from different species associate. In this instance, along one stream as the habitat became drier harlequin frogs (*Atelopus varius*) gathered near waterfalls, so increasing the probability of their being attacked by parasitic flies, and this in turn resulted in increased mortality (Pounds et al., 1999). The researchers concluded that the aforementioned population crashes belong to a constellation of demographic changes that have altered communities of birds, reptiles and amphibians in the area and which are clearly linked to regional climatic warming. Later work more clearly linked *Atelopus* population crashes with climate change. There were some 110 species of *Atelopus* endemic to the New World tropics, a number that made it possible to make a statistically valid local extinction study.

The idea was put forward that temperatures at many highland habitats of *Atelopus* spp. were shifting towards the growth optimum of a chytrid fungus, *Batrachochytrium dendrobatidis*, which is known to infect many species of frog. In short, in this instance climate change was promoting infectious disease in highland areas. Looking at the records of when species were last recorded in highland areas in the wild and comparing these to climate records the researchers were able to show that most extinctions (78–83%) occurred in years of higher tropical temperature and that this was statistically significant to 99.9%. Further, the extinctions varied with altitude (which is also related to temperature) and it is thought that this reflected the optimal conditions for *Batrachochytrium* to flourish (Pounds et al., 2006). This means that not only does global warming reduce individual mountains habitat islands as habitat zones migrate upwards, but that the environmental conditions of the complex mosaic of habitats in highland areas can change so that some areas see an increase in disease organisms.

There is a further twist to this story. Around a third (1856) of amphibian species in the Global Amphibian Assessment are classified globally as threatened and 427 species as critically endangered. One question therefore is whether the *Atelopus* studies meaningfully relate to the global amphibian picture. The earliest known hosts of *Batrachochytrium* were a continent away, the African clawed frog (*Xenopus*) of South Africa. These frogs became economically important in the 1950s as their tissue was used in some pregnancy-testing kits. Museum records also suggest that *Batrachochytrium* achieved a global distribution after the 1960s. The possibility exists therefore that the expansion in one frog species due to commercial activity might have led to the extinction of other amphibian species (Blaustein and Dobson, 2006).

The recent decline in amphibians is not just due to the chytrid fungus *B. dendrobatidis* and climate change (not to mention the interaction of the two), but also land-use change leading to loss of habitat. And, of course, land-use change involving deforestation and drainage releases carbon and so is a factor in climate change. In 2011, Christian Hof, Miguel Araújo, Walter Jetz and Carsten Rahbek attempted to map these three factors of amphibian decline. They found that the areas harbouring the richest amphibian faunas are disproportionally more affected by more than one of these factors than areas of low amphibian diversity. They therefore predicted that amphibian declines are likely to accelerate in the 21st century. (See also Alford, 2011.)

This is not the only interaction known to take place between climate and disease. Similarly the warmer climate conditions in upland regions of the western USA favour the mountain pine beetle (*Dendroctonus ponderosae*), allowing it to complete its life cycle in 1 year rather than 2 years. These beetles act as a vector, enabling the transmission of pine blister rust (*Cronartium ribicola*), and this is having a serious impact on pine trees on some of the Rockies' highest mountains (Blaustein and Dobson, 2006). The issue of economic activity, globalisation, human population growth, disease species and climate interactions is a theme that we shall return to when we look at human ecology in Chapter 7.

The early-21st-century outbreak of the mountain pine beetle in western Canada is an order of magnitude greater in area than previous outbreaks. Beginning just before 2000, it was facilitated by higher minimum winter temperatures, higher summer temperature and reduced summer rainfall. By the end of 2006 the outbreak had covered some $130\,000$ km^2, with commercial timber losses estimated to be more than 435 million m^3 and additional losses outside commercial areas. But in addition to the commercial costs arising from the biological impact of this climate-facilitated outbreak, there were also changes to the forests' ecosystem function with regards to carbon balance. In 2008 researchers primarily from the Canadian Forest Service (Kurz et al., 2008), reported that between 1990–2002 the average net biome production (net primary and secondary production[3]) for the south central region of British Columbia (an area of approximately $660\,000$ km^2 and the region most extensively affected by the outbreak) was 0.59 Mt C year^{-1}: that is to say, it was acting as a carbon sink. The researcher used actual ecological data from the area between 2000 and 2006 and then modelled the outbreak's likely course through to 2020, with an expected peak around 2008. They found that net biome production was changing and, including the modelled future prognosis for the outbreak, concluded that the forest would be a net source of carbon of some 17.6 Mt C year^{-1} until 2020 but with peak carbon release years seeing between just below 20 to just under 25 Mt C year^{-1}. Again, this is another example of the many biosphere feedback cycles in the global climate system discussed in section 1.5. The researchers also predict that outbreaks of other forest insects are likely to affect forest carbon balance, including the eastern spruce budworm (*Choristoneura fumiferana*) and forest tent caterpillar (*Malacosoma disstria*), by reducing tree growth and increasing tree mortality. With tree mortality comes the risk of increased forest fires, and the research team postulated that such insect outbreaks together with fire could turn current North American forest carbon sinks into carbon sources. Their study is one example of how climate change biology has an impact on the global carbon cycle providing a positive feedback to the global climate system. (Of course, climate change can also work the other way to increase other carbon sinks so it is the overall global balance between sources and sinks that will determine the trajectory of the global system.)

Returning to climate change and altitude, an example of a plant species restricted by altitude and which may see a population decline with warming is Britain's Snowdon lily (*Llodia sevotina*).

In this section on species' climate-induced range shifts we have noted that some high-latitude species have nowhere to go with warming that takes the Earth above the temperature of the warmest times of the current and past interglacials (above the warmest global climate level of the Quaternary to date, the past 2 million years): they have run out of biome in which to live. We have also noted that high-altitude species on the tops of mountains face a similar problem. However, in the tropics in addition to

---

[3] Gross primary production is the dry biomass produced (mainly) by photosynthetic organisms and secondary production is the dry biomass by those species living on primary producers. Net primary production is gross primary production minus biomass lost by respiration.

species with nowhere to go with further warming – tropical lowland species that do not have an upward corridor to higher elevations or latitudinally from where they are now through which they can migrate – there is another ecological problem. This is, what replaces the world's most tropical species that already live in tropical lowlands and that have either migrated away or died *in situ*? There is no terrestrial biome warmer than the tropics. So in the event of further warming above today's levels we can expect increased tropical lowland ecological attrition in addition to those at the highest of tropical altitudes. This tropical lowland ecological attrition will be more severe than the temperate lowland attrition in developed nations discussed at the start of this section.

The seriousness of this tropical ecological attrition may not at first seem obvious; after all, in theory a lowland tropical species should be able to migrate latitudinally towards the poles or upwards. In the tropics, and indeed elsewhere, a species responding to climate change will find it easier to shift its range vertically than horizontally as the vertical migration required to compensate for, say, half a degree warmer climate necessitates a far shorter vertical migration than a horizontal one. Indeed, in the tropics this difference between the more easy vertical and the harder horizontal latitudinal migration to compensate for temperature is greater than in temperate zones: vertical temperature gradients in most places on Earth are steep, around a 5.2–6.5°C decrease per 1 km elevation but this distance is far less than the horizontal distance required, which is more than 1000 times more: you need to leave the tropics first before there is cooling with increased latitude. Once beyond the tropics in the temperate zone, latitudinal cooling is around 6.9°C per 1000 km at latitudes around 45° south or north: these are the latitudes of places such as Bordeaux in France, Minneapolis in the USA, or Oamaru in New Zealand. In addition, there is the aforementioned landscape fragmentation problem that is both natural (the Earth's geography being not one where all lowland elevations are connected by corridors that gradually rise because many places, especially of mid-elevations, see their route to higher elevations blocked by intervening warmer lowlands) and anthropogenic, with intervening human settlements or agricultural lands (i.e. the landscape fragmentation already mentioned a number of times) inhibiting migration. But there is the lowland tropical ecological problem: irrespective of whether there are appropriate migration corridors and species migrate, or they locally go extinct *in situ*, what replaces them? At the tops of tropical mountains, species that go extinct are replaced by others tolerant to slightly warmer conditions migrating from below, but what replaces those current tropical species that will no longer live in the even warmer tropical lowlands: tropical lowlands are the warmest biomes on Earth (not including extreme niches such as volcanic springs). So, we are faced with this prospect of lowland tropical ecological attrition in addition to the extinction of alpine-allied species at the tops of tropical mountains. All this leads us to consider the seriousness of this problem.

In 2008 the US ecologist Robert Cowell, the German zoologist Gunnar Brehm and colleagues attempted to get a handle on this issue. They analysed altitudinal range data for four large survey data sets of plants and insects: epiphytes (plants that live on the surface of other plants but do not derive nourishment from them), understorey Rubiaceae (the family that includes blackberry and raspberry), Geometridae (a large family of large-winged moths with distinctive larvae) and Formicidae (ants). The data for all 1902 species were taken between 2001 and 2007 from the Barva Transect, a continuously forested corridor ascending 2900 m up an elevation gradient from La Selva Biological Station, near sea level, to the top of Volcán Barva, in Costa Rica. To explore the question of tropical lowland ecological attrition we must ask whether tropical lowland species are already living near the thermal optimum of their climatic niche, above which fitness would decline in the absence of acclimation or adaptation. For plants, especially, because temperature and precipitation interact strongly through transpiration water loss, the answer includes precipitation change as well that of temperature. With climate change, as some parts of the tropics will see more rainfall, and others less, the picture becomes less clear. Current evidence, however, suggests that contemporary warming of around $0.25°C$ decade^{-1} since 1975 in the tropical lowlands has already driven global mean temperature to within approximately $1°C$ of the Earth's maximum temperature over the past couple of million years. The Cowell–Brehm team used a simple graphical model that relied on temperature to assess potential altitude range shifts for their fauna and flora transect data. Assuming an IPCC 2007 warming of $3.5°C$ by the year 2100, a 600 m vertical compensatory shift in their Barva Transect species' range would be required. About half (53%) of the 1902 species in their study (those that are currently in the lowest elevations) are candidates for lowland biotic attrition, and about half (51%) may be faced with range-shift gaps. The potential for mountains-top extinctions for these groups on the Barva Transect is minimal for a 600 m shift in isotherms but begins to be more significant at about a 1000 m range shift (which could occur by the early to mid-22nd century if the IPCC 2007 longer-term scenario forecasts are correct). So, although mountain-top extinction has received quite a bit of recent attention in the ecological literature, this study suggests that in the near term a far greater proportion of the tropical species are threatened with lowland attrition or are challenged by landscape fragmentation (natural and anthropogenic). Many face more than one challenge.

A broadly similar result came in 2010 from Kenneth Feeley and Miles Silman, at Wake Forest University, Northern Carolina. They looked at more than 2000 plant species from tropical South America and their thermal niches to see how they might respond to $1–5°C$ of warming. Looking at the results on the basis of observed thermal niches, they predicted an almost complete loss of plant diversity in most South American tropical forests due to $5°C$ warming, but correcting for possible niche truncation they estimated that most forests will retain 50–70%, or more, of their current species richness. This Feeley–Silman result, and the Cowell–Brehm et al. result, both very broadly suggest that lowland tropical areas in the warmest part of the Amazon might see an attrition of roughly between a third and half of species with warming of $3.5–5°C$. This is the temperature rise that the central Amazon is likely to experience by the end of the 21st century under the 2007 IPCC A2 B-a-U emissions scenario. (Of course the above does not include precipitation change, hence forecast summer drought, in some parts of a warmer Amazonia, which would provide additional plant stress.)

Finally, before leaving the subject of ecological and species response to climate change in the tropics it is worth recalling some of the ground previously covered when looking at the last glacial (section 4.5). Because tropical lowlands were refugia during glacials to what are now tropical montane species, tropical mountains are today home to species with a long genetic history and this has contributed to their high biodiversity.[4] However, in the past the global climate has alternated with glacial–interglacial cycles from present-day to cooler temperatures. Yet today we are seeing temperatures going the other way, into warmer times that are unprecedented for the Quaternary, and this means extinctions that are unprecedented in the Quaternary. So not only will there be extinctions associated with species at very high latitudes but we will also lose tropical montane biodiversity.

Of course, species' shifts across landscapes are not just associated with future climatic warming; we have already discussed (Chapter 4) how species relocated with past climate change. Notably, this took place to and from refugia where a species (or species community) survived either a glacial or an interglacial.

One significant concern in all of this is the plight of our nature reserves. These sit mainly among landscapes that, as has been noted (and will be again), have become highly fragmented by human use and transformation. This fragmentation impedes species migration and so means that many nature reserves can be considered to be ecological islands: an idea that sprang from two North American ecologists, MacArthur and Wilson, in their *Theory of Island Biogeography* in 1967. The problem is that because the climate is changing these island nature reserves are in the wrong place. Alternatively, they could be considered as being in the right place but not featuring the communities of species likely to flourish under future climates. Either way, these nature reserves will not, in a climatically changed future, sustain the species they have helped conserve in the past. Species will be stranded. So the current places of refuge will not be sufficient for our present conservation goals in the future. What is to be done?

One solution could be to ensure that there are ecological corridors to allow species to migrate. Of course, a highly fragmented landscape tends not to lend itself to such corridors. In such circumstances it may just be possible to create a sufficient number of scattered small reserves or mini-communities of species in the landscape. These might be carefully managed hedgerows, strips of green along motorways, parks (and especially managed areas within parks) and even clusters of urban gardens. Such areas may provide stepping stones for some species. Altogether the landscape, though still fragmented, could feature escape routes. The landscape, with its scatter of mini and temporary refugia, could be viewed as a matrix. (Although many may associate the term matrix with mathematics, it actually comes from the Latin meaning a breeding animal or mother but with particular respect to propagation, and womb. Nonetheless, this conservation use does chime with its other use to denote a complex assemblage.)

Of course, as part of creating an assemblage of mini-communities of species, many of these communities will need to be managed to encourage diversity. This is not just for the obvious goal of maximising biodiversity within the managed ecosystem (which may in some cases is not be ecologically desired), but within the assemblage of ecosystems within the reserve, as there are often knock-on effects between ecosystems. Effects can cascade across ecosystems. For example, the presence of fish in a

---

[4] Readers who are not life scientists may care to note that biodiversity does not just relate to the diversity of different species but also the genetic diversity within a species.

pond can have the effect of increasing plant diversity on the surrounding land. As has been shown in a Florida reserve, fish can reduce, through predation, the abundance of dragonfly larvae (such as *Anax junius* and *Erythemis simplicicollis*) in ponds, which leads to fewer adult dragonflies in the air that would in turn prey upon pollinator insects. Fewer predating adult dragonflies means more pollinator insects and so more plants pollinated, hence, in theory, more plants (Knight et al., 2005). So not only does conservation management need to be conducted with regard to the landscape level to facilitate migration between sites, it also needs to be done at the ecosystem level, not just for the ecosystem but to enable trophic cascades between ecosystems to realise biodiversity synergies.

With climate change, managers of such sites will need to consider a number of parameters to their local conservation strategy. The first is 'ecosystem resilience'. This can be defined as the ability of an ecological system to absorb disturbances while retaining the same basic structure and ways of functioning, and to have the capacity for self-organisation and the capacity to adapt to stress and change. This definition differs from another ecological definition of the same term as being the speed with which a community returns to its former state after it has been disturbed. With current climate change, globally the IPCC anticipate a long-term warming trend and so in this light absorbing disturbance is more appropriate than returning to a former state. Having said that, global climate change is one thing, but local vagaries of weather and climate are another. Here the ability to return to a former state is of value.

The importance of these two views of ecosystem resilience is highlighted by understanding the manner of species' range shift with climate change. In 2011 ecologists Regan Early and Dov Sax highlighted this when looking at 15 amphibian species in the western USA, making the following predictions. First, interdecadal variability in climate change can prevent range shifts by causing gaps in climate paths, even in the absence of geographic barriers. Second, the somewhat unappreciated trait of persistence during unfavourable climatic conditions is critical to species' range shifts: a key predictor of success was a species' ability to live with unfavourable climate for – in the case of their amphibian study – up to a decade. Third, climatic fluctuations and low persistence could lead to endangerment even if the future potential range size is large. These considerations may render habitat corridors ineffectual for some species, and conservationists may need to consider managed relocation (species translocation).

Ecosystem resilience should not be confused with ecological resistance, the ability of a community to avoid displacement from its present state by a disturbance. What ecological conservationists need to consider is how to build in ecosystem resilience so as to manage ecological resistance because we do expect communities to migrate with long-term climate change. This problem of terminology is common in science, not to mention other walks of life. (Earlier there was the example of the misleading term medieval climatic optimum and we shall in the coming chapters see others, such as population 'fertility', climate 'tipping point' and low 'carbon' economy.) While it is difficult if not impossible to impose vocabulary (living languages are dynamic), we can and need to be clear about what we mean.

Then there is 'ecological accommodation'. This relates to allowing ecological change and is important in the context of global climate change as ecological change across the planet is a virtual certainty. This modifies our local ecological conservationist brief when formulating site-management strategies to that of considering how to build in ecosystem resilience to manage ecological resistance in order to allow accommodation. In practical terms this means considering a number of scenarios, especially as the IPCC warn us to be wary of climate surprises.

Yet even with the most astute conservationists, because human pressures are so great and becoming greater, the management of reserves and the landscape to create a matrix needed to facilitate species' climate change-driven range migration will be difficult (as we shall see in Chapters 7 and 8). Even so, this is a possible conservation tool currently being advocated by a number of ecologists (Lovejoy and Hannah, 2005). Certainly, the trend towards the end of the 20th century among ecologists was to have regard for managing ecology on the scale of the landscape, rather than purely site by site.

The biological impacts of climate change discussed in this section have been thematic: generalities that apply to many places. To re-cap, these include phenological changes, species migration (and limitations), changes in ecological communities (recombinant ecology) and the problems of an increasingly anthropogenically fragmented landscape. But because the determinants of biological effects of climate change are so multi-dimensional, they can only really be considered in the context of specific regional climate change: one size does not fit all. The next three sections provide case studies of likely change and impacts for three areas: North America, the UK and Australasia.

6.2 Case study: climate and natural systems in the USA and Canada

The year 2001 saw a major policy study entitled *Climate Change Impacts on the United States* by the National Assessment Synthesis Team from the US Global Change Research Program (USGCRP), although it was actually released in 2000 with a draft for consultation also released earlier that year. A second report was published by the USGCRP in 2009 entitled *Global Climate Change Impacts in the United States*. The first 2000–1 (hereafter 2001) report was, in effect, a stock-take of US scientific understanding at the turn of the millennium of how climate change might affect the USA. The second report confirmed and extended the first's conclusions, and affirmed them with updated scientific evidence.

The 2001 report concluded that the USA would very probably get substantially warmer, with temperatures rising in the 21st century more rapidly than in the previous 10 000 years. There would also be more rain, some of it in heavy downpours, even though some parts of the USA would become more arid. Drought in some parts may become more common as the effects of temperature rise on water evaporation outstrip increases in precipitation. It noted that reduced summer run-off coupled with increased summer demand for water would probably stress some parts of the US water-supply system, while increased heavy precipitation would increase the need for flood management. These changes would disturb freshwater ecosystems. Some species (both aquatic and terrestrial) would in all probability adapt to climate change by shifting their ranges, but invasive species and land use would together undermine this, resulting in a risk to biodiversity.

There were particular worries expressed for 'rare ecosystems', but not so for managed ones. Forest growth would possibly increase due to the increase in carbon dioxide and temperatures, spurring growth; however, some forests would probably become more susceptible to fire and pests. Then in 2009 the second USGCRP report added that forest productivity is projected to increase in much of the east, while it is projected to decrease in much of the west where water is scarce and projected to become more so. Wherever droughts increase, forest productivity will decrease and tree death will increase. In addition to occurring in much of the west, these conditions are projected to occur in parts of Alaska and in the eastern part of the south east.

With regards to agriculture, the first 2001 report concluded that there was no concern for any decline in food security and the assessment considered that agricultural profit margins would continue to decline for non-biological reasons. The 2009 report was less sanguine. It warned that the projected climate change is likely to increasingly challenge the USA's agricultural capacity to efficiently produce food, feed, fuel and livestock products. Water supply in some regions was a key concern.

Human health impacts were also noted in the 2001 report, especially in the summer with heat stress. The 2009 report went further, introducing concerns as to climate changes' overseas human health impacts possibly threatening US security. It also emphasised the health impacts of extreme weather events. Now, the USGCRP's 2001 assessment did warn of the likelihood of extreme events but this became more of a recurring theme in the 2009 report.

The importance of the USGCRP's National Assessment Synthesis Team 2001 report was that not only was it the US Government's first major attempt to have its science agencies assess the likely impacts of climate change on the nation, but that it also underpinned the US Department of State's 2002 *US Climate Action Report* (covered in Chapter 8).

The 2001 assessment painted a very much broad-brush picture and its 2009 follow-up corroborated and added detail. This first report's coarseness was due to the coarse nature of (the then) cutting-edge computer models, which by 2000 had not the sufficient agreement in terms of forecasting the future to portray a spatial picture with much certainty. However, the two models (the Canadian model from the Canadian Climate Centre and the Hadley Centre model from the UK) that the report frequently referred to did, on a state-by-state scale, give a broadly similar pattern of temperature change throughout the 21st century, even if in some places they disagreed by a few degrees Celsius. Conversely the 2009 report's climate projections were more coherent and broadly speaking chimed with the regional projections given by the 2007 IPCC's own assessment.

The 2001 assessment had noted that in the past century temperatures across much of the USA have risen, but the states of, and around, Mississippi and Alabama had

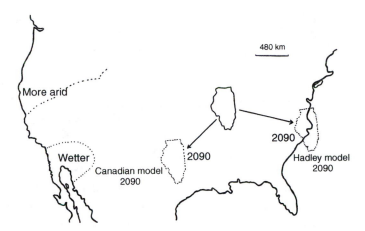

Fig. 6.4 Difficulties in predicting 21st-century change in the USA, in this case by the Hadley and Canadian climate models (National Assessment Synthesis Team, 2001). Both models suggested that southern California would become wetter and that parts around Colorado and Kansas would become drier. However, differences between the models include whether Illinois's climate will become like North and South Carolina's or like Oklahoma's.

cooled. With regards to the 21st century, both the Canadian and the Hadley models agreed that the USA would warm up but disagreed as to the exact intensity of this warming on a state-by-state basis. The Canadian model predicted warming across the USA but with a large centre of stronger warming in the central states of South Dakota, Wyoming and Nebraska. Conversely, the Hadley model predicted an almost similar, but marginally milder, background of warming compared to the Canadian model, without a central island of stronger warming and with slightly milder warming of the south-eastern states around Florida, Alabama, Georgia and North and South Carolina. Although computer models have improved since then, model improvement is still an ongoing enterprise and likely to be so for decades to come. So it is perhaps worth noting (as we shall shortly see, hence Figure 6.4) some of the areas of debate and uncertainty as to computer model output at the beginning of the 21st century, because the issues then are more or less the same as the issues now, only starker: subsequent model improvement has, and will, reduce uncertainty and provide additional detail.

With regards to changes in precipitation, the 2001 USGCRP report again presented outputs from both the Canadian and Hadley models. These suggest that much of the USA will see an increase in precipitation of 20–40% and that California (especially southern California) might experience more than this. However, they differed in that the Canadian model attributed far more rainfall to the states to the east of California and also up to 20% less rain in the central and eastern-seaboard states. Of course, as previously noted, increased rainfall does not necessarily mean more water for plants. If evaporation due to a temperature rise is more than the extra rainfall then soils become drier. Both models forecast that the area around the Colorado/Oklahoma border will experience a decline in soil moisture and that in California there will be an increase. Other than this, the models gave contrasting forecasts. With reference to the medieval climatic anomaly (MCA; see Chapter 4) being an analogue to a degree or so of medium-term future climatic warming that we would expect, only the Hadley model suggested that part of the Sierra Nevada will experience drought. Both models do indicate that Montana, Wyoming and North and South Dakota may retain less soil moisture, as would be expected from looking at past climates when it was a little warmer than today.

These temperature and rainfall factors suggest increases in vegetation with climate change across much of the USA, with the exception of the Mississippi floodplain and Florida. However, the Hadley model suggested a marked decline in vegetation in Arizona. This was not as big a difference between the two models as it seems. If an area is already vegetation-poor, in this case due to aridity, then only a small proportional change in an already small biomass per unit area equates to a big difference in the ecosystem. So much depends on which parameter is represented more in the output.

At the moment one of the larger features in an ecosystem map (technically, a biome map) of the USA is the eastern states, with their temperate deciduous and mixed forest, and the grasslands of Oklahoma, Kansas, Nebraska and South Dakota, with a border between these two ecosystems running through eastern Texas, and up to the Iowa/Illinois boundary. This boundary line lies north by north-east. However, at the end of the 21st century both models suggest that this will become north–south, still beginning in mid-Texas but instead ending close to the Minnesota and North Dakota border (see Figure 6.4). These last two states, together with Iowa and Missouri, will see considerable ecosystem change.

These are the broad changes across large areas that both models predict: there will also be climate change elsewhere. However, they also predict much change (but differently) on smaller scales. The Rocky Mountains, already a highly fragmented natural landscape in terms of ecosystem type, are predicted to change. Nonetheless, it is possible that the small lateral distances of such change might be accommodated through vertical migration, but the picture is not that clear for definitive predictions, let alone the national determination of ecosystem-management policy at the local level.

The agreement, or lack of, between the two models used in the first USGCRP report from 2001 is aptly illustrated by how they each predict the summer climate of Illinois will change by 2090 (see Figure 6.4). Both models predict that the climate approximately along the latitude 35°N will shift 5° further north by the end of the 20th century. However, they disagree as to whether Illinois's climate will become like North and South Carolina's or will be more like Oklahoma's. The main difference between the two is the amount of summer rainfall: Oklahoma gets less.

With regards to ecosystems, the 2009 report gave a little more detail. Of possible note, while not included in the chapter's key messages summary, the 2009 report's main text did refer to the possibility of thresholds being crossed whereby an ecosystem stops changing gradually but instead moves quickly to a new state. Climate-related thresholds are also referred to elsewhere in the report, such as in relation to health issues.

The 2001 USGCRP report did not just look at possible US climate and ecosystem change but also how its population and economy would develop throughout the 21st century. It predicted that the population would increase by between just 20% and almost a doubling, depending on various emigration scenarios. Employment forecasts were also variable, from barely rising at all through to rising up to 2070 but then levelling off or slowing down in the century's final quarter. Either way, the US gross domestic product (GDP), it was thought, would increase, although there was considerable uncertainty about how much. Meanwhile the second report noted that the US population had been estimated to have grown to more than 300 million people, nearly a 7% increase since the US's 2000 census, and that then current Census Bureau projections were for this growth rate to continue, with the national population projected to reach 350 million by 2025 and 420 million by 2050. The highest rates of population growth to 2025 were projected to occur in areas such as the south west that are at risk for reductions in water supplies due to climate change.

My purpose in describing the first 2001 USGCRP report and simply noting just some of the principal points of the second one in 2009, rather than concentrating on the latter, is to demonstrate two things. First, much is old news and little is new, even if the updates are extremely useful in their extra detail. All too often in both science and policy development there is focus on current developments and not how we got there. In one sense this is regrettable because we then do not see failure in addressing matters previously raised. Second, by concentrating purely on the latest reports, those doubtful (or even sceptical) of the science can claim that somehow this is a re-writing, and that because a re-writing was required it may well be again in the future, so concerns do not need to be raised seriously. Such a perspective is, of course, obfuscation. Those beyond science need to appreciate that science really is all about 're-writing', but in an evolutionary sense of it building upon past understanding, and that it does need to be taken seriously.

Returning to climate change and its consequence for the USA, putting the above population and wealth forecasts into the context of extreme climatic events such as hurricanes (which in a warmer world we might reasonably expect to see more frequently), the 2001 report noted that although the number of hurricanes per year has been constant in the last century, the number of deaths by hurricanes has declined (presumably through better facilities affordable in a richer society). Yet, conversely, property losses from hurricanes increased (presumably because a wealthier society has more to lose).

That there is a link between climate and the destructiveness of cyclones has become increasingly apparent. In 2005 Kerry Emanuel from the Massachusetts Institute of Technology published results based on surface wind measurements taken during storms over the past 30 years. He found a high correlation between the destructive potential of storms and sea-surface temperature. He suggested that future warming might lead to an upward trend in destructive potential and predicted a substantial increase in hurricane-related losses in the 21st century. Meanwhile, in the same year Peter Webster, of the Georgia Institute of Technology, and colleagues, including one from the National Center for Atmospheric Research in Colorado, published a study covering a 35-year period of cyclone activity. They found no statistical increase in the global number (frequency) of tropical storms and hurricanes but, in a way affirming Emanuel's results, a large increase was seen in the number of the most powerful storms (categories 4 and 5 on the Saffir–Simpson scale).

Over the 35-year study period the number of the most powerful hurricanes had increased by 80%. Interestingly, the survey showed no statistical increase in the actual intensity of the most intense hurricanes: the maximum intensity appeared static but the number and proportion of the most powerful hurricanes have risen. In this precise sense the two reports disagree: although there was agreement that climate change was bringing more hurricane destruction, whether it was frequency and/or intensity of the more intense events was unclear. And so the first years of the 21st century saw much debate about the exact link between climate change and the nature and/or frequency of hurricanes even though it was agreed that the frequency spectrum of hurricanes in a given year was shifting in the direction of greater intensity. Things were to become a little clearer in 2008 when Jason Elsner, James Kossin and Thomas Jagger had their analysis of 1981–2006 Atlantic tropical hurricanes published. While previous studies had looked at hurricanes' average strength, this new work looked at the maximum strengths reached by hurricanes. Here they found that in 1981–2006 the maximum strength of hurricanes was increasing and suggested that as seas warm the ocean has more energy to convert to tropical cyclone wind. This conclusion appeared in the 2009 USGCRP report.

Both the 2001 and 2009 USGCRP reports looked at each region. Alaska, with its ecologically important Arctic tundra and sub-Arctic environments, is likely to see much change in its wildlife conservation and biodiversity concerns. This applies to much of Canada too. Elsewhere in the USA the question largely becomes one of having to manage change. However, in some places the change is not one of movement along a broad ecotone (a gradual environmental gradient, such as a hill) but catastrophic change on a comparatively isolated ecosystem type. For example, the encroachment of salt water due to sea-level rise on the coastal forests around the Gulf of Mexico has led to trees dying. This problem is particularly acute in parts of south Florida and Louisiana, with the creation of dead or dying areas called ghost forests. In other areas there is the problem of increased summer heat which, despite increased rainfall, is likely to lead to greater aridity that will raise the likelihood of wildfires. This again is likely to be of particular concern to those in the south-east USA. The reports' regional conclusions are summarised in the following paragraphs.

The midwest, which contains Chicago, the nation's rail hub, is likely to see transportation problems due to extensive flooding, which might in turn result in nutrient enrichment (eutrophication) of aquatic systems. Ironically, despite increased rainfall (much of which will be in seasonal bursts), it is anticipated that there will be a reduction in river and lake levels. These in turn will result in transport impacts as docks, locks and associated navigation structures will require re-engineering. The reports cite the anticipated reduction of water levels in the Great Lakes and the 2009 report forecasts that by the end of the century, under IPCC Business-as-Usual (B-a-U) scenarios, heatwaves will occur ever other year.

The west, the USGCRP reports conclude, will continue to see an increase in precipitation, as it has already done during the 20th century. In some parts river flow may double by the end of the 21st century. However, some parts of the west are arid and it is not clear whether increased temperature would offset any extra precipitation. Droughts are likely to increase; a conclusion in line with palaeoclimatological evidence.

The Pacific north-west is likely to see far greater temperature changes, at 4–4.5°C by the end of the 21st century, than the 1.5°C rise experienced in the 20th century. Rainfall across the region is likely to increase, although decreases in some areas may be experienced. Much of the north west's extra rainfall is likely to take place in winter months. There is considerable uncertainty associated with this extra rain, which could be as little as a few per cent or as much as 50%. The overall concern is about prospective deficiencies in water supply during the summer and this is likely to impact the region's important agriculture sector. In Columbia, where water-allocation conflicts are already acute, matters are likely to become increasingly difficult. Erosion from sea-level rise in areas already threatened (such as in low-lying communities of southern Puget Sound) will need to be addressed.

Alaska, one of the states with the lowest population density, is an area of important wildlife resources. It is also a large state, with an area nearly a fifth of the size of the entire southern 48 states. It contains 75% (by area) of US National Parks, 90% of wildlife refugia and 63% of wetlands. It also has more volcanoes and glaciers than the other states combined. Alaska has already warmed considerably: by 2°C between 1950 and 2000. This considerable warming, far more than the global average, is likely to continue and is part of a trend seen over much of the Arctic. Permafrost thawing and sea-ice melting are likely to be at the heart of many of the climate change problems and Arctic summer sea ice is unlikely to exist by the end of this century. Meanwhile, the precipitation/warming duality occurring elsewhere in the USA is also likely to apply in Alaska, with soils (without permafrost) becoming easier to drain. Together this adds up to considerable environmental change.

The changes to the Alaskan soils, particularly those of tundra, have potentially extreme carbon cycle implications (see Chapter 1). They are also likely to have profound ecological consequences regarding their flora; hence there are probable nutritional consequences for caribou (*Rangifer arcticus*) and its more domesticated counterpart, the reindeer/caribou (*Rangifer tarandus*). Thawing permafrost is likely to undermine the stability of Alaskan oil pipelines. The loss of sea ice will have a dramatic effect on polar bears (*U. maritimus*), which rely on ice rafts from which they hunt (see the previous section). The loss of protective sea ice combined with sea-level rise is likely to result in increased erosion, with consequences for some Alaskan settlements.

The south-east USA is already vulnerable to weather, let alone climate change. During the last century's final two decades the south-east saw half of the nation's costliest weather-related disasters, mostly floods and hurricanes. These had cost some US$85 billion. Over the past century precipitation has increased and this is forecast to continue. Then there is sea-level rise. The USGCRP anticipates increased flooding in low-lying counties from the Carolinas to Texas and this is likely to increase the area of the aforementioned ghost forests, and have impacts on human health and property. Because coastal flooding concerns in the region were already an issue (for example, with Hurricane Betsy in 1965), some protection proposals have already been made. In 1998, prior to the 2001 USGCRP report, a 30-year restoration plan was proposed, called *Coast 2050*. However, to date (2012) there has been political stagnation on its implementation due to Congressional disagreement about who should pay the estimated bill of $14 billion.

One proposal has been to restore the wetlands around New Orleans, one of the biggest property flood risks in the south-east USA, due to both sea-level rise and increased precipitation, as a result of being at the head of one of North America's largest rivers. New Orleans lies in a basin, much of which is below sea level, and which at its lowest is 6 m below sea level. Past water management is in no small part to blame. Since 1930 much of the lower Mississippi has had extensive embankments to control flooding. These prevent water and sediment from reaching the delta, which is steadily eroding. Not surprisingly the 2009 USGCRP report emphasised these concerns.

Events subsequent to 2001 make it worthwhile to look at this last USGCRP conclusion more closely. It was thought that New Orleans could survive surges caused by a category-3 hurricane, but not by any higher categories. Yet on 16 September 2004 Hurricane Ivan (a category-5 hurricane with winds in excess of 240 km/h) hit the coast some 160 km to the east of New Orleans. Despite this narrow escape a serious flood incident appeared to be just a question of time. Then, on the 29 August 2005, the category-5 Hurricane Katrina struck. (Bearing in mind the possible relationship between hurricanes and sea temperatures discussed above it is interesting – but not scientific – to note that Gulf of Mexico sea temperatures at the time were, for August, the highest ever recorded.) Aside from the immediate wind damage caused by Katrina, sea surges flooded much nearby coastal property. The levy defences around New Orleans itself were breached so that most of the city was flooded: the flood area exceeded 400 km^2. The city effectively shut down and nearly all its inhabitants either left or subsequently were evacuated. President George W. Bush declared a state of emergency for Louisiana and surrounding states. The economic losses associated with Katrina were in excess of $125 billion. Many would have wanted the authorities to heed the warnings given years in advance in reports such as those from the USGCRP (see also 8.5.1).

The USGCRP's 2001 and 2009 reports also looked at commercial and infrastructure sectors, and, although there is not the space here to review all of them, the section on agriculture deserves special attention, as the USA is a major supplier of food globally. In 2000 US products accounted for more than 25% of the global trade in wheat, maize, soya beans and cotton. Furthermore, cropland occupied some 17% of the total US land area and in addition pastures and grazing lands accounted for a further 26% of land area. Economically, agricultural commodities annually exceeded $165 billion in 1999 at the farm gate and over $500 billion (10% of US GDP) after processing and marketing. By the time of the 2009 report the gross value of the US agricultural sector was in excess of $200 billion.

Historically, US agricultural productivity has improved by roughly 1% per annum since 1950. This success could cause an economic problem for the sector, as profit margins are tight with little scope for further efficiency gains. The 2001 USGCRP report focused primarily on plant crops and not animal husbandry, even though (as the report itself acknowledged) climate change will also effect this dimension of food production.

The USGCRP reports note that region-by-region climate change will have both positive and negative effects on different types of agricultural activity. For instance, yields in the northern midwest are currently limited by temperature, so agricultural production, it is thought, would increase with warming. Indeed, the anticipated longer thermal growing season (TGS) would allow double cropping. The carbon dioxide fertilisation effect would also be an additional positive influence on yields. However, these would be balanced by yield decline in the southern part of the region by as much as 10–20% due to a lack of sufficient water that would be needed in the warmer climate.

Much of the nation's meat, wheat and fibre is produced on the Great Plains, including 60% of the country's wheat, 87% of sorghum and 36% of its cotton. In the 20th century temperatures across the northern and central Great Plains increased by more than 1°C and by up to 2°C in parts of Montana and North and South Dakota. The eastern portion of the Great Plains has seen precipitation increased by more than 10% and Nebraska and Texas both had higher levels of rainfall at the century's end compared to its beginning. Climate models predict that in the 21st century temperatures would rise throughout the region, with the largest rise in the western parts of the Great Plains. Precipitation generally is likely to continue to increase, except in the lee of the Rocky Mountains. Furthermore, the effects of temperature increases on potential evapotranspiration (the theoretical maximum level of evaporation and plant transpiration with unlimited water and optimal nutrients) are likely to outstrip the extra supply of water received through increased precipitation. The result is likely to be a net decline in soil moisture from levels in 2000, especially in Louisiana and Mississippi. Maintaining water supplies through supply management throughout the year and ensuring water quality are likely to become key issues for the region, especially in the south.

The USGCRP 2001 report (relying, as it did, on the sometimes contradictory Hadley and Canadian models of the late 1990s) did not raise concerns as to lack of water in some parts of the west. Instead, it concluded that more atmospheric carbon dioxide, higher temperatures and increased precipitation are likely to increase crop productivity and extend the TGS. However, it did warn that the patterns of potential crop productivity are likely to shift. It notes that fruit and nuts make up 32% of the west's crop value (with a third of that from grapes alone) and that changing the principal locations of such crop production may well be problematic, bearing in mind that such crop plants can take decades to become established. Here, if anywhere, there is an urgent need for accurate local forecasting of likely climate change. This changed with the 2009 report concluding that drought frequency and severity are projected to increase in the future over much of the USA, particularly under higher-emissions scenarios. Increased drought will be occurring at a time when crop water requirements also are increasing due to rising temperatures. Water deficits are detrimental for all crops.

The USGCRP's report in 2001 concludes that without adaptation many crops will grow better by the end of the 21st century, especially cotton and grapefruit (but not oranges), and pastures will improve. However, the 2009 report mentions that water shortage, especially in the summer, is a cause of great concern. Barley and oats are unlikely to improve or will fare badly under future conditions. Potato and orange production could also see some experience decline. *If* adaptation is made then crops could see a 10–100% increase in productivity. Such adaptations importantly will necessitate changing which crop grows where and changes in water management. This last will often involve interstate cooperation and considerable cost, whereas a minority of farmers will have to accept abandoning their land.

Adaptations needed, and again echoed in the 2009 report, are likely to include the following:

- changing sowing dates and other seasonal activities in response to phenological changes;
- introducing new cultivars (crop varieties), including those created through biotechnological methods as opposed to by conventional breeding;
- improved water-supply management and water use;
- changes in tillage practice to retain organic matter in soils;
- increase use of short-term (6-month) forecasts relating to key climatic cycles, such as El Niño events;
- tweaking of management systems from plough to gate, and even plough to market.

The USGCRP's reports also have a sector report on the related topic of forestry. Many of the concerns focus on the Pacific north west, which is likely to suffer from changes in water availability throughout the year. The Cascades coniferous forest, in the west of the Pacific north west, covers about 80% of the land and, more importantly, this accounts for about half the planet's temperate rainforest. Because of the length of the forest crop cycle (decades rather than within a year for most agriculture) the need is to plant species adapted to the forecasted climate rather than the climate as it is now. Careful management is also required to reduce water stress and the risk of forest fires. The net economic effect of the forecast climate change is thought to be marginal but minor economic improvement is anticipated, with most of the gains relating to hardwood as opposed to softwood. However, this does necessitate the adaptations being undertaken in a timely manner and so climate-driven agricultural change is unavoidable.

The other important sector the USGCRP's reports covered was water. As seen from the above summary of the report's key points, water in the USA was considered a key dimension to the biological impacts of climate change on both natural and managed systems. The report notes that population in the USA increased by nearly two-thirds (well above 60%) in the latter half of the 20th century. However, whereas water consumption did increase between 1950 and 1979 by around 60%, it stabilised thereafter to the century's end, at around 1.5 billion m^3 day^{-1}. Farming today, at 85%, dominates water use, with 82% being used for crops and 3% for livestock (see Figure 6.5). This highlights the importance of water management in food production. However, water management throughout the year will become harder with warming in those areas relying on seasonal mountains snow melt. With the amount of snow due to decline as snow lines rise, and there being warmer years, snow melts will occur earlier in the year and summer melt-fed streams will be drier. Models suggest that the Southern Rocky Mountain, Sierra Nevada and central Rocky Mountain snowpack volumes are likely to decline by between 50 and 90% by the end of the 21st century. Given this, and that a number of groundwater aquifers (such as the Great Plains Ogallala) are already diminishing due to extraction rates exceeding replenishment rates, water storage combined with increased efficiency of use will be the important management measures required by the main users of water in the USA, the farmers. The 2009 report concludes that the past century is no longer a reasonable guide to the

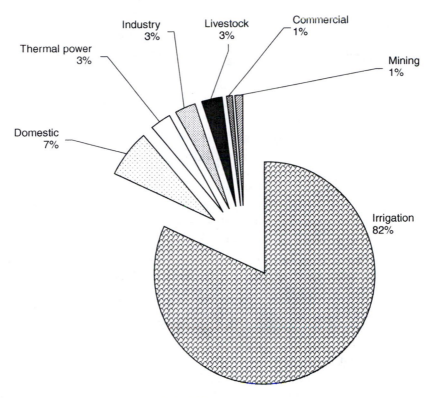

Fig. 6.5 A breakdown of US freshwater consumption in 1995.

future for water management and that climate change will place additional burdens on already stressed water systems.

The 2001 USGCRP report concluded that climate change impacts 'will be significant for Americans', although 'for the nation as a whole the direct economic impacts are likely to be modest'. There were many uncertainties regarding the precise nature and locality of specific impacts, even though themes such as the need for water management were clear.

We will turn to the human ecology needed to underpin climate policy matters briefly in Chapter 7 and policy itself in more depth in Chapter 8. For now it is worth noting that the USGCRP's 2001 report underpinned the 2002 *US Climate Action Report* from the US Department of State, which outlines what the USA must do to address greenhouse (energy-policy and climate change) concerns. But this Department of State report was light on biological specifics and stated that 'one of weakest links in our knowledge is the connection between global and regional projections of climate change'. This is now changing. And with regard to potential impacts the 2001 report said that we have a 'lack of understanding of the sensitivity of many environmental systems and resources – both managed and unmanaged – to climate change'. This too is changing, although we have still much to discern.

Whereas the USGCRP reports effectively represented a stock-take of their respective years of US climate change science understanding as to the likely future climate of the USA, research naturally is still continuing. Of note among more recent contributions was an assessment as to whether the current multi-year drought being experienced in the western USA has any precedent. The research team was led by Edward Cook of the National Oceanic and Atmospheric Administration (NOAA) and their work was published in *Science* in 2004. The drought experienced by the western USA at the end of the 20th century and now is both severe and unprecedented over the period of detailed hydroclimatic measurements, which mainly covers the 20th century. If the early 21st century has its palaeoclimatic analogue back in the medieval warm period (MWP) or medieval climatic anomaly (MCA), to see whether the current drought is a likely response to global warming what is needed is to ascertain conditions during and prior to the MWP, so as to capture all of the MWP episode. Edward Cook and colleagues looked at palaeoclimatic dendrochronological indicators in a geographical grid across the USA and southern Canada. They found that the current drought does appear to be analogous to one that took place in the western half of the USA and western half of southern Canada during the MWP. However, the recent drought up to July 2004 (when the paper was submitted) was not as severe as the one during the MWP between AD 900 and 1300. They concluded that the western US drought is likely to get worse with continued global warming.

Further corroboration of analogous droughts in the region, but much earlier still, comes from Peter Fawcett, Josef Werne, Scott Anderson et al. in 2011 and reviewed by John Williams in the same issue of *Nature*. They noted the above concerns of past mega-droughts and particularly in the southwestern USA. Multi-year droughts during the instrumental period and decadal-length droughts of the past two millennia were shorter and climatically different from the future permanent, 'dust-bowl-like' mega-drought conditions, lasting decades to a century, that some predicted as a consequence of warming. They used molecular palaeotemperature proxies to reconstruct the mean annual temperature in earlier Pleistocene (the rest of the Quaternary prior to our interglacial) sediments from the Valles Caldera, New Mexico, that were laid during previous interglacials. They found that the driest conditions occurred during the warmest phases of interglacials, when the mean annual temperature was comparable to or higher than today's, and that these droughts lasted over a thousand years. It seems that with south-western North America, water variability seems to be the rule rather than the exception during interglacial periods and that during the warm parts of previous interglacials there were extended mega-droughts that lasted far longer than those occurring during human history.

In 2010 the National Research Council – whose members are drawn from the councils of the National Academy of Sciences, the National Academy of Engineering and the Institute of Medicine in the USA – provided their perspective on US climate change and impacts via their Committee on Stabilization Targets for Atmospheric Greenhouse Gas Concentrations in the report *Climate Stabilization Targets: Emissions, Concentrations, and Impacts over Decades to Millennia*. The ground it covered overlapped much of that the USGCRP and IPCC Working Group I assessments already covered (and so will not be reiterated here).

In particular, the report demonstrated that stabilizing atmospheric carbon dioxide concentrations will require deep reductions in the amount of carbon dioxide emitted. Because human carbon dioxide emissions exceed removal rates through natural carbon sinks, keeping emission rates the same will not lead to stabilization of carbon dioxide. Emissions reductions larger than about 80%, relative to whatever peak global emissions rate may be reached, are required to approximately stabilize carbon concentrations for a century or so at any chosen target level. It also concluded that certain levels of warming associated with carbon dioxide emissions could lock the Earth and many future generations of humans into very large impacts; similarly, some targets could avoid such changes. It made clear the importance of our 21st century choices regarding long-term climate stabilization. Among the biological and human impacts it noted that nearly 40% of global maize production occurs in the USA, much of which is exported. It concluded that the most robust studies, based on analysis of thousands of weather station and harvest statistics for rain-fed maize (>80% of US production), suggest a roughly 7% yield loss per °C of local warming. It also warned that changes in climate and carbon dioxide beyond 2100 will probably be sufficient to cause large-scale shifts in natural ecosystems (see section 6.1.5) and that the Initial Eocene or Palaeocene–Eocene Thermal Maximum (IETM or PETM) provides one of the most worrying indicators of Earth system sensitivity and is the closest analogue to what might happen to the carbon cycle in a climate substantially warmer than the present (see section 3.3.9). The report noted that that the Earth system is subject to feedbacks amplifying polar warming, which are not adequately represented in first-decade 21st-century computer models. It also considered very long-term (in the human sense) consequences, and considered that over the coming millennia some impacts of climate change may settle into new patterns of climate variability with the successful implementation of stabilization policies that cap cumulative emissions and therefore limit increases in global mean temperature. They thought that it is possible that societies could become accustomed to these new environments. Even so, that future world would be different from today, but new conditions could become routine to people living on Earth one or two thousand years from now. Other impacts, though, could continue for many centuries past the date of temperature stabilization some time following our ceasing to add greenhouse gas to the atmosphere.

So, what will the North American climate be like towards the end of the 21st century? Readers are urged to seek out the climate change forecasts for the USA in the USGCRP 2009 report and compare them with the IPCC's 2007 forecasts (and, when published, the 2013 IPCC forecasts), not to mention the National Research Council's 2010 report. Nonetheless, readers might like at least a broad-brush picture of what may be expected and so the maps in Figure 6.6 very roughly portray some of the salient points of these projections.

Turning to Canada, we have already covered some of the issues affecting natural systems in the course of this section's largely-US focused discussion. However, Canada's high-latitude environment deserves special attention, not withstanding that Arctic amplification (see section 5.3.1) is seeing greater climate change in those latitudes. In November the 2005 the *Arctic Climate Impact Assessment*, or *AICA* (ACIA, 2005), prepared by more than 250 scientists for the eight-nation Arctic Council, affirmed predictions that anthropogenic warming would hit the Arctic earlier and more significantly than other parts of the planet. It reiterated the IPCC (2001b) scientific consensus that human emissions of greenhouse gases are causing much of the warming observed in recent decades. However, the President George Bush

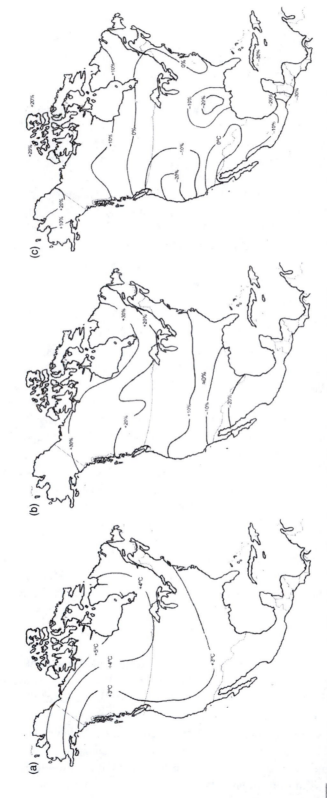

Fig. 6.6 Rough sketch maps of key aspects of North American climate change. (a) Mean annual temperature change for the 21st century under B-a-U scenarios. (a) Mean annual temperature change over the 21st century. Everywhere in North America warms, with warming greater at higher latitudes. This does not mean it will be warmer there, just that the warming is greater there over the 21st century. (b) Changes in winter precipitation. −20%, −10%, +10%, +20% and +30% contours are shown. (c) Changes in summer precipitation. −30%, −20% −10%, 0%, +10 and +20% contours are shown. North of the 0% change zone increased summer precipitation is expected. Again, note that the *change* in precipitation is mapped (not actual precipitation). The south west and parts of the western USA are expected to see much greater drought than at present. Parts of central North America are expected to see net drying compared to the present. Even if precipitation in parts of North America remains the same at the end of the 21st century compared to the present, the warmer end-of-21st-century temperature will result in a dryer environment. Compare these sketch maps with the IPCC 2007 assessment (WG1, chapter 11) and the USGCRP 2009 report (section on national climate change).

Jnr's administration downplayed the second finding, and opposed strong emissions controls. The report had been due out earlier, and there was speculation that Bush had delayed the *ACIA*'s publication until after the US presidential elections. Not surprisingly, the report was not only in line with the IPCC view, but also drew on the IPCC's 2001 assessment's scenario forecasts and actually referred in the text to 'anthropogenic climate change', which at the time the President did not accept was happening.

With regards to ecological impacts, the *ACIA* said that observations had provided evidence of nutritional stresses on many animals that were indicative of a changing environment and changes in food availability. New species, never before recorded in the Arctic, had also been observed and the distribution ranges of some species of birds, fish and mammals now extended further to the north than in the past. These observations were significant for indigenous communities since changes likely to occur in traditional food resources will have both negative and positive impacts on the culture and economy of arctic peoples. The *ACIA* considered some Arctic species, especially those that are adapted to the cold Arctic environment (e.g., mosses, lichens and some herbivores and their predators) especially at risk of loss in general, or displacement from their present locations. Present species diversity is more at risk in some of Arctic regions than others; for example, Beringia has a higher number of threatened plant and animal species than any other *ACIA* region. Overall it concluded that while there will be some losses in many arctic areas, movement of species into the Arctic is likely to cause the overall number of species and their productivity to increase, so overall biodiversity measured as species richness is likely to increase along with major changes at the ecosystem level. However, some high-latitude species were at risk of extinction. For example, as forests expand and in some areas, where present-day tundra occupies a narrow zone, are likely to reach the northern coastline. The expansion, the assessment considered, would be slowed by increased fire frequency, insect outbreaks and vertebrate herbivory, as has already been observed in some parts of the Arctic. With regards to carbon, the assessment concluded that methane fluxes were likely to increase as vegetation grows in tundra ponds, and as wetlands become warmer (until they dry out). Methane fluxes were also likely to increase when permafrost thaws (see section 6.1.5 for carbon loss in peri-Arctic Siberia).

In 2008 Canadian scientists Donald Lemmen, Fiona Warren, Jacinthe Lacroix and Elizabeth Bush produced *From Impacts to Adaptation: Canada in a Changing Climate 2007*. It too drew on the IPCC and noted that climate change was affecting every region of the country. Loss of permafrost was considered one of the more serious natural system issues, in addition to impacts on water supply in some regions such as the Prairies. It was also noted that the warming of the Prairies would mean less harsh winters, which some agricultural systems rely on to remove pests. However, the Canadian Prairie productivity gains through warming were not signalled in the report and this was probably related to concerns about water availability. This was due to the report's focus on adapting to impacts. The report was one of the first comprehensive, and indeed first Canadian governmental, assessments of Canadian climate change impacts.

6.3 Case study: climate and natural systems in the UK

It was the UK Prime Minister, Margaret Thatcher (herself inspired by the UK's former UN ambassador Crispin Tickell), who in the 1980s did much to promulgate climate change concerns among international politicians. Whereas researchers in the USA had done important climate work (for example at the National Center for Atmospheric Research in Boulder, Colorado), it is probably fair to say that US research into likely impacts was not pursued as vigorously as in the UK. Already by the late 1980s (before the first IPCC assessment) UK state-funded science had climate change as a research priority and was communicating research outputs to a broader audience, such as with the Natural Environment Research Council's (NERC's) former Institute of Terrestrial Ecology (ITE) booklets (the ITE subsequently became the NERC's Centre for Ecology and Hydrology, or CEH). The first of these booklets was *Climate Change, Rising Sea Level and the British Coast* (Boorman et al., 1989). This considered the likely impact of a sea-level rise of up to 1.65 m over the next 100 years. It noted that estimates for maintaining sea defences to withstand (as opposed to retreat from) this sea-level rise would cost some GB£5 billion in the money of the day. It cited stretches of the coast and land deemed at risk from sea-level rise (see Figure 6.7). It also considered what would be the general ecological impacts of various response options, from raising present sea walls and building storm-surge barriers to new sea walls constructed landward, as well as abandonment and retreat. Each measure would have an effect on biomes; specifically, changing their size. One of the first significant events testifying to the ITE's concerns came in November 2007 when a storm surge hit the coast of East Anglia. A flood alert was issued for Norfolk, Suffolk and Essex as well as East London downstream of the River Thames flood barrage. The flood risk was deemed to be the worst since 1953 and police personally called on some 7500 households recommending temporary evacuation. Fortunately the defences held, but the Environment Agency (the UK Government agency responsible for flood defence) Chief Executive Barbara Young said that East Anglia had come within a 'whisker' of widespread flooding. With climate change, and its related sea-level rise, such incidents will increase.

Of greatest wildlife concern are the likely effects on salt marshes and mudflats. Here the effects depend on which coast-management option is chosen. However, in addition to whether the UK adopts protection or retreat, there are likely to be transition problems which could greatly restrict a biome before expansion as coastal biotic and abiotic systems adjust to new sea levels. In addition to biome-related flora and invertebrates, bird populations were considered to be at greatest risk. Unfortunately the likely impacts are too numerous (due to many possible management options) for them to be summarised here.

Turning to subsequent research outputs, early in the 21st century there were two major reports on climate change impacts on British natural systems. The first was the 2000 Government departmental report *Climate Change and UK Nature Conservation: a Review of the Impact of Climate Change on UK Species and Habitat Conservation* (or UKCIP, UK Climate Impacts Programme, 2011), published by the

Fig. 6.7   Coast and land in the UK deemed to be at risk from sea-level rise.

then Department of the Environment, Transport and the Regions (DETR) in 2000 as part of the UK Climate Impacts Programme. The second was *Climate Change and Nature Conservation in Britain and Ireland* for the UK Climate Impacts Programme (or UKCIP; UK Climate Impacts Programme, 2001) compiled by leading government departmental agencies and leading UK wildlife charities led by the agency English Nature, again as part of the UKCIP, and also known as MONARCH I (Modelling Natural Resource Responses to Climate Change).

The DETR report based its assessment on the UK's Biodiversity Action Plan (or BAP), which the UK Government developed as part of the nation's commitment to the 1992 UN Conference on Environment and Development (held in Rio de Janeiro) and its Convention on Biological Diversity, but not its Framework Convention on Climate Change (which in the USA prompted that country's *Climate Action Report* [Department of State, 2002]; see the previous case study).

The UK departmental report (DETR, 2000) addresses many of the points regarding species impact covered in this and the preceding chapter. However, it also notes that the UK, being comprised of islands, has restricted population pools and that the sea inhibits species migration (unlike the USA, which has Canada to the north and Mexico to the south). The report also comments on human-imported species, currently residing in parks and gardens, noting that some of these may find future conditions favourable for expansion.

Landscape fragmentation is also an issue highlighted by the report. It noted that some species, which otherwise might benefit from a warmer climate, such as the dormouse (*Muscardinus* spp.), may have their expansion hindered by landscape fragmentation.

The report's consideration as to the vulnerability of UK biomes to climate change is summarised below.

Low vulnerability

Arable and horticulture: the change of species will need to be managed and a management regime for existing species in existing locales established.

Low-to-medium vulnerability

Mixed broad-leaved and yew woodland: damage will occur due to extreme weather events. Climatic effects may become manifest in 100–150 years. There is a particular risk to lowland woodland in southern England due to dry summers.

Fen, marsh and swamp: there will be a change in species composition and a risk of soligenous (communities relying on water that has drained through geological strata) fens drying out in summer. Drought may exacerbate air-pollution impacts. Coastal fens will be affected by sea-level rise and a sea-defence response to sea-level rise will be necessary.

Standing water and canals: isolated high-altitude lakes are vulnerable to temperature increase. High-summer temperatures could restrict cold-water species. There could be changes in the frequency of algal blooms in eutrophic (nutrient-rich) waters. Warmer winters could increase populations of overwintering plankton.

Medium vulnerability

Improved grassland: there may be changes in the thermal growing season (TGS). Risks include an increase in moisture-loving species (such as rushes; *Juncus* spp.) and surface damage to soil and vegetation due to hoof action on wetter ground (called poaching). Coastal grazing marshes will be affected by sea-level rise and changes in sea defences.

Acid grassland: there may be possible changes in competitive ability with other grassland types and the potential for increased bracken invasion in the north (with a possible reduction in the south). Soil-nutrient limitations may restrict adaptations to changes in the TGS.

Dwarf shrub heath: a decrease in summer rainfall may change community types in the south. There may also be an increase in insect damage.

Medium-to-high vulnerability

Coniferous woodland: there may be extreme weather events, encroachment by southerly species and limitation in altitudinal migration due to land-use change and specifically grazing.

Calcareous grassland: well-drained chalk soils increase the likelihood of drought stress in dry summers. Species community change and/or structure alteration are likely.

Bogs: there may be possible peat erosion and vulnerabilities due to changes in the water table. Lowland bogs will be at risk from decreased summer rainfall, whereas upland bogs will face erosion pressures from increased winter rainfall.

High vulnerability

Rivers and streams: low summer and high winter flows, together with high summer temperatures, are likely to have major impacts on species and the river environments.

Montane habitats: species migration is likely but some altitudinal limitations may result in population declines/extinction. Less snow may affect snow buntings (*Plectrophenax nivalis*), which rely on insects in snow patches, and some bryophyte species (a group of small, rootless, non-vascular plants), which rely on snow cover for insulation and/or spring moisture. An increase in average climate of just 0.5°C would eliminate many montane biomes in the UK.

Unknown vulnerability

The effects of climate change on the following biomes are unknown: inland rock, built-up areas and gardens, maritime cliffs and coastal sand dunes.

Finally, the DETR report highlighted the need for climate impacts to be considered proactively within UK conservation policy.

The second UK assessment was the UKCIP (2001) report on *Climate Change and Nature Conservation in Britain and Ireland*, or MONARCH I. It was a suitable follow-on from the 2000 departmental report. It linked established environmental impact models to a national bioclimatic classification. As such, the report took on board climatic changes but not impacts arising from elevated levels of carbon dioxide. (You will recall that increased carbon dioxide levels have an effect separate to that of climate change, even if there may be interactions and synergies between the two.) MONARCH I was a detailed report and its overall conclusions were that a more proactive and flexible management of ecological sites is required. It recognised that many existing protected sites were not in the right place to fulfil current protection goals in the future but that such sites will continue to be important for maintaining certain species populations.

Virtually echoing the US GCRP report discussed in the previous section, the MONARCH I report's authors also pointed to the difficulty in developing response strategies when the regional and local nature of future climate change was so uncertain. They noted that, up to the report's publication, the only detailed monitoring of biological climate indicators had been those commissioned by the UK Government's environment ministry, the then DETR. There the focus had been change across a wide range of sectors and not just natural ecosystems. However, MONARCH I also made research recommendations, including the following:

- refining models predicting change and the use of field studies for model validation;
- applying the MONARCH methodology to different scales;
- improving understanding of ecosystem dynamics;
- identifying possible natural adaptations to climate change.

The MONARCH report noted that for wildlife conservationists to address climate change their response must lie somewhere between two extremes. At one end there is a 'King Canute' (*sic*)-type approach, resisting change and trying to maintain existing natural communities. (An historical analogy that does not bear too close a scrutiny as Canute 'attempted' to hold back the tide to show that even he, a king, was not impervious to natural order and not, contrary to popular belief, in a bid to demonstrate his regal power.) At the other end of the spectrum is 'wildlife gardening', with species translocation and habitat-type creation in suitable areas, or areas due to become suitable through change climate.

The MONARCH authors also drew attention to the plight of the natterjack toad (*Bufo calamita*) as an example of special conservation interest, as its habitat area will become reduced before climate change creates new areas for it. The authors therefore recommended consideration to purchasing likely areas to facilitate such species translocation, and even provide 'holding bays'.

A second report, MONARCH II, was published in 2005 (Berry et al., 2005). It took the MONARCH I climate space approach but downscaled it for use at the local level. A dispersal and ecosystem function model was also developed to run alongside the climate space models. MONARCH II focused on methodological development and testing in four case-study areas. A third study, MONARCH III (Walmsley et al., 2007), began in April 2004 and was published in May 2007. It refined and applied the output from MONARCH II in the context of the UK Biodiversity Action Plan. (Biodiversity Action Plans themselves were a result of the aforementioned 1992 international Biodiversity Convention – see Chapter 8 – and so relate MONARCH III to Britain's UN obligations.)

MONARCH III's species range map scenario forecasts up to the 2080s did not attempt to simulate the future distribution of species in response to climate change (it left forecast scenarios to UKCIP). However, in presenting likely shifts they did broadly show where there could be suitable climate space. Overall they forecast a northward shift in suitable climate space for many species, some of which have the potential to extend their range within Britain and Ireland as northern limits shift farther than their southern limits. Fifteen species were projected to gain substantial potential climate space, with no significant UK range loss. They were: the birds stone curlew (*Burhinus oedicnemus*), corn bunting (*Emberiza calandra*) and turtle dove (*Streptopelia turtur*); the butterflies pearl-bordered fritillary (*Boloria euphrosyne*), marsh fritillary (*Euphydryas aurinia*), silverspotted skipper (*Epargyreus clarus*), heath fritillary (*Melitaea athalia*) and Adonis blue (*Polyommatus bellargus*); the mammals greater horseshoe bat (*Rhinolophus ferrumequinum*) and lesser horseshoe bat (*Rhinolophus hipposideros*); the plants stinking hawk's-beard (*Crepis foetida*), red-tipped cudweed (*Filago lutescens*), broad-leaved cudweed (*Filago pyramidata*), red hemp-nettle (*Galeopsis angustifolia*) and small-flowered catchfly (*Silene gallica*). Eight were projected to lose significant UK range due to climate change with no significant gains. They were: the birds skylark (*Alauda arvensis*), common scoter (*Melanitta nigra*), black grouse (*Tetrao tetrix*), capercaillie (*Tetrao urogallus*) and song thrush (*Turdus philomelos*); and the plants Norwegian mugwort (*Artemisia norvegica*), twinflower (*Linnaea borealis*) and oblong woodsia (*Woodsia ilvensis*). Three indicated no significant gain or loss of climate space, the tree sparrow (*Passer montanus*), linnet (*Carduelis cannabina*) and shepherd's needle (*Scandix pecten-veneris*). Six were projected to both gain and lose potential climate space, resulting in a northward shift. These included a beetle, the stag beetle (*Lucanus cervus*), a mammal, the Barbastelle bat (*Barbastella barbastellus*), and the plants tower mustard (*Arabis glabra*) and cornflower (*Centaurea cyanus*).

MONARCH outputs were valuable as broad signposts to help develop policies for nature conservation in a changing climate. The broad adaptation measures MONARCH III recommended conservation managers should adopt included conserving and restoring the existing biodiversity resource; reducing other sources of harm such as pollution and inappropriate habitat management; and developing ecologically resilient landscapes through reducing habitat fragmentation.

As in the USA, the British series of climate reports tended to leapfrog each other. This was particularly clear with the MONARCH reports as effectively this was an evolution of specific policy concerns; those of wildlife conservation and climate change. Understanding in both areas evolved in parallel so that any reference from one to the other tended to become a little dated in any given report. (This is an almost inevitable problem common to a number of areas of cutting-edge science.) Just as the USGCRP recognises the limitations of computer models, hence the need for their development and to use the latest available, so MONARCH used the most recent models available in each of its successive reports. The need to do this is aptly demonstrated when looking at the anticipated change in precipitation: as with the USA, water issues are of concern when considering likely climate change. The climate model used in MONARCH I suggested that by the middle of the century the south of England (particularly the central south) would become drier in the summer compared to the 1961–90 average, by up to 110 mm, and the north west of Scotland slightly wetter, by 60 mm. Conversely, in winter Scotland and western England down to and including Wales would get wetter by up to 60 mm whereas the south of England

and East Anglia would only get wetter by 40 mm. This was based on the UKCIP model UKCIP98. Compare this to the UKCIP02 model forecasts used in 2002, which predicted the opposite for winter 2050, that the south east of the UK would have more rain and the north west of Scotland would remain within the natural variability of 1961–90. It is because of such vagaries that successive MONARCH reports, and other UK policy reports, did not build on the results of previous models but built on policy methodologies and understanding, while starting afresh each time using the latest computer climate models.

In 2002 the UKCIP updated its first (1998) climate change scenarios for the UK (Hulme et al., 2002). That in Britain (more so than in the USA) climate change issues were of considerable political concern, both by the Government and the opposition parties, is indicated by the Secretary of State Margaret Beckett personally endorsing the report with an introduction. UKCIP02 took four carbon-emission scenarios (low, medium-low, medium-high and high), which were based on the IPCC Special Report on Emission Scenarios (or SRESs) (IPCC, 2000). These in turn were related to change from a 1961–90 baseline to 2011–40 (labelled the 2020s), 2041–70 (the 2050s) and 2071–2100 (the 2080s).

It concluded that by the 2080s:

- temperatures across the UK may rise by between 2 (low-emission scenario) and 3.5°C (high-emission scenario), with the greatest warming being in the south and east and greater warming in the summer than winter (parts of the south east may be up to 5°C warmer in the summer);
- high summer temperatures will become increasingly frequent, and cold winters rare;
- winters will become wetter and summers drier. Overall the UK will become drier as the winters are expected to become only 30% wetter while the summers may have only 50% of the precipitation;
- days of heavy rain are likely to increase in frequency;
- the sea level will rise between 26 and 86 cm in the south east of England but, due to geological (isostatic) rise, the sea level may fall by 2 cm (low-emissions scenario) or rise by 58 cm (high-emissions scenario) in western Scotland;
- the Gulf Stream is unlikely to weaken its warming of the UK, but there is uncertainty due to lack of detailed knowledge of interactions between ocean circulation and climate.

(Note: the above UKCIP 2002 scenario forecasts were changed slightly in 2009, as we shall shortly see.)

In 2007 UKCIP produced the first of the UKCIP08 reports, *The Climate of the United Kingdom and Recent Trends* report (Jenkins et al., 2007). It looked at historic trends and showed that in the 19th century through to the early years of the 21st century there had been clear warming and changes in precipitation. It noted that Central England Temperature (CET) had risen by about a degree Celsius since the 1970s, with 2006 being the warmest year on record to date. Sea-surface temperatures around the UK coast had risen over the past three decades to 2007 by about 0.7°C. It noted that regions of the UK had experienced an increase over the past 45 years in the contribution to winter rainfall from heavy precipitation events; in summer all regions except north-east England and northern Scotland showed decreases. It concluded that it is likely that there has been a significant influence from human activity on the recent warming.

The 2009 UKCIP report was branded as UKCP09 (not UKCIP09 as had been expected prior to 2008) and had the full title of *UK Climate Projections* (Jenkins et al., 2009). Its scenario forecasts were only slightly different to those given by UKCIP02. (The model resolution for UKCIP02 was 50 km² whereas for UKCP09 it was 25 km².) It gave a range of probabilities: 10, 50 and 90%, with the middle 50% being the most likely (the mean values) and the 10 and 90% probabilities being the high and low extremes (the two tails to the binomial bell-shaped probability distribution) that were thought possible but only with low probability. The principal scenario forecast differences between UKCIP02 and UKCP09 were as follows.

- In the case of mean temperature, projected changes in UKCP09 for were generally somewhat greater than those in UKCIP02.
- In the case of the summer reduction in rainfall, the UKCP09 prediction was not as great as that projected in UKCIP02. UKCP09 had UK annual rainfall in the 2080s largely unchanged compared to the beginning of the century but with central England and most of East Anglia receiving slightly less rain and the rest of the UK slightly more rain across the year.
- The overall outcome under the medium emissions scenario (equivalent to IPCC scenario A1B) was that all areas of the UK will become warmer, more so in summer than in winter. Changes in summer mean temperatures are greatest in parts of southern England (up to 4.2°C; 2.2–6.8°C) and least in the Scottish islands (just over 2.5°C; 1.2–4.1°C).

The key scenario forecast points are outlined below.

- Mean daily maximum temperatures will increase everywhere. The summer average will increase by 5.4°C (2.2–9.5°C) in parts of southern England and 2.8°C (1–5°C) in parts of northern Britain. Increases in winter will be 1.5°C (0.7–2.7°C) to 2.5°C (1.3–4.4°C) across the country.
- Changes in the warmest day of summer will range from +2.4°C (−2.4 to +6.8°C) to +4.8°C (+0.2 to +12.3°C), depending on location, but with no simple geographical pattern.
- The biggest changes in precipitation in winter, increases up to 33% (9−70%), will be seen along the western side of the UK. Decreases of a few per cent (−11 to +7%) will be seen over parts of the Scottish highlands.
- The biggest changes in precipitation in summer, down to about −40% (−65 to −6%), will be seen in parts of the far south of England. Changes close to zero (−8 to +10%) will be seen over parts of northern Scotland.

However, we must bear in mind that water demand in the summer is higher than that in the winter and that the bulk of the UK's population lives in the southern half of England and Wales, and that the above refers to future *change* whereas *currently* the east and south east of England have lower rainfall that the west and north of Great Britain. So water shortages in the south east of England are a concern. For a summary overview of the temperature and precipitation changes anticipated by the 2009 report see Figure 6.8.

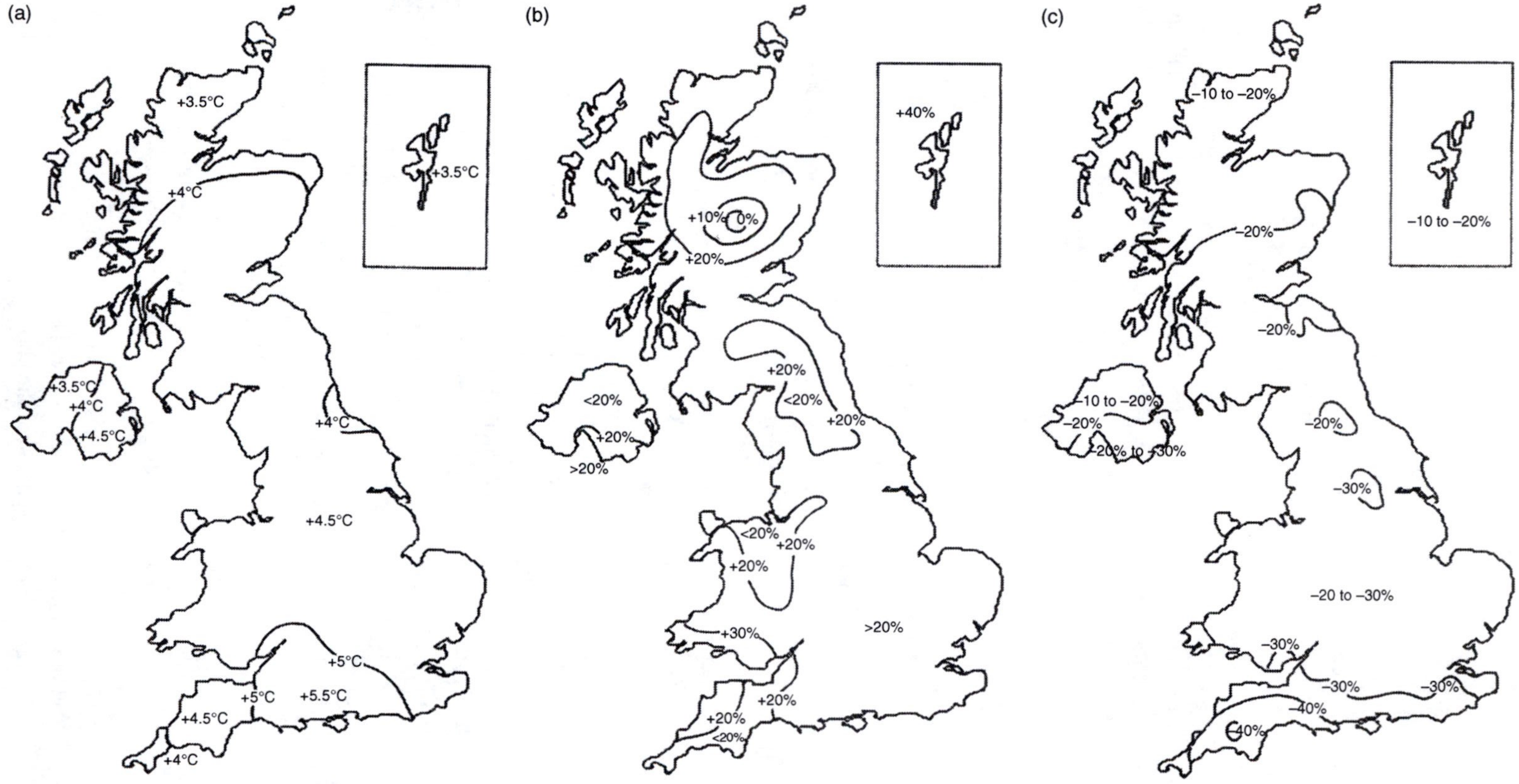

Fig. 6.8 Forecasted key aspects of climate change for the British Isles up to the 2080s. (a) Summer mean daily land temperatures under the high-emissions scenario to the nearest half degree Celsius. (b) Winter precipitation changes under the high-emissions scenario. (c) Summer precipitation under high-emissions scenario.

The UKCIP and MONARCH reports are being used to underpin national policy. In the UK, as in the USA, government-funded research is supported both by scientific agencies (such as NERC, the semi-autonomous agency that distributes public money for environmental research) and governmental departments such as DETR and now its successor the Department for Environment, Food and Rural Affairs (DEFRA). The former DETR also had agencies to undertake environmental monitoring and protection work and, of these, English Nature (which was integrated in 2006 with the Countryside Agency) was responsible for wildlife and the Environment Agency was responsible for waterways and coastal protection. UKCIP scenarios have alerted the Environment Agency to the problem of winter flooding due to excessive rain bursts becoming a chronic condition (see also Chapter 7), although flooding in the late 1990s and at the beginning of the 21st century effectively ensured that the agency would develop a long-term strategy. Some of the areas at specific risk are shown on Figure 6.9. However, there are other impacts on society. As already seen in Chapter 5, there have been past climate change impacts on human activities and Chapter 7, on human ecology, will look at likely future impacts.

One of the indicative ways in which climate change will affect UK natural systems was discussed in 2005 at a symposium at Sussex University run jointly by the Royal Horticultural Society, Forest Research and UKCIP, entitled 'Trees in a Changing Climate'. There are some 4 billion trees in the UK. In summary, it was noted that trees would thrive in the north and west of the UK, where mainly there is little water limitation. Conversely, as the south and east of the UK is likely to become drier, it is likely that some species will suffer. Sycamore (*Acer pseudoplatanus*) is one species at risk and on well-drained soils birch (*Betula* spp.) in particular will suffer. Finally, beech trees (*Fagus sylvatica*) generally do not fare well in hot summers.

In a hypothetical warming Britain without humans, the mix of species in woodlands would change, with tree species moving to the north and west. Furthermore, woodland understorey species would also change and so with them the mix of invertebrate systems. However, as previously noted, in the real world many parts of developed and even less-developed nations have an increasingly fragmented landscape that does not facilitate species migration. Britain is no exception. Biologists have for many years argued that biological conservation should not solely be carried out on a site-by-site basis, laudable as such initiatives are. What is needed is active management at the landscape level. Present-day concerns about climate change underscore such needs. Even so, one confounding factor may well be the speed and magnitude with which anthropogenic climate change is taking place. This, as has been noted, is greater than anything seen for the past 10 000 years of the Holocene and is likely, over the coming two centuries, to cause a return to a warmer world that has not existed for tens of millions of years.

Shifts in species ranges are not just confined to well-known birds, animals and butterflies, although naturally these form the basis of many studies. Many other invertebrates are also affected. In 2006 a telling study of many UK species was published (Hickling et al., 2006), demonstrating that current climate change is impacting on many classes and families of species. It looked at dragonflies and damselflies (Odonata), grasshoppers and related species (Orthoptera), lacewings (Neuroptera), butterflies (Rhopalocera), spiders (Arachnida), herptiles (Amphibia and Squamata),

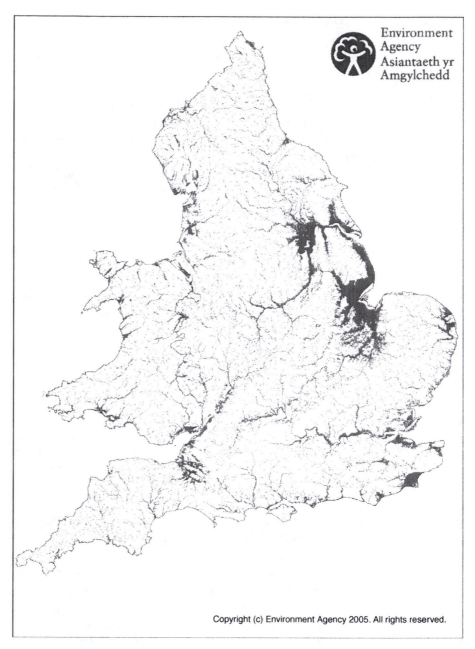

Fig. 6.9   Principal areas of England and Wales at risk from river flooding as a result of an extreme weather event with a 1% chance of an event happening in any given year, and from the sea with a 0.5% chance in any year, if there were no flood defences. These data are for 2005 and do not include the effects of future climate change. Compare this with Figure 6.7, which shows just the principal areas prone to flooding. The value of this map is to illustrate that there are many small areas at risk of flooding in extreme events which, in a densely populated island, poses considerable economic risk. © The Environment Agency, reproduced with permission.

freshwater fish (Teleostei), woodlice (Isopoda), ground beetles (Carabidae), harvest-men (Opiliones), millipedes (Diplopoda), longhorn beetles (Cerambycidae), soldier beetles and related species (Cantharoidea and Buprestoidea), aquatic bugs (Heterop-tera) as well as birds (Aves) and mammals (Mammalia). For each group, species were only included in the analyses if they were southern/low-elevation species, as these species would be expected to increase their range sizes, move northwards and/or shift to higher elevations if they were responding solely to temperature. Northern species were excluded from the analyses because of a lack of data, with the exception of birds. The results revealed that out of a total of 329 species analysed across 16 taxa, 275 species shifted northwards, 52 species shifted southwards and two species' range margins did not move, with an average northwards shift across all species of 31–60 km. Similarly, with regard to altitudinal shifts, 227 species moved to higher altitudes and 102 species shifted to a lower altitude, resulting in a mean increase in altitude of 25 m overall. Work such as this fed into MONARCH.

In addition to biological change, change in the soil carbon in England and Wales has also been detected. Indeed, in 2005 the results of a detailed DEFRA-funded survey was published (DEFRA was formerly the DETR, and before that it was the Ministry for Agriculture, Fisheries and Food, or MAFF). The survey was conducted by Pat Bellamy and colleagues and covered the 25-year period from 1978 to 2003. Its importance was that up to then evidence from temperate climates had virtually entirely come from small-scale field experiments, laboratory studies and computer modelling. However, the Bellamy team looked at samples of 15 cm depth from 5662 sites across England and Wales that had originally been taken as part of the National Soil Inventory. They then revisited sites 12 or 25 years later. This was the only soil inventory to have included resampling on such a scale anywhere on Earth. In addition to soil analyses they also examined land use at each site. They found that carbon loss from soils across England and Wales over the survey period took place at a mean rate of 0.6% a year. They also found that the relative rate of carbon loss increased with a soil's carbon content and was more than 2% a year for soils with greater than 100 g of C per kg of soil. Their findings, which relate to ecological carbon leakage (as opposed to economic carbon leakage, see sections 8.3.1 and 8.3.3), suggested that the losses of soil carbon in England and Wales were irrespective of land use, which suggests a link to climate change. However, they could not say where the carbon had gone but likely destinations are a combination of transfer to the atmosphere as carbon dioxide and leeching deeper into the ground. Nonetheless, considering the area of temperate terrestrial ecosystems internationally, the implications for changes in carbon cycling are significant.

Putting this top-soil carbon loss into the context of Britain's anthropogenic emissions of carbon dioxide, the carbon loss from England and Wales is around 13 million t year^{-1}. This is equivalent to about 8% of UK carbon dioxide emissions (which includes those from Northern Ireland and Scotland, the soils of which were not included in the survey). It is also more than equivalent to the past reductions in UK carbon dioxide emissions between 1990 and 2002 (12.7 million t of C year^{-1}).

In 2010 DEFRA sought an appraisal of England's wildlife sites specifically with regard to climate change, and this resulted in an assessment called *Making Space for Nature: A Review of England's Wildlife Sites and Ecological Network* (Lawton et al., 2010). The problem Britain has compared to North America is that it has a far higher population density. England and Wales have a greater population density than Scotland, and the most populous area is south-east England. This means that the competition between humans and wildlife is a major ecological issue. Wildlife in many parts of the UK, and especially in the areas of high human population, is restricted to small pockets, or islands of semi-natural (ecological and amenity-managed) environments, as well as applied ecological environments, primarily agricultural land. When DEFRA Secretary of State, Hilary Benn, briefed John Lawton to chair the review, Benn said:

> With the effects of climate change and other pressures on our land, now is the time to see how we can enhance ecological England further. Linking together areas to make ecological corridors and a connected network, could have real benefits in allowing nature to thrive. (Lawton et al., 2010)

The report concluded that it was not all bad news. Targeted conservation efforts have improved the fate of many British species and extensive new areas of habitat have been recreated. In other words, given resources, determination and skill, British ecologists know what to do, and how to do it. The report argued that we need a step-change in the approach to British wildlife conservation, from trying to hang on to what we have, to one of large-scale habitat restoration and recreation as part of a UK-wide ecological network, underpinned by the re-establishment of ecological processes and ecosystem services, for the benefits of both people and wildlife. This addresses the problem that our nature reserves are in the wrong place, or rather have the wrong species given inevitable climate change. The report also challenged the practical conservation value (as distinct from their theoretical scientific value) of Sites of Special Scientific Interest (SSSIs): perhaps this should not have come as a surprise because SSSIs were not designated with this aim in mind. Another problem was that many wildlife sites needed to be larger, and that more connecting corridors and/or stepping-stone sites were required with, preferably, surrounding buffering zones. Finally, the report emphasised that it only provided a strategic vision and not a tactical plan as to how to realise it. The ball was very much back in the Government's court. However, in 2010 Britain was in the middle of an economic slump and governmental cutbacks for the following years were expected.

6.4 Case study: climate and natural systems in Australasia

Unlike the previous two case-study regions, Australasia is located at the edge of the El Niño Southern Oscillation (ENSO) migration of warm water. The ENSO is the largest of the climate cycles and its effects are near global. Agriculture is at least four times more important to New Zealand's and Australia's economy than it is to the UK or the USA: Australia has 3.9% of GDP coming from agriculture and New Zealand has 4.7% of GDP (data from 2010). Consequently climate variability, and hence the impact of extreme events, is arguably of greater economic concern.

Eighty per cent of Australia's agricultural output by value comes from, and 70% of its irrigated cropland and pasture is found in, the Murray-Darling basin that covers much of New South Wales and Victoria. Much of New Zealand's agriculture is located on or around its South Island Canterbury Plain. Both these areas are increasingly seeing climate variability (specifically with respect to water supply), which is of central concern regarding agricultural productivity. The impact of such variability can be considerable. In New Zealand in a bad year agricultural output can decline by over a fifth and knock the best part of a percentage point off the nation's annual GDP. Bearing in mind that if a developed nation's GDP grows by 1–2% then its economy is considered to be doing well, losing a percentage point is a major sociopolitical concern.

The significance of climate variability and extreme events came home to Australians in the State of Victoria on 'Black Saturday', 7th February 2010, and the subsequent month, when there was a record-breaking heatwave that followed 2 months of drought. Fires destroyed 2030 domestic dwellings and over a thousand other types of building, and displaced more than 7500 people. Ecologically, other than the fires, the most dramatic event was the grey-headed flying fox (*Pteropus poliocephalus*) 'mass-death event'. This flying fox is a large fruit bat whose range in Australia has already begun to move poleward. Being a flying creature, they have an active metabolism which means that they often need to lose excess heat to their surroundings. However, if the surrounding environment is too warm then the first law of thermodynamics comes to the fore and the bat is unable to lose heat: 43°C is the critical temperature. As the heatwave progressed the bats first unfurled their wings to maximise their surface area and therefore heat loss. When this did not work they flew, with air passing over their wings to cool them. However, when it is too hot even this strategy will not work. On Black Saturday many flying foxes could be seen falling out of the sky, dying from heat exhaustion. One of Victoria's largest grey-headed flying fox colonies, with a population of around 10 000, saw 20% lost in the mass-death event. Prior to the 2010 heatwave, between 1994 and 2007, more than 24 500 grey-headed flying foxes have been recorded as dying from extreme heat events. One of the largest losses of foxes prior to 2010 was on 12 January 2002 when New South Wales temperatures exceeded 42°C, which resulted in the loss of 3500 individuals in nine mixed-species colonies (Welbergen et al., 2008). To put these losses into context, the species population was thought to be around 300 000 at the turn of the millennium and it is listed as vulnerable by the International Union for the Conservation of Nature (IUCN) Red List.

The IPCC (2007a) anticipate that all of Australia and New Zealand is likely to get warmer this century: somewhat more so than the surrounding oceans, but comparable to the overall mean global warming. The concern is with climate variability and water deficit. It is thought that precipitation over Australia will not change that much over the 21st century but, factoring in warmth, moisture balance deficits are expected. Given all of the above the very low stream flow in the River Murray, in Victoria, Australia, for the 1998–2008 period is very rare – about a 1-in-1500 year event (Climate Commission, 2011) – and of some concern. Equally of concern for New Zealand agriculture is the anticipated increase in temperature and rainfall in the eastern half of the South Island over the 21st century (IPCC, 2007b).

The flora and fauna of Australia and New Zealand have a high degree of endemism (80–100% in many taxa). Many species are at risk from rapid climate change because they are restricted in geographical and climatic range. Most species are well-adapted to short-term climate variability, but not to longer-term shifts in mean climate and increased frequency or intensity of extreme events. Many reserved areas are small and isolated, particularly in the New Zealand lowlands and in the agricultural areas of Australia (IPCC, 2007b). Major changes are expected in all vegetation communities. In the Australian rangelands (75% of total continental land area) shifts in rainfall patterns are likely to favour establishment of woody vegetation and encroachment of unpalatable woody shrubs.

The following are examples of key climate-induced ecological impacts likely in the 21st century:

- bleaching of the Great Barrier Reef,
- decrease in Australian wet tropical habitat,
- risk of 200–300 indigenous New Zealand alpine plant species becoming extinct,
- montane tropical rainforest impacts in northern Australia.

Then there are the species migration and phenological issues, which I have covered previously in reference to other parts of the world. An example of the latter from Australasia, one of a number, is that of a shift in the emergence date of the common brown butterfly (*Heteronympha merope*) in response to regional warming in the south-east Australian city of Melbourne. The mean emergence date for the common brown has shifted 21.5 days decade^{-1} over a 65-year period with a concurrent increase in local air temperatures of approximately 0.16°C decade^{-1} (Kearney et al., 2010).

This chapter's UK, Australasian and North American case studies illustrate the types of terrestrial biological change that are taking place and the likely threats due to current climate change. For a summary map of key Australasian areas of climate change see Figure 6.10. Other than temperature, central to the changes in climate factors affecting biology is water availability. This also applies to many other terrestrial areas subject to climate change outside these global regions. For instance, one example is that the arid desert biome of northern Africa may extend and jump the Mediterranean into Spain. In 2005 the Spanish Government reported, when announcing a £50 million programme to combat desertification, that more than 90% of the land between Almeria in the south to Tarragona in the north east is considered at risk. This includes the provinces of Murcia, Valencia and southern Catalonia. Over-grazing and development pressures have exacerbated the problems.

Spain, though, is not alone in facing both climate and development pressures given both a warming world and growing global population. Access to water will become a far bigger issue as the century progresses. Indeed, not only will African hot dry biomes probably spread across the Mediterranean, deserts are likely to spread within parts of Africa. This is even though the balance between deserts spreading and areas of high rainfall spreading, or rain getting more intense over a similar land area, remains uncertain. Nonetheless, there is some evidence to suggest that areas that were previously drier and more sandy during (geologically) previous climatic regimens may once more become desert. In 2005 researchers from Oxford and Salford in Britain looked at southern Africa's dune systems, much of which today

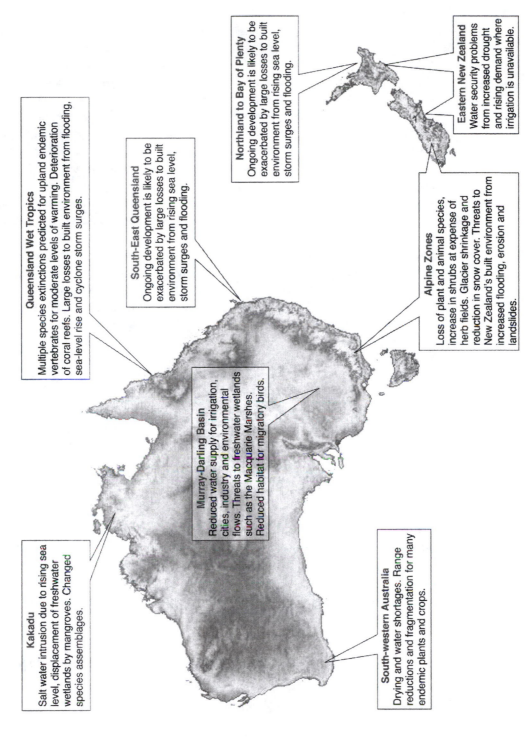

Kakadu
Salt water intrusion due to rising sea level, displacement of freshwater wetlands by mangroves. Changed species assemblages.

Queensland Wet Tropics
Multiple species extinctions predicted for upland endemic vertebrates for moderate levels of warming. Deterioration of coral reefs. Large losses to built environment from flooding, sea-level rise and cyclone storm surges.

South-East Queensland
Ongoing development is likely to be exacerbated by large losses to built environment from rising sea level, storm surges and flooding.

Murray-Darling Basin
Reduced water supply for irrigation, cities, industry and environmental flows. Threats to freshwater wetlands such as the Macquarie Marshes. Reduced habitat for migratory birds.

South-western Australia
Drying and water shortages. Range reductions and fragmentation for many endemic plants and crops.

Northland to Bay of Plenty
Ongoing development is likely to be exacerbated by large losses to built environment from rising sea level, storm surges and flooding.

Eastern New Zealand
Water security problems from increased drought and rising demand where irrigation is unavailable.

Alpine Zones
Loss of plant and animal species, increase in shrubs at expense of herb fields. Glacier shrinkage and reduction in snow cover. Threats to New Zealand's built environment from increased flooding, erosion and landslides.

Fig. 6.10 Key hotspots identified for Australia and New Zealand, assuming a medium emissions scenario for 2050. Reproduced with permission from IPCC (2007b), originally figure 11.5.

are covered with grass and, indeed, trees (Thomas et al., 2005). This is an important biome because desert dunes, whether dry and active, such as in much of the Sahara Desert, or inactive, covered by vegetation north of the Kalahari Desert, account for 5% of the global land surface and 30% in Africa. Knowing how moisture levels determined by precipitation and plant evapotranspiration, together with wind speeds, combine to affect dune mobility, they were able to apply climate models to see the effect on sand dunes. If, due to global warming and regional climate change, temperatures rise without a counteracting increase in rainfall, and factoring in wind speeds, then vegetation that currently stabilises dunes could die, and the dunes could mobilise. This would also kill off other vegetation previously sheltered by the dune, and so the sandy desert would return. Such an analysis suggests that the current situation is precarious. It would appear that regardless of which IPCC future scenario is considered, the southern dunes bordering the dry Kalahari Desert that are already mobile will become prone to greater movement by 2040. Meanwhile, the eastern and northern dune fields that are currently covered by trees and/or grass could become mobile by 2070. By 2100 all dune fields from northern South Africa to Angola and Zambia may become mobile. The authors conclude, 'There are uncertainties within the modelled Kalahari scenarios but the general trend and the magnitude of possible change in the erodibility and erosivity of dune systems suggests that the environmental and social consequences of these changes will be drastic' (Thomas et al., 2005).

In 2011 a large international team of researchers, led by Michael Burrows, David Schoeman and Lauren Buckley, published research in the journal *Science* that mapped the distances that species range would have to move based on actual climate change in the past half century. They used two measures of thermal shifts that species would have to (try to) track: the velocity of climate change as in the geographical movements of isotherms (temperature), and the phenological shift (seasonal timing) of temperatures. These indices resulted in a complex mosaic of predicted range shifts and phenology changes that deviated from a straightforward poleward migration that a simple analysis based on the global temperature and average latitude temperatures would have given.

Be it vegetation changes in sand dunes, species range migration or many of the other responses to climate change discussed in this chapter, the likelihood with confounding factors is that change will not be smooth but at some point will move rapidly to a new state. Such points are critical in the transition and thresholds to new environmental, climatic and or ecological conditions. We will return to this later this chapter.

6.5 Biological responses to greenhouse trends beyond the 21st century

Biome shifts and ecosystem change, with their consequential impacts on human activities, have been a theme of this chapter. As we shall see when looking at the human ecology of anthropogenic global warming, it is most likely (some other global catastrophe aside) that fossil fuel burning will continue throughout this century, so raising atmospheric carbon dioxide levels from around 400 ppm (soon to be reached early in the 21st century) to 700–1000 ppm during the early 22nd century.

There is therefore little doubt that biome shifts and ecosystem change of the sort described so far will continue to be seen both while atmospheric carbon dioxide continues to increase and also after carbon dioxide levels have stopped increasing and while ecosystem communities stabilise, adapting to the new conditions. However, there will be other changes. As noted in Chapter 2 with, for example, leaf shape and climate, species themselves will change. Some of these transformations will manifest themselves genetically and others will be physiological, with regards to the way carbon is metabolised (see Chapter 1).

A recent instance of genetic change under circumstances analogous to 22nd-century carbon dioxide conditions (namely, carbon dioxide at 1050 ppm) was found in 2004 when Canadian biologists Sinéad Collins and Graham Bell grew a protoctist species of Chlorophyta (green algae) called *Chlamydomonas reinhardtii* over about 1000 generations. *C. reinhardtii* was used because it has been studied extensively to elucidate photosynthetic mechanisms. Like most other eukaryotic algae, it is known to have a carbon-concentrating mechanism (CCM) that increases the concentration of carbon dioxide near the enzyme Rubisco (see Chapter 1). After 1000 generations at elevated carbon dioxide levels there was less need for a CCM. So the CCMs in *Chlamydomonas* became less efficient and its cells became smaller but with higher rates of both photosynthesis and respiration. The researchers tentatively concluded that as carbon dioxide levels increase so many Chlorophyta will have less-effective CCMs and changes in photosynthetic and respiratory rates. This, they note, would affect global processes by changing the rate of carbon turnover in aquatic and perhaps terrestrial ecosystems. It needs to be remembered that very roughly a quarter of the Earth's photosynthetic primary production is undertaken by algae, so even if other photosynthetic species do not see such a radical response to elevated carbon dioxide (although some change is likely), there will be a major change in carbon transferences around sources, sinks and reservoirs and not just changes part-mediated by alterations in physiology. These unknowns further confound those developing computer models that forecast climate change from the effects of greenhouse gases. They will, though, be important should we enter a warming period analogous to the Initial Eocene Thermal Maximum (IETM).

6.6 Possible surprise responses to greenhouse trends in the 21st century and beyond

The IPCC recommendations that provided the scientific consensus on climate change were summarised in the last chapter (see section 5.3). Despite the broad picture (greenhouse gases are increasing, the Earth is warming up and sea levels are rising) the IPCC acknowledge that there are uncertainties and that in addition to heeding their main conclusions we also need to be aware of surprises. As we have seen from the above three case studies, at the present time the global IPCC broad picture at best translates to very basic scenarios, as do other perspectives that are able to draw on additional regional research. These regional scenarios will undoubtedly improve with further research both in the field and with the development of more sophisticated models. But what of the IPCC warning about surprises? This section looks at some key areas.

6.6.1 Extreme weather events

One swallow does not make a spring, and so a single extreme weather event (be it an abnormally heavy downpour or a particularly dry month) does not necessarily signify climate change. But how many swallows do? An increase in the frequency of weather abnormalities will, at some point, signify that the climate has altered. Recent years have seen an increasing frequency of strong hurricanes (it is thought that the total number of all hurricanes has broadly remained the same) and record-breaking global warmth. This last is only to be expected with (indeed, it is inherent in) global warming. There have also been times of unusually intense local and regional rainfall.

Such events are surprising, especially for those who experience them. When a 'hurricane' hit southern England one night in 1987 it caught the population by surprise and disrupted lives of virtually everyone for at least a day, with power cuts, telephones not working and roads closed due to fallen trees. Not least surprised was one meteorologist working for BBC Television as a weather presenter, Michael Fish, whose evening broadcast prior to the storm was erroneously reassuring. Now, whereas this particular single swallow cannot itself be attributable to climate change, it was the sort of event that climatologists were contemplating. Indeed, the incident was referred to by one UK Governmental department as climate change-related, even though that claim could not be substantiated. Nonetheless, the Department of Energy used the 1987 storms in an energy-conservation promotion, with the slogan 'Global warming, we have been warned'. The department was subsequently reprimanded in March 1992 by the Advertising Standards Authority (Cowie, 1998). Today there is arguably less chance that such a reprimand would be issued.

Having discussed increases in the number of strong hurricanes earlier in the chapter, let us now focus on temperature. A warmer world will, by definition, hold more thermal energy in the atmosphere and ocean circulation systems. Some of this energy will become manifest in the form of greater evaporation (hence rainfall) and some in wind speeds or storm size. Therefore, as the world warms so the frequency of extreme weather events is likely to increase. Although worldwide ocean evaporation and hence precipitation are two of the key indicators of global warming, arguably the factor of greatest psychological impact is the intensity and frequency of heatwaves. These too are likely to become more intense and frequent as the world warms. For example, in 2004 Gerald Meehl and Claudia Tebaldi of the National Center for Atmospheric Research in Boulder, Colorado, published the results of a coupled model together with field observations to see whether the extreme heatwaves of Chicago and the south-western area of the Great Lakes (1995) and London/Paris region of Europe (2003) were unique events or, alternatively, more likely to occur in a greenhouse future. Their conclusions were that the heatwaves over these parts of the USA and Europe coincided with a specific atmospheric circulation pattern that is intensified by ongoing increases in greenhouse gases; this suggests more intense and severe heatwaves in these regions in the future.

Do not be surprised by extreme weather events in our greenhouse future. But equally do not be quick to attribute them to climate change as there is a difference between an anomaly within background noise and a statistical trend. To overcome this, we need statistical analyses of what might be expected (1) with natural variability assuming no carbon dioxide forcing and (2) with climate forcing from additional anthropogenic greenhouse gases. Both scenarios need to be compared with reality as revealed by environmental measurements.

In 2004 a small UK team led by Peter Stott performed such an analysis to examine whether the 2003 western European heatwave was due to either (1) natural variability or (2) global warming, as per the previous paragraph. These two scenarios were derived using one of the more sophisticated climate models available at the time, the HadCM3 climatic model developed by the UK's Meteorological Office at the Hadley Centre for Climate Prediction and Research. The model took into account the impact of actual volcanic eruptions and also solar variability specifically for the south and western part of the European peninsula. It was run three separate times simulating June–August temperatures as if there were human additions of greenhouse gases and over the period from 1900 to 2003 and then on to 2100: these are type 2 scenarios as per the previous paragraph. It produced three temperature lines with a typical annual variation of about ± 1 K and this followed both actual meteorological measurements and also a smoothed, few-year-filtered, line to 2000. The actual meteorological line and, consequently, the smoothed line both lay within the model's annual variability band through to the 1990s. Consequently, the smoothed line also rose, albeit erratically due to actual annual, volcanic and solar variation, by about a degree along with the simulated lines, so suggesting that the model effectively captured anthropogenic greenhouse warming.

The model was run again, but under a scenario assuming no additional greenhouse forcing over this time – that is, as per scenario 1 in the above paragraph – and so only included anthropogenic greenhouse forcing up to 1900 (and not beyond that year). This non-global warming line largely followed (intertwined with) both the natural and smoothed natural lines for the 20th century's first three quarters. However, for the century's final quarter this simulation continued horizontally, albeit erratically as if with natural variation, so that by the century's end there was nearly a 1 K difference between it and both the actual instrumental record as well as the anthropogenic greenhouse simulation. The inference clearly was that this 1 K difference at the end of the 20th century was due to anthropogenic emissions of greenhouse gas.

However, looking at the actual meteorological line, not only did it broadly follow the model's global warming scenarios but summer 2003 stood out as being about 0.75 K warmer than any other previous June–August period on record and is thought to have given parts of western Europe the warmest summer for more than 500 years. Indeed, the next warmest summers have all taken place since the mid-1980s and this too was reflected by the three global warming model scenarios. It suggests that the model successfully captures the annual *likelihood* of an exceptional summer heatwave in western and southern Europe even if it cannot predict the *actual year* in which such a heatwave might take place.

Having established that the HadCM3 climatic model captures both global warming and the annual likelihood of extreme events, what did the model predict for the 21st century? The indication is that for 2000–2100 the summers in this part of Europe will gradually warm by 6.5 K on average. In addition, extremely hot summers will continue, at least once or twice a decade, above this average warming. However, because the average summer temperature is set to increase during the 21st century, an exceptionally warm summer up to 2020 will become a normal summer by the 2040s. From the 2040s exceptionally warm summers will all be far warmer than summer 2003, so that by the end of the century exceptional heatwaves can be expected to be around 6.5 K warmer that those at the century's beginning. The researchers' conclusion was that it seems likely that past human influence (greenhouse emissions) has more than doubled the risk of European mean summer temperatures as hot as those in 2003 for the early 21st century, and further that the likelihood of such warmth is projected to increase 100-fold over the next four decades. They said, 'It is difficult to avoid the conclusion that potentially dangerous anthropogenic interference in the climatic system is underway' (Stott et al., 2004).

In 2011 two Swiss researchers (Fischer and Schär, 2010) attempted to model the likely nature of European heatwaves for the 21st century. Their results suggested that whereas extreme heatwaves are expected to increase across all Europe they will be both more frequent and more intense in southern Europe and particularly Spain and Portugal. For the Iberian peninsula and the Mediterranean region the frequency of heatwave days is projected to increase from an average of about 2 days per summer for the period 1961–90 to around 13 days for 2021–50 and 40 days for 2071–2100.

In a warmer world one might also expect fewer very cold winters. Although this may be true there are likely to be exceptions, for a number of reasons. First, the shifting of weather systems in a warmer world may result in the blocking of the jet stream, such that cold Arctic air is drawn south (as happened in western Europe in the winter of 2010–11). In winter, north-eastern North America and north-eastern Asia are both colder than other regions at similar latitudes. In 2011 Yohai Kaspi and Tapio Schneider, from the California Institute of Technology, showed that this anomalous winter cold can result in part from westward radiation of large-scale atmospheric waves – nearly stationary Rossby waves (large-scale meanders in high-altitude winds caused by Coriolis force sheer) – generated by heating of the atmosphere over warm ocean waters. They showed that warm ocean waters contribute to the contrast in mid-latitude winter temperatures between eastern and western continental boundaries, not only by warming western boundaries but also by cooling eastern boundaries. Such reasons mean that cold weather extremes are likely to persist well into the 21st century even as – in fact, because – the Earth warms.

Other work looking at the likelihood of winter extreme cold events as the 21st century progresses still has them taking place by the end of the 21st century, albeit with decreasing frequency, and by then on average 1 winter in 10 will be cooler than the winter mean of the 20th century (Kodra et al., 2011).

In 2011 the IPCC published an advance copy of its *Summary for Policymakers* of its 2012 *Special Report on Managing the Risks of Extreme Events and Disasters to Advance Climate Change Adaptation (SREX)*[5] (IPCC, 2011, 2012). It concluded that

---

[5] The SREX (2012) was due to be published after the submission of the manuscript for this book's second edition.

it was *virtually certain* that increases in the frequency and magnitude of warm daily temperature extremes and decreases in cold extremes will occur in the 21st century on the global scale. Also, it is *very likely* that the length, frequency and/or intensity of warm spells, or heatwaves, will increase over most land areas. Based on the IPCC's A1B and A2 emissions scenarios, a 1-in-20-year hottest day is *likely* to become a 1-in-2-year event by the end of the 21st century in most regions. Additionally, average tropical cyclone maximum wind speed is *likely* to increase, although increases may not occur in all ocean basins. However, it is *likely* that the global frequency of tropical cyclones (hurricanes and typhoons) will stay roughly the same or decrease. As for the damage that climate extremes will bring, the *SREX* concluded with *high confidence* that disaster losses associated with weather, climate and geophysical events are higher in developed countries. Fatality rates and economic losses expressed as a proportion of GDP are higher in developing countries. Here the IPCC have a specific meaning for the words 'likely', having a notional probability of 66–100%, and 'high confidence', that there is high agreement from a body of robust evidence. Of course, it should be remembered (in case you are reading this in the decades to come) that what is considered an extreme event early in the 21st century will be more normal (frequent) towards the century's end when the characteristics of extreme events will be even more extreme. Finally, the IPCC are careful to stress that they draw the above conclusions in general and not specifically: that is to say, it is impossible to distinguish the cause of a specific extreme event between climate change on one hand or the vagaries of weather on the other. So, from the IPCC perspective in 2011 it would be impossible to sue a fossil fuel company for the damages caused by a heatwave or a hurricane.

The agricultural and health impacts of such extreme weather events will be considered in the next chapter.

6.6.2 Greenhouse gases

The IPCC scenarios are based on a number of global economic possibilities ranging from business as usual to prudent policies curbing emissions and introducing new low-fossil and/or alternative technologies. These, from the later 2001 reports, result in a range of predictions (see Figure 5.7). Fortunately, for simplicity the clear majority of these fall within the high and low estimate boundaries of the IPCC's original 1990 Business-as-Usual (B-a-U) scenarios (Figure 5.5 and Table 5.1). Yet, even so, there may be surprises. For example, these models assume that the biosphere's response tomorrow to additional atmospheric carbon dioxide will be proportionally similar to the response today. But what if some threshold were reached? It may be that circumstances are such that the oceans cease to absorb as much of this additional carbon as they have done in the past. (Perhaps because increasing acidity means that the oceans are unable to buffer so much carbon dioxide, as will be discussed shortly.) In such a case the build-up of atmospheric carbon dioxide would be greater than expected and, as a result, so would climate change. Alternatively, if natural (as opposed to human) systems do operate as expected then there may be unforeseen developments in human systems. Then again, what would happen if China's population of 1.3 billion people were to start increasing their fossil fuel consumption at rates previously not contemplated? If either natural or human systems do not behave in the future as anticipated then there will be unforeseen consequences.

Such unforeseen circumstances could take us outside of the IPCC forecasts in either a positive or a negative way. At the most fundamental level, with regard to natural systems, we know that in the past atmospheric carbon dioxide has been both markedly higher and lower than the IPCC window. However, since the Industrial Revolution the trend has been upwards, so it would be prudent to consider surprises that increase climate change. Indeed, in 2004 it was realised from direct measurements that atmospheric carbon dioxide was increasing faster around the turn of the millennium than had previously been contemplated. The current question revolves around whether this is an atypical blip or part of an unforeseen trend.

6.6.3 Sea-level rise

The IPCC predictions for sea-level rise have, if anything, become more conservative with successive reports (see the summary in Table 5.1). To some, in the UK scientific community at least, this has been itself surprising as a body of research since the late 1980s (the time from which the IPCC first drew on existing peer-reviewed academic literature) has pointed to higher sea levels of up to 6 m during the warmest part of the last (Sangamon or Eemian) interglacial, even though much of that interglacial was climatically comparable to the present one and only part of it marginally warmer than today. Further, there is recent evidence that Antarctic ice flows in some places are faster than thought and that Greenland's ice cap is melting more rapidly than predicted. Having said this, there are also possible counteracting factors but there is uncertainty, and hence the possibility of surprise.

It has to be said again that the IPCC provide a consensus view; it came to its conclusions through a committee process and that its committees were not always as unified as might be preferred, even if a good proportion of its members were in general accord. It is not appropriate, or possible, for us to explore exactly how they came to their conclusions, although we can be certain that their conclusions do arise from a variation in opinion and/or confidence in forecasts (hence one reason for the upper and lower estimates in addition to the best estimates the IPCC provide). We also have to acknowledge, with a significant proportion of the global population living in areas likely to be affected by sea-level rise, that flooding attributable to climate change could have major political implications. In other words, irrespective of the risk (probability), the potential hazard is high. So, given the comparative reassurance of a low forecast on one hand, and potentially high costs in the event of high sea levels on the other, it may be tempting for policy-makers to play down sea-level concerns when in fact there should be a greater apprehension and attention to the precautionary principle. Psychological perceptions and political drivers are one thing but scientific understanding is another, even if the latter can contribute to the former (which is what was hoped with the formation of the IPCC).

We know that during the last interglacial sea levels were up to 6 m higher than today. This therefore might be considered a realistic scenario of concern that would be *currently* foreseen to take place *beyond* the 2100 horizon frequently used by policy-makers. Conversely, the total sea-level rise due to mass contributions if all

Table 6.3 Potential sea-level rise from all terrestrial ice melt from estimates by the US Geological Survey (1999)

Source	Volume (km^3)	Percentage	Level rise (m)
Glaciers, etc.	180 000	0.55	0.5
Greenland	2 600 000	7.90	6.5
Antarctic[a]	30 109 800	91.49	73.5
Total	32 889 800	100	80.5

[a] Of which the West Antarctic Ice Sheet would contribute 5–6 m. Note: this excludes any rise from thermal expansion of the oceans. Also note that Antarctic land ice is largely stable and unlikely (unless there are surprises) to make a major contribution to sea-level rise, if at all, in the 21st century. However, since 1999 new evidence suggests that terrestrial ice melt may be faster, especially from Greenland (see text). The US Geological Survey data have been rounded to the nearest half metre.

terrestrial ice melted would be somewhere around 80 m. (This melt does not include the sea-level rise from thermal expansion, and the melt would probably take a few thousand years.) In 1999 the US Geological Survey published the estimates shown in Table 6.3 for the breakdown of this 'all-melt' potential.

Those concerned that the IPCC consensus view of sea-level rise might be too conservative can point to the work of marine biologists from Mexico and Germany (Blanchon et al., 2009). Corals track sea level and so these researchers isotope-dated corals from the last interglacial using ^{230}Th. They not only confirmed what had been known for a few decades, that sea levels at times during the last interglacial were 6 m higher, but also found that there were rapid jumps in sea level of 2–3 m, during which the rate of sea-level rise was more than 36 mm year^{-1}. This rate is faster than the IPCC's forecast based on models.

On the one hand a worst-case 6 m rise (as in the last interglacial) over a few (two, three or four) centuries seems modest against a very long-term potential rise of around 80 m in the event of *all* terrestrial ice melting over very many centuries. The long-term all-melt scenario is far greater that the IPCC's 2007 forecast for the end of the 21st century, but may occur should an event like the Initial Eocene Thermal Maximum (IETM) be triggered (see Chapter 3). The 6 m rise seen in the last interglacial took place at a time when temperatures were not far from what they are anticipated to become later in the 21st century; although, of course, due to thermal inertia it will take a long time for Greenland, let alone Antarctica, to melt. On the other hand, at more than 392 ppm, current atmospheric carbon dioxide levels are already exceeding the peak level of just over 300 ppm seen during the last interglacial.

The reason for the disparity of the last interglacial sea-level peak (approximately 5–6 m) and the potential rise if all our planet's ice melted (approximately 80 m) is due to the aforementioned huge thermal inertia inherent in very large ice caps: it takes a while for ice caps of a very large mass to adjust to extra warming. Conversely, if we stopped further global warming geo-biosphere inertia would ensure that some melting would continue through to the end of the century due to the increased warmth already in the system.

In addition to the above terrestrial ice-melt factors, roughly a quarter to a third of recent sea-level rise has not been due to melt but thermal expansion of the oceans. As the oceans warm, they expand and the sea-level rises. This means that the sea-level rise in a warmer world thousands of years from now, with all terrestrial ice melted, would exceed the 80 m or so anticipated rise from melting alone.

Before dealing with possible surprises, it is worth reminding ourselves that our current understanding of historic, let alone possible near-future, sea-level rise is fraught with uncertainties (see the previous chapter). Uncertainties are not the same as surprises, even if one can lead to, or spring from, the other. As previously mentioned our estimates for the various components of historic sea-level change include a considerable margin of error. So, in one sense we could say that we can only explain about half the change seen over the past couple of centuries. In addition, we not only do not understand how Antarctica is likely to behave in a warmer future, we do not know entirely what is going on today.

What we do know is that the marine ice shelves of the Antarctic Peninsula are vulnerable to collapse, but because the bulk of these is already under water their future melting will not contribute as much to sea-level rise as the continental ice shelf (grounded on rock). The Antarctic Peninsula is particularly sensitive to climate change as it juts away from the main body of the continent and the South Pole. We also know that much of West Antarctica consists of floating marine shelves and so will certainly contribute to sea-level rise in a warmer future, long before East Antarctica. It appears that parts of the West Antarctic Ice Sheet (or WAIS) contributed to the (5–6 m) higher sea levels seen in the previous interglacial. Conversely, East Antarctica is largely composed of grounded continental ice shelves and is higher (hence cooler) than West Antarctica. On one hand these ice shelves are more stable and so will not melt, except with warming over a considerable period of time (hundreds of thousands of, if not one or two million, years) or unless global warming is really exceptional, such as perhaps with an IETM analogue event.

Yet uncertainty abounds as to what is currently happening in much of Antarctica (let alone what might happen in the future). There were in 2006 18 long-term monitoring stations in Antarctica. Eleven of these show warming over the past few decades and seven show cooling. This may suggest that Antarctica is warming up. Yet, it must be remembered that Antarctica is a continent the size of Europe and the USA combined, and so 18 stations are not sufficient for detailed monitoring. This insufficiency is even more evident when one considers that 16 are arranged around the coast and just two (at the South Pole and at Vostok) are in the continent's interior. Indeed, even the South Pole station is not particularly useful for getting an idea of what is happening in East Antarctica, as it is not that far from the West Antarctic Ice Sheet, in the transition border zone between the part of the continent that is thought to be losing water and the part that is accumulating it. Consequently, that station does not tell us as much about the continent's interior as we would like. Remote sensing via satellite has also been limited as no satellite has passed directly over the South Pole (see Figure 5.9). However, this will change with results from the second European Space Agency CryoSat satellite. Following the first CryoSat's crash due to a launcher failure in 2005, CryoSat-2 was launched in April 2010 and, following in-orbit testing, became operational in October that year. This will help corroborate results from the Gravity Recovery and Climate Experiment (GRACE) satellite missions (see section 5.3.2).

It is worth briefly summarising our current knowledge of the situation in Antarctica. This is very important to sea-level rise and will affect low-lying regions worldwide, although it is decidedly geology, not biology. There are broadly three Antarctic zones. The East Antarctic (see Figure 5.9) contains some 90% of the continent's ice and this is virtually (apart from a few small coastal areas) all land, and which is reasonably stable but not immune to climate change. It is thought that this ice is getting thicker because snowfall (precipitation) – although low – exceeds snow melt. The second area is the Antarctic Peninsula which, as mentioned, sticks out away from the continent and is sensitive to warming, with its marine shelves (resting on, or floating over, the sea floor); this is particularly so on its eastern side. These include the Larsen ice shelves that are being undermined by submarine currents from the east that have warmed a little in recent decades (at a rate of around $0.01–0.02°C$ $year^{-1}$) and have recently been subject to warmer surface air. Some of the peninsula's ice-shelf retreat is unique to the current Holocene interglacial, although melt appears to have previously been slower but continual over much of the Holocene. What is now known is that the collapse of Antarctica's Larsen B ice shelf from the mid-1990s appears to have been a unique event. One possibility was that the ice shelf may have collapsed previously during other warm periods in the Holocene and then returned in cooler times. After all, we know that the Holocene has had warm periods (such as the MCA or MWP) and cooler times (such as the Little Ice Age); see Chapters 4 and 5. However, in 2005 the results were published of research by geoscientist Eugene Domack and colleagues. They analysed core samples taken in 2002 from various sites that had been beneath the former ice shelf that had existed up to 2001. They also looked at a core taken where the ice shelf had existed up to 1995. A few large boulders were also recovered and only some showed evidence of recent colonisation by organisms. Carbon analysis of these organisms gave modern ^{14}C ages (it was possible that older carbon could have been recycled). Their conclusions, based on the biological and geological evidence, was that the Larsen ice shelf had been in continual existence throughout the Holocene interglacial. Its disintegration from the mid-1990s was therefore an event that had not occurred since the last interglacial or earlier. Domack and colleagues said that the ice shelf's demise was probably the consequence of long-term thinning (by a few tens of metres) over thousands of years and of recent short-term (multi-decadal) cumulative increases in surface air temperature that exceeded the Holocene's natural variation in regional climate.

The third zone is the West Antarctic Ice Sheet, other than the Peninsula, which consists of the Ross, Amundson and Weddell Zones. Of these, some of the Amundson ice shelf has melted before in previous interglacials and so will play an important part in near-future sea-level rise. Indeed, we know from foram remains under some of the Amundson shelves that these have retreated in the past and reformed during glacials. (The foram remains detected were from species that evolved in the past 600 000 years and so could not have been left over from before Antarctica began to freeze a few million years ago.)

In 2009 the Scientific Committee on Antarctic Research (based at the Scott Polar Research Institute, Cambridge) of the International Council for Science expressed concern about possibly large contributions to sea level from the dynamic instability of Antarctic ice sheets during the 21st century. These were not included by the IPCC's 2007 assessment. They recommend that we prepare for a maximum sea-level rise of 2 m (part of which will come from Antarctica) by 2100 on the basis that it may be less than this.

Where does this leave us with regards to surprises? The IPCC's 2007 best estimate for sea-level rise from 1990 to the end of the 21st century is 37 cm (A2 median value), in the range 23–51 cm. The 2007 IPCC's A1F1 scenario gives a higher estimate of 59 cm (see Table 5.1). For the IPCC, and hence policy-makers, the surprise would be if we were taken outside this window. Is this likely? The reasons to suggest that it might be are 3-fold.

1. Some of Greenland's margins and Antarctica's ice shelves appear to be disintegrating faster than anticipated.
2. Antarctic and Greenland interior ice is now moving faster to the coast.
3. Mass additions to past 20th-century sea-level rise appear to be greater than thought.

First, with regard to points 1 and 2, even though the potential sea-level rise from Antarctica greatly exceeds that from northern hemisphere ice caps (see Table 6.3), recent research suggests that during the last glacial the sea-level rise of 30 m (when the sea was much lower than now) over the period 65 000–35 000 years ago was due to equal contributions ($\pm 10\%$) of melt from northern hemisphere ice caps and Antarctica, and that these were pulsed with 5000–8000-year spacing between events. The suggestion is that while the then glacial Earth warmed 2–3°C the sea-level rise of 30 m took place at a rate of 2 m per century (Rohling et al., 2004). This is over three times the IPCC's best estimate (IPCC 2001a, 2001b) of the rate of rise forecasted for the 21st century.

Furthermore, many had assumed that just 25 m of the total 120 m glacial–interglacial difference in sea level came from Antarctica. Consequently this more recent research suggests much of the past 25 m Antarctic (100 000-year glacial–interglacial) melt was released in episodes within a 30 000-year period in the depth of the ice age. Whereas *past* melt is not necessarily a guide to *future* melt (the 21st century does not have large Laurentide or Fennoscandian ice sheets, which accounted for much of the rest of the 120 m glacial–interglacial rise), it does suggest that past melt was not smooth, and so future melt might not be as linear or smooth as the IPCC predict. In 2011 an international team, including Europeans but primarily from Canada, led by Alex Gardner and Geir Moholdt noted that the Canadian Arctic Archipelago, located off the north-western shore of Greenland, contains one-third of the global volume of land ice outside the ice sheets, but its contribution to sea-level change remains largely unknown. Their estimates were based on three independent approaches: surface mass-budget modelling plus an estimate of ice discharge, repeat satellite laser altimetry (ICESat) and repeat satellite gravimetry (GRACE). All three approaches showed consistent and large mass-loss estimates. They showed that the Canadian Arctic Archipelago (the two largest islands of which are Baffin and Ellesmere) has recently lost 61 ± 7 Gt year^{-1} of ice, contributing 0.17 ± 0.02 mm year^{-1} to sea-level rise in 2007–9. Between the periods 2004–6 and 2007–9 the rate of mass loss increased sharply from 31 ± 8 to 92 ± 12 Gt year^{-1}. The duration of the study was too short to establish a long-term trend, but for 2007–9 the increase in the rate of mass loss makes the Canadian Arctic Archipelago the single largest contributor to eustatic sea-level rise outside Greenland and Antarctica. This sea-level rise compares with the global rate the IPCC 2007 assessment gives for the decade 1993–2003 of 3.1 ± 0.7 mm year^{-1}.

We should not allow current slow sea-level rise to lull us, as a period of faster rise may yet take place. Of course, while comparisons of a present interglacial Earth warming of 2–3°C are not directly analogous to a past glacial Earth warming 2–3°C, we might at least bear in mind the past when considering possible future surprises of which the IPCC warn.

Second, with regards to point 2 above, Antarctic (specifically West Antarctica) and Greenland ice may be less stable than previously thought. There has already been much concern over the stability of Antarctic ice sheets, especially those grounded beneath the sea along the coast (continental ice shelves). (Those ice sheets that are already floating – marine ice shelves – will not contribute to sea-level rise when they melt.) In 2004 concern was raised that the Greenland ice cap was melting faster than thought. More evidence came to light in 2006 (to which we will return shortly). Indeed, around the turn of the millennium large parts of marine ice shelves in the Antarctic Peninsula broke free. Among the most spectacular collapses were those of the Larsen A and B ice shelves in 1995 and 2002. In 2004 it was reported that a survey in West Antarctica, using ice-penetrating radar, had revealed a change in ice flows compared to that about 1500 years ago, as deduced from geological evidence (Siegert et al., 2004). It is now thought that coastal ice was/is physically supporting ice further inland and that with this coastal ice gone ice inland was/is more free to flow to the coast. It is separately known that some Antarctic glaciers more than 100 km inland are now accelerating (Kerr, 2004). If ice caps are moving faster than thought then IPCC estimates of sea-level rise (which are based solely on work published more than a year prior to each report) will need to be revised. At the moment the IPCC consider that, as previously mentioned, the Antarctic contribution to sea-level rise in the 21st century will be negligible due to ocean evaporation falling as Antarctic snow, which will then become trapped as ice. This provides a countering effect to Antarctic ice melt. The question is whether Antarctica is taking more water from the oceans, via the evaporation/snowfall route, than it is contributing to the oceans through melt. This is a current research priority.

With regards to estimates of the mass of meltwater contributing to sea-level rise, the IPCC (2001a, 2001b) refer to 20th-century sea-level measurements to assist them in forecasting the future. The IPCC do warn that their 21st-century sea-level forecasts have considerable uncertainty to them (a warning not often highlighted, or even mentioned in much other documentation). Indeed, there has been a discrepancy between direct estimates of 20th-century rise compared to estimates based on contributions of mass (meltwater) and volume (thermal expansion). Research published subsequent to the IPCC's 2001 assessment looked at temperature and salinity close to sea gauges. Sea temperature affects the amount of thermal expansion and salinity is an indication of how the sea is being diluted by fresh water. They found that the mass (melt) contribution to 20th-century rise was greater than previously estimated (Miller and Douglas, 2004). This picture has been reinforced by other research published just prior to the IPCC's 2007 report (Overpeck et al., 2006). This compared sea-level rise during the last interglacial with computer models and current trends. This again suggests that the IPCC's 2001 estimates for sea-level rise may need to be revised upwards and further that during the 21st century we may pass thresholds that enhance Antarctic warming as well as – as others have thought (such as Toniazzo et al., 2004; referred to in the previous chapter) – Greenland. For example, if the collapse of West Antarctic Ice Sheet and other melting changes albedo sufficiently then atmospheric circulation patterns may also change, so altering East Antarctic warming, not to mention that of the whole continent. This would increase the likelihood of sea-level rise reaching more quickly that of the last interglacial 127 000–130 000 years ago: 6 m above that at the beginning of the 21st century. The big question is how quickly would this happen and so how great a revision of the IPCC's 2001 sea-level rise estimate would be required?

As shown in Table 5.1, in 2007 the IPCC revised their upper and median sea-level rise estimates for the 21st century downwards but raised their lower estimates for all scenarios. This, they said, was due to reduced uncertainty of estimates compared to their earlier reports and because the latest report did not include a small contribution from the thawing of permafrosts. Climate system surprises were not included but, as the IPCC themselves say, we should be wary of them.

Again, if there is more melt, and if we have yet to experience much of the thermal expansion anticipated from global warming, then both these factors have the possibility of increasing potential 21st-century sea-level rise. This has conceivably serious consequences for the resulting impacts on human activity. Policy-makers and planners need to take this certain uncertainty, if you will, into account.

Finally, what of beyond the 21st century? Whereas the 6 m sea-level rise during the last interglacial may arguably be a very (possibly unconceivable) worst-case scenario for the 21st century, it is not so in the longer term. Conversely, it is most unlikely (indeed, virtually impossible), for *all* Antarctic, Greenland and glacier ice to melt by the end of the 21st century or even the next few centuries: so unlikely, in fact, that it can be discounted. However, unless greenhouse emissions are curbed considerably in the near future, and indefinitely, on the scale of many centuries if not millennia hence, a significant proportion of such melt becomes far more probable. Indeed, if current global warming triggered an event analogous to the IETM then such melt would become more likely. If the Earth warmed sufficiently over the millennia sea levels could well rise by more than 80 m (Table 6.3). This would have a profound effect on the global civilisation. All current ports, and many capital cities, would be under water. It would also re-shape coastlines in a manner close to (because of plate tectonic movement) that not seen since the Pliocene. Such extremely long-term thinking is not currently on the political agenda. It is also beyond the IPCC's current agenda. Yet at the moment, with greenhouse gas emissions currently growing (and not being curbed), this is the long-term future into which we are probably heading.

Turning away from Antarctica, the climatic monitoring of the Greenland ice cap is far better. Furthermore, computer modellers seem to have greater confidence in their forecasts of possible future Greenland melt under a variety of IPCC SRES scenarios. Taking the high IPCC climate change scenario, with carbon dioxide rising to some 1000 ppm up to the year 2200 and then stabilising (assuming it is not pushed higher still by other surprises from other – non-fossil – carbon pools), the Greenland ice cap will disappear over the next 2000 years (although there might be a small amount of ice remaining on top of its eastern mountains). This melt alone will raise sea levels by 6.5 m; although should this happen then sea-level rise will be higher than this due to thermal expansion and other contributions from Antarctica (Alley et al., 2005).

Although our understanding of the Greenland ice cap is better than that of Antarctica, our knowledge is still far from complete. The above 2000-year scenario is largely based on melt factors. However, there are other mechanisms that serve to release water from Greenland ice. One of these is increased glacier movement. In 2006 it became apparent from satellite images that glacier movement had increased more than had previously been thought. It was already known that glaciers near the coast had been moving faster, but the new satellite data showed that there had been a marked increase some way inland and so the flow rates of Greenland's largest glaciers had doubled in the 5 years since the beginning of the 21st century. This in turn means that Greenland's estimated loss of ice mass has increased from a little more than 50 km^3 $year^{-1}$ to in excess of 150 km^3 $year^{-1}$. The mechanisms behind this change are thought to include removal of the formerly stable glacier ends (so reducing the glacier ends' buttressing effect) and percolation of meltwater from the top of glaciers through to their bases, so lubricating them (Dowdeswell, 2006; Rignot and Kanagaratnam, 2006). Consequently the IPCC 2001 and the US Geological Survey estimates (Table 6.3) will need revising and so the IPCC's next assessment report (with AR5 Working Group I reporting in 2013 and Working Groups II and III reporting in 2014) may well contain increased estimates for sea-level rise. These will be at the high end of the predicted range, albeit with similar or lower estimates for sea-level rise at the low end. (The latter, possibly lower, AR5 sea-level rise estimates are probable due both to scientific uncertainty and the realization that the global economic slump beginning in 2009 could be part of a longer-term trend that reduces greenhouse emissions and hence 21st century warming.)

Such a remote future may seem to have little bearing on present-day life. Yet even a small proportion of this long-term rise would in the shorter term have a significant effect on humans and human-managed ecosystems in the present, and on land that we currently consider that we will utilise indefinitely (from our human perspective). What we have inherited from the past we largely expect to either use, or have its use evolve, in the future. Today much of our infrastructure had its practical day-to-day foundations laid centuries ago. For example, much of Europe's rail network was established well over a century ago; in Britain Brunel's bridges are still being used today. Much of Europe's principal road networks were established centuries earlier still. Yet again, the geographical location of many of humanity's major settlements were established thousands of years ago. But consider a future a thousand years hence. London, for instance, would either be submerged or protected, either as an island or as part of the south-east region of England, needing hundreds of kilometres of towering sea defences scores of metres high. London would not be alone. New York, Washington and indeed many of the Earth's major urban centres would need similar protection. Similarly, low-lying countries, such as Bangladesh, would be submerged, so they would effectively vanish. Even low-lying developed nations, such as The Netherlands, with a reputation for excellent sea protection, would have to bolster these several-fold or be submerged. The big question is what will happen between now and this far future? When exactly will a 1 m rise become manifest, let alone a 2 or 3 m rise?

Sea-level rise is a recognised dimension to global warming. Yet, other than to very low-lying states (such as the Maldives), it is all too often either not considered at all by planners above and beyond immediate risks (for instance, storm surges based on historic levels), or proper investment fails to follow policy initiatives (such as in the southern USA). Indeed when it is considered by planners, the IPCC summary position is the one taken into account without allowance for IPCC main-text caveats, let alone climate surprises. Such is the way of likely climate impacts that future sea-level rise will almost certainly receive more serious attention as this century progresses.

6.6.4 Methane hydrates (methane clathrates)

Methane is a far more powerful greenhouse gas than carbon dioxide (see Chapter 1) and releases from the dissociation of semi-stable methane hydrates in marine sediments have been thought to contribute to past warming and some mass-extinction events (see Chapter 3). So what are the prospects of such a methane release today? There is still considerable uncertainty, which is why the IPCC include methane releases as a possible surprise in their forecasts. There is still much to learn and research continues to point to this being a risk warranting attention.

In 2004 Hornbach et al. reported the discovery of methane gas being trapped in a layer below some methane hydrate provinces. The researchers state that *if* such a layer is common, and thick along passive continental slopes, then a 5°C increase in sea temperature at the sea floor could release as much as around 2000 GtC. This is the same order of magnitude of carbon as was involved in the Eocene thermal maximum. Whereas estimates as to the global inventory of methane clathrates seem to preclude this, this newly discovered layer of trapped methane may be a climate risk factor. Furthermore, it now seems that carbon from other biosphere pools was also involved in the Initial Eocene Thermal Maximum/Palaeocene–Eocene Thermal Maximum (IETM/PETM) (see section 3.3.9) and so peri-Arctic soil carbon is also an area of risk (see section 6.1.5). Then again, methane clathrate dissociation itself may well have been one of the triggers for further carbon release.

Whereas a 5°C rise is unlikely during the 21st century (but is likely for the early 22nd under the IPCC 2001 Business-as-Usual [B-a-U] scenario), a release of methane from clathrates, and subsequently other biosphere pools of carbon, would have a major (high-hazard) warming effect. The climate effect of 2000 Gt of methane would be approximately equivalent to that of 120 000 Gt of carbon dioxide on a 20-year time horizon (see global warming potentials in Chapter 1). Given that in 2001 fossil fuel burning additions of carbon dioxide to the atmosphere were just over 8 GtC year^{-1} with a further 2 GtC from land use and industry, the climate change associated with a methane release equivalent to 120 000 Gt of carbon dioxide (37 000 GtC) would be dramatic.

There could be between 2000 and 4000 times as much methane locked up as hydrates as is currently in the atmosphere, although it is unlikely that all of this would be released without a far greater warming of the ocean than is anticipated in the worst IPCC forecasts, even beyond the 21st century. However, the release of a proportion of this is likely if not inevitable with sufficient warming. The question is, how much, especially as a small proportion of even a few per cent would present a major change in atmospheric methane concentration and a major climate hazard.

Low-probability, high-hazard surprises pose great difficulty for policy-makers. Investment in monitoring a low-probability event is – virtually by definition – unlikely to see a return, yet should the event actually happen the costs would be high. A recent illustration of failing to address such a low-probability, high-hazard event was that of the 2004 Indian Ocean tsunami. Prior to 2004 it was not considered worthwhile investing in a seismic-event monitoring and alarm system in the Indian Ocean. After the 2004 tsunami, a warning system was developed. The difficulty of investing in low-probability events is also highlighted by some failures to invest in countering likely probability, high-hazard events. A recent example of this is the flooding of New Orleans in 2005 (see also section 6.2). This risk was previously identified and the economic value of New Orleans was also (obviously) known to be high, yet proposed new flood defences were not built.

So what are the prospects of a mega-large release of methane hydrates? As previously mentioned, the big concern is whether global warming beyond anything known in the Quaternary will trigger something analogous to the IETM or, worse, the Toarcian event (see section 3.3.7). Methane releases from the ocean are considered a key source (along with some terrestrial sources of carbon) of the Eocene carbon isotope excursion (CIE) and the warming that took place then (55 mya). It is thought that the release was at least triggered by volcanic action (from what is now a large igneous province) heating organic-rich (fossil fuel-type) strata, so releasing greenhouse gases.

From the CIE, we have a very rough estimate of the total ^{12}C involved. What we do not know is how much came as a result of the volcanic action on organic strata and how much came from marine methane hydrates and other sources (such as wetlands). Yet estimates are that during the 21st century, through fossil fuel and land-use change, we are likely to release a similar amount of carbon from the atmosphere that took place during the IETM. Two possibilities, which are not mutually exclusive, therefore present themselves.

The first possibility is unlikely but it frames the problem. It is that *all* the Eocene ^{12}C came as a result of volcanic action warming organic-rich strata (and none initially from methane hydrates, but only far later in the event as the oceans warmed along with other carbon). Then, because the amount of carbon involved is similar to what we are releasing now, in a worst-case scenario we would be looking at triggering an early-Eocene-like thermal event sooner rather than later. This is so even if its effects from further carbon release came much later: an analogy would be an early lighting of a fuse with detonation taking place later.

The second possibility is if *just part* of the Eocene ^{12}C came as a result of volcanic action on organic strata early in the event (during the first few thousand years), and the remainder early on from methane hydrates, then the current prospect for future warming would be worse. We already know that the *total* ^{12}C during the Eocene event is broadly comparable to the amount we expect to release over the next century under IPCC B-a-U scenarios. This then can be considered to be analogous to the carbon released due to volcanic action from the early Eocene organic-rich strata. Subsequent carbon from marine methane hydrates and terrestrial soils would be extra to this carbon. If this was so, then as today we are already in the process of releasing carbon from an Eocene-comparable fossil fuel burden, it means that any future release from hydrates and other natural biosphere reservoirs of carbon *must be above and beyond this*. In short, we could be heading for a climatic event that is greater than that during the Eocene. Equally we might perhaps see a bit of both possibilities: an event sooner rather than later, and one that is greater rather than comparable to the early Eocene temperature rise.

Whatever transpires, it is worth remembering that currently the prospect of such a mega-release is considered low even if of high hazard. Such mega-releases are only likely *if* we continue (as we currently are) down the B-a-U route. There are other options to B-a-U, but as we shall see in Chapter 8 realising these will be difficult.

Finally, it needs to be emphasised that Eocene warming and the Eocene levels of atmospheric carbon built up over several thousands of years. In the present day the same amount of carbon will be added in just a century or so. Whereas the ocean warming will be far slower, the terrestrial warming will be far faster than during the Eocene. The Eocene thermal maximum cannot therefore be considered as analogous as might be thought to a B-a-U scenario for the late 21st and early 22nd centuries. Our current predicament is worse, in that terrestrial biological systems will not have the time to adapt as they had in the Eocene: that release of carbon took centuries, not decades. Further, even though the oceans will take longer to warm because *complete* circulation-driven ocean turnover takes the order of 500–1000 years, there are also bound to be surface-ocean acidity concerns due to the same, slow, complete ocean turnover. This may well be why we are already seeing the early signs of surface-ocean acidification that also occurred during the early Eocene.

Leaving aside the above risk of a mega-release of methane, what are the chances of smaller releases from semi-stable marine methane hydrates that might drive up atmospheric methane by, say, 50–150 parts per billion by volume (ppbv), as happened during the last glacial? (For comparison the concentration in 2005 was 1774 ppbv.) This could happen as a result of either decreased water pressure or increased temperature. The chances of it being due to sea-level reduction are quite small (if not virtually zero as sea levels are rising, not falling). However, the chance of a release due to temperature increases is greater. The Earth is getting warmer and warmth increases methane hydrate instability. Methane releases during the last glacial have been deduced for interstadials (short, comparatively warm periods within glacials), which have atmospheric concentrations that are increased by 100 ppb or more (Kennett et al., 2000). Here it is not so much that the Earth is warmer during interstadials, but that the now warmer interglacial Earth may see different atmospheric and ocean circulation patterns. These could suddenly bring warmer water to an area that was previously cooler, which is conducive to hydrate dissociation. This relates to another possible surprise in the change in circulation patterns (see below).

A small methane release (boosting atmospheric methane by only 50–150 ppbv) may change global circulation patterns that themselves might greatly change the climate in a few regions. However, the overall effect of such a methane release itself in terms of raising the average global temperature would be not that great, and far smaller than the several degrees difference between a glacial and an interglacial. It might, though, be as much as the temperature difference between the depth of the Little Ice Age and today, or possibly more. Even so, the concern would be that such a sudden but smaller release would still have a rapid effect on the global climate, possibly equivalent to a few decades of IPCC-anticipated, 21st-century B-a-U warming impacting within a year or so.

Whether a big release or a small one, from our understanding of past events it does appear that methane releases from marine hydrates can either trigger climate change or exacerbate the episode of climate change that has already begun. Consequently, it is not surprising that some concerned with gathering science to underpin policy are looking seriously at such risks. Indeed, in 2004, when the UK Department of Environment, Food and Rural Affairs (DEFRA) was considering its science requirements over the next decade, it specifically included research into reasons behind sudden climate change, such as release of methane hydrates (Department of Environment, Food and Rural Affairs, 2004), as is the IPCC for its forthcoming 2013 assessment.

So what do we need to look for? Methane hydrates form at depths roughly exceeding 400 m in some places and exceeding 1 km and more in others. Aradhna Tripati and Henry Elderfield (2005) of the University of Cambridge, using fossil forams and employing a temperature proxy based on magnesium/calcium ratios, ascertained the (sea) bottom water-temperature rise at the Palaeocene–Eocene boundary. Their conclusion was that the water temperatures then rose by 4–5°C and that this was caused by a change in ocean circulation (see section 6.6.6). To put this into perspective in terms of temperature, since 1960 to the early years of 21st century's first decade much of the ocean surface rose by 0.2°C. One of the places where surface warming has been greatest is the North Pacific (0.25°C). As such, this is broadly in line with the IPCC estimate of global surface warming and so greater warming should be expected if we go down the IPCC B-a-U route. However, this warming fingerprint has only penetrated about the first 75 m of the sea column and this penetration seems to reasonably reflect current computer models (Barnett et al., 2005b). Even with further warming, as the IPCC B-a-U scenarios suggest, it is likely that a change in ocean circulation to bring warm surface waters to a depth that had previously remained cool will be necessary to trigger a major methane pulse in this century. Nonetheless, this possibility cannot be dismissed.

Finally, what of the global estimates as to the amount of marine methane clathrates present? First, no detailed mapping has been done, so what we have are estimates based on scaling up those areas surveyed. Bruce Buffett and David Archer of Chicago University in 2004 estimated that there was 3000 GtC in clathrates and 2000 GtC in methane bubbles, with 85% being released if there was to be 3°C warming (see also Hornbach et al. 2004, mentioned earlier). Second, there is much about clathrate chemistry we do not know. Indeed, in 2007 a new clathrate structure, previously discerned only in a laboratory, was found in the marine environment. Furthermore it is a fairly stable clathrate structure suggesting that it can be found under a wide range of temperature and pressure regimens (Lu et al., 2007).

Taking all this together, it may initially appear that marine methane releases are unlikely: the amounts in official estimates are not high and the ocean takes a long time to warm up. However, we do know that the Earth is warming up (even if we cannot forecast future warming exactly). Further, as shall be discussed in later sections, with warming there is an increased chance of changes in circulation patterns. The question is not of whether there will be such methane releases, but when and by how much? To address this, research on Earth system thresholds is a current priority.

6.6.5 Volcanoes

Major volcanic activity has been associated with mass-extinction events in the past (see Chapter 3). Specific events are the end-Permian or Permo-Triassic extinction 251 mya and the Cretaceous–Tertiary extinction 65.5 mya. In the former some 90% of ocean species vanished and 70% of vertebrate families on land became extinct, and with the latter 75–80% of terrestrial species died out. We have to remember that asteroid impacts have also been associated with both these extinction events (especially the Cretaceous–Tertiary extinction) but it is thought that the impacts provided the *coup de grâce* and that many species were in decline beforehand, with vulcanism implicated in the cause.

So what are the chances of a major or super-volcanic event today? There are a number of places where such significant volcanic events might take place (albeit smaller than those that created igneous provinces in either the Permian or Cretaceous periods). These include Long Valley in eastern California, Toba in Indonesia (which last erupted 74 000 years ago, and which is thought to have lowered temperatures 3–5°C for up to 5 years) and Taupo in New Zealand. However, one of the sites of greatest concern is at Yellowstone National Park in the USA. Starting 2.1 mya, Yellowstone has been on a regular eruption cycle of 600 000 years. The last big eruption was 640 000 years ago. Yellowstone has continued in a quieter fashion since, with a much smaller eruption occurring 70 000 years ago, so a large one is overdue. Indeed, today it is clear that volcanic activity beneath the park is taking place and this is signalled by marked surface deformation.

The next question is whether a Yellowstone eruption would affect the global climate and have a global biological impact. There are two ways that we might begin to make a rough estimate of the impacts. First, the Yellowstone eruption 640 000 years ago takes us to long before the Vostok ice-core record and to seven or eight glacials ago. However, the marine ^{18}O isotopic evidence does go back this far and it suggests that the eruption was roughly contemporary with the height of a particularly cold glacial (there has been none colder since). Milankovitch factors (see Chapter 1) would have made it a cold glacial anyway so the question remains as to whether that eruption made that glacial even colder than it would have otherwise been, and if so by how much and for how long?

Second, the last great Yellowstone eruption is thought to have released about 1000 km^3 of lava. This compares with a release of 1–4 million km^3 of lava in Siberia at the time of the Permo-Triassic extinction and some 512 000 km^3 from the Deccan Traps in north-west India with the Cretaceous–Tertiary extinction. As stated, whereas these eruptions were not the sole cause of the mass extinction, it is now well established (especially with regards to the Cretaceous–Tertiary event) that species were declining at the time of the eruptions. So, by this standard alone a large Yellowstone eruption may have roughly a five-hundredth the impact of Cretaceous–Tertiary vulcanism. Current global warming might actually help a little in offsetting the cooling effect of the eruptions. Even so, an eruption at Yellowstone would most likely have a climatic impact far greater than that of large eruptions of normal volcanoes that do not have large volcanic traps (a trap is the magma chamber of a volcano).

So, what of a large eruption of a standard-type volcano? The eruption of Huaynaputina in February 1600 (see Chapter 4 and Figure 2.1) was not the biggest eruption in the Little Ice Age. Nonetheless, the following few years saw very low temperatures. In Europe the summer of 1601 was cold, with freezing weather in northern Italy extending into July and overcast skies for much of the year, while in parts of England there were frosts every morning throughout June. Yet despite this the economic impact of this climate change was marked, although not as clear cut as one might expect. The eruption happened to follow some years of the late 1590s that also saw climate-related food shortages. By 1601 people were adapting (as they will in the future once they have experience of global warming in their area). In England in 1600–10 documented wheat prices were not high. However, the purchasing power of non-agricultural workers, as revealed by English builders' wages, not only dipped due to the earlier shortages of the 1590s but failed to rise significantly throughout much of the first half of the 17th century, after which there was a steady recovery and upward trend lasting a century (Burroughs, 1997). Builders' wealth is a useful indicator of societal economic well-being, as building is inevitably associated with a wealthy society, but builders themselves do not generate food. Consequently builders' wealth is in no small part an indicator of a society's agricultural surplus. So the Huaynaputina eruption during the Little Ice Age, while not having a devastating effect on the economy, does coincide with a period of little food surplus (if anything, a marginal food deficit).

While the economically discernable effect on agricultural systems from standard (as opposed to super-) volcanoes lasts only a few years, their thermal impact on the biosphere lasts longer. Evidence comes from today's computerised climate models that are sufficiently good to simulate at least most continental-scale climate events of which we have knowledge (be it through direct measurement or from several sets of palaeoclimatological data). Of course, there is still work to be done regarding modelling of high latitudes and increasing the level of spatial resolution, not to mention in reducing uncertainty relating to, and increasing the number of, the various climate forcing factors. Nonetheless, that many of today's global models do successfully capture past climatic change suggests that, even if much refinement is still possible, they do have a very meaningful role to play in informing us about likely climate factors (and indeed policy). There is some evidence that volcanic eruptions, such as Krakatau in 1883, have a small, but clearly discernable, impact on deep-ocean temperature that lasts for between 50 and 100 years.

Twelve state-of-the-art global climate models were analysed, both including and omitting the impact of the Krakatau eruption and its injection of particulates, gases and aerosols into the atmosphere. Each of these models differed in its physics, resolution, initialisation and ocean–atmosphere coupling as well as different combinations and strengths of forcing factors. However, each successfully captured the principal climate features of the past couple of centuries, albeit coarsely and within certain margins of error. However, the analysis did show that simulations that included a Krakatau-type eruption noticeably differed from those that did not. The eruption's thermal fingerprint is clearly seen in the simulated oceans, with a small but clear temperature difference of around $0.01°C$ in the top kilometre of the ocean column that lasted for about 50–100 years. This analysis also revealed that present-day (non-super-) volcanoes have a smaller cooling event. The simulated heat-content recovery after the 1991 Pinatubo eruption (which was comparable to Krakatau in radiative climate forcing terms) was far quicker: this was actually measured by satellite monitoring data. The reason for this more speedy oceanic thermal recovery was that at the time of Pinatubo (and more so now) there were rising greenhouse gas emissions and global warming taking place. Anthropogenic greenhouse gas emissions at the time of Krakatau were roughly a tenth of those at the time of Pinatubo. The thermal impact on the oceans of each volcano may be small, and shorter-lived in today's faster-warming planet than the past, but taken together over time they do have a more pronounced effect. The importance of this work is that it helps validate components of climate models; in this case the inclusion of volcanic activity (Gleckler et al., 2006). Including volcanic eruptions in models also helps current forecasts, as there are different likelihoods of global warming depending on whether or not there is a run of volcanic activity. Remember that much of the 20th century up to the 1960s was largely free of major (standard) volcanic eruptions, whereas the period from the 1960s to the mid-1990s saw considerable climate forcing volcanic activity.

Although standard-sized volcanic eruptions do seem to impart a discernable thermal fingerprint on the oceans and large volcanoes may impact on food prices for a couple of years, a super-volcanic eruption has a different order of impact that noticeably affects biological systems. By comparison today, should the Yellowstone trap erupt, it would impart a much greater level of impact than a standard Huaynaputina-type volcano, so that there would almost certainly be significant food-security implications. These would be global, so there would be no major bread-basket country from which to import food. Furthermore, eruption size aside, the lack of food security would be far worse than the European food shortages experienced periodically during the Little Ice Age, due to the 21st century's high population density (and the concomitant smaller areas of woodland in which to forage for alternatives) and the lack of recent experience of such conditions.

In 2005 the Geological Society of London established a working group to examine the probability and likely global effects of a super-eruption. Its report's conclusions called for the completion of a database of super-eruptions over the past 1 million years. It also recommended that the UK Government establish a task force to consider the environmental, social and political consequences of large-magnitude volcanic eruptions and that this involve international collaboration (Geological Society, 2005).

6.6.6 Oceanic and atmospheric circulation

As the Earth warms it is possible that new oceanic and atmospheric circulation patterns may become manifest. For example, the Broecker thermohaline ocean circulation (see Chapter 4) is one of the key mechanisms by which heat is transported away from the tropics to higher latitudes. It is driven by warm water from the tropics moving northwards and becoming saltier due to evaporation. At higher latitudes this denser water becomes even more dense through cooling and so sinks. A current in the ocean depths then carries it between oceans, where it surfaces elsewhere. One of the key places where a warm current from the tropics sees considerable evaporation and then sinking at high latitudes is the Gulf Stream in the North Atlantic, with sinking taking place around Iceland and Greenland. However, freshening of this water, such as by iceberg discharges, Arctic ice melt or changes in regular precipitation patterns, would make the current less dense and so reduce sinking. If the reduction was sufficiently great it could shut the conveyor down. This might block the Gulf Stream's northern route. Of course, the Coriolis force from the Earth's rotation would still continue, as would trade winds, so a warm current would still leave the Gulf of Mexico, but it would not travel so far north, but rather circle around at a lower latitude. Without the Gulf Stream off north-western Europe, the British Isles, the North Sea continental nations and the western edge of Scandinavia would see far harsher winters, just as are today experienced in Labrador, which is on the same latitude as Great Britain.

Although we do not have a particularly thorough understanding of the conveyor's precise operation, the current understanding is that a complete shut-down of the North Atlantic driver of the conveyor is unlikely (but not impossible). However, a weakening of the conveyor by around 30% is (again, currently) considered a reasonable possibility. If this happened then seasonality in north-west Europe is likely to increase as winters would not warm in pace with summers due to long-term climate change.

Whereas such a change in ocean circulation would result in agricultural dislocation and the need for infrastructure investment, the effect would be regional. Overall the Earth would still be experiencing global warming, it is just that the climatic patterns would change more markedly on a regional basis in many places. This would have profound consequences for north-west Europe.

It is known from salinity studies that more fresh water has been added to the North Atlantic in the 20th century. In particular, during the late 1960s a large pulse of fresh water entered the Nordic Seas (between Greenland and Scandinavia) through the Fram Strait. This freshwater pulse was of the order of around $10\,000$ km^3 at a rate of about 2000 km^3 year^{-1}, implying a net flux anomaly of approximately 0.07 Sverdrups (Sv; where 1 Sv $= 10^6$ m^3 s^{-1}) during a 5-year period. It is known as the Great Salinity Anomaly (GSA). Naturally there is always fresh water being added (early during interglacials) into the Arctic from rivers and ice melt, but the GSA is thought to represent roughly around a 40% increase of normal flow. In 2005 Ruth Curry, from the Woods Hole Oceanographic Institution, and Cecilie Mauritzen, from the Norwegian Meteorological Institute, used these long-term salinity data sets to ascertain the likelihood of an effect on the Broecker thermohaline circulation. In a carefully worded conclusion they said that, assuming the rates of fresh water observed between 1970 and 1995 continued, it would take a century for enough fresh water (9000 km^3) to accumulate to affect the ocean that is exchanged across the Greenland–Scotland ridge and two centuries to stop it. Consequently disruption of the ocean thermohaline circulation did not appear imminent. However, they noted that this assumption would not hold if enhanced high-latitude northern hemisphere precipitation and/or glacial melt increased. That these are predicted in many climate-model forecasts for the 21st century means that disruption of the thermohaline circulation cannot be ruled out.

If the currently prevailing view is that the North Atlantic driver of the conveyor (Atlantic meridional overturning circulation) will only modestly weaken – a view of the 2007 IPCC assessment – what is the evidence for this and for the notion of perhaps an even greater conveyor shutdown with further warming? As stated above, it is known that in the 20th, but also early in the 21st, century key parts of the North Atlantic have been freshening. Part of this is thought to be due to Greenland ice-cap melt. However in 2005 research, funded by the UK Department for Environment, Food and Rural Affairs (DEFRA) and carried out by the Met Office's Hadley Centre for Climate Prediction and Research in Exeter by Peli Wu and colleagues, showed that the Arctic Ocean is freshening due to an increasing hydrological cycle (Wu et al., 2005). Remember, in a warmer planet there is more ocean evaporation and hence rainfall. What Wu and colleagues did was to look at data from the six largest Eurasian rivers flowing into the Arctic Ocean since the mid-1930s, namely the Yenisey, Lena, Ob, Pechora, Kolyma and Severnaya Dvina. This showed that river flows had increased by about 2 ± 0.7 mBe year^{-1} (1 Be or Bering $= 10^3$ km^3) over a 140-year historic average of 2.3 mBe year^{-1}. This was a very good agreement with the predictions from runs of the HadCM3 climate model of 1.8 ± 0.6 mBe year^{-1}. With this qualified validation of the model, they then ran the model to forecast the future to the end of the 21st century assuming standard IPCC predictions. It showed that the discharge from Eurasian rivers may increase by between 20 and 70% during the course of this century. This is quite a range and so the modelling still needs to be improved. Nonetheless, such results have implications for the Broecker circulation and this line of research is among that being considered by compilers of the fifth IPCC assessment due in 2013. However, it is likely that we will still need to accrue at least a decade or so of data before we gain a clear view based on field evidence rather than a perspective derived from theoretical computer models.

The UK's Natural Environment Research Council (NERC) Rapid Climate Change programme used a North Atlantic array of sensors to measure the circulation in the field. It found that between March 2004 and September 2006 the Atlantic meridional overturning circulation went through an annual cycle varying in strength from winter weaks of less than 10 Sv to summer strengths of more than 30 Sv. The success of the NERC pilot scheme (2004–8) ensured the future of the RAPID-Watch array until 2014 (Natural Environment Research Council, 2008). So the question is, will increased peri-Arctic melt run-off alter this, and if so by how much?

Changes in atmospheric circulation are also possible and again these would mainly have regional consequences and be less likely to have as pronounced an effect on the Earth's overall temperature. Nonetheless, regional consequences could be severe, especially if climate change disrupted water-distribution events such as the monsoon. Even if major individual water-distribution events remained unaltered, globally water distribution about the planet via the atmosphere will almost certainly change. Indeed, this is connected to the predictions of regional precipitation change and water availability noted earlier in the case studies of climate and natural systems of the USA and UK. Water availability is not only important for ecosystem function and geophysical processing but (as shall be explored in the next chapter) it also fundamentally affects human ecology and therefore economic activity. Further to the above work on the hydrological cycle as it affects Arctic rivers, a global assessment of stream flow and water run-off suggests that there will be considerable change in many areas by the mid 21st century compared with the 20th century.

In 2005 an ensemble of 12 climate models was used by Milly and colleagues from the US Geological Survey to see how coherently they forecast changes in run-off (run-off is defined as precipitation less evapotranspiration). The picture emerging is that as the climate warms there is more ocean evaporation and hence precipitation, but also that this precipitation is subject to more evaporation. The picture is therefore somewhat complex, with some areas having more precipitation but paradoxically less run-off due to this increased evapotranspiration. Overall, most areas are likely to see change in run-off and this change is likely to continue as climate change (in the thermal sense) progresses through to the middle of the 21st century (which is as far as this analysis forecasted). Initial increases of run-off seen in the 20th century are projected to reverse in some regions in this century, namely eastern equatorial South America, southern Africa and the western plains of North America. The modelled drying of the Mediterranean extends north from Spain and Greece (which we were beginning to see by the end of the 20th century) to south-eastern Britain, France, Austria, Hungary, Romania and Bulgaria as well as the sub-Russian states. In many regions *all* the models in the ensemble agreed on the likely mid-21st-century change. These agreements include increases of typically 10–40% in the high latitudes of North America and north Eurasia (as in Wu et al., 2005), in the La Plata basin of South America and in eastern equatorial Africa as well as some of the major islands of the eastern equatorial Pacific. Agreements on decreasing run-off include southern Europe, the Middle East, mid-latitude western North America and southern Africa (Milly et al., 2005).

And then there is the weather factor, and specifically how individual weather events (such as cyclone/hurricane strength) are altered by global climate change. In 2007 US climatologists Ryan Sriver and Matthew Huber calculated the effect of tropical cyclones on depth of sea-water mixing. Their results indicated that tropical cyclones are responsible for significant heat transport by mixing heat from the air into the sea. They concluded that around 15% of peak ocean heat transport may be associated with vertical mixing by tropical cyclones and that the magnitude of this mixing is related to sea-surface temperature. This phenomenon was not at the time included in computer models (and so is another example of new discovery driving ongoing model improvement).

The possibility of changes in oceanic and atmospheric circulation is one current focus of climate research. It appears that circulation systems interact with the global climate system through a number of feedback cycles (see Chapter 1). These result in semi-stable global climate modes, with each mode having its own characteristic regional distribution of local climate types. Consequently the stability of a global climate mode is related to the stability of circulation patterns such as ocean thermohaline circulation (Clark et al., 2002). Perturb the climate a little too much and this may affect thermohaline circulation (as discussed above). Perturb the thermohaline circulation and this can affect ocean circulation, creating new regional patterns of warm and cool localities. Create new regional temperature patterns and this will affect atmospheric circulation, and so the Earth system will settle into a new semi-stable state until further climatic perturbation takes place. Note that in addition to immediate regional climatic impacts a change in the thermohaline circulation could possibly disrupt marine methane hydrates (see section 6.6.4) and these would have an additional and global climatic effect.

So, how likely is a circulation change? The IPCC's warming projections for the 21st century are fairly linear. Most of their scenarios result in near-linear projections; projections that do not appear to contain a quantum jump. Yet the palaeoclimatic record is full of sudden jumps and this alone suggests that we need to take changes in oceanic and atmospheric circulation seriously.

If the question of rapid change deviating from the linear has any credence, and if such fluctuations are discernable at the global level, then they should be seen within major biosphere subsystems (remembering that the biosphere includes the geosphere). In 2005 a team led by Chih-hao Hsieh of the Scripps Institution of Oceanography in San Diego, California, published work suggesting that large-scale marine ecosystems (such as oceans) are dynamically non-linear. As such they have the capacity for dramatic change in response to stochastic fluctuations in basin-scale physical states. This, they say, is exemplified by fish-catch data and larval fish abundance that are prone to non-linear responses and do not simply track a dimension of physical change such as sea-surface temperature. In short, ecological catastrophes (such as a species' regional population crash) may be the result of a modest change in, say temperature, at a critical point. This has implications for adopting the precautionary principle when estimating ecosystem sustainable yields in a time of change.

Of course, as noted a number of times, biological systems feed back into climate. The area of wetlands affects atmospheric methane, which in turn affects climate. Vegetation cover, be it a change in or decline of cover, affects regional albedo and so local climate. So, if the biological response to modest linear environmental change is not necessarily itself linear, and biology affects environment, then environmental change such as climate change will also be non-linear. On one hand this does not lend confidence to the IPCC's near-linear forecasts. On the other, the IPCC do provide high and low near-linear forecasts and so can be said to provide a window in which non-linear responses can take place. This theoretical line of argument for non-linear change is all very well, but what is happening in the real world at the macro scale?

There was much concern with regard to changes in atmospheric and ocean circulation in the late 1990s and early 21st century. Computer modellers have undertaken much of the work in possible atmospheric circulation changes. Conversely, work on ocean thermohaline circulation has been undertaken more equally by those making field observations whose results then feed into modellers' work. Changes in thermohaline circulation are of particular concern as field observations have shown that some factors potentially affecting the circulation are beginning to take place. For example (just one of many), there has been a decrease since 1950 of cold, dense water from the Nordic Seas across the Greenland–Scotland ridge. The decrease is of the order of 20% and if this reduction is not compensated for by increased cold, dense water from elsewhere then the thermohaline circulation could be weakening (Hansen et al., 2001). What is known is that the North Atlantic around Iceland is freshening. The thinning of Arctic ice (itself a result of climate change) could account for part of this freshening. Sea-level rise has not been as great (considering contributing factors such as ocean thermal expansion and mountains glacier melt) yet glaciers almost everywhere around the world are in retreat so some of this freshening could come from North Atlantic continental glacier melt (with the missing sea-level rise perhaps being accounted for by increased evaporation and precipitation as snow over Antarctica taking water out of the ocean system). If Arctic melt accounts for much of this freshening then the problem will be comparatively short-lived as it is likely that the Arctic could become largely ice-free during the summer in the course of the 21st century. Ice-free summers mean that the Arctic annual carry-over store of freshwater ice has been depleted and so major freshening of the North Atlantic from this source would cease.

Meanwhile, in the southern hemisphere the Ross Sea indeed became fresher during the late 20th century, which itself has thermohaline implications for one of the drivers of the Broecker conveyor in that hemisphere (Jacobs et al., 2002). Of course, a change in any single driver of the conveyor has implications for the conveyor as a whole; hence, remote impacts even in another hemisphere are important. The possibilities of changes in the Broecker circulation in the North Atlantic are, as mentioned, the subject of research initiatives in the UK and Norway and these nations have engaged in joint research programmes, whereas European, North American and Australasian researchers have undertaken work on the thermohaline circulation in the southern hemisphere.

6.6.7 Ocean acidity

Human additions of carbon dioxide to the atmosphere affect flows to other carbon cycle reservoirs and notably flows into the ocean (see section 5.4). This has resulted in increased ocean acidification, much like that thought to have happened early in the Eocene. However, there are limits as to how acid the ocean is likely to become; after all, there have been times both tens and hundreds of millions of years in the past when atmospheric carbon dioxide was higher than today and calciferous sediments such as chalk did not dissolve. In part this is because should the ocean (or rather the ocean/atmosphere interface) become saturated then sea uptake of carbon dioxide would be reduced. This would mean that the atmospheric increase in carbon dioxide associated with the burning of a unit of fossil fuel would rise and with it the climatic impact. Also, acidity would be stabilised by buffering.

With little other than clues from the Eocene thermal maximum and Toarcian events (see Chapter 3), we do not know exactly what will happen to ocean pH if an IPCC B-a-U scenario were followed for just three centuries to the year 2300, but a handful of researchers are beginning to look at the problem. In 2003 Ken Caldeira and Michael Wickett of the Lawrence Livermore National Laboratory ran the IPCC

1992a scenario (IS92a) through Livermore's ocean general-circulation model. This assumed that fossil fuel consumption (and greenhouse gas emissions) would continue to grow much as they have from 1950 to date through to the year 2100 and then begin to stabilise logistically to 2300 as the economically recoverable reserves were used up. The IS92a scenario is, in terms of the IPCC Special Report on Emission Scenarios (SRESs) used for their 2001 assessment (and with respect to their 21st-century forecasts), a middle-of-the-road B-a-U scenario.

Caldeira and Wickett's results suggest that the atmospheric carbon dioxide level would exceed 1900 ppm by 2300. At first this might seem acceptable; after all, some 500 mya the biosphere had atmospheric concentrations up to some 7500 ppm (see Figure 3.1) and calcareous marine species survived that. However, the current anthropogenic carbon dioxide pulse is different from that early Palaeozoic rise. That peak was reached after many millions of years of carbon dioxide build-up; conversely, the current human B-a-U rise would be literally millions of times faster. When an atmospheric carbon dioxide change of such magnitude takes place in less than 1000 years then geological weathering and the resulting buffering cannot take place. For greater buffering it is thought that such a fossil-fuelled carbon dioxide increase would need to take place over tens of thousands of years or longer. Consequently, Caldeira and Wickett's simulation predicts that the pH reduction would be 0.7. Currently there is no evidence that ocean pH has been more than 0.6 pH units lower than today. Their conclusion is that unabated carbon dioxide emissions over the coming centuries may produce changes in ocean pH that are greater than any experienced since 300 mya, with the possible exception of those resulting from some mass-extinction events. This suggests that a marine mass-extinction event, certainly of calcareous forams, would be a possibility.

Then in 2005 an international team led by James Orr used 13 ocean carbon cycle models to assess ocean chemistry, especially with respect to calcium carbonate and its associated ions. As with the Caldeira and Wickett team, the Orr team were again relating ocean chemistry to the IPCC's IS92a scenario. Using pteropods, a group of species of gastropod mollusc that have calcareous shells, in tanks of water matching the chemistry of the IS92a scenario they attempted to see whether there were any adverse effects. They found that water simulating the IS92a scenario up to the year 2100 did affect species. This was particularly a problem at high latitudes when the sea water becomes undersaturated with aragonite (a metastable form of calcium carbonate). According to the models, aragonite undersaturation is likely to take place around the middle of the century at some high latitudes and by 2100 throughout the entire Southern ocean and in the sub-Arctic Pacific. The researchers exposed a small population of a live pteropod (*Clio pyramidata*) to water simulating IS92a-forecast, high-latitude ocean waters. They observed 'notable dissolution' of their calcareous shells (see section 5.4).

The ecological consequences of ocean acidification are likely to be significant. The researchers suggested that 'conditions detrimental to high latitude ecosystems could develop within decades [of 2005], not centuries as [previous work] suggested' (Orr et al., 2005).

Pteropods, especially at high latitudes, form integral components of food webs. Species high in the food web also rely on pteropods and include North Pacific salmon (*Oncorhynchus*), mackerel (family Scombridae), herring (*Clupea*), cod (family Gadidae) and baleen whales (Cetacea of the suborder Mysticeti). In the Ross Sea, for example, the prominent sub-polar pteropod, *Limacina helicina*, sometimes replaces krill as the dominant zooplankton. Indeed, in this sea pteropods account for the majority of the carbonate and organic carbon-export flux. If pteropods cannot grow their shells then they are unlikely to survive. If carbonate and organic carbon fluxes are altered then this could also conceivably have climatic repercussions.

Calcite is another form of calcium carbonate that is a little more stable than aragonite. According to the ocean-chemistry models that the researchers developed from the IS92a scenario, calcite undersaturation will lag that for aragonite by some 50–100 years. Species that use the calcite form of calcium carbonate include Foraminifera and coccolithophorids. Moreover, some Antarctic and Arctic species use magnesium calcite, which can be more soluble than aragonite. These include sea urchins (echinoderms). Although all these species may survive in warmer latitudes, nearer the poles ecosystems are likely to be severely disrupted. Further, should acidification continue well beyond this century then oceanic refugia for vulnerable species will become more scarce and the likelihood of an extinction event more probable. Finally, the IPCC 2007 assessment notes that the net effect on the biological cycling of increased ocean acidity is not well understood.

6.6.8 Climate thresholds

Sudden jumps in the global climate, or regional conditions and processes, at first might seem surprising given that the IPCC forecasts depict gradual change. Yet the IPCC do warn of surprises and sudden[6] jumps can take place as the biosphere system, or components thereof, crosses a critical transition. In such circumstances the Earth system, or components thereof, can be said to have passed a critical threshold and entered a new state. A good example of a climate threshold at the global scale and associated critical transition is that of those between the current Quaternary Earth's glacial and interglacial states (see Chapter 4); the last of these occurred before 10 000 years ago when in a few thousand years the Earth system flipped from the glacial mode and crossed the climate threshold into the Holocene interglacial mode (section 4.6.3) that we have today.

The phenomena of critical transitions may seem new but readers will undoubtedly have come across them without realising many times before. For example, supply heat at a steady rate to water and the temperature rises in a near-linear way. But continue to supply heat and suddenly the system flips to a new state: the water boils. Indeed, at the point near boiling it takes extra energy (latent heat) to get the water to actually vaporise. The water has reached a critical threshold and is transitioning to a new state.

---

[6] 'Sudden' might be a year in which a region's seasonal rains, which had failed periodically, fail and then do not return in subsequent years. It might be just a few decades within a century of warming regarding the disappearance of end-summer Arctic ice. Sudden might also be in the *geological* sense of a few centuries or even a few thousand years (a very brief time, geologically speaking) but involve a major change to the entire global system.

(a)

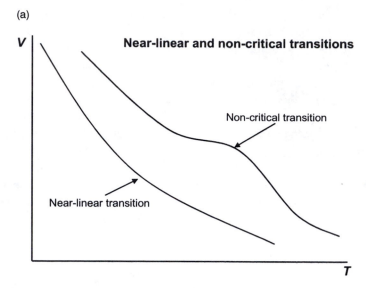

(b)

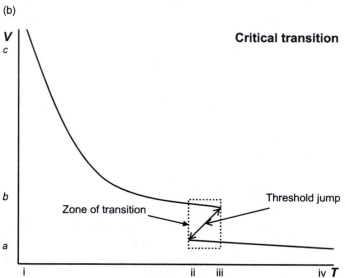

Fig. 6.11   (a) A near-linear relationship between two variables and a relationship with a non-critical transition. (b) A relationship between two variables that includes a critical transition. See text for discussion.

The idea of critical transitions in complex systems came more to the fore in popular science in the 1980s with catastrophe theory and bifurcation theory and in the 1990s with chaos theory.

Examples of near-linear and non-critical (sometimes called non-catastrophic) transitions are given in Figure 6.11a. (The example of a linear transition is not given, as this is simply a straight line.) Figure 6.11b shows a critical transition (or catastrophic transition in the parlance of catastrophe theory). This shows two states that are not directly connected. Consider two variables, V and T. If it helps, as this is a book about global warming you might consider T to be temperature and, to take a purely hypothetical example in the form of a component of the Earth system, V might be the annual end-summer volume of ice on some high-latitude island. To continue this purely invented analogy, as the temperature warms the ice volume on the island decreases. However, as warming continues we may get more ocean evaporation and hence precipitation as snow, and so the rate of ice depletion on the island slows. But a point will come with continued warming when the extra snowfall cannot counteract the effect of the warming and suddenly all the ice on the low-lying area of the island melts. Indeed, that this can happen quite fast may not come as a surprise given local albedo effects: less ice and less sunlight reflected by the landscape and so forth and so even more warming (see Figure 1.8). Snow and ice on the island's mountains are all that remains and even these continue to decline with even further warming and snow lines rise up our hypothetical island's mountains.

There are two important points about critical transitions. First, the transition does *not* take place about a specific point, but in a zone. This is why the term 'tipping point' is not appropriate. There is instead a zone of transition. Second, there is a large, sudden change in the zone of transition. (This is why some do call it a tipping point, as it conveys the notion of a large change in one variable for a very small change in another variable.) It is as if the system has two overlapping states: between the area encompassed by i–iii/*b–c* on Figure 6.11b and the area encompassed by ii–iv/near-horizontal line level with *a*. The states of the system within these areas might be considered analogous to areas of comparative stability around a 'Lorenz attractor', in the parlance of chaos theory, outside of the transition zone.

Other examples of components of the Earth system thought to potentially exhibit critical transitions include the Meridional Atlantic Turnover, Greenland ice-cap melt and the south-east Amazon ecosystem. Other examples where the whole Earth system went through a critical transition include the onset and termination of the Toarcian and Eocene carbon isotope excursions.

The question that arises is whether or not it is possible to predict that changes in a system are driving it towards a critical transition. In 2008 a team of German and Dutch researchers, led by Vasilis Dakos and Marten Scheffer, examined data from eight ancient abrupt climate change events. They showed that all were preceded by a characteristic slowing down of climate fluctuations and that this slowing down started well before the actual shift across the threshold. By slowing down they mean that far away from the transition zone and abnormal event (be it a drought or whatever) sees the system quickly return to normal. However, near the transition zone the system only returns to past normality slowly (e.g. the drought persists for a while). Such slowing down, measured as what mathematicians call increased 'autocorrelation', Dakos et al. showed to be a hallmark of climate critical transitions. The following year this question was reviewed in *Nature* with the inclusion of other possible mathematical diagnostics (Scheffer et al., 2009).

The problem is that a long-term environmental database needs to be analysed to see this change in autocorrelation and other diagnostic signals. But long-term research projects are time-consuming and so comparatively expensive. Consequently considerable foresight, as well as the willingness to invest in the research, is needed to determine in good time which key long-term monitoring databases need to be established.

6.6.9 The probability of surprises

It may seem oxymoronic to discuss the probability of surprises, for if it were possible to calculate any such events with accuracy then strictly the surprise would not be a surprise. However, we do know a few things about the surprises discussed so far.

First, regarding vulcanism, we know that over a period of centuries there is a certain level of volcanic activity that includes large eruptions of standard-type volcanoes and that over far longer timescales the eruption of volcanic traps occurs. That, for example, a Yellowstone eruption is overdue on a geological timescale of hundreds of thousands of years translates to such an eruption in the next 100 years as being not particularly surprising. We also know that in the past other surprises, such as methane hydrate releases, have happened. With further elucidation of their mechanisms we may be able to make better estimations of the probability of future happenings.

Second, we know we are not in a steady-state situation. We are not just talking about the early 21st-century world being warmer than the 19th century, but a world that is continuing to get warmer.

If the second point was not significant enough there is a third, that anthropogenic releases of carbon have been rising almost exponentially since the Industrial Revolution (a graph of historic carbon releases on a logarithmic scale against time is nearly linear). We are therefore steadily accelerating the greenhouse forcing of climate year on year: the increase in warming noted in the previous paragraph is not linear but has an exponential component.

Fourth, some greenhouse surprises (for example ocean circulation changes and methane from hydrate releases) may be linked. Surprises can compound and sometimes in unexpected ways, resulting in a critical transition and a climate threshold being crossed.

Finally, we are reaching the limitations of Quaternary experience. The 20th century saw the global climate warm to levels not seen for more than 500 years. Early in the first quarter of the 21st century we are likely to see temperatures not previously reached for 750 years since the MCA (MWP). Sometime before the middle of the 21st century could well see temperatures rise to those of the Holocene maximum or above, which took place some 8000–5000 years ago (Chapter 4). Shortly after that, sometime around the middle of the century, we are likely to reach a temperature comparable to that experienced at the height of the last interglacial. Some decades before the end of the 21st century (should the IPCC forecasts be largely correct) the Earth is likely to experience a greenhouse regime broadly analogous to a time at least 1.0–3.5 mya, and by the century's end there could be conditions not seen since the Pliocene. ('Broadly analogous' because these palaeo-analogues did not have quite the same greenhouse forcing as today due to the circulation reasons discussed in Chapter 4.) The passing of each of these successive stages is likely to increase the prospects of a climatic surprise.

One purpose of current climate research is to try to ensure that IPCC surprises don't surprise us. Climate-related policy needs to reflect this. Furthermore, just because the IPCC focused on a 100-year forecast does not mean that greenhouse problems will cease or become static in the year 2100. Global fossil fuel reserves (especially dirty oils such as orimulsion, shale gas and lower-grade coal) could continue to provide a significant proportion of civilisation's energy budget well into the 22nd century. This may seem like a long time hence and not worth worrying about but it is not an especially long time in geological and evolutionary terms. Most nations' capital cities have many buildings and infrastructure that were built two or more centuries ago. In London commuter trains use bridges built by Brunel in the 19th century, so it would not be ridiculous – indeed it would be most prudent – to plan with this time frame in mind at the very least, rather than the shorter-term considerations all too often seen.

Of course, encouraging policies that are likely to offset the worst effects of future climate surprises is easier said than done. After all, by definition, surprises are surprising. However, a clear priority to inform such policies is research into natural processes that may become the drivers of abrupt climate change. Here special attention needs to be given when these drivers function close to sensitive thresholds that may become critical. So, identifying such thresholds is equally important.

To engage in a longer-term perspective the likely climatic impacts on human society, and civilisation's probable energy requirements for the remainder of this century, need to be assessed. This is not a straightforward task as the human ecology of the 21st century will inevitably be as different from that of the 20th century as that century was from the 19th. Human ecology as it relates to climate change and greenhouse gas emissions in the 21st century is outlined in the next chapter.

6.7 References

ACIA (2005) *Arctic Climate Impact Assessment*. Cambridge: Cambridge University Press.

Alford, R. A. (2011) Bleak future for amphibians. *Nature*, 480, 461–2.

Alley, R. B., Clark, P. U., Huybrechts, P. and Joughin, I. (2005) Ice-sheet and sea-level changes. *Science*, 310, 456–60.

Atkinson, A., Siegel, V., Pakhomov, E. and Rothery, P. (2004) Long-term decline in krill stock and increase in salps within the Southern Ocean. *Nature*, 432, 100–3.

Baldocchi, D. (2011) The grass response. *Nature*, 476, 160–1.

Barber, V. A., Juday, G. P. and Finney, B. P. (2000) Reduced growth of Alaskan white spruce in the twentieth century from temperature-induced drought stress. *Nature*, 405, 668–72.

Barbraud, C. and Weimerskirch, H. (2001) Emperor penguins and climate change. *Nature*, 411, 183–5.

Barnett, T. P., Adam, J. C. and Lettenmaier, D. P. (2005a) Potential impacts of a warming climate on water availability in snow-dominated regions. *Nature*, 438, 303–9.

Barnett, T. P., Pierce, D. W., AchutaRao, K. M. et al. (2005b) Penetration of human-induced warming into the World's oceans. *Science*, 309, 284–7.

Behrenfeld, M. J., O'Malley, R. T., Siegel, D. A. et al. (2006) Climate-driven trends in contemporary ocean productivity. *Nature*, 444, 752–755.

Bellamy, P. H., Loveland, P. J., Bradley, R. I., Lark, R. M. and Kirk, G. J. D. (2005) Carbon losses from all soils across England and Wales 1978–2003. *Nature*, 437, 245–8.

Berry, P. M., Harrison, P. A., Dawson, T. P. and Walmsley, C. A. (eds) (2005) *Modelling Natural Resource Responses to Climate Change II (MONARCH II): a Local Approach*. Technical report. Oxford: UKCIP.

Bertrand, R., Lenoir, J., Piedallu, C. et al. (2011) Changes in community composition lag behind climate warming in lowland forests. *Nature*, 479, 517–20.

Blanchon, P., Eisenhauer, A., Fietzke, J. and Liebetrau, V. (2009) Rapid sea-level rise and reef back-stepping at the close of the last interglacial highstand. *Nature*, 458, 881–4.

Blaustein, A. R. and Dobson, A. (2006) A message from the frogs. *Nature*, 439, 143–4.

Boorman, L. A., Goss-Custard, J. D. and McGrorty, S. (1989) *Climate Change, Rising Sea Level and the British Coast*. ITE Research Publication no. 1. London: HMSO.

Both, C., Bouwhuis, S., Lessells, C. M. and Visser, M. E. (2006) Climate change and population declines in a long-distance migratory bird. *Nature*, 441, 81–3.

Buffett, B. and Archer, D. (2004) Global inventory of methane clathrate: sensitivity to changes in the deep ocean. *Earth and Planetary Science Letters*, 227, 185–99.

Burroughs, W. J. (1997) *Does the Weather Really Matter? The Social Implications of Climate Change*. Cambridge: Cambridge University Press.

Burrows, M. T., Schoeman, D. S., Buckley, L. B. et al. (2011) The pace of shifting climate in marine and terrestrial ecosystems. *Science*, 334, 652–5.

Caldeira, K. and Wickett, M. E. (2003) Anthropogenic carbon and ocean pH. *Nature*, 425, 365.

Carpenter, K. E., Abrar, M., Aeby, G. et al. (2008) One-third of reef-building corals face elevated extinction risk from climate change and local impacts. *Science*, 321, 560–3.

Chen, I.-C., Hill, J. K., Ohlemüller, R., Roy, D. B. and Thomas, C. D. (2011) Rapid range shifts in species associated with high levels of climate warming. *Science*, 333, 1024–6.

Clark, P. U., Pisias, N. G., Stocker, T. F. and Weaver, A. J. (2002) The role of the thermohaline circulation in abrupt climate change. *Nature*, 415, 863–9.

Climate Commission (2011) *The Critical Decade: Climate Science, Risks and Responses*. Canberra: Department of Climate Change and Energy Efficiency.

Collins, S. and Bell, G. (2004) Phenotypic consequences of 1,000 generations of selection at elevated CO_2 in a green alga. *Nature*, 431, 566–9.

Cook, E. R., Woodhouse, C. A., Eakin, C. M., Meko, D. M. and Stahle, D. W. (2004) Long-term aridity changes in the western United States. *Science*, 306, 1015–18.

Coulson, T. and Malo, A. (2008) Case of absent lemmings. *Nature*, 456, 43–4.

Cowell, R. K., Brehm, G., Cardelús, C. L., Gilman, A. C. and Longino, J. T. (2008) Global warming, elevational range shifts, and lowland biotic attrition in the wet tropics. *Science*, 322, 258–61.

Cowie, J. (1998) *Climate and Human Change: Disaster or Opportunity?* London: Parthenon Publishing.

Crimmins, S. M., Dobrowski, S. Z., Greenberg, J. A. et al. (2011) Changes in climatic water balance drive downhill shifts in plant species optimum elevations. *Science*, 331, 324–7.

Curry, R. and Mauritzen, C. (2005) Dilution of the northern North Atlantic Ocean in recent decades. *Science*, 308, 1772–4.

Dakos, V., Scheffer, M., van Nes, E. H. et al. (2008) Slowing down as an early warning signal for abrupt climate change. *Proceedings of the National Academy of Sciences USA*, 105(38), 14308–12.

Department of Environment, Food and Rural Affairs (2004) *Evidence and Innovation: DEFRA's Needs from the Sciences over the Next 10 Years*. London: DEFRA.

Department of Environment and Transport for the Regions for the UK Climate Impacts Programme (2000) *Climate Change and UK Nature Conservation: a Review of the Impact of Climate Change on UK Species and Habitat Conservation Policy*. London: DETR.

Department of State, US (2002) *US Climate Action Report: the United States of America's Third National Communication Under the United Nations Framework Convention on Climate Change*. Washington DC: US Department of State.

Devictor, V., Julliard, R., Couvet, D. and Jiguet, F. (2008) Birds are tracking climate warming, but not fast enough. *Proceedings of the Royal Society of London Series B*, 275, 2745–8.

Doak, D. F. and Morris, W. F. (2010) Demographic compensation and tipping points in climate-induced range shifts. *Nature*, 467, 959–62.

Domack, E., Duran, D., Leventer, A. et al. (2005) Stability of the Larsen B ice shelf on the Antarctic Peninsula during the Holocene epoch. *Nature*, 436, 681–5.

Doney, S. C. (2006) Plankton in a warmer world. *Nature* 444, 695–6.

Dowdeswell, J. A. (2006) The Greenland ice sheet and global sea-level rise. *Science*, 311, 963–4.

Early, R. and Sax, D. F. (2011) Analysis of climate paths reveals potential limitations on species range shifts *Ecology Letters*, 14 (11), 1125–33.

Edwards, M. and Richardson, A. J. (2004) Impact of climate change on marine pelagic phenology and trophic mismatch. *Nature*, 430, 881–3.

Elsner, J. B., Kossin, J. P. and Jagger, T. H. (2008) The increasing intensity of the strongest tropical cyclones. *Nature*, 455, 92–5.

Emanuel, K. (2005) Increasing destructiveness of tropical cyclones over the past 30 years. *Nature*, 436, 686–8.

Fawcett, P. J., Werne, J. P., Anderson, R. S. et al. (2011) Extended megadroughts in the southwestern United States during Pleistocene interglacials. *Nature*, 470, 518–21.

Feeley, K. J. and Silman, M. R. (2010) Biotic attrition from tropical forests correcting for truncated temperature niches. *Global Change Biology*, 16, 1830–6.

Field, D. B., Baumgartner, T. R., Charles, C. D., Ferriera-Bartrina, V. and Ohman, M. D. (2006) Planktonic foraminifera of the California current reflect 20th-century warming. *Science*, 311, 63–6.

Fischer, E. M. and Schär, C. (2010) Consistent geographical patterns of changes in high-impact European heatwaves. *Nature Geoscience*, 3, 398–403.

Gardner, A. S., Moholdt, G., Wouters, B. et al. (2011) Sharply increased mass loss from glaciers and ice caps in the Canadian Arctic Archipelago. *Nature*, 473, 357–60.

Geological Society (2005) *Super-Eruptions: Global Effects and Future Threats*. London: Geological Society of London.

Gleckler, P. J., Wigley, T. M. L., Santer, B. D., Gregory, J. M., AchutaRao, K. and Taylor, K. E. (2006) Krakatau's signature persists in the ocean. *Nature*, 439, 675.

Gunnarsson, T. G., Gill, J. A., Sigurbjornsson, T. and Sutherland, W. J. (2004) Arrival synchrony in migratory birds. *Nature*, 431, 646.

Hansen, B., Turrell, W. R. and Østerhus, S. (2001) Decreasing overflow from the Nordic seas into the Atlantic Ocean through the Faroe Bank channel since 1950. *Nature*, 411, 927–30.

Harsch, M., Hulme, P., McGlone, M. and Duncan, R. (2009) Are treelines advancing? A global meta-analysis of treeline response to climate warming, *Ecology Letters*, 21(10), 1040–9.

Hickling, R. W., Roy, D. B., Hill, J. K., Fox, R. and Thomas, C. D. (2006) The distributions of a wide range of taxonomic groups are expanding polewards. *Global Change Biology*, 12, 450–5.

Hill, J. K. and Fox, R. (2003) Climate change and British butterfly distributions. *Biologist*, 50(3), 106–10.

Hoegh-Guldberg, O. and Bruno, J. F. (2010) The impact of climate change on the world's marine ecosystems. *Science*, 328, 1523–8.

Hof, C., Araújo, M., Jetz, W. and Rahbek, C. (2011) Additive threats from pathogens, climate and land-use change for global amphibian diversity. *Nature*, 480, 516–19.

Hornbach, M. J., Saffer, D. M. and Holbrook, W. S. (2004) Critically pressured free-gas reservoirs below gas-hydrate provinces. *Nature*, 427, 142–4.

House of Commons Science and Technology Select Committee (2011) *The Reviews into the University of East Anglia's Climatic Research Unit's E-mails*, vol. 1, HC444. London: Stationery Office.

Hsieh, C., Glaser, S. M., Lucas, A. J. and Sugihara, G. (2005) Distinguishing random environmental fluctuations from ecological catastrophes for the North Pacific ocean. *Nature*, 435, 336–40.

Hulme, M., Turnpenny, J. and Jenkins, G. (2002) *Climate Change Scenarios for the United Kingdom: The UKCIP02 Briefing Report*. Norwich: Tyndall Centre.

Intergovernmental Panel on Climate Change (2000) *Emission Scenarios – IPCC Special Report*. Geneva: World Meteorological Organization.

Intergovernmental Panel on Climate Change (2001a) *Climate Change 2001: Impacts, Adaptation and Vulnerability – a Report of Working Group II*. Cambridge: Cambridge University Press.

Intergovernmental Panel on Climate Change (2001b) *Climate Change 2001: the Scientific Basis – Summary for Policymakers and Technical Summary of the Working Group I Report*. Cambridge: Cambridge University Press.

Intergovernmental Panel on Climate Change (2007a) *Climate Change 2007: the Physical Science Basis – Working Group I Report*. Cambridge: Cambridge University Press.

Intergovernmental Panel on Climate Change (2007b) *Climate Change 2007: Impacts Adaptation and Variability – Working Group II Contribution to the Fourth Assessment of the IPCC*. Cambridge: Cambridge University Press.

Intergovernmental Panel on Climate Change (2011) *IPCC SREX Summary for Policymakers*. Geneva: IPCC.

Intergovernmental Panel on Climate Change (2012) *Special Report on Managing the Risks of Extreme Events and Disasters to Advance Climate Change Adaptation (SREX)*. Geneva: IPCC.

Jacobs, S. S., Giulivi, C. F. and Mele, P. A. (2002) Freshening of the Ross Sea during the late 20th century. *Science*, 297, 386–9.

Jenkins, G. J., Perry, M. C. and Prior, M. J. O. (2007) *The Climate of the United Kingdom and Recent Trends*. Exeter: Met Office Hadley Centre.

Jenkins, G. J., Murphy, J. M., Sexton, D. M. H. et al. (2009). *UK Climate Projections: Briefing Report*. Exeter: Met Office Hadley Centre.

Kaspi, Y. and Schneider, T. (2011) Winter cold of eastern continental boundaries induced by warm ocean waters. *Nature*, 471, 621–4.

Kausrud, K. L., Mysterud, A., Steen, H. et al. (2008) Linking climate change to lemming cycles. *Nature*, 456, 93–7.

Kearney, M. R., Briscoe, N. J., Karoly, D. J. et al. (2010) Early emergence in a butterfly causally linked to anthropogenic warming. *Biology Letters*, 6(5), 674–7.

Kennett, J. P., Cannariato, K. G., Hendy, I. L. and Behl, R. J. (2000) Carbon isotope evidence for methane hydrate instability during Quaternary interstadials. *Science*, 288, 128–33.

Kerr, R. A. (2004) A bit of icy Antarctica is sliding toward the sea. *Science*, 305, 1897.

Kinnard, C., Zdanowicz, C. M., Fisher D. A. et al. (2011) Reconstructed changes in Arctic sea ice over the past 1450 years. *Nature*, 479, 509–14.

Knight, T. M., McCoy, M. W., Chase, J. M., McCoy, K. A. and Holt, R. D. (2005) Trophic cascades across ecosystems. *Nature*, 437, 880–3.

Kodra, E., Steinhaeuser, K. and Ganguly, A. (2011) Persisting cold extremes under 21st-century warming scenarios. *Geophysical Research Letters*, 36, L047103.

Kurz, W. A., Dymond, C. C., Stinson, G. et al. (2008) Mountain pine beetle and forest carbon feedback to climate change. *Nature*, 452, 987–9.

Laurance, W. F., Oliveira, A. A., Laurance, S. G. et al. (2004) Pervasive alteration of tree communities in undisturbed Amazonian forests. *Nature*, 428, 171–4.

Lawton, J. H., Brotherton, P. N. M., Brown, V. K., Elphick, C. et al. (2010) *Making Space for Nature: a Review of England's Wildlife Sites and Ecological Network*. Report to DEFRA. London: DEFRA.

Lemmen, D. S., Warren, F. J., Lacroix, J. and Bush, E. (2008) *From Impacts to Adaptation: Canada in a Changing Climate*. Government of Canada: Ottawa, Canada.

Lenoir, J., Gégout, J. C., Marquet, P. A., de Ruffray, P. and Brisse, H. (2008) A significant upward shift in plant species optimum elevation during the 20th century. *Science*, 320, 1768–71.

Lewis, S. L., Lopez-Gonzalez, G., Sonké, B. et al. (2009) Increasing carbon storage in intact African tropical forests. *Nature*, 457, 1003–6.

Lovejoy, T. E. and Hannah, L. (eds) (2005) *Climate Change and Biodiversity*. Yale: Yale University Press.

Lu, H., Seo, Y.-t., Lee, J.-W. et al. (2007) Complex gas hydrate from the Cascadia margin. *Nature*, 445, 303–6.

MacArthur, R. H. and Wilson, E. O. (1967) *The Theory of Island Biogeography*. Princeton, NJ: Princeton University Press.

Malhi, Y., Roberts, J. T., Betts, R. A. et al. (2007) Climate change, deforestation, and the fate of the Amazon. *Science*, 319, 169–72.

Meehl, G. A. and Tebaldi, C. (2004) More intense, more frequent, and longer lasting heat waves in the 21st century. *Science*, 305, 994–7.

Miller, J. B. (2008) Sources, sinks and seasons, *Nature*, 451, 26.

Miller, L. and Douglas, B. C. (2004) Mass and volume contributions to twentieth century global sea level rise. *Nature*, 428, 406–9.

Milly, P. C. D., Dunne, K. A. and Vecchia, A. V. (2005) Global pattern of trends in streamflow and water availability in a changing climate. *Nature*, 438, 347–50.

Monson, R. K., Lipson, D. L., Burns, S. P. et al. (2006) Winter forest soil respiration controlled by climate and microbial community composition. *Nature*, 439, 711–14.

Morgan, J. A., LeCain, D. R., Pendall, E. et al. (2011) C_4 grasses prosper as carbon dioxide eliminates desiccation in warmed semi-arid grassland. *Nature*, 476, 202–5.

National Assessment Synthesis Team (2001) *Climate Change Impacts on the United States: the Potential Consequences of Climate Variability and Change*. US Global Change Research Program. Cambridge: Cambridge University Press.

Natural Environment Research Council (2008) *Rapid Climate Change: The Rapid Climate Change Programme*. Swindon: Natural Environment Research Council.

National Research Council (2010) *Climate Stabilization Targets: Emissions, Concentrations, and Impacts Over Decades to Millennia*. Washington DC: The National Academies Press.

Nemani, R. R., Keeling, C. D., Hashimoto, H. et al. (2003) Climate-driven increases in global terrestrial net primary production from 1982 to 1999. *Science*, 300, 1560–3.

Orr, J. C., Fabry, V. J., Aumont, O. et al. (2005) Anthropogenic ocean acidification over the twenty-first century and its impact on calcifying organisms. *Nature*, 437, 681–6.

Overpeck, J. T., Otto-Bliesner, B. L., Miller, G. H., Muhs, D. R., Alley, R. B. and Kiehl J. T. (2006) Palaeoclimatic evidence for future ice-sheet instability and rapid sea-level rise. *Science*, 311, 1747–50.

Ozgul, A., Childs, D. Z., Oli, M. K. et al. (2010) Coupled dynamics of body mass and population growth in response to environmental change. *Nature*, 466, 482–5.

Parmesan, C. and Yohe, G. (2003) A globally coherent fingerprint of climate change impacts across natural systems. *Nature*, 421, 37–42.

Parmesan, C., Ryrholm, N., Stefanescu, C. et al. (1999) Poleward shifts in geographical ranges of butterfly species associated with regional warming. *Nature*, 399, 579–83.

Peck, L. S., Barnes, D. K. A., Cook, A. J. et al. (2009) Negative feedback in the cold: ice retreat produces new carbon sinks in Antarctica. *Global Change Biology*, 16(9), 2614–23.

Phillips, O. L., Malhi, Y., Higuchi, N., Laurance, W. F. et al. (1998) Changes in the carbon balance of tropical forests: evidence from long-term plots. *Science*, 282, 439–42.

Piao, S., Ciais, P., Friedlingstein, P. et al. (2008) Net carbon dioxide losses of northern ecosystems in response to autumn warming. *Nature*, 451, 49–52.

Pounds, A. J., Fogden, M. P. L. and Campbell, J. H. (1999) Biological response to climate change on a tropical mountain. *Nature*, 398, 611–15.

Pounds, A. J., Bustamante, M. R., Coloma, L. A. et al. (2006) Widespread amphibian extinctions from epidemic disease driven by global warming. *Nature*, 439, 161–7.

Rignot, E. and Kanagaratnam, P. (2006) Changes in the velocity structure of the Greenland ice sheet. *Science*, 311, 986–90.

Rohling, E. J., Marsh, R., Wells, N. C., Siddall, M. and Edwards, N. R. (2004) Similar meltwater contributions to glacial sea level changes from Antarctic and northern ice sheets. *Nature*, 430, 1016–21.

Root, T. L., Price, J. T., Hall, K. R. Schneider, S. H., Rosenzweig, C. and Pounds, A. (2003) Fingerprints of global warming on wild animals and plants. *Nature*, 421, 57–60.

Rosenzweig, C., Karoly, D., Vicarelli, M. et al. (2008) Attributing physical and biological impacts to anthropogenic climate change. *Nature*, 457, 353–7.

Scheffer, M. Bascompte, J., Brock, W. A. et al. (2009) Early-warning signals for critical transitions. *Nature*, 461, 53–9.

Schiermeier, Q. (2010) Ecologists fear Antarctic krill crisis. *Nature*, 467, 15.

Scientific Committee on Antarctic Research (2009) *Antarctic Climate Change and the Environment*. Scientific Committee on Antarctic Research: Cambridge.

Siegert, M. J., Welch, B., Morse, D. et al. (2004) Ice flow direction change in interior west Antarctica. *Science*, 305, 1948–51.

Smith, L. C., Sheng, Y., MacDonald, G. M. and Hinzman, L. D. (2005) Disappearing Arctic lakes. *Science*, 308, 1429.

Sparks, T. and Collinson, N. (2003) Wildlife starts to adapt to a warming climate. *Biologist*, 50(6), 273–6.

Sriver, R. L. and Huber, M. (2007) Observational evidence for an ocean heat pump induced by tropical cyclones. *Nature*, 447, 577–80.

Stine, A. R., Huybers, P. and Fung, I. Y. (2009) Changes in the phase of the annual cycle of surface temperature. *Nature*, 457, 435–40.

Stott, P. A., Stone, D. A. and Allen, M. R. (2004) Human contribution to the European heatwave of 2003. *Nature*, 432, 610–11.

Thackeray, S. J., Sparks, T. H., Frederiksen, M. et al. (2010) Trophic level asynchrony in rates of phenological change for marine, freshwater and terrestrial environments. *Global Change Biology*, 16(12), 3304–13.

Thomas, D. S. G., Knight, M. and Wiggs, G. F. S. (2005) Remobilization of southern African desert dune systems by twenty-first century global warming. *Nature*, 435, 1218–21.

Todd, B. D., Scott, D. E., Pechmann, J. H. K. and Gibbons, J. W. (2011) Climate change correlates with rapid delays and advancements in reproductive timing in an amphibian community. *Proceedings of the Royal Society of London Series B*, 278, 2191–7.

Toniazzo, T., Gregory, J. M. and Huybrechts, P. (2004) Climatic impact of a Greenland deglaciation and its possible irreversibility. *Journal of Climate*, 17, 21–33.

Tripati, A. and Elderfield, H. (2005) Deep-sea temperature and circulation changes at the Paleocene-Eocene Thermal Maximum. *Science*, 308, 1894–8.

UK Climate Impacts Programme (2001) *Climate Change and Nature Conservation in Britain and Ireland*. Oxford: UK Climate Impacts Programme.

USGCRP (2009) *Global Climate Change Impacts in the United States*. Cambridge: Cambridge University Press.

US Geological Survey (1999) *U.S. Geological Survey Professional Paper – Estimated Present-Day Area and Volume of Glaciers and Maximum Sea Level Rise Potential*. Reston, VA: US Geological Survey.

Walmsley, C. A., Smithers, R. J., Berry, P. M. et al. (eds) (2007) *MONARCH – Modelling Natural Resource Responses to Climate Change – A Synthesis for Biodiversity Conservation*. Oxford: UKCIP.

Walther, G.-R., Post, E., Convey, P. et al. (2002) Ecological response to recent climate change. *Nature*, 416, 389–95.

Warren, M. S., Hill, J. K., Thomas, J. A. et al. (2001) Rapid response of British butterflies to opposing forces of climate and habitat change. *Nature*, 414, 65–8.

Walter, K. M., Zimov, S. A., Chanton, J. P., et al. (2006) Methane bubbling from Siberian thaw lakes as a positive feedback to climate warming. *Nature*, 443, 71–5.

Webster, P. J., Holland, G. J., Curry, J. A. and Chang, H.-R. (2005) Changes in tropical cyclone number, duration and intensity in a warming environment. *Science*, 309, 1844–6.

Welbergen, J. A., Klose, S. M., Markus, N. and Eby, P. (2008) Climate change and the effects of temperature extremes on Australian flying-foxes. *Proceedings of the Royal Society of London Series B*, 275, 419–25.

Westgarth-Smith, A. R., Leroy, S. A. G. and Collins, P. E. F. (2005) The North Atlantic Oscillation and UK butterfly populations. *Biologist*, 52, 273–6.

Williams, J. (2011) Old droughts in New Mexico. *Nature*, 470, 473–4.

Wu, P., Wood, R. and Stott, P. (2005) Human influence on increasing Arctic river discharges. *Geophysical Research Letters*, 32, L02703.

Yvon-Durocher, G., Montoya, J. M. Trimmer, M. and Woodward, G. (2010) Warming alters the size spectrum and shifts the distribution of biomass in freshwater ecosystems. *Global Change Biology*, 17(4), 1681–94.

Zwiers, F. and Hegerl, G. (2008) Attributing cause and effect. *Nature*, 453, 296–7.

7 The human ecology of climate change

Given that climate change has been part of, and has helped shape, the biosphere's development, it should not have been surprising that climate change affected human evolution and humanity's historical affairs (see Chapter 5), or that it will do so in the future. That humanity itself has affected the climate – modern technological civilisation significantly so – all the more demonstrates the interconnections between climate and our species. If we are to begin to assess the future of this relationship it is necessary to understand the relevant fundamentals of human ecology (the way our species as a population develops and its relations with other species). This includes population demographics (our numbers influence the amount of biomass of, and hence impact on, the species to which we relate), energy supply (which relates to carbon cycle short-circuiting), health (mainly with regard to those species that prey on us) and food supply (those species that we harvest). Finally, given all of this, what are the prospects for limiting or ameliorating our short-circuiting of the biosphere's carbon cycle?

7.1 Population (past, present and future) and its environmental impact

7.1.1 Population and environmental impact

Generally, if one person has an environmental impact of some arbitrary quantification, x, then it is not unreasonable to suppose that two people will have an impact of $2x$, three people an impact of $3x$, etc; in other words, that environmental impact is proportional to population and so may be represented in equation form.

One of the first environmental-impact equations widely considered, where impact is proportional to population, was formulated by Paul Ehrlich and John Holdren (1971). It was expressed as:

$$I = PAT$$

where I is the impact on the environment resulting from consumption, P is the population, A is the consumption per capita (affluence) and T is a technology or, as it has more recently sometimes been known, an inefficiency (inverse efficiency of use) factor.

Today, environmental impact is considered in a more complex and multifactorial way. Most often impact is not considered in terms of arising from a human population, but occasionally it is considered in this way: one example would be the concept of a community's, or a nation's, ecological footprint. More often than not, impact is considered as relating to a development project, be it an airport or the construction of a road, when an environmental impact assessment is carried out.

Policy-makers rarely consider environmental impact in human population terms. Instead, much of their focus is on either the affluence or technological dimensions; that the policy in question will have to generate this much wealth for that much cost, incidentally generating so much pollution. Attempts to curb population through lowering fecundity[1] are rarely considered as a means to curb impact or frame economic and sustainability policies. There are exceptions, the most notorious being China in the late 20th century; 'notorious' because China's draconian measures greatly infringed individual freedom and distorted its population's gender balance. However, in the main, up to the early 21st century addressing population issues has been low on policy agendas.

More recently there has been some interest in population, P in the above equation; population concerns have slowly begun to come to the fore. Politicians in developed nations perceive that current population demographic changes (the changes in a population's composition, such as by age) mean that there will not be enough younger people to care for the increasing number of elderly (and especially partially infirm elderly, for we are living longer more in an unhealthy state). The early 21st century has seen a number of developed nations, such as the USA, UK, Italy and Australia, actively introduce policies to address this problem, be it through immigration or fiscal encouragements to increase fecundity for couples wanting children. Consequently, following the Ehrlich–Holdren equation, such policy trends are likely to increase environmental impact rather than reduce them. Indeed, because the citizens of developed nations are more affluent the environmental impact arising from each one is greater than in their less-developed counterparts, so actively encouraging a growing developed nation's population will often (but not always) tend to increase that population's environmental impact. However, some academics do present detailed arguments as to why population growth might be seen in a positive light (Simon, 1996) and, as we shall see, the addition of an extra person to a population can result in either an improvement or a decline in the wealth-to-impact ratio.

As per the Ehrlich–Holdren impact equation carbon dioxide emissions do broadly (but again, not always) relate to affluence in terms of gross domestic product (GDP; the sum total of that population's earnings) and GDP per capita. In the late 20th century, affluent North America (Canada and USA), with approximately 2% of the global population, was responsible for about 25% of greenhouse gas emissions. Conversely, less wealthy China, with some 20% of the Earth's population, then generated just 12% of global emissions. The carbon emissions per capita from energy use for selected countries in 2002 and 2009 are depicted in Figure 7.1. From this it can be seen that

[1] Until recently many demographers used fertility to mean both the actual level of human reproduction and the potential for human reproduction, even though these concepts are quite different. More exact terminology is required (especially when communicating across disciplines) and increasingly the ecological term fecundity is used to relate to the amount of offspring a human couple actually have, while fertility is used to reflect the potential to have offspring. This is how these terms are used in this book.

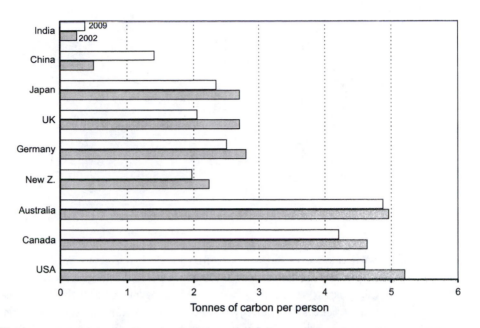

Fig. 7.1 Carbon emissions per person from energy production in selected countries in 2002 (shaded bars) and 2009 (white bars). Data for 2002 from the Worldwatch Institute (2003), which in turn obtained it from the International Energy Agency (IEA); 2009 data from the IEA, 2011 (www.iea.org); both data sets were converted from tonnes of carbon dioxide to tonnes of carbon. (See also Appendix 3.)

post-industrial developed areas such as from Europe, North America and Australasia have decreasing per-capita emissions between 2002 and 2009, whereas India and China have growing emissions.

Affluence does not vary just between national populations but within them, and it varies over time. Overall, citizens of the Earth are getting wealthier (see Figure 7.2), although some remain more wealthy than others and for a minority wealth is decreasing. The differences in wealth across the planet do have a bearing on environmental impact but as this relates to international development it is a subject too great to discuss here, other than as a brief reference. Nonetheless, broadly speaking there is some correlation between financial wealth and carbon emissions. For example, as noted, the USA is the biggest economy on the planet and also, with less than 5% of the global population, generates nearly a quarter of all greenhouse gas emissions, so making it the largest fossil fuel emitter on both a per-capita and a national basis. Whether such correlation will continue remains to be seen. This is because the per-capita carbon emissions of a number of less-developed nations are also increasing, even if they are currently low. Here the concern is that the populations of countries like India and China are large, so that if their per-capita emissions continue to rise then on a national basis this will have a major impact in the way global trends in atmospheric carbon dioxide concentrations change.

Notwithstanding all this, according to our current global society's accounting system (there are others) we are getting wealthier and, for the moment at least, our global greenhouse gas emissions are similarly increasing. This overall increase in

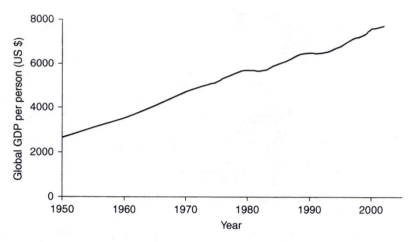

Fig. 7.2 Worldwide average wealth (annual income) per person in 2011 US dollars. Wealth per capita is expressed here not as world gross domestic product (GDP) in a traditional sense but as gross national income (formerly gross national product) per person. It includes income from investments. It is calculated here on the exchange-rate basis as opposed to the purchasing parity basis. These caveats are mentioned in case you come across other global wealth indicators. The data from the graph was downloaded from the World Bank's website at http://data.worldbank.org/indicator/NY.GDP.PCAP.CD in November 2011.

affluence is reflected in parameters such as household size (number of occupants), which is one of the key indicators of a populations' environmental impact, although the relationship is not a linear or a proportional one. The broad reason for this is that a building or apartment consists of a fixed mass, volume and area so that even if the food and water and other consumables used by a household of three are approximately half those of a household of six, the same volume of air and bricks needs to be warmed with both households, irrespective of the number of people living there. In short, there is a decrease in energy intensity per person as the number of occupants in a household increases.

Furthermore, a halving of household size for a fixed population means that the number of houses required for that population doubles. Take both factors (energy intensity and number of households) together and it can be seen that a halving of household size more than doubles the environmental impact. This decrease in household size is manifest in both less-developed and developed nations, albeit more markedly in the latter (see Figure 7.3).

Decline in household size is just one aspect of the way affluence generates environmental impact irrespective of population size. Affluence, among other things, also affects annual per-capita distance travelled and consumption of material goods, water and food. In terms of greenhouse impact, goods and services each have an inherent energy content (energy intensity) associated with their production and delivery. If this energy comes from fossil fuel then there will be an associated greenhouse cost. This cost might be met (say by chemically capturing the carbon dioxide emitted from the power station [see Chapter 8], through planting new woodland, etc.) and the cost covered by the price of the energy sold. Or, alternatively, this greenhouse cost may not be met, in which case the greenhouse gas escapes, to warm up the planet. In this

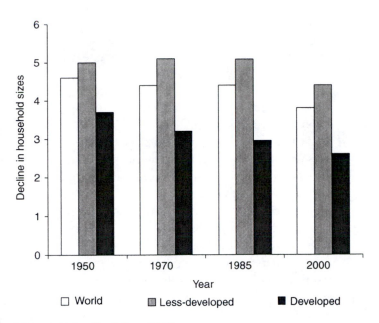

Fig. 7.3   The decline in household size. After Keilman (2003).

instance the environmental cost is not covered by energy sales and so is external to the price of energy. In economic terms such environmental impacts are known as environmental externalities.

Accurately quantifying environmental externalities globally is extremely difficult. There are two main reasons for this: assessing the cost of the impact and assessing affluence. Assessing the cost of an impact is difficult because, for example, how does one price the loss of a species, let alone an ecosystem? The financial value of a species that has no known economic value is obviously going to be hard. Valuing an ecosystem in one sense is a little easier. Ecosystems have function. For example, a wooded valley will retain rainwater falling on it and release it slowly, so providing a more steady water supply through the catchment. The same valley devoid of life with just bare rock will see the water run off immediately, so scouring the lower catchment and exposing it to periods of drought. This particular function of the ecosystem can be part quantified in the value of the water that would have been purchased commercially to replace that lost in the drought as well as the erosion-protection mechanisms to prevent scouring.

This idea of ecosystem services is not purely theoretical. Furthering the above ecosystem function example, in 1996 New York City invested between US$1 and 1.5 billion in ecosystem management in the Catskill Mountains, which provide its main water supply. Up to the 1990s plant root systems, soil structure and its ecology largely served to purify the water to US standards. However, sewage, fertilizer use and pesticide use caused water standards to fall. New York either had to restore the Catskill ecology and ecosystem function to what it had been or to build a water-purification plant at an estimated capital cost of $6–8 billion and annual running costs of $300 million. By comparison the ecosystem-management option was more attractive (Chichilnisky and Heal, 1998).

However, valuing ecosystems can be harder than for species because it is difficult to quantify all the benefits arising out of an ecosystem's function and not every ecosystem has a human population with an easily identifiable ecosystem demand and associated opportunity cost (see Glossary, Appendix 1). This does not mean that some far-flung piece of tropical forest has no ecosystem service to offer, because vast tracts of tropical forest, for instance, affect the regional climate and the (fast) global carbon cycle.

In short, even though many developed nations theoretically embrace the polluter-pays principle, the reality is that much of such costs remain environmental external-ities if only because we are not, when undertaking economic transactions, valuing all existing ecological services.

Nonetheless, rough valuation attempts have been made to value ecosystem services globally. These suggest that the global value of services provided by ecosystems, but external to our human economy, is of the same order of magnitude as the 'real' human economy, if not greater. One 1997 estimate was in the range of US$$(16–54)\times10^{12}$ and this range was considered by the researchers to be a minimum one. It does, though, compare with the then global annual economy of $\$18\times10^{12}$ (Costanza et al., 1997) even if some of the assumptions and methods caused some debate in the ecological economics community. Subsequently, in the run-up to the 2002 UN Conference on Environment and Development +10 (UNCED+10) summit (see section 8.1.9), another assessment concluded that the benefit/cost ratio of conserving the Earth's then remaining wildlife and natural systems was at least 100/1 (Balmford et al., 2002).

Although attempts to quantify the (largely unrecognised) value of ecological ser-vices have proven difficult – other than to say that they probably contribute roughly as much value as the conventional economy – it is possible to identify how such services might be affected in a region due to climate change. One such assessment of change in ecosystem service supply due to climate change in Western Europe identi-fied both positive and negative service changes (Schroter et al., 2005). Positive ones included increases in forest area and productivity. Negative ones included declin-ing soil fertility (due to temperature and extreme rain-burst erosion) and increased risk of forest fires. These positive and negative service changes respectively provide opportunities (biofuel production or increased agriculture) or remove options (such as traditional water use). This assessment particularly noted that among the Western European countries, those around the Mediterranean appeared to be most vulner-able, with multiple projected impacts primarily related to increased temperature and reduced precipitation with resulting loss of agricultural potential.

Despite difficulties in assessing the value of ecosystem services, there is increasing interest in this approach by policy-makers. One project, supported by the European Commission, the United Nations Environment Programme and government depart-ments responsible for the environment in the UK and Germany, is The Economics of Ecosystems and Biodiversity (TEEB) study project that came out of a G8 summit in 2007. It published a relevant report, *Climate Issues Update*, in 2009 (TEEB, 2009). This concluded that there is a compelling cost-benefit case for public investment in ecological infrastructure (especially restoring and conserving forests, mangroves, river basins, wetlands, etc.), particularly because of its significant potential as a means of adaptation to climate change. This included, among other things, an assessment of the ecosystem service value of coral biomes at on average US$13 541 ha^{-1} year^{-1} but noted that in some instances this might be as high as \$57 133 ha^{-1} year^{-1}.

Of biological note, it seems that high biodiversity is needed to maintain ecosystem services as many species contribute to these services at different times of the year and in different years due to climate variability (see, for example, Isbell et al., 2011). However there frequently appears to be a mismatch between the conservation value placed on some ecosystems and their ecosystem service value (Turner et al., 2007), which of course is why the ecosystem service approach is useful. Of special relevance to global climate change is the ecosystem service value of some systems' role (ecosystem function) in carbon cycling and/or storage.

Assessing affluence (A in the equation at the start of this section) can be hard because, in the sense of Ehrlich–Holdren impacts, it does not equate with money, although frequently it can approximate to it. Yet the currency of money is at the heart of our economic system. A person's affluence, for example, might be considered in terms of the size of their home (harking back to a previous example), together with diet, material possessions and so forth. Yet property and even food prices across a nation vary, and between nations they vary considerably. The price of a four-bedroomed house in, say, Washington DC, USA, is different from the price of a similar-sized and -constructed house in Timisoara, Romania. The same applies to a loaf of bread in the two places. Yet an American living in an analogous dwelling and consuming analogous goods and services to a Romanian will be paying far more in US dollar terms than a Romanian. The Romanian may even pay the same proportion of his or her financial income for a loaf of bread as the American, and so have similar so-called purchasing parity. The problem comes when the Romanian wishes to purchase goods sold only in US dollars, as the US dollar is stronger than the lei (the unit of Romanian currency). To get around this problem economists sometimes use parity purchasing prices.

Even parity purchasing prices are not a perfect way of measuring affluence and all monetary, indeed all other, means have their own limitations and this subject has been the focus of considerable debate among economists for many decades. Any monetary measure of a society's GDP, even compared with making parity-pricing allowances with another country's GDP, does not accurately reflect that society's affluence, let alone enable true comparisons with other societies both economically and in terms of environmental impact. This is in part because it is possible to employ someone to create something, or provide a service, that enhances a citizen's well-being as well as someone who does not. For example, those involved in oil-spill accidents as well as those who clean them up are both paid a salary, but all these citizens' salaries contribute to their nation's GDP. However, clearly their actions, taken together, have a zero effect on well-being and a negative effect on the environment, which has been disrupted. This is somewhat analogous to the problem (see below) as to what size of population the Earth can support. Using genuine progress indicators (such as literacy levels, longevity, health, etc.), as opposed to purely GDP-type indicators (of economic wealth and growth), does lead to a different perspective. For example, one such indicator suggests that genuine progress in the USA remained broadly static between 1970 and 2000, even though total US GDP more than doubled and GDP per capita increased by more than 60% (Worldwatch Institute, 2003).

Purely in greenhouse terms (which is not satisfactory for a broad range of other economic purposes), using the emissions of carbon per person, as shown in Figure 7.1, is one way to begin to address the difference in the monetary value of goods globally (note, it is important to check whether it is carbon by weight or carbon dioxide being used, as the latter is 3.666 times heavier than the former and yet refers to the same amount of fossil fuel energy consumed). However, because there is no one perfect measure of affluence, its inclusion in the Ehrlich–Holdren equation must be notional, and therefore so must the equation itself. Even so, notions and theories do have merit and can help illuminate issues – in this case that of environmental impact – and so help frame solutions.

The final component of the Ehrlich–Holdren equation is T, the technological factor. This too is difficult to quantify. Ehrlich and Holdren meant it to reflect the way different technologies impart different environmental impacts. For example, generating a kilowatt hour of electricity from coal will impart a greenhouse impact as well as involve part of the environmental impact of the coal mine. Whereas generating the same electricity from a hydroelectric scheme will have a smaller greenhouse impact (there will be a greenhouse impact from creating the cement and the energy used in the dam's construction), there will also be the environmental impact of the volume of dammed water on land use and so forth. So the different technologies have different impacts and so T remains hard to quantify. More recently T has been replaced by some users with a term reflecting the inverse of environmental efficiency of that population. Energy that is produced with high environmental efficiency will have a low inverse value and so contribute to a low environmental impact.

All the above demonstrates that quantifying environmental impact and hence the cost of global warming is difficult, if not impossible, to assess with any accuracy. Indeed, there is an entire sub-discipline of environmental economics that attempts to come to grips with such problems. As we shall see, this problem affects one type of greenhouse policy whereby permits to release carbon dioxide can be sold by those who have invested in low-fossil fuel technology to those whose investments are already tied up in high-fossil fuel technology. Because the estimation of impacts, identifying of externalities and allocation of costs and benefits to energy technologies is so fraught with difficulty, any system of greenhouse permit trading is likely to be very complex and fail to reflect properly the impacts. Such systems also can easily lead to gamesmanship whereby energy firms, and countries, play the system, so undermining their contribution to greenhouse solutions (we shall return to this).

These difficulties aside, the consequences are that the impact of two populations of similar size and affluence can be quite different. Similar ecosystems can support different numbers of people depending on what they do. This notion did not begin with, but was developed by, the American demographer Joel Cohen (1995). Specifically, Cohen suggests that it is the choices a society (or a population) makes that affect the carrying capacity of that nation's environment. Consider two self-sufficient populations of the same size. The birth of an extra person (in the absence of any deaths) in each will increase those populations by one. Yet that extra person in one population might go on to invent greenhouse-efficient systems or something else that lowers environmental impact without reducing the goods and services generated by that population. In the other population that person might become, to take an extreme example, a tyrant bent on domination through aggression. This population will divert resources to those ends rather than improving environmental efficiency. Such acts lower the carrying capacity of the environment. The inescapable conclusion in greenhouse terms is that it is the choices we make as a global society that determine our ability to handle climate change concerns, and that there is not some fixed global level of fossil fuel consumption that can be considered ideal. This may seem counterintuitive but can be illustrated by considering a power station that burns fossil fuels whereby all the carbon dioxide generated is isolated through carbon capture (see Chapter 8). Such a station will not affect the climate through its emissions (though admittedly it may in other ways) as a normal power station would.

To sum up, although it is extremely difficult to quantify the costs of the global population's greenhouse impact, the levels of population and affluence do very broadly relate to the level of impact. However, above and beyond this, it is environmental efficiency, and specifically the choices made to realise this efficiency, that affects climate change. This last lies at the heart of much of the next chapter.

7.1.2 Past and present population

The global population of *Homo sapiens sapiens* (hereafter referred to simply as *H. sapiens*) is currently growing and for many (there are exceptions) this is an accepted and often unremarked-upon fact of modern life. This is in no small part because everyone alive today has lived in a world that has seen continuous growth, so that such growth seems unremarkable: after all, the continuation of the line above and below one point on an exponential curve looks just like elsewhere on the same line above and below any other point on that curve. A person today in their late middle age will have seen the global population grow from a little over 2 billion to more than 7 billion, while a typical undergraduate student aged 19 or 20 will have seen the population rise some 25% from over 5 billion (see Figure 5.10). Yet our post-Industrial Revolution period of population growth is not typical.

Joel Cohen (1995) cites the independent estimates of early human population size for over a million years ago of demographers Edward Deevey Jr and Michael Kremer. They both estimated the global population at the time to be approximately an eighth of a million (although Kremer based his estimates on those of Deevey Jr for global population up to 25 000 years ago). After that time their estimates for population differed substantially up to the beginning of the Common Era. From AD 1650 their figures are joined by similar estimates from five other demographers and together these present a coherent picture.

If these estimates are to be believed then it would appear that from 1 million to 300 000 years BC the Earth's population of *Homo* spp. grew slowly from an eighth of a million to a million. The rate of increase required for this is little over one person a year. However, 1 million to 300 000 years ago was a period of Pleistocene Quaternary glacial–interglacial cycles so fairly linear population growth is unlikely to have occurred. More probable is that the global population was reduced in times of extreme climate change (such as at the beginning and end of glacials and interstadials) and grew when the climate was stable. Furthermore, this meta-population of *Homo* spp. saw its various component populations (of *Homo erectus*, *Homo sapiens* and

Homo floresiensis) fluctuate and, of course, there was an overall trend of succession from one to the other, albeit with considerable overlap, before *H. sapiens* rose (see Chapter 4).

By the time of the domestication of barley and wheat, early in the current Holocene interglacial some 10 000–9000 years ago, *H. sapiens* was the sole hominid extant with a global population probably somewhere between 3 and 5 million. This increased over the next 7000–8000 years to the start of the Common Era. It is not known for certain whether or not this growth was slow and steady, at a fraction of a per cent a year, or whether it increased dramatically when and where agriculture replaced hunting and gathering as the primary food source, but the latter is more likely. Either way, this length of time was sufficient for the global population to grow to somewhere between 100 and 300 million.

The next one and a half millennia to AD 1500 saw the population roughly treble to below half a billion. But it was after AD 1700, and the Renaissance and then the Industrial Revolution, that the population began to grow at a previously unprecedented rate.

In terms of human population the development of science and its resulting technology have proven hugely successful. Indeed, many of the problems we (as a species and the planet) now face are the problems of success. Namely, how do we maintain a large population that is used to harnessing considerable environmental resources? This success in population is demonstrated by the leap in population that largely began with the Renaissance and the Industrial Revolution, and can be best seen by looking at a graph of the estimated global population over a few millennia (see Figure 7.4a). Much of the past couple of millennia has been characterised by low population and low population growth by today's standards. But it is clear that since the Industrial Revolution the global population has experienced a radical demographic change. This has been fuelled by

- lower mortality, especially child mortality,
- increased longevity (by around 50% since the onset of the Industrial Revolution) and
- a lag in the decline of fecundity (which counters the lower mortality).

The population curve looks almost exponential in nature. (That is to say, the curve is similar to an equation of the type of $y = x^n$, where n is a constant.) So, another way to look at how the global population has evolved is to plot it on a logarithmic scale. If population growth was truly exponential then such a graph (as in Figure 7.4b) would be a straight line, the slope of which would reflect the rate of growth. Indeed, up to the Industrial Revolution the population on a logarithmic scale does approximate a *shallow* straight line, which suggests a slow but steady rate of compound growth. However, the time after the Industrial Revolution to shortly after the mid-20th century sees this line become an upward curve, indicating that the rate of population growth was itself growing. That this upward curve itself looks as if it has an exponential shape signifies that for a time the population was growing at a super-exponential rate (see Figure 7.4b). Not surprisingly, population demographers of the 1960s realised this and it is from this time that modern concerns as to the population explosion (or bomb) came to the fore. Finally, the latter part of the 20th century saw a return

(a)

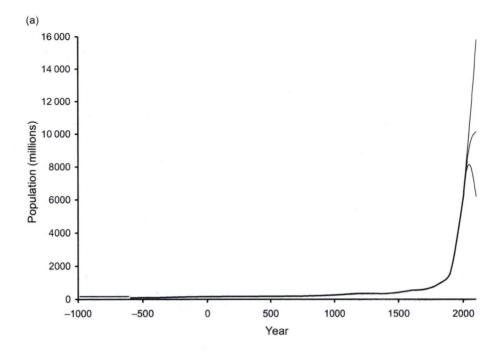

(b)

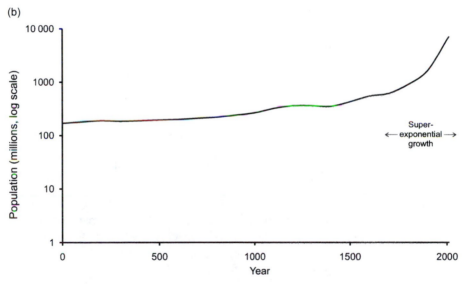

Fig. 7.4 (a) World population over 3000 years compiled from various demographic, historic estimates presented in a table by Cohen (1995) and future forecasts from the UN Population Division of the Department of Economic and Social Affairs (2010). The thick line depicts the actual population up to the early 21st century. Thin lines depict the UN's future 2010 low, medium and high projections to 2050. (b) World population on a logarithmic scale. This indicates the rate of exponential growth varying over time. A straight line would represent a constant rate of proportional growth and its steepness indicates the rate of growth. Note that the graph is not a straight line but an upward curve in the late 20th century, reflecting super-exponential growth for part of that century, after which it becomes less steep, indicating that population growth is slowing down.

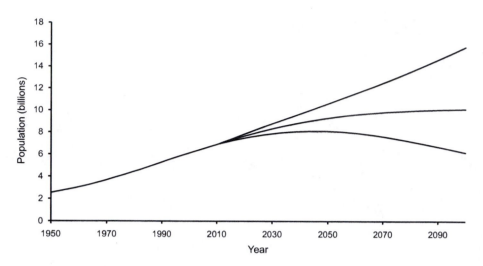

Fig. 7.5 The last 150 years of the 3000 years represented in Figure 7.4(a). This details the Earth's population of late 20th century to 2010 and also forecasts (in high, medium and low scenarios) the future up to 2100 (UN Population Division of the Department of Economic and Social Affairs, 2010).

from super-exponential growth to a more logistic rate of growth, when the population stabilised.

Figure 7.4a helps explain a good part of the reason why human impact on the global commons (the atmosphere and oceans) has been most marked in recent decades, rather than in pre-industrial times.

From the perspective of the Ehrlich–Holdren equation, the post-Industrial Revolution escalation in human population and industrially born wealth has had a profound effect on likely environmental impact. Indeed, future population growth and wealth will be drivers of future greenhouse gas emissions.

7.1.3 Future population

The UN forecast for the future of the Earth's population is based on three scenarios: high, medium and low. These are shown at the right-hand end of Figure 7.4a, which covers 3000 years, but are seen more clearly when the Earth's population for the last decades of the 20th century are plotted together with the UN forecast for the 21st century from 2010 onwards (see Figure 7.5).

With increased individual wealth a range of factors come into play that affect likely future population growth. These include increased control over personal reproduction (with late 20th-century biomedicine), lower childhood mortality (so increasing future parental security), increased aspirations for offspring and, from this last, increased cost of raising offspring to realise such aspirations. Each of these factors serves to lower the need and/or desirability for offspring and so lower the rate of population growth. However, past population growth still impacts on the present and the future because typically a birth in one decade will see a contribution to the population for a number of decades (through the individual's lifespan and own reproductive contribution).

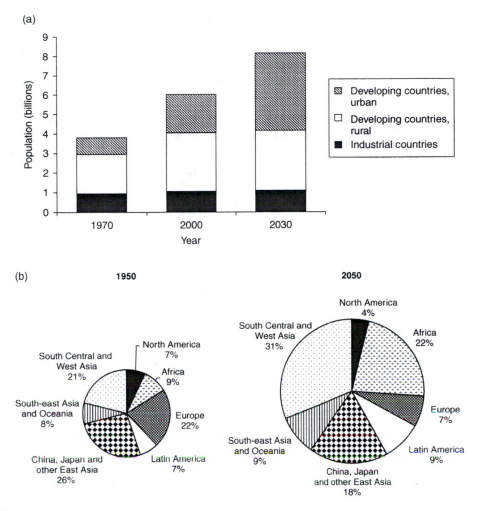

(a) Developing and industrial nations' contributions to global population growth (past actual values for 1970 and 2000 and UN medium-scenario future forecast for 2030). (b) Geographical distribution of the global population in 1950 and as predicted for 2050. After Raleigh (1999).

Such individual wealth factors show up as drivers of past actual and near-future population estimates in wealthy and poor countries (see Figure 7.6a). The traditional view has been that poorer nations have greater fecundity in part because of higher infant mortality and in part because children can contribute to their family income. Conversely richer countries, with lower infant mortality and children needing expensive education and/or wanting material goods such nations offer, see lower fecundity and smaller families. This brings us to the UN's estimates of the future population of the 21th century.

When the UN first estimated likely future population in 1980, it gave a high figure due to the particularly strong super-exponential growth seen early in the 20th century. However, over the next three decades it reduced the size of its forecast population for the 21st century as super-exponential growth declined (even though the population was still rising). With its 2010 forecast, the UN's Population Division of the Department of Economic and Social Affairs increased its 21st-century population projections (Figure 7.5). This was in no small part because the mantra that increasingly wealthy families have fewer children was found to be incorrect, due to a ground-breaking paper by Mikko Myrskylä, Hans-Peter Kohler and Franceso Billari in 2009. They found that while fecundity (which being demographers they called 'fertility' even though couples' potential to have children was not in question) declined with wealth in 1975, by 2005 some nations had become richer than the richest in 1975. The average life expectancy in countries with low infant mortality was 75 years or more and per-capita GDP was in excess of US$25 000 (in terms of purchasing power parity in 2000). These richer countries had larger families than those of the richest in 1975. In short, once wealth exceeds a certain point, even though having children is expensive they become more affordable.

This clearly is a double blow regarding anthropogenic climate change, as not only do larger families generate more greenhouse gas than smaller ones, but affluent families also produce more than their less-wealthy counterparts.

Figure 7.6b compares the geographic locations of the actual global population in 1950 with its anticipated locations in 2050. It is expected that by 2050 North America and Europe will have a markedly smaller proportion of the total world population, although in absolute numbers their populations will only fall marginally in the first half of the 21st century. The largest increase in world population will take place in regions with high poverty and unemployment (Raleigh, 1999). Irrespective of climate change concerns, this means that resources will need to be used more effectively and sustainably if human well-being is to be maintained without increasing environmental impacts.

Meanwhile, in a number of developed nations the decline in fecundity of the aging population has raised concerns about the size of the future economically active cohort necessary to sustain a larger elderly population. In other words, the ratio of active workers to retired people is decreasing. Just one instance of the policy response to such concerns came in 2003 in France, where the government announced an €800 grant to each baby born to a mother from 2004.

The growth in developing nations' population has driven, and is expected to continue to drive, an increase in their contribution to global greenhouse gas emissions. This will be explored in section 7.2, on energy.

7.1.4 Food

Agriculture enabled early civilisations to secure a regular surplus of food. Even so, the negative pressures of disease on the population remained, albeit in a changed manner: the gathering of humans into larger local population centres increased the opportunities for disease transmission even if an improved diet reduced disease associated with poorer nutrition. Consequently, despite an agricultural surplus there was not entirely a commensurate increase in the population of early civilisations. Indeed, the agricultural surplus of early civilisations was not great, and although more of this surplus could have supported more children, hence greater population growth, it was instead used largely for other purposes. For instance, Egypt during the time of the Pharaohs (2780–1625 BC) had a population of around three million people, of whom 95% were engaged directly in food production, so being able to support the remaining 5%, which consisted largely of slaves, some military and a small aristocracy. Calculations suggest that at that time the 20-year construction of the Cheops pyramid absorbed nearly all of this agricultural surplus. In later periods the emphasis of support from this proportionally small agricultural surplus shifted from slaves to the military. At other times there was little agricultural surplus and these are known to have been times of social strife (Cottrell, 1955).

The historic period (the time since written records began) has seen a gradual accumulation of knowledge and technical ability. This enabled there to be a similar development of the aforementioned small agricultural surplus and so gradually this freed more of the population for other endeavours. Virtually all these early societies relied on (non-fossil) biological energy and as such their impact on the biosphere's global commons was negligible: climate variation then was determined by non-anthropogenic factors such as circulation change, solar and volcanic climate forcing.

For thousands of years plant and animal breeding contributed hugely to the success of early agrarian societies. Prior to the Industrial Revolution, understanding the need to lay fields fallow and the use of animal and vegetable wastes as fertilizer improved agricultural productivity wherever such practices were conducted. The Industrial Revolution itself enabled agriculture to slowly become more mechanised, so reducing the need for such a large rural workforce. It also enabled agricultural goods to be shipped in bulk over long distances. Finally, the mid-20th-century development of fertilizers and pesticides in the Green Revolution further increased agricultural productivity to levels comparable with the present day.

By the end of the 20th century in industrialised countries the proportion of the population directly employed in agriculture had become a small minority. In the UK, which then produced nearly three-quarters of its own food, just 2% of the population was directly employed in agriculture, contributing to about 1% of GDP (Cabinet Office, 1994), with food processing, packaging and transport contributing an additional 2% of GDP. Allowing for the overseas contribution to current UK food supply, a reasonable estimate for the total equivalent proportion of the UK population involved in food production, processing and transport is 7–8%. Although this excludes retail and catering commercial activities, it is still a dramatic contrast to the involvement in food production of people in ancient civilisations. Today 92% or more of the population in many industrialised nations are free to engage in activities other than basic food supply, which is almost a complete reverse of the situation in ancient Egypt.

What enabled this radical change from ancient to modern societies to take place was the increased contribution made by energy. Indeed, with regards to nitrogen fertilizer (which commonly comes in the form of urea), fossil fuel is consumed in two ways. The first way is in the energy-intensive fixing of nitrogen from the atmosphere. This has in the past mainly been done using energy derived from fossil fuels, which are finite or so-called fund resources (the fund being finite). However, in the future nitrogen fixation might be done by energy from renewable (flow resources) and/or nuclear alternatives. Second, fossil fuel is used to provide urea's carbon content (mainly from natural gas, methane). With regards to this last, the prospects of not using fossil carbon for fertilizers are limited, although biogas may provide part of the solution.

Today's copious energy supply resulted, in the industrialised world, in our becoming used to supermarket goods that are largely independent of the seasons and consisting of produce from around the planet (i.e. with high 'food miles'). Furthermore, as the global economy becomes wealthier, so the meat content of the average human diet increases. The point here is that meat is an inefficient food source in terms of both energy and land – feeding grain to animals that we in turn consume – compared to humans consuming grain directly. World meat production per person increased more or less linearly throughout the 20th century through to the present day, from 17.2 kg per capita in 1950 to 39.0 kg in 2002 (FAO, 2002).

Consequently, the food dimension is not just a fundamental part of human ecology and the way we as a species relate to others, it is also an important driver of human activity – which depends on a societal food surplus above subsistence – as well as being a sector that consumes energy.

This energy relationship is two-way. Not only does the agricultural production of food – its processing, storage and delivery – consume energy, which in turn, if fossil fuel energy, has a climate impact. These climate impacts also impact back on agriculture. For example, the vagaries of the Holocene climate have affected European agricultural activity (see Chapter 5), and so future climate change is also likely to affect agriculture (see Chapter 6 and the case studies on the probable US and Australasian impacts, sections 6.2 and 6.4).

If human food supply is fundamental to the nature of human activities, if it is dependent on fossil energy and if food supply is shaped by climatic patterns, then what about food security against a backdrop of an increasing and unsurpassed global population? Will future climate change improve or weaken the global security of food supply? This question will be addressed in section 7.4.

7.1.5 Impact on other species

Because the Earth's population is now so large, and the commensurate need for food is so great, food production dominates much of the most biologically productive temperate land. Together with the land required to house, transport and provide a location for economic activity, this has seen our species take much land away from purely natural systems. Indeed the (perceived market) value of economic activity can be such that it surpasses that of purely agricultural value, which in turn leads to housing and urban settlements being constructed on fertile land. As mentioned in Chapter 4, the extent of this domination over natural systems is such that it is estimated that just under a third of global terrestrial net primary production (the biomass arising from the energy trapped by photosynthesis) is appropriated for human use (Imhoff et al., 2004). Another perspective on the degree of human management of global ecosystems is provided by the amount of nitrogen fixed by humans (primarily for fertilizers), which already exceeds that fixed globally by natural means. In addition, more than half of all accessible fresh water (i.e. excluding sources such as the Antarctic ice cap) is put to use by humanity (Vitousek et al., 1997), and 10–55% of terrestrial photosynthetic production is managed by humans (which includes not just food for humans or farm animal feed, timber, biofuels, etc., but also amenity value and active wildlife conservation management; Rojstaczer et al., 2001). Such natural resource management has necessitated the modification of many ecosystems and this in turn has resulted in a high degree of fragmentation of the natural landscape (see the previous chapter). Again it is important to remember that this fragmentation is unique to modern times and so will greatly exacerbate the impact of climate change on biodiversity compared to nearly all other times in the Earth's history.

At this point it is worth considering the benefits of intensive agriculture. Intensive agriculture has had some bad press, and although some bad reports may be justified there are also clear benefits. Of course, use of land to produce food intensively for agriculture is almost certain to have a negative effect on biodiversity, if only because primary production will be geared to that of nurturing food species at the expense of others. What is less considered is the effect of a plot of intensive farming on landscape biodiversity as a whole. Which would be better: 1 ha of natural environment next to 1 ha of highly intensively farmed land or 2 ha of far less intensively farmed land? The exact answers tend to be site-specific within the broader landscape and in no small part dependent on the natural environment and the type of intensive farming being considered.

In 2011 biologists from the University of Oxford and the Royal Society for the Protection of Birds (Phalan et al., 2011) noted that the question of how to meet rising food demand at the least cost to biodiversity requires the evaluation of two contrasting alternatives: 'land sharing', which integrates both objectives on the same land, and 'land sparing', in which high-yield farming is combined with protecting natural habitats from conversion to agriculture. To test these alternatives they compared crop yields and densities of bird and tree species across gradients of agricultural intensity in south-west Ghana and northern India. More species were negatively affected by agriculture than benefited from it, particularly among species with small global ranges. For both these types of species in both countries, land sparing is a more promising strategy for minimizing the negative impacts of food production, at both current and anticipated future levels of production.

In short, if 20th-century farming had not become more intensive then inevitably more land would have been switched from being habitat for wildlife to land for agriculture. This would have made the (potential) impact of climate change on natural systems more acute (see Figure 7.7).

The impacts of this competition on other species are many and are mainly negative, covering competition for and with natural systems, and land and related landscape fragmentation. This topic itself is so great that it is the focus of a considerable body of work that is far too great to meaningfully summarise here. However, it is worth citing some recent work that suggests the scale of the problem.

In 2005 a team of researchers led by Gerado Ceballos of the National Autonomous University of Mexico conducted the first global analysis of the conservation status of all known land mammals with regard to their spatial distribution. The International Union for the Conservation of Nature (and Natural Resources; IUCN; now also known as the World Conservation Union) and the United Nations Environment Programme (UNEP) have for many years identified and/or sponsored the identification

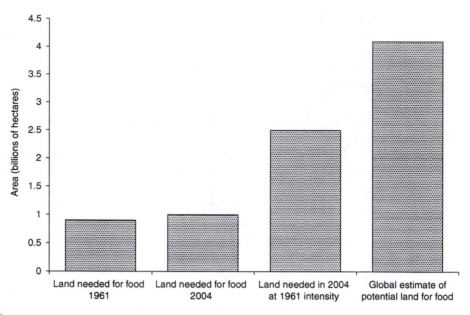

Fig. 7.7 Land for global food production; cropland versus intensification. After Mooney et al. (2005).

of endangered species and listing protected areas, but this is different to a global analysis of the spatial distribution of all mammals. The Ceballos team compiled the geographic ranges of all 4795 known species of land mammal. They then linked this to cells, rectangles of 10 000 km² each. They found that 1702 cells of land would conserve at the very least 10% of the range of each species; this 10% is a critical threshold that many conservation biologists use as the minimum of a species' natural range needed to conserve a viable population. The spatial detail of the 1702 cells is, unfortunately, too coarse for it to be used in practical conservation management. However, 1702 cells is roughly equal to 11% of the Earth's ice-free land surface and provides an indication of the scale of the problem. That around 80% of the critical 1702 cells are already used for agriculture demonstrates the imposition of human food demands on conservation. Conversely, looking at some 250 conservation scenarios, 95% of these cells can be replaced by other cells without loss of conservation value, and only 135 cells (<1%) are irreplaceable; that is, were required in all 250 conservation scenarios. Of these, those located in wealthy countries (which have the resources for implementing developed, sustained conservation programmes) are most likely to succeed, whereas extinction risks are higher for cells in less-developed nations.

Another key question is how, as opposed to where, should finite conservation effort be focused? A long-running debate in conservation biology has been whether efforts should be area-based (say, conserving a region as a national park or World Heritage Site) or based on a species: such are often called flagship species (a famous example is the giant panda, *Ailuropoda melanolueca*). Marcel Cardillo, Georgina Mace and colleagues (2005) looked at 4000 species of terrestrial mammal. They looked at three principal groupings of predictors of extinction risk: (1) environmental factors such as size and range of species, (2) species' ecological traits such as population density and (3) species' life-history traits such as gestation period and weaning age. They found that extinction risk exhibited a positive relationship with a species' body mass. Importantly, they noted that extinction risk increases sharply when a species' adult body mass exceeds 3 kg. Second, whereas extinction risk in small species is driven by environmental factors, in larger species it is driven by both environmental factors and the species' intrinsic traits. Consequently, they conclude, 'the disadvantages of large size are greater than generally recognized, and future loss of large mammal biodiversity could be far more rapid than expected' (Cardillo et al., 2005). Interestingly, this is in line with separate work done on mass-extinction events demonstrating that it is the large species that are most vulnerable in such events (Hallam, 2004).

Linking such research to the spatial manifestation of climate change is likely to receive greater attention in the coming years and no doubt will continue to do so as geographic climate models become more detailed with regards to both climate and geography. Meanwhile, competition for land for humans as opposed to habitat for wildlife continues irrespective of climate change, which is a compounding factor.

To practically address this, biological conservation managers are developing ecological networks of habitat with corridors between reserves, small 'stepping stone' reserves and permeable areas (either low-intensity human use or a landscape with semi-natural features) through which species can migrate and buffer zones of semi-natural landscape to protect the key elements. There are now around 250 ecological networks globally. In 1995, 53 European countries agreed to the establishment of the Pan-European Ecological Network (PEEN). The European Centre for Nature Conservation (ECNC) coordinates this work in collaboration with the Council of Europe (Parliamentary Office of Science and Technology, 2008).

7.2 Energy supply

Over half of global warming in recent decades is attributable to anthropogenic carbon dioxide (see Figure 1.2) and the clear majority of this comes from the burning of fossil fuels (see Chapter 1). Consequently, the fuels we use to supply commercial energy are central to climate change policies.

7.2.1 Energy supply: the historical context

Ultimately nearly all energy on Earth – except for hydrogen (and perhaps theoretically potential exotic energy such as zero-point energy) – comes from the Sun and other stars. Most mechanical energy on the Earth arises in one, or a combination, of two ways: first, from the solar energy driving movement in the atmosphere, wind and in turn waves, waving tree branches and so forth; second, gravitationally through Earth–Moon interactions and, of course, motion within the Earth's gravitational field.

Solar energy is also trapped by photosynthesis and stored chemically. Chemically stored energy may be harnessed by living things through the various trophic levels of ecosystems or geologically trapped as fossil fuels. Geothermal energy (in no small part) comes from the decay of radionuclides deep in the Earth: the Earth would have cooled to a solid long ago if its internal heat had arisen solely from its formation, notwithstanding the energy arising from the Moon's tidal drag. These radionuclides were formed from supernovae and so they too have stellar origin.

Humans as a species are unique in that it appears that we learned quickly to use fire to supplement the biological energy of our own bodies. However, for many millions of years for this supplement to our energy we were reliant on basic biofuels such as wood and animal oils and dung. Even so, the use to which this energy was put was almost invariably confined to cooking, space heating and illumination, and not motion, or information processing and transference. As discussed in the previous section, this meant that the energy surplus which early societies had above the level of subsistence was small and largely dependent on bio-energy in the form of the excess food that early agrarians could provide to support others engaged in non-food tasks. Mechanical energy was harnessed from animals, as well as wind and water, and this made was a substantial contribution to the energy surplus enjoyed by early agrarian *H. sapiens* over hunter–gatherers. The benefits of harnessing energy are clear. This leads to two ultimate societal consequences, as I shall describe.

First, to realise these benefits there was and still is a drive to increase energy consumption. If such consumption increases globally in an uninterrupted fashion over time then sooner or later the magnitude of this energy surplus will affect the biosphere's thermodynamic balance. This is inescapable. It is exactly what is happening with anthropogenic global warming from fossil fuel burning and its resultant climate change, although – almost ironically – it is not the energy surplus itself that is causing the problems but the thermodynamics of its chemical carriers. Prior to this, energy-related environmental impacts existed but were not climatically significant. For instance, the early human use of wood as a fuel and construction material impacted on ecosystems and species as woodland was cleared and replaced by fields, but did not impact as it does today on biosphere energy flows.

Second, given clear benefits, it follows that there must be equally clear disadvantages to a society that has enjoyed a certain level of energy surplus if that surplus is reduced. We tend to think of energy crisis as a modern problem but this is not so. Early societies were constrained by the limitations they had in harnessing and deploying energy resources. Consider what the Egyptians would have done had their society's surplus been 10 or 15% rather than just 5%. More recently, in 17th-century England there was an acute shortage of fuel wood and charcoal. This led to more than a century of iron shortage. By the late 19th century the use of wood as a major fuel had declined to negligible levels and coal dominated. However, coal had its limitations and, while suitable for steam locomotives, was not nearly as suitable as liquid fossil fuel, oil, for personal transport. That fuel made a major contribution to society's global energy budget in the 20th century.

Our species' use of energy has therefore been characterised by fuel-switching at an ever faster rate. Aside from the various early biofuels and the early use of wind and water, this fuel-switching is clearly seen from the advent of the Industrial Revolution to the present day in the country where the revolution began. Figure 7.8a depicts the proportional contribution of various fuels made to the UK's energy budget since

(a)

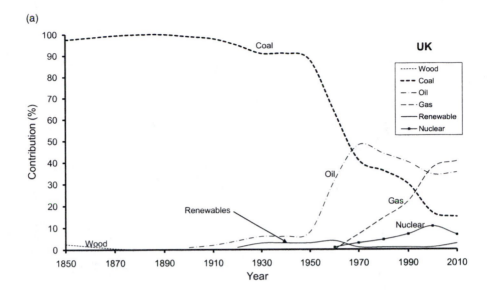

(b)

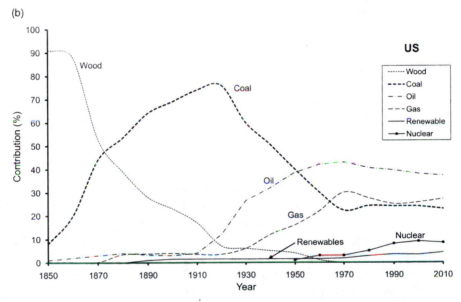

Fig. 7.8 The proportional contribution (as a percentage) from various fuels made to the energy budget of the (a) UK and (b) USA. (c) Summary of the proportional contribution (percentage) from various fuels to global energy supply. Before 1950 theoretical estimates are used and after 1950 data are based on estimates derived from commercial figures with a 10-year filter; see Appendix 3 for sources.

1850. It can also be seen in other developed nations outside of Europe but with a delayed start and hence in a contracted form. Figure 7.8b relates to the US energy budget. We see that in the 20th century it took the order of three or four decades to significantly develop a new fuel stream for a nation's energy budget. Such lead times need to be considered when contemplating future energy resources.

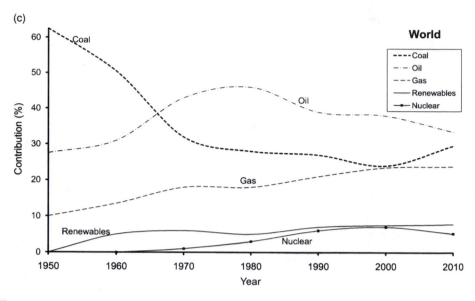

(c)

Fig 7.8 *(continued)*

The global energy picture that emerges has been one of diversification. There was a proportional decline in the importance of coal over the last half of the 20th century (see Figure 7.8c), before a resurgence in the early 21st century. The proportion of global energy demand met by oil over this time increased to around 46% by 1980 when price and energy-security concerns began to cause a shift to other energy sources such as nuclear power and gas. Most of the global energy diversification has been at the expense of coal, much in the way that fossil fuels replaced wood in the USA and much of the New World, and a little later less-developed nations, in the late 19th and early 20th centuries. However, longevity of individual energy resources (the economically recoverable ones) will, as we shall see in the next chapter, determine the economic sustainability of respective fuels.

At the end of the 20th century nearly all countries were dependent on fossil fuels for the majority of their energy needs, although nuclear and renewable energies, as a proportion of the total, were beginning to grow.

Although a look at the breakdown of a nation's (or the global) commercial energy budget informs as to fuel-switching and is suggestive of new fuel-introduction lead times, it does not indicate how much carbon dioxide will be generated in the future: it all largely depends (higher energy per carbon methane notwithstanding) whether the total fossil energy consumed remains constant, increases or decreases. In fact, commercial *energy* consumption overall has been increasing globally with time, and markedly so (see Figure 7.9). This exerts a pressure to increase *fossil fuel* consumption in the absence of new non-fossil energy resource deployment. Indeed, by the end of the 20th century most of our global commercial energy consumption (around 87%) came from fossil fuels, this having increased by more than 350% since the middle of the century. Even though the contributions from nuclear and renewable energies have grown slowly, their growth – as a proportion of the global commercial energy budget – has increased the annual non-fossil contribution only

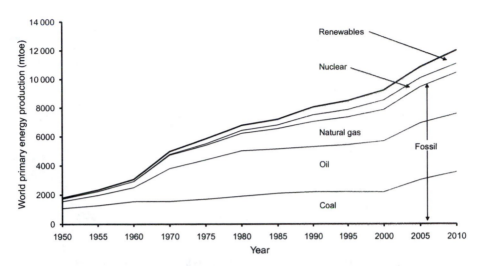

Fig. 7.9   Broad summary (with a 5-year filter) of global primary energy production/consumption in million tonnes of oil equivalents (mtoes). Because of the 5-year filter short marginal dips in production/consumption (such as 1979–82 and 2008–9) do not show. See Appendix 3 for sources.

marginally compared to fossil exploitation in absolute terms. Consequently there has been a commensurate rapid increase in carbon dioxide emissions in the latter half of the 20th and early part of the 21st centuries compared to earlier (see Figures 7.9 and 7.10b). Note, from Figure 7.9, how fossil fuel use has changed (increased) since the first Intergovernmental Panel on Climate Change (IPCC) assessment in 1990. This has caused some academic commentators to say that we are not embarked on a 'business-as-usual' but 'burn baby burn' scenario (Lenton and Watson, 2011).

Reflecting the nature of human population growth (and the Ehrlich–Holdren equation, linking environmental impact to population), most of the 19th- and early 20th-century carbon emissions were made by industrialised countries, with little from the less-developed nations. In the 20th century the populations of the developed industrialised countries had largely ceased to grow (Figure 7.10a), whereas those of the less-developed nations grew markedly. Furthermore, as the population grew in less-developed nations their energy usage was developing and increasing on a per-capita basis. Figure 7.10b looks at global carbon emissions from fossil fuel in more detail on an annual basis since the mid-20th century and shows the relentless increase of annual global emissions.

As we shall see in the next chapter, this has coloured international policy discussions. On one hand some have argued (as did the USA around the turn of the millennium) that, as the less-developed nations were contributing more carbon than the developed nations, these countries needed to either demonstrate a commitment to cutting emissions or make some other contribution to the international effort to curbing greenhouse gas emissions. On the other hand (argued at the time mainly by the less-developed, or developing, nations) the industrialised countries need to carry the brunt of costs for international greenhouse controls as, on a per-capita basis, each person in an industrialised developed nation contributes far more than a person in a less-developed nation (see Figure 7.1 and relate that to Figure 7.6). Remember that

(a)

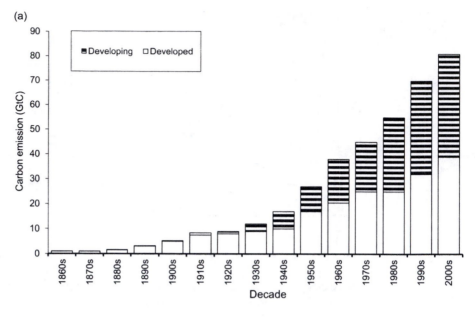

(b)

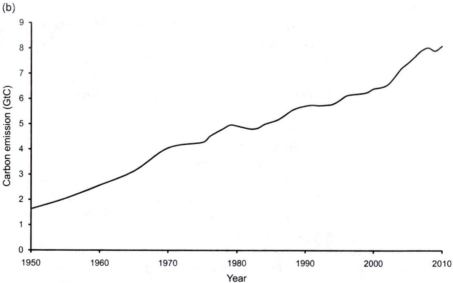

Fig. 7.10 (a) Outline of developed and developing nations' emissions of carbon (GtC) by decade from fossil fuel burning and cement production for each decade. (b) Annual (not decadal) global emissions of carbon (GtC) from fossil fuel burning (excluding other minority industrial sources) since 1950, the period of most emissions. A 5-year filter applies to years before 1975. Sources: see Appendix 3. Fossil fuel annual emissions prior to 1990 were less than 6 GtC and emissions prior to 1850 less than 0.1 GtC (IPCC, 1990).

North America (USA and Canada), with roughly 5% of the global population, around the turn of the millennium was contributing about a quarter of global emissions. The argument here is that greenhouse gas emissions need to be tackled on a per-capita basis. Such policy concerns will be discussed further in the next chapter.

Finally, the IPCC (IPCC, 1990) estimated that the total emissions of carbon from fossil fuels between 1850 and 1987 were around 200 GtC ($\pm 10\%$). This is nearly double the IPCC's estimate of 115 GtC ($\pm 10\%$) for carbon release due to historical changes in land use (deforestation, etc.). This, and that two-thirds of fossil fuel emissions over this 120-year period took place in its last three decades, explains the importance of recent fossil fuel carbon releases to climate change and why climate forcing has dominated matters only towards the end of the 20th century. With the prospect of ever larger fossil fuel releases in the 21st century (as demonstrated by each subsequent IPCC assessment since 1990), the implication is that the thrust of 21st-century policies to control atmospheric concentrations of carbon dioxide need to focus on the energy dimension (see Chapter 8). It also explains why the political manifestation of climate concerns became more marked at the 20th century's end, with more severe impacts anticipated in the 21st century, even though Svante August Arrhenius identified the problem as far back as 1896.

7.2.2 Future energy supply

Nobody can predict the future. However, we do know with certainty the historic pattern of energy production and the present situation. We also know, as described earlier in this chapter, the likely future population and economic trends. Overall this suggests that, barring major disruption of global society, demand for energy will increase. How will the increased demand for energy be met and what will this mean in terms of carbon dioxide emissions?

The IPCC's 2007 report (IPCC, 2007a) and their forecasts for 21st-century global warming are based on a series of future scenarios developed in 2000: the *Emission Scenarios – IPCC Special Report* (IPCC, 2000). These scenarios vary from their lowest (which forecasts marginal growth in emissions to the middle of the century followed by a decline to just below 1990 levels) to their highest (which predicts an increase in emissions up to five times 1990 levels by the end of the century). Again remember, carbon dioxide accounts for more than half of anthropogenic warming (see Figure 1.2).

The pattern that the IPCC emission scenarios outline for methane, the next major greenhouse gas contributing to anthropogenic warming (about 18% in 2005; Figure 1.2b), proportionally follows that of their carbon dioxide scenarios. This is because methane too is either population-driven (needing an increased area of rice paddies and number of enteric fermenting animals) and/or energy-related (needing more gas drilling and venting). However, in the future this may not be so, as the technical and energy options can vary.

Whatever the future, the IPCC predict a range of possible futures and it is this that drives their core warming forecasts (see Table 5.1). So, what are the energy possibilities and options? Excluding environmental costs, the cheaper options (excluding long-term environmental and social costs) include the Business-as-Usual (B-a-U) scenario; that is to say, humans broadly continue the present pattern of energy-resource consumption. The reason for this is that no extra cost is required to develop and implement new energy technologies. In this scenario we would continue to rely heavily on fossil fuels as long as reserves will allow throughout the 21st century. The caveat about reserves allowance is an important one. For whereas there is plenty of economically recoverable coal worldwide, over two centuries' worth at current rates of production, there is less oil and gas. BP Economics Unit (2011) estimated that there were enough commercially available proven reserves of gas to last nearly six decades at 2010 levels of production, and only enough oil for very roughly around four and a half decades. It needs to be remembered that the pressures for global consumption of these resources to increase are plainly evident, as discussed earlier in this chapter.

Of course, should prices change (as they will), today's economically recoverable reserves may not be so tomorrow. Indeed, with increasing scarcity prices will rise, so enabling reserves that are currently either harder to reach, or more expensive to refine (such as sulphur-rich deposits), to be economically recovered. Then again, future innovation and new technology may enable reserves that were previously uneconomical to recover to become economical. Furthermore, as prices rise so non-fossil technologies, formerly considered too expensive, will become economically viable. Finally, proven reserves are a subset of total reserves, which includes deposits yet to be discovered as well as exotic sources (possibly such as marine methane hydrates) that are not usually considered. In short, estimates as to ultimate economically recoverable reserves vary.

To see how these reserve caveats work, consider the analogous question of how long US food stocks may last. One might say that adding up all the food in people's homes, shops and warehouses, and dividing it by the rate of consumption, might give a figure of, say, x months. However, in addition to this there will be crops growing in the fields not yet harvested. There will also be the food wasted due to, say, combine harvesters missing some of the crop, mishandling or poor storage. Then there are species of food plant outside of the economic system, from apple trees in people's gardens to raspberries growing in the wild. There are also exotic food species, that is to say perfectly edible plants and animals that do not normally feature in most Americans' cuisine. Finally, there will be species not normally edible that will require some sort of processing before they can be used. In short, what can be considered the total US food stock depends on what exactly is being considered. So, such an estimate, when expressed in terms of a reserve/production (r/p) ratio, is therefore not as exact a figure as one might think. Food in the USA will not run out on the last day of the xth month. Quantifying oil reserves is fraught with analogous difficulties and so estimates have to be viewed with extreme caution.

An example of how reserve assessments can be misleading comes from the Kern River oil field. It was originally discovered in California in 1899. Calculations in 1942 suggested that 54 million barrels of economically extractable oil remained. However, in subsequent years some 736 million barrels were recovered and a new estimate later in the century suggested that 970 million barrels remained. What had happened was not that the oil field had changed but that knowledge and technical ability had increased (Maugeri, 2004). Having said this, it does not mean that oil is an inexhaustible resource but this example does demonstrate that estimates have their limitations.

From this it can be seen that r/p ratios are at best only a broad indication of future potential. However, with regard to fossil fuels, they do suggest a magnitude of resources for continued, if not increasing, exploitation throughout the 21st century irrespective of whether or not they fully meet energy demand. They can tell us whether the resource is likely to be a major economic one for years, decades or longer, after which it will become more scarce/expensive, or be less economically competitive, and perhaps indicate a time when other (possibly more expensive) energy resources may become more competitive. In greenhouse terms r/p ratios are suggestive of the *minimum* time in which under B-a-U fossil resources are likely to make major contributions to atmospheric carbon dioxide. Of course, 'business' is unlikely to remain 'as usual' the further into the future one projects a forecast and so no meaningful estimate can be made as to fossil fuel consumption in the 22nd century or beyond. Among the various wild cards are the roles the various alternative resources will play.

Alternative (non-fossil) energy resources are in the main currently more expensive than fossil fuels and/or have other problems associated with them. Renewable energy from so-called flow resources such as tidal, wind and wave power ('flow' because unlike 'fund' resources, there is a continual energy flow to be harvested) has a lower energy density and so requires a larger area for its exploitation, hence a large land-use impact. Conversely, finite fund resources such as fossil fuels and uranium have high energy intensities. Biofuels, although carbon-based, do not short-circuit the deep carbon cycle, and do take up productive land that could be used for food crops. This needs to be taken into account. We need to remember, against a backdrop of a growing global population, that land for which there will be an increasing food-crop demand has competing use options. Biofuels have a significant biological opportunity cost, being the cost of a venture in terms of lost opportunity of another venture. In short, agricultural land can be used for food or energy crops, or even a mix, but one will always tend to offset the yield from another. There are also non-biological opportunity costs, for the agricultural land might be sold and used in a non-biological way such as to provide a site for a factory.

Other than sources of renewable energy, the other principal energy alternative to fossil fuels is nuclear power. The cost of nuclear generation is generally considered at best comparable with the cleanest and most efficient of coal stations but is usually more expensive. Its advantages are that it furthers a nation's energy security and, of course, has a small greenhouse impact. Its disadvantages (other than cost) are that it has other environmental impacts, chief among which is the problem of long-term disposal of radioactive waste. This is a financial unknown and has not generally been completely solved to public and political satisfaction. Further, much of the unknown component of the cash cost of long-term disposal will not be incurred until many years into the future so that, unless a provision is made today to put aside some of the income from the sale of nuclear-generated electricity, there will be an intergenerational transfer of externality ('externality' because this cost is external to the price of the electricity sold). In this sense nuclear power is similar to power arising from fossil fuels in that the consequential global warming generates an environmental impact with economic consequences for future generations to bear. Economists therefore say that both fossil fuel use and nuclear power currently have environmental externalities associated with them, which in turn have intergenerational transfer consequences. (The issue of nuclear waste will be touched upon briefly in the next chapter.)

Then there is the question of how much fossil carbon is used by nuclear power. It has been said that nuclear power is not a zero-carbon energy source as fossil fuel energy is required to mine, transport, process and refine (enrich) uranium ore. This is true. While it is not this book's purpose to look at energy resources in detail, because of the biological and human-ecology consequences of climate change it is necessary to at least examine them in a broad way to ascertain the approximate greenhouse gas-emitting potentials of differing energy resources. Fortunately, comparing UK and France's energy consumption and their respective fossil fuel/nuclear balance can allow us to make a very simplistic but informative assessment. This comparison is all the more valuable because France and the UK are geographically close, share the same European economic grouping, have a similar population size (in 2004, 60.6 million and 60.4 million respectively) and have a similar annual total (fossil and non-fossil) energy consumption, which for 2004 was 262.9 and 226.9 million t of oil equivalent (mtoe) respectively (BP Economics Unit, 2005). Yet France consumes some 29% (60.4 mtoe annum^{-1}) less fossil fuel energy than the UK. This fossil saving comes from France's nuclear power programme which is far bigger than the UK's.

That there is a fossil saving means that France itself is not expending more fossil energy (offsetting the fossil savings) in processing or enriching uranium ore or processing nuclear waste. It can be argued that there is a considerable amount of energy expended in mining and transporting uranium ore. Looking at nations from which uranium ore mining takes place it is difficult to discern any major difference in fossil fuel consumption on a per-capita basis above that of their economically comparable neighbours. This is not to say that there is no energy expended in uranium extraction and refining but that it does not put anywhere near as heavy a fossil fuel burden on a nation as nuclear power saves. Indeed, one estimate, admittedly assuming efficient fast-breeder reactors, is that a low-grade uranium ore (Chattanooga black shale, with just 60 g of uranium per t of shale) has an energy equivalent to 1000 t of coal (McMullan, 1977). Higher-grade ore is better still. So, even without using fast-breeder reactors, we can see that a tonne of higher-grade uranium ore is worth thousands (or tens of thousands for even higher grades of ore) of tonnes of coal, hence the energy used for mining and refining it is nearly proportionally less per tonne of fuel (hundreds or thousands of times less, as uranium is more energy-intensive to refine than oil). This was reflected in a 2006 estimate provided by a UK 'green' policy forum, the Sustainable Development Commission. It cited an estimate for carbon emissions from (non-fast-breeder) nuclear power as being 4.4 tC GWh^{-1} compared to 243 tC GWh^{-1} for coal. In short, while nuclear power does not have a zero-fossil carbon cost it does have a very significant net fossil carbon-saving impact.

The exact magnitude of such savings will become more certain in the coming decade as assessments are made prior to the second phase of the Kyoto Protocol (see Chapter 8) that is due (at the time of writing) to begin in 2012.

The other difficulty with the nuclear option is that civil nuclear power can be used to further military nuclear technologies that in turn undermine international stability. This is likely to become an increasingly pertinent issue as on one hand developing nations legitimately seek to become more energy-sustainable, and so explore the nuclear option, while on the other they may be tempted by the international leverage that a nuclear military option confers. (Iran in the years subsequent to 2003 is but one example.)

There is also another similarity between nuclear fission and fossil fuel use, which is that strictly both are finite or fund resources. This is not always appreciated. Currently the contribution of nuclear power is such that there are enough known uranium-containing geological strata to fuel our current nuclear power industries for over a century, and more considering that uranium prospecting has not been as intensive as that for oil. This broad estimate does not take into account fast-breeder options that would greatly enhance uranium's longevity as a resource. But equally, as mentioned, this comes with greater technological as well as military security challenges (see section 8.2.5 on the prospects for nuclear power). Furthermore, since the mid-1980s (specifically the Chernobyl power-plant disaster in 1986) the growth of the nuclear industry has slackened markedly. If there was a major international resumption of the nuclear industry then global uranium resources would need to be reassessed.

With regards to the global commercial energy budget, the factors of resource availability (or scarcity) and cost compete. Together these largely determine the global energy resource mix, the fossil contribution of which will determine anthropogenic carbon releases from energy (sequestration technologies aside). Although overall reserves of fossil fuels (coal, oil and gas) are enough to theoretically enable carbon emissions to increase throughout the 21st century, the low r/p ratios for gas and oil suggest that the consumption of at least oil and gas might peak early in the 21st century. Estimates that are not based on r/p ratios vary but there is a rough consensus from a number of analysts (for example Bentley, 2002) suggesting that the peak consumption of oil and gas might be around the middle of the 21st century. The options for coal are less restricted and this could make up for any shortfall in gas or oil should additional alternatives fail to be developed.

In short, such is the likely magnitude of future energy demand that unless non-fossil energy dominates the global energy mix fossil fuels are likely to continue to add significantly to the atmospheric reservoir of carbon throughout the 21st century and well into the next century. Given that to stabilise temperatures we would need to lower greenhouse gas emissions by around 60% of their 1990 level, and that the atmospheric residence time of a carbon dioxide molecule is between one and two centuries, it seems unlikely that global warming will cease over the next few centuries. Further, in all likelihood human society will have to live with the impacts of further global warming beyond the 21st century. (Indeed, there will be many centuries' worth of impact if global warming causes carbon to cascade from biosphere pool to biosphere pool via the atmosphere in an early Eocene-like Initial Eocene Thermal Maximum/Palaeocene–Eocene Thermal Maximum (IETM/PETM) event.

In terms of human biology and ecology these climate impacts can largely be grouped into the related categories of health and food supply.

7.3 Human health and climate change

Extreme weather events, such as the heatwaves of north-east North America and Western Europe in August 2003, according to some climate models, are thought to be associated with global warming (see section 6.6.1; Meehl and Tebaldi, 2004). The weather, it is said, is a traditional topic of conversation associated with the British. It should therefore not be surprising that when extreme weather events take place the

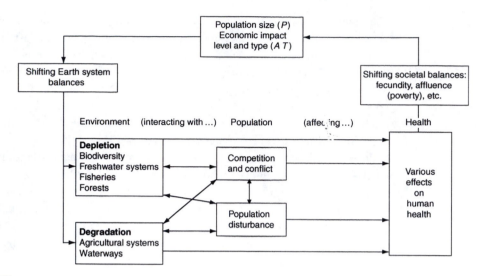

Fig. 7.11 Global environmental change, development and health impacts. Further developed from McMichael and Haines (1997).

British press write some colourful copy. And so *The Times* informed us that during heatwaves the clock of the Palace at Westminster in London (colloquially known as Big Ben) slows down but also that there were 5–15% more recorded sexual offences during the unusually warm summers of 1976, 1989 and 1990 (Coates and Cohen, 2003). While the psychology of climate change is itself a fascinating subject, and has been outlined elsewhere (Cowie, 1998), global warming has more immediate effects on human biology. Indeed, during the 2003 heatwave it was reported that the UN Early Warning and Assessment Centre was considering concerns that global warming would impact on human health through changes in the geographical distribution of disease, flooding and sea-level rise. This prompted the UK Environment Minister at the time, Elliot Morely, to comment that climate change 'is a major threat to our global community and one that has to be taken seriously'. The relationship between climate change, development and health impacts is summarised in Figure 7.11.

Climate change affects human health through processes mediated by disturbances of ecological systems (Chapters 4 and 5). The degree and nature of recent and future health effects from climate change is related to the nature of the said climate change (Chapters 5 and 6). However, in terms of energy, climate and human health, these dimensions have only recently come to the fore in academic discussions even though there were earlier greenhouse warnings arising from fossil fuel use (Chapter 1).

In the early 20th century the concern of industrial countries, regarding energy and environment matters, focused on ensuring that people were warm enough in the winter and, to a lesser degree, cool enough in the summer. Here the energy conclusions of Sir (Sydney) Frank Markham's sociological analysis of 1942 are typical and are reflected in its title, *Climate and the Energy of Nations*, which he further developed in a 1944 edition in the middle of World War II (Markham, 1944). Sir Markham had an interest in history, having a humanities degree from the University of Oxford and being President of the Museums Association, and also in politics and science, being a Parliamentarian for 35 years, during which his duties included terms as

Parliamentary Secretary to the Prime Minister and Chairman of the Parliamentary Science Committee. His original goal was to discern what made nations great. He was prompted, one day in 1933, when despondently Prime Minister Ramsay MacDonald had said to him that 'Government has become no more than an ambulance, and its success is measured by the speed with which it deals with disaster.' (This was during the global economic depression of the 1930s but is relevant to the 2009 depression as well as current climate concerns.) After years of on and off contemplation Markham concluded that climate and climatic independence was central to national success. His assessment focused on how nations through the ages had striven to become more independent from the environment and he argued (albeit obliquely) for the need to address what is known today as fuel poverty. The one area of environmental degradation from energy and health consequences he touched on concerned the urban smogs caused by the prodigious domestic coal consumption at the time. Nonetheless, his overall conclusion was that, to be successful, nations need to ensure (through coal, oil and electricity consumption) that their citizens' indoor environment should be 'ideal'. What he would have made of the indoor environment of citizens' homes in the industrialised nations in the 21st century we will never know, as the norms three-quarters of a century later have changed considerably. However, it is likely that he would consider today's greenhouse and health concerns somewhat ironic.

Nearly a quarter of a century on from Markham's book, in the 1960s, all but the poorest of people in Britain had warm homes in winter. British biologists' concerns arising from climate were not so much focused on human health due to temperature but on climate and Britain's ability to feed its population; remember that rationing from World War II only completely ended in 1950. In the 1960s scientists were aware that the climate had changed in the past, so when in 1963 the Institute of Biology (the professional body for UK biologists now re-branded as the Society of Biology) in its annual major symposium, looked at the biological significance of climatic change in Britain, the human health dimension was not considered. Instead the Institute's focus was on the impact of climate change on natural and agricultural systems (Johnson and Smith, 1964). Apart from a few specialists, only since the late 1980s have health concerns from climate change (as opposed to weather) had serious attention.

Contemporary concerns as to climate change and health impacts have been summarised by the IPCC (2001a, 2007b) and the World Health Organization (WHO, 1999, 2001a), and the impact of climate change on human health was even the theme of WHO's World Health Day in 2008 (WHO, 2008). Indeed, the WHO recognises that not only are health impacts from climate change per se important, but also ecosystem degradation. A significant cause of this is climate change, as the WHO acknowledges in its 2005 report, *Ecosystems and Human Well-Being* (WHO, 2005). The emerging picture is that some populations will experience health impacts from climate change due to

- permanent or intermittent changes in disease range due to shifting climate zones and/or extreme weather events (such as the El Niño),
- flooding, due to either sea-level rise or extreme weather events,
- changes in food security, and
- ecosystem degradation.

Finally, both the IPCC and WHO emphasise that the people most vulnerable to health impacts arising from climate change will be the poor: those in the less wealthy nations and the poor in wealthier nations.

But it is not just climate changes' impacts that have health consequences; there are positive health benefits to policies that promote climate change mitigation. In 2010 in a joint editorial, published simultaneously by the *British Medical Journal*, *The Lancet* and *Finnish Medical Journal*, Ian Roberts and Robin Stott urged clinicians, and those in professions allied to medicine, to put health at the heart of climate change negotiations. They argued that moving to a low-fossil economy 'could be the next great public health advance'. For example, a low-fossil economy would mean less pollution and a need for more physical activity. A low-fossil diet (especially eating less meat) that relies on less fossil energy in the provision of food, and taking more exercise will mean less cancer, obesity, diabetes, heart disease and even depression. Further, a reduction in car use and meat consumption would also cut world food prices.

The question of how the developed and less-developed nations may respond to climate change will be explored in the next chapter, which will review climate change and sustainability policies.

7.3.1 Health and weather extremes

Weather extremes affect health in a number of ways. Besides affecting crops and facilitating outbreaks of disease, even the annual seasonal cycle of weather change outside of the tropics affects health. Human physiology can handle most variation in weather, but only within certain limits. Marked short-term fluctuations in weather that take the human body beyond these limits can cause adverse health effects that lead to a greater number of hospital admissions (so straining health services) and even to increased death rates. Depending on location, hence climate and the nature of typical weather extremes, regional mortality can deviate from the annual average (particularly in winter and summer, the ends of the annual temperature cycle outside of the tropics), and in particular during extreme weather events.

An analysis of the Central England Temperature record (see Chapter 5) between 1665 and 1834 (when the British population was far less independent from the climate) shows that mortality increased above the annual average in warm summers and cold winters. The main mortality effect of extreme winter cold was immediate but in the summer the effect of excessive heat was delayed by a month or two. This is because in the winter the elderly died of influenza and bronchitis but in the summer a broader section of the population died from food poisoning, which took longer to have an effect. Broadly, in England in the 17th to 19th centuries, a 1°C warming in winter reduced annual mortality by about 2%. Conversely, a 1°C cooling in the summer reduced mortality by around 4%. There was also a synergistic correlation with the economy and prices, which suggests that poverty was a supplementary factor (Burroughs, 1997). However, care needs to be taken when correlating extreme weather events and mortality. An extreme event occurring once a decade among normal years will show a clear correlation, but a run of, say, three severe winters shows the greatest excess in mortality in the first year but there may even be no excess in the third consecutively bad winter (as happened in Britain in 1940–2). The correlation therefore is not so much between an extreme event and a population's mortality rate but the extreme event and the absolute number of susceptible individuals within that population. If that number is high then the mortality will be high but if that number is low then it is impossible for the mortality to be high, even if the extreme event is a record-breaking one.

The most susceptible people in a population are likely to be

- the elderly,
- the infirm (which will not only include the elderly),
- the poor (biologically due to poor diet, health care) and
- the poor (due to lack of insulation from environmental vagaries).

Because the most susceptible within a population will suffer, although the figures in terms of number of deaths may be high, the number of person-days of quality life lost may actually be low. This is another manifestation of the declining mortality from a run of extreme events, described above, that shows that only the susceptible tend to be vulnerable to extreme winters. If someone is susceptible then they could already be close to death, so the extreme event is bringing their death forward by weeks or months or a year or two. Those that would in any case die within a few days will die first, then later those that would otherwise have lived for weeks will die, and so forth.

Although it may seem to be humanly cold and calculating, this is how rigorous quantitative analysis needs to be conducted when attempting to calculate the costs and benefits of climate change. It is not mortality per se that should be counted but the number of person-days of quality life lost. For example, consider a population whose society can afford either air conditioning for all its elderly citizens or bush fire-fighting equipment (or flood protection or another such opportunity cost). Which opportunity should they choose assuming the numbers of anticipated deaths from heat stroke or fires (or whatever) were equal? The number of anticipated deaths does not help provide an answer but the number of person-days of quality life lost often provides a clear indication. Of course, in developed nations this dilemma need not arise (although sometimes it does with tight civil budgets). Conversely, such quandaries are more common in less-developed nations.

Surprising as it may seem, climate change may cause health effects simply because the new climate is not within a population's perceptual history, and so the risks and avoidance measures may not be recognised. If they were, adverse health effects may be avoidable. This was brought home with the publication of research in 1998 by a team co-operating between London and Moscow (Donaldson et al., 1998). On balance, conventional wisdom had it that the more affluent Western European society would have lower mortality than their Russian counterparts. However, when winter temperatures reached $0°C$ excess deaths were not seen in the elderly cohorts of the Russian population, although they are known to exist in Western European populations. The reason given is that the Russian population was used to such winters and was prepared. Their houses were warmer and they wore more clothes. However, for many Western European population winters of 0 to $-6°C$ were extreme events. The research suggests that *some* deaths from extreme events in affluent societies are avoidable (Donaldson et al., 1998).

That excess mortality in heatwaves, or cold spells, is dominated by what a population is used to, and how it copes, as well as physiological acclimatisation rather than intrinsic physiological failings, was highlighted by a study that compared summer mortality in a hot part of Europe (Athens, Greece), a median part of Europe (London, Britain) and a cool part of Europe (north Finland). Heat-related mortality occurs at higher temperatures in hotter regions than in colder regions of Europe but does not account for significantly more deaths in hotter areas. Surveys indicate that people in cold regions of Europe protect themselves better from cold stress at a given level of outdoor cold. A similar explanation, better protection from heat stress in hot than cold regions, also helps account for these findings (Keatinge et al., 2000).

Of the possibly avoidable deaths from extreme heat or cold events, it is the excess mortality above the norm in winter that can more easily be negated. Heating to combat cold need not rely on advanced technology and basic heating technology is as old as the camp fire. Conversely the least avoidable deaths will be those from excess summer heat, because to negate this usually does involve the higher-level technology of air conditioning. Before the use of air conditioning in the developed industrialised countries, major heatwaves would typically increase mortality in cohorts over 50 by several-fold. In 1936, one of the hottest summers on record in the USA, some 4700 people were estimated to have been killed directly by heat stroke. This represented an increase of several per cent above the numbers of deaths expected over the same period. Again, in Melbourne, Australia, in 1959, there was a 4-fold increase in mortality (145 excess deaths) that was directly attributable to a prolonged heatwave, although most of these occurred in the first 4 days of the event, for reasons presumably akin to those where mortality from a successive run of bad winters occurs early in the run (see above). A similar pattern of early-run mortality occurred during the run of summer heatwaves in Los Angeles in 1939, 1955 and 1963 (McMichael, 1993). At the end of the 20th century in the USA some 1000 died each year from extreme cold events, but twice as many died from extreme warm events. This last might in all probability be an underestimate, as a proportion of death certificates made out during heatwaves cite respiratory disease or some other cause rather than heat stroke.

The UK provides an example of a policy response to the likelihood of increasing heatwaves. The August 2003 heatwave in north-west Europe is estimated to have caused some 15 000 extra deaths in France, 2000 in Spain, 1300 in Portugal and around 2000 in England and Wales. In 2004, from concerns that climate change will increase the frequency of heatwaves, the UK Government's Department of Health and the National Health Service published a *Heatwave Plan for England* (National Health Service, 2004). As outlined in the box, the plan has four levels of response for when threshold temperatures are forecast or exceeded. This plan is to be further developed over the years.

The Heatwave Plan for England

The plan details a heat-health watch from the beginning of June to mid-September each year. It consists of four levels of response based on threshold temperatures. These vary with region but are approximately 30°C in the daytime and approximately 15°C at night.

Level 1 (awareness) The minimum constant level of heat-health awareness among heat-health stakeholders and the general public.

Level 2 (alert) To be triggered as soon as the UK Meteorological Office forecasts that threshold temperatures are to be exceeded for 3 days or that there will be two consecutive days of extremely high temperatures near the threshold.

Level 3 (heatwave) Triggered as soon as threshold temperatures are reached.

Level 4 (emergency) When a heatwave is so severe that it threatens the integrity of health and social care systems.

It is the responsibility of the government-funded health and social services to identify groups and individuals likely to be adversely affected by heatwaves, provide advice and implementing appropriate measures. At-risk groups include the elderly, babies and the very young, those with mental health problems, those taking certain medication, those with certain chronic conditions such as breathing problems, those with a disease/infection induced by high temperature and the physically active, such as manual workers and athletes. It is the responsibility of the UK Meteorological Office to provide the health and social services with 3-day forecast warnings.

Climate change due to greenhouse warming for most (but not all) places on the planet will mean an increase in extremely warm summer events (above the mid-20th-century average). It will also for most (but again, not all) places outside of the tropics mean a decrease in extremely cold winter events. (See the discussion on extreme winters at the end of section 6.6.1). The overall mortality of a population throughout the year may (or may not) increase, but what almost certainly will change will be the pattern of mortality within the year. There will in all likelihood be an above-average level of mortality in the summer and a lower level in the winter. Consequently hot countries are likely to see a net increase in annual mortality but countries with a mild climate (such as the UK) may see hardly any change in overall annual mortality and instead a decline in winter mortality in the more frequent years without an extreme winter and an increase in summer mortality. The winter exception to the above is that in a warmer world it is expected that the strength of severe storms would increase. Injury from debris and destruction could potentially increase public awareness, and enable local resource provision, of what to do in the event of severe storms.

The WHO recognises that global warming comes with health impacts and it has been working with the World Meteorological Organization (WMO) and the UNEP (both of whom in turn formed the IPCC; see Chapter 5) since 1999. Its work in this area has three main strands: capacity building (informing to help nations develop the necessary services to address climate change-related health impacts), information exchange and research promotion (WHO, 2001a).

Extreme events are also associated with climatic cycles. There are three or four principal climatic cycles: two encircling the poles (one each), a cycle related to northern mid latitudes and the El Niño Southern Oscillation (ENSO). The North Pole cycle is the Arctic Oscillation or Northern Annular Mode, while the northern mid-latitude cycle is known as the Pacific-North American pattern (Kerr, 2004). Of these cycles, the best studied is the ENSO. The ENSO has a period of 2–7 years and within these years warm water periodically forms in the eastern Pacific to create an El Niño event. The El Niño begins with a weakening of the prevailing winds in the Pacific and a change in rainfall patterns that create extreme-flood and drought events in countries surrounding the eastern Pacific and climate-related events further afield. Prolonged dry periods may occur in south-east Asia, southern Africa and northern Australia, as took place in 1991–2, together with heavy rainfall, and sometimes flooding in Peru and Ecuador. During a typical El Niño event the Asian monsoon usually weakens and is pushed towards the equator, often bringing summer drought to north-west and central India and heavy rainfall in the north east. The regions where the El Niño has a strong effect on climate are those with the least resources: southern Africa and parts of South America and south-east Asia.

The WHO has noted that the risks of natural disasters are often highest in the years during and immediately after the appearance of an El Niño, and lowest in the years before. During the 1997 El Niño central Ecuador experienced rainfall 10 times higher than normal, which resulted in flooding, extensive erosion and mudslides with the loss of lives, homes and food supplies. During the same El Niño there were droughts in Malaysia, Indonesia and Brazil, exacerbating the extensive forest fires taking place at the time. The El Niño cycle is also associated with increased risks of disease range (see section 7.3.2).

The El Niño not only serves to illustrate the type of health impact an extreme weather event may impart; the impacts themselves may change with global warming. It has been suggested that the El Niño may become more intense and/or more frequent with warming and so will the potential health impacts (WHO, 1999, 2001b).[2]

In addition to the direct physiological effects of heat, extreme climatic events also have indirect effects that are no less serious. A drought increases the likelihood of bush and forest fires which are not only dangerous in themselves but destroy homes and property, which can also have health consequences, including loss of life. For example, in August 2003 a heatwave in Portugal resulted in the nation's worst forest fires for 20 years, killing six people. Meanwhile, at the same time Canada was experiencing the worst fires for 50 years, which necessitated the evacuation of some 8500 people. Consequently those concerned with the biological management of woodland are having to come to grips with climate change and fire hazards. For example, since 1990 the western USA has seen some extreme fires that have been notable for their size and severity. The annual US cost of fire suppression exceeded US$1.6 billion in 2004 and the figure is rising. In the absence of large fires, much of the western USA during the 20th century saw many forests fill with a dense understorey of shrubs and small trees that provide so-called ladder fuels that set crowns of trees alight: crown flares are among the most destructive of forest conflagrations. To address this the US Healthy Forest Restoration Act (2003) determined programmes of aggressive thinning, burning and replanting to create fire breaks and more open conditions by removing ladder fuels (Whitlock, 2004). However, there is evidence to suggest that this may not be appropriate as a generic strategy.

In 2004 Jennifer Pierce and colleagues from the universities of New Mexico and Arizona published results indicating that a one-size-fits-all woodland strategy may not be pragmatically effective. They came up with a way of looking at the long-term

---

[2] See also discussion of the El Niño in the warmer Pliocene in section 4.3.

fire record in low-altitude pine forests of the western USA. These forests are prone to drying. There are also open forests that lie above the steppe and grasslands of the valleys and basins. At higher altitudes there is more moisture and the forest becomes enclosed and subalpine. The low-altitude forests have been affected by human land use (hence fires there have a potential health impact) and there is some controversy as to how best to return some areas to their 'natural' state. One question is whether the recent severe fires are atypical. Another concerns the role that climate change may (or may not) be playing. Conventional wisdom has also been challenged, in that fires have affected both open and closed forests: open forests were thought to be more fire-resistant.

Our current understanding of past low-altitude forest fires is based largely on two records. The first record is the fire scars of living and dead trees. As such, this only provides a record within about 500 years, and a record that is increasingly biased with age towards fires that scar but do not kill. Second is charcoal records from lake sediments. These can often provide a record going back a few thousand years but only in the main relating to montane and subalpine forests: only when a forest is sufficiently higher than a lake is any charcoal transferred by watercourses.

Conversely, Pierce and colleagues (2004) looked at fire-related sediments at low altitudes. Severe fires can remove surface litter and small plants, so enabling soil movement (especially following storms or snow melt) that result in debris-flow deposits. These accumulate in alluvial fans, and other places in valley floors, and can be carbon-dated. Pierce et al. found that over the past 8000 years in xeric (that is, characterised by dry conditions) ponderosa pine (*Pinus ponderosa*) forests, fire-related debris flows were associated with warm, dry periods such as the medieval climatic anomaly (MCA) around AD 1100–1300 (see Chapter 4). This contrasts with cooler, wetter periods such as the Little Ice Age (AD 1550–1850) that saw fires from the dense grass cover, but only minor episodes of sedimentation. Adding this knowledge to the previous picture it becomes clear that strategies of forest management need to recognise that the frequency and nature of fires varies with location (hydrology and geology) and climate regimen. So, one-size-fits-all forest management may not be appropriate in that we may not be currently minimising risks to human settlements that are mainly associated with lower-, rather than higher-, altitude forests.

Forest fires also affect whole ecosystems and the humans associated with them. Conversely, the physiological effects of extreme heat (and cold) are a manifestation of effects on the whole organism, be it plant or animal (including humans), as discussed above. However, there are also effects of extremely hot weather at the cellular level. Hot weather events tend to be associated with sunnier days, which combined with the human behavioural response of spending more time outdoors during such events can result in an increased exposure to ultraviolet (UV) radiation. Sunburn (erythema) and skin cancer are largely due to UV-B radiation, which has a spectral range centring around 300 nm. But there are other cellular degenerative effects more related to UV-A radiation, which has a spectral range centring around 340 nm.

The risk of skin cancer is greatest among fair-skinned populations that evolved in northern latitudes that are exposed to lower levels of intense sunshine. It is thought that the evolutionary driver for pale skin was to enable the skin to continue to produce sunlight-induced vitamin D. Further, sunlight-induced skin cancers are the common-est form of cancer among light-skinned populations: for example, they account for an estimated 70% of all skin cancers in the USA. Exposure to sunlight is the major environmental cause of these cancers. The least frequent, but most serious, of the three skin cancer types is malignant melanoma. If not detected and treated early, this cancer will spread via the bloodstream to internal organs and this is often fatal, with mortality of around 25% (McMichael, 1993). In the UK the Department of Health assessment is that by 2050 there will be a possible 5000 extra cases of skin cancer a year due to climate change (Department of Health, 2002).

UV light also damages the eyes in a variety of ways, which include encouraging cataract formation. The Department of Health estimate an increase of some 2000 cataract cases a year by 2050 (Department of Health, 2002). In its second *Health Effects of Climate Change* assessment (an update of the 2002 report but jointly published with the Health Protection Agency; Department of Health and the Health Protection Agency, 2008) the warnings about UV radiation were strengthened but it also noted that quantitative estimates remain elusive.

As with the extreme cold events discussed above, the health effect of increased surface UV light on a population is strongly affected by behaviour. Avoiding sun-bathing (and especially by avoiding sunburn) greatly reduces risks, although not sunbathing does not eliminate risk because skin exposure to UV radiation takes place outdoors on nearly all uncovered parts of the body. Wearing glasses with glass lenses (as opposed to plastic-lensed glasses transparent to UV radiation) helps reduce the risk of cataract formation. However, small-lensed, dark sunglasses allow sunlight in around the rim while the dark lens allows the iris to dilate, so allowing in extra UV light, even if it is not focused on the fovea of the retina, the most sensitive part of the visual field. Large-lensed and only slightly tinted sunglasses give the best protection. (Glass wearers take note when next visiting your optician.)

Extreme hot-weather events can also affect microbial populations and this is dis-cussed in section 7.3.2. Finally, extreme weather events can also result in flooding due to rain bursts and the synergies between sea-level rise and storm surges that are associated with global warming. These too have health impacts and are covered in section 7.3.3.

7.3.2 Climate change and disease

Pathogens are disease-causing living organisms that are predominantly protoctistans (single-celled eukaryotes, meaning with internal membranes), prokaryotes (such as bacteria, which have no internal membranes) and viruses (genetic material – DNA or RNA – with a protein and/or glycoprotein coat; although there is debate as to whether viruses should be considered as living). There are also metazoan (multicelled animal) parasites. The ability of all these to flourish is affected by their environmental conditions, and hence climate change. Even those pathogens that rely on a vector (an organism capable of transmitting a disease from host to host) are affected by climate change, in that the vector itself will flourish best under certain environmental conditions. In short, just as a species' range and/or population will vary with a combination of climate change and other environmental factors (Chapter 5 and 6),

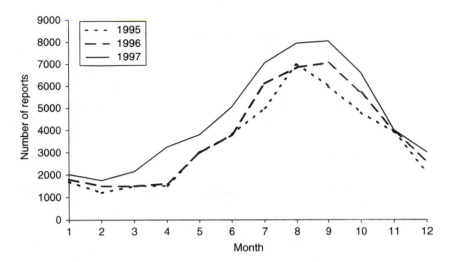

Fig. 7.12 Reported cases of *Salmonella enteritidis* by month in England and Wales in 1995–7. After Stanwell-Smith (1999).

so a disease's range will also change, as will the disease organism's population if not controlled.

As with humans and the annual temperature cycle outside of the tropics, so microbial populations vary. The incidence of reported cases of food poisoning in developed nations is illustrative of such annual cycles, with cases peaking in the summer months. In the UK the number of officially notified food-poisoning cases was rising in the late 20th century, so that by 1998 it had reached about 100 000 per year. However, notifications do not always strongly correlate with monthly temperature and sometimes are more strongly associated with the previous month's temperature. This is not due to incubation period, as this is usually short (typically less than 2 days), and it is not the result of delays in notification, because the UK system is designed for prompt reporting. It is possibly to do with the level of infections in animals prior to slaughter, as meat is an important source of food poisoning. Consequently global warming does not just present a challenge to consumers but also to those responsible for food production, processing and transportation (Bentham, 1999).

The annual pattern of reported cases of *Salmonella enteritidis* by month in England and Wales is typical (see Figure 7.12). *Campylobacter* species show a similar pattern with a summer maximum. *Campylobacter enteritis* has been linked to barbecues, particularly those involving chicken. Nonetheless, this only explains a small proportion of UK outbreaks and a very few of the sporadic cases. However, that cases of *Campylobacter*-related illness are rising has made this species of interest to those engaged in climate change and microbial-disease research in the UK.

This pattern of summer increase in temperate-zone incidence of food poisoning suggests that in a warmer world there will be more such cases. Worldwide, the WHO and the UN Food and Agriculture Organization (FAO) estimate that there are at least two billion people affected by food poisoning each year, and many die. For example, in Asia alone some 700 000 people die due to food- and water-borne (contaminating food) disease (WHO/FAO, 2004).

The seasonality of Lyme disease is also likely to be affected by climate change. It was first identified in the USA in 1975 and then noticed in much of Europe. Its name comes from a district in Connecticut, which is one of the many areas in which it is endemic. The disease is caused by the spirochaete bacterium *Borrelia burgdorferi*, with mammals acting as reservoirs, and is transmitted by ixodid ticks such as *Ixodes ricinus*. Humans are not the preferred host, but nonetheless Lyme disease is a public-health concern in a number of rural areas in Europe and especially the USA, where there are many thousands of cases a year. In the UK and Ireland the incidence of the disease has been sporadic with typically just a few scores of cases a year. It results in arthritis of large joints and may also inflame the brain and nervous system. Transmission occurs during the spring, summer and early autumn months, which is also when humans are likely to be outdoors in rural areas and exposed to the vectors. Because of climate change both the seasonality of the disease and also its geographic distribution may be extended.

Climate change not only affects the propensity of a disease to affect a human population but can also change the species range of disease vectors. Of the major vector-borne diseases, malaria is the most significant. In the 1990s it had the potential to have an impact on more than 2.1 billion people and successfully infected some 300–500 million, of whom some 90% are in tropical Africa. Malaria currently kills more people than any other infection except tuberculosis, and has an annual death toll of 1–2 million people, mainly young African children. It possibly accounts for 24% of child deaths. Annual deaths from malaria rose steadily from around a million in 1980 to about 1 900 000 in 2005. Following substantial investment in control (development assistance for tackling malaria increased from US$149 million in 2000 to almost $1.2 billion in 2008) deaths from malaria declined to around 1 300 000 in 2010 (Murray et al., 2012). This peak-and-decline pattern of mortality is reflected in WHO data, although the WHO's figures (WHO, 2011) are broadly around 25% lower than this. Having said that, the methodology of Murray et al. has caused some debate, relying as it does on interviews with next of kin and friends of the deceased (Anon, 2012).

The range of malaria, among other diseases, has also changed with climate, and will undoubtedly continue to do so. The disease itself is caused by protoctistan parasite species of *Plasmodium*, namely *Plasmodium ovale*, *Plasmodium vivax*, *Plasmodium malariae* and *Plasmodium falciparum*, which are transmitted to (and between) humans by mosquitoes belonging to the genus *Anopheles*. *P. ovale* results in a less-severe form of malaria compared with the other three species, and Hypocrites (*c.* 2500 years ago) divided malaria into three different types that, with the benefit of subsequent knowledge, largely appear to relate to three of the different *Plasmodium* species. As for the vector, there are some 380 species of *Anopheles*, of which about 60 transmit malaria.

Genetic studies suggest that the virulent form of malaria evolved some 10 000 years ago just after the last glacial. This broadly coincides with the development of agriculture and the supposition has been posited that water held for irrigation provided an ideal habitat for the *Anopheles* mosquito to evolve into its current form, but this has not been proven.

Malaria evolved in the tropics and subtropics of the Old World (Africa) and it did not arrive in the New World (the Americas) until around 1500 with European settlers and African slaves. Malaria was not controlled with pesticides in the USA until the 1940s, but eradication has proven difficult and there are still sporadic outbreaks. Eradication programmes elsewhere have been more successful, such as in Brazil and Egypt.

The disease used to be found in Europe, including in southern England. Historically, in England malaria was known as ague and was frequently found around coastal salt marshes. Indeed, 200 years ago it was thought to be a leading cause of death in many marshland communities including the Fens (in Norfolk), the Thames estuary and further along the Kent and Essex coasts, and the Somerset Levels. It declined in the 1840s but there were still pockets of the disease in Kent and in other low-lying areas through to the early 1900s. The last major UK outbreak took place in Queenborough on the Isle of Sheppey (in Kent). Soldiers returning from World War I carried the malaria parasite, and they were bitten by local mosquitoes that went on to transmit the disease to 32 people over several years. However, indigenous malaria remained and from 1917 to 1952, 566 people contracted the disease in Britain. The indigenous disease died out in 1953 with two final cases in London: mosquitoes (probably *Anopheles plumbeus*) were found breeding in a tree hole near one of the victims' houses.

Twentieth-century England typifies the problem of malaria in many temperate zones in the 21st century. First, we currently have a far more internationally mobile population than ever before. This increases the chance of re-introducing malaria. Indeed, in 1971 Europe as a whole saw around 1400 cases of imported malaria. The subsequent years saw Europeans travel further outside the continent so that by 1986 cases of imported malaria had increased to around 6800 (Martens, 1999). By the turn of the millennium the number importing malaria into the UK alone approximated 2000. Fortunately, only 25% of these had *P. vivax*, which used to be common.

Second, the disease's range is complex and the areas in which it may exist are determined by many factors. Just as climate change threatens many wildlife species (and note that the malaria parasites and vectors are wildlife species) so climate change may not necessarily serve to help promulgate the disease as might be thought. The actual theoretical areas where malaria may exist are determined by a number of factors, of which climate is one. Just as human land use restricts wildlife species and impedes their migration/expansion (see Chapter 6), so it does with malaria. In the UK the extent of many wetlands near to large centres of population have declined. Climate itself is multifactorial and malaria, or rather the parasite in the *Anopheles* vector, not only requires a specific temperature range but also humidity and duration of these conditions, as well as the aforementioned land-use criteria. In 2000 David Rogers and Sarah Randolph showed, using climate change models, that during the rest of this century – even under the more extreme climate change scenarios – *P. falciparum* malaria might move into new areas, but that more or less equally it would move out of old ones. The overall potential health effects of climate change appear to be marginal; it is just the distribution that will change. This is further complicated by whether the new areas contain more humans to infect compared to the areas vacated. The picture is complex and again is not untypical of the way ecosystems might respond to climate change: species assemblages may not always shift uniformly with climate change (see Figure 6.2b). The aforementioned forecast is also supported by limited observations in four high-altitude sites in East Africa where increases in malaria have been reported and then present and past conditions compared using 95-year climate records. Looking at the combination of required temperature, rainfall, humidity and number of months suitable for *P. falciparum*, conditions have not changed significantly either over the 95-year period or in the two decades of malaria increase (Hay et al., 2002). In short, there is a natural variation in the disease in addition to anything that might be connected with climate change.

Does this mean that there will be no future health impacts from malaria due to climate change? Certainly not. Notwithstanding surprises, the distribution of malaria will change but it may not result in huge increases in numbers of cases due to climate change alone, although the places where infections take place are likely to alter. However, climate change in combination with other factors, such as continued increasing international mobility, might result in sporadic localised outbreaks in parts of Western Europe and North America. By the end of the 21st century malaria might be found in some parts of mid South America, parts of southern Africa and further east of its existing range in Asia. Equally, parts of mid South America, eastern Africa, eastern India and parts of east Asia that are presently suitable for malaria may become unsuitable. Consequently, the deployment of anti-malaria measures will need modifying.

Here climate change science is set to help. The marked increase in sophistication of computer models of the global climate, especially their regional-level spin-offs, are beginning to be used to develop forecast systems of malaria outbreaks. (Here, as elsewhere in this book, the term model means a mathematical simulation as opposed to how epidemiologists sometimes use the word as a simple correlation or statistical relationship between two things; Palmer, 2006.) One such system has already had practical success in semi-arid Botswana, adding up to 4 months' extra lead time in advance of existing warnings of outbreaks. It predicts regional climate patterns about 6 months in advance. In Botswana the rainy season, hence the malaria season, begins in December but health officials have to wait until February or March before they can make an accurate assessment. The new model enables them to make the prediction months before the rainy season starts (Thomson et al., 2006).

Then again, some non-climatic factors can aggravate matters. For example, *Anopheles* pesticide resistance was first noticed in the 1950s and resistance to chloroquine (an anti-malaria pharmaceutical) was first seen in the 1960s and has become an increasing problem since. Evidence of resistance to the modern anti-infective drugs atovaquone and proguanil was noted in 2003. In China artemisinin-based combination drugs derived from the sweet wormwood plant (*Artemisa annua*) have yet to encounter resistance and they are increasingly being used in Africa. Consequently, demand for these drugs is rising but even with them the evolution of resistance will continue to be of concern, so increasing dependence on artemisinin-based pharmaceuticals. Indeed, in 2004 the WHO announced a shortage of artemisinin-based combination drugs and forecast a demand of 10 million treatment courses, predicting that this would rise to 60 million in 2005 (WHO, 2004).

Finally there is the question of how long artemisinin-based drugs will continue to avoid resistance by malaria parasites. One option to delay the onset of resistance would be to avoid the use of single-drug treatment (monotherapy) and instead use cocktails of pharmaceuticals or artemisinin combination therapies (sometimes known as ACTs). So, in January 2006 the WHO appealed to all pharmaceutical companies to stop marketing oral artemisinin monotherapies and 13 countries over the subsequent 3 months agreed to withdraw marketing authorisation for such monotherapies (WHO, 2006).

The ranges of many other diseases are likely to change with climate. Among those of major concern is Chagas' disease or American trypanosomiasis, a South American variation of the sleeping sickness (trypanosomiasis) that affects central Africa. Chagas' disease is named after the doctor who researched the affliction in northern Brazil during the building of a railway in 1909. About 18–20 million people in Central and South America and the USA are at risk of contracting Chagas' disease. Its geographical distribution ranges from the southern USA to southern Argentina and Chile. It is transmitted by the blood-sucking insect *Triatoma infestans*, and other species, which carry the protoctistan *Trypanosoma cruzi*. It originally affected rodents and marsupials but has adapted to human encroachment. Unlike African sleeping sickness, the trypanosomes do not multiply in the blood but become localised in various organs. The disease causes long-term, essentially untreatable debilitation and, by affecting heart muscle, it is a major cause of chronic heart disease and sudden death in apparently healthy young people.

Conversely, with the related African sleeping sickness the brain and the meninges (the covering of the brain and spinal cord) become infected with *Trypanosoma brucei* and, if treatment is delayed, it usually proves fatal. The WHO estimates that between 300 000 and 500 000 people are affected worldwide, and that the disease threatens 60 million people in the poorest underdeveloped countries. As for treatment, relatively drastic measures are required. In the acute stage of the disease pentamidine is given. If the parasite is also in the central nervous system, the drug melarsoprol is used, so exposing patients to serious risk, especially of arsenic encephalopathy, of which almost 1000 people die each year. Both substances are administered directly into the bloodstream. There is no vaccine or chemical prevention due to the relative toxicity of the drugs, so it is necessary to prevent bites from the tsetse fly (of the genus *Glossina*) and control the places where the fly lives.

Dengue, which covers both the milder 'breakbone fever' and the more serious but rarer haemorrhagic fever, is caused by arboviruses that are transmitted by *Aedes* mosquitoes. Dengue fever can be caused by any one of four types of dengue virus: DEN-1, DEN-2, DEN-3 and DEN-4. A person can be infected by at least two if not all four types at different times during their lifetime, but only once by the same type. Symptoms of typical uncomplicated dengue usually start with fever of up to 40.5°C within 5–6 days of a person being bitten by an infected mosquito. Other symptoms include severe headache, retro-orbital (behind the eye) pain, severe joint and muscle pain, rash, nausea and vomiting. Of interest to clinicians is that most infected children never develop typical symptoms. There is no specific treatment for dengue fever, and most people recover completely within 2 weeks. Dengue is of particular concern because the vectors are associated with human environments and, unlike many anopheline mosquitoes, they prefer to feed on humans. Consequently, when epidemics take place they do so rapidly.

Currently milder dengue has all-year transmission, but mostly during and shortly after the rainy season around the tropics, between about 25°N and 25°S, where the principal vector is *Aedes aegypti* (which is also the vector for yellow fever arbovirus, now largely eliminated from the USA, Central America and most of South America and still present in much of tropical Africa). The WHO estimates that there are 50 million cases of dengue infection each year in more than 100 countries. This includes 100–200 cases reported annually to the US Centers for Disease Control and Prevention (CDC), mostly in people who have recently travelled abroad. From 1977 to 1994, US health-care workers reported to the CDC 2248 cases of dengue that had been imported into the USA. Many more cases probably go unreported because some doctors do not recognise the disease. Cases of dengue began to increase in many tropical regions in the late 20th century. Epidemics also began to occur more frequently, and to be more severe. The concern for Europe is that with climate change another dengue vector, *Aedes albopictus*, is currently increasing its range and can already be found in Italy and parts of the Balkans.

In North America West Nile fever has become a concern in recent years. It is caused by a member of a group of viruses called flaviviruses, which are transmitted mainly by birds to humans via mosquitoes. It results in a mild, influenza-like illness that is often called West Nile fever. More severe forms of the disease, which can be life-threatening, may be called West Nile encephalitis, West Nile meningitis or neuroinvasive disease. Although many people are bitten by mosquitoes that carry West Nile virus, most do not know that they have been exposed. Few people develop severe disease or even notice any symptoms at all. It is thought that only one in five infected people develop mild illness, and that only 1 in 100 of fever-infected people go on to develop brain inflammation (meningitis or encephalitis). By 2000 the disease was considered firmly established in the USA and it is of concern in Canada. In 2002 and 2003 in the USA it caused the death of 284 and 264 people, respectively. By 2002 in Canada several populations of birds were found to have the disease. It was originally found throughout much of Africa, from the Cape of Good Hope to Cairo, and is also widespread in the Middle East and India. The disease occasionally appears in parts of southern Europe, such as Italy, and in Belarus, Ukraine and southern Russia. In 1997 some 500 in the Bucharest region of Romania caught the disease, with 50 people dying.

In 2004 there was sufficient concern over West Nile fever that the UK Department of Health issued a short bulletin alerting clinicians to be vigilant. The document described a plan to control key mosquito populations and a contingency plan for when cases were diagnosed (Department of Health, 2004).

In North America, West Nile virus (which emerged in New York city in 1999 before spreading across much of the continent) has been implicated in large-scale declines of bird populations. In 2007 Shannon LaDeau and colleagues from the Smithsonian Migratory Bird Center used a 26-year bird breeding survey to demonstrate that the populations of seven species from four families of birds had declined, coincident with the predicted areas of the pathogen. These were American crow (*Corvus brachyrhnchos*), blue jay (*Cyanocitta cristata*), tufted titmouse (*Baeolophus bicolour*), American robin (*Turdus migratorius*), house wren (*Troglodytes aedon*), chickadee (*Poecile* spp.) and the eastern bluebird (*Sialia sialis*). The American crow population was severely hit, with its population declining by 47% since the arrival of West Nile virus and by 2005 only two of the seven species had recovered to pre-West Nile outbreak levels: blue jay and house wren.

That West Nile virus hit the headlines in North America and Europe at the beginning of the 21st century suggests that flavivirus zoonoses (virus diseases of animals that affect humans) have only recently presented a serious health impact on humans. However, Japanese encephalitis (found in Asia) is a virtually identical bird-borne disease caused by a related virus. It affects more than 50 000 people a year in Asia, and kills 15 000 of them. It currently has a far greater health impact, but is not of such concern in North America or Europe because it has not touched the western hemisphere (yet). But there are occasional incidents. For example, one such illness has already appeared in Britain, louping-ill fever. It affects a variety of animals and in sheep is known as ovine encephalomyelitis or infectious encephalomyelitis of sheep, for which the main vector is the sheep tick *Ixodes ricinus*. Other tick vectors include *Rhipicephalus appendiculatus*, *Ixodes persulcatus* and *Haemaphysalis anatolicum*. Infection has been demonstrated in other domestic species and wildlife; namely, cattle, horses, pigs, dogs and deer, as well as in a range of species of small mammals such as shrews, wood mice, voles and hares, the latter being considered an important reservoir species for the virus. It can decimate grouse populations but humans are only a tangential host and there have only been occasional isolated outbreaks of non-fatal meningo-encephalitis. Louping-ill is endemic in rough upland areas in Scotland, northern England, Wales and Ireland. A disease of sheep very closely related to louping-ill has been reported in Bulgaria, Turkey, the Basque region in Spain and Norway.

The aforementioned are just some of the diseases with an incidence in part related to climate change. Of course, these diseases would still be with us without climate change and in one sense virtually all diseases are in some way or other influenced by climate and/or weather. Even common influenza viruses tend to have an annual cycle. As a very broad generality within the range of environments commonly colonised by humans, microbial organisms prefer warm, moist conditions as opposed to dry, cold conditions. Consequently, as global warming will result in an increase in precipitation globally, and as (again a generality) temperate-zone winters will become less cold, the balance is such that a warmer world is likely to see more, rather than less, microbially caused disease. This increased likelihood is synergistically increased by non-climate factors such as the increasing global population and international mobility. Indeed, this last is causing a noticeable increase in tropical infections among travellers from temperate, developed nations returning home.

Leaving aside the above spatially leptokurtic[3] outbreaks, the early signs that global warming is causing a change in a disease's range will be found at nations on the edge of their range. Italy, mentioned earlier in this section, is one such critical area.

---

[3] A probability distribution with a tail that has a kink in it at the end is known as leptokurtic. The relevance here is that normally tropical disease cases will be found in the tropics with low to zero probability outside of the tropics. Yet air travellers may take a disease (or their vectors) outside the tropics, causing a small outbreak of the disease outside of the (tropical) zone of main probability. This results in a leptokurtic distribution.

Although Italy was declared free of malaria in 1970, by 2007 there was concern that it might become re-established there. Then there is tick-borne encephalitis, only 18 cases of which had been reported in modern times before 1993. By 2007 there had been around a hundred cases. In the UK, the *Health Effects of Climate Change in the UK 2008* report (Department of Health and the Health Protection Agency, 2008) notes that while outbreaks of malaria in the UK are, for the time being, likely to remain rare, health authorities (the regional administrative bodies responsible for the nation's health concerns) need to remain alert to the possibility of outbreaks of malaria in other European countries and to the possibility that more effective vectors (different species of mosquito) may arrive in the UK, necessitating rapid response. It pointed out that UK Climate Impact Programme (UKCIP) scenarios for warming in the UK will make re-establishment of the disease more likely, especially in the south east with risk areas including the Thames Valley, Norfolk Broads, and fenland near Cambridge and south west of The Wash. Meanwhile tick-borne diseases are likely to become more common in the UK, but this will more likely be due to changes in land use and leisure activities than, in the short-to-medium term, to climate change. So, the likelihood that tick-borne encephalitis will become established in the UK is very low. The report also carried area of malaria suitability maps and these showed that in North America the Sacramento and San Joaquin valleys, the Rio Grande, and coastal wetlands of Texas, Louisiana and Florida are potential risk areas by the 2080s.

In December 1997, the first workshop for international agencies was held on climate change and human health monitoring when the UNEP, the WHO and the UK Medical Research Council joined forces to assess the problem. This paralleled a WHO, UNEP and WMO initiative to establish an interagency network on climate-induced health impacts. The workshop helped set the international research agenda and allowed agencies to pool resources to begin to tackle this issue. Worldwide health efforts to address these diseases have increased in recent years, although a lack of resources in many less-developed nations has hindered (and in some instances reversed) progress. Consequently, where climate change is a significant epidemiological factor, once again it is the wealthy countries generating the most greenhouse gases that can afford to take remedial action, while poorer nations tend to remain exposed.

7.3.3 Flooding and health

Sea-level rise (see Chapter 5) and increased precipitation on a warmer planet are likely to increase the prospect of flooding. Importantly, the volume of water that can be held in the air increases with temperature. As this goes up so the possible maximum precipitation (or PMP) increases. With higher PMP in a warmer world storm bursts could become more extreme. In short, a warmer world can increase hydrological cycling.

In itself, will this seriously impact on people's lives, their well-being and health? A brief consideration as to how floods have affected people worldwide in the last century gives us an indication as to their human impact in the more crowded and flood-prone 21st century. The WHO (2001b) stated that 'almost 2 billion people – one-third of humanity – were affected by natural disasters in the last decade of the 20th century. Floods and droughts accounted for 86% of them.' Floods are the second most frequent cause of natural disaster after windstorms but affect more regions and more people than any other phenomenon (WHO, 2001b). Against this background any factor, such as climate change, likely to increase the frequency of floods (not to mention severe windstorms) needs to be examined so that the appropriate measures required can be anticipated.

The WHO consider the most vulnerable victims of flooding globally to be the poor and the marginalised, most of whom live in low-quality housing, in flood-or drought-prone regions. Those fleeing floods often drink unclean water. If the drinking-water supply and sanitation systems are already inadequate, flooding poses a major health risk. Flooded areas that have industrial waste, such as used engine oil, and refuse dumps add to health risks. The WHO says that, 'people who have lost everything in the flood – their homes, their food, their livelihood – are all the more prone to disease'. It also noted that floods are becoming more frequent: from 66 major floods in 1990, the number rose to 110 in 1999. Further, the number of people who died in floods in 1999 was more than double that in any other year of the 1990s (WHO, 2001b).

In tropical countries receding waters provide an ideal breeding ground for mosquitoes and other insects, so creating an increased risk of diseases such as malaria and dengue (see the previous subsection) as well as Rift Valley fever viral infections that affect both humans and animals. Flood water combined with the effects of open sewage and reduced opportunities for good personal hygiene lead to cholera, diarrhoea and gastrointestinal viruses, and displaced rodent populations can themselves cause outbreaks of leptospirosis and hanta virus infection. Flooding may also encourage *Cryptosporidium* infections. *Cryptosporidium* is a widespread protoctistan found in both tropical and temperate climates, the infection of which causes cryptosporidiosis, a diarrhoeal disease. *Cryptosporidium* oocysts (the cysts formed around two conjugating gametes) can survive for long periods and are capable of penetrating groundwater sources. In addition *Cryptosporidium* spp. are readily spread from person to person. Infections have also been linked to swimming pools. *Campylobacter* is similarly associated with water.

As with disease impact and the resistance to anti-infectives (antibiotics and anti-virals), the health impact from floods can be exacerbated by such non-climate-related factors, and others. For example, regional development increased the impact in 2004 of heavy storms that hit the Philippines early in December of that year, which resulted in exceptional floods along rivers. By 3 December more than 650 people were dead and some 400 were missing. Thousands more were made homeless. Extensive logging was blamed as a compounding factor. The next day the government announced that all logging was to be banned, subject to a review.

Another example of a major flood imparting the aforementioned health impacts is that of the Bangladesh flood in July and August 2004. Aside from this low-lying area having a large population, it has also been identified by the IPCC (1990) as likely to have a higher summer rainfall in a warmer world. In 2004 the region experienced heavy monsoon rains, the worst on record. It was an extreme weather event, again of the kind that the IPCC anticipate. In this instance 1 million people were displaced from their homes, out of the 10 million in the nation's capital, Dhaka, mosquitoes, escaped sewage, disrupted food supplies and disease (for example, more than

270 000 cases of diarrhoea) compounded the problem. The year 2004 was particularly bad but floods in this region are common and in 1998 about 70% of the country was under water for nearly 3 months. With around 65% of the population deemed by the International Red Cross to live below the poverty line, the people affected were typical of those the IPCC consider to be most vulnerable to extreme weather events and other climate change impacts. The IPCC cites Bangladesh as an example of a risk area because it is low-lying, hence the potential for exposure to sea-level rise, and because its climate naturally features weather extremes (often monsoon-related). Factors conspired synergistically to make the 2004 floods in the north-eastern region of Bangladesh impact heavily on people. First, four flash floods came in quick succession and, second, unusually high tides slowed the water's discharge to the sea. But, in addition, population pressures, to farm and other land development, had constrained the waterways, so increasing their chance of failing during extreme events (Nature News, 2004).

In 2007 some 20 million were displaced in monsoon floods in northern India, Bangladesh and Nepal. In 2008 the monsoon rains were so severe that the number of people displaced necessitated the establishment of 119 relief camps. In 2010 Pakistan monsoon floods, the worst for over 80 years, displaced 20 million people with a death toll of close to 2000, and saw outbreaks of cholera and other diarrhoea-related disease. Some 6.9 million ha of farmland was flooded and the loss of the cotton crop alone was valued at US$1.8 billion. In 2011, by comparison, just two million were affected by monsoon floods in India. Across the border in Pakistan some 2600 villages were flooded, more than 400 people died, with 5.3 million people displaced, and millions more otherwise affected. The District Badin in Sindh province saw record rainfall of 615.3 mm during the monsoon spell, breaking the earlier record of 121 mm in Badin in 1936. The area of Mithi too had record rainfall of 1290 mm during the spell, where maximum rainfall previously had been 114 mm in 2004. The 2011 monsoon season also saw record flooding in Thailand, Cambodia and Myanmar and heavy flooding in Vietnam. In China rain-fed floods affected more than 36 million people and caused a direct economic loss of some $6.5 billion.

But are rain-fed floods actually getting worse? The IPCC (2001a) stated that there were no observable changes in tropical cyclones (hurricanes and typhoons), and peak precipitation intensities were not found in the few analyses available to it. In the 21st century it forecast that increases would in the future be likely in some areas. This all changed with the IPCC's 2007 assessment that noted that *intense* tropical cyclones have increased with warming since around 1970, and that there have been increases in heavy precipitation events.

Indeed, research is revealing increases in rain-fed floods, although it is still unclear whether these will persist: a longer time frame is required as 20th-century analyses reveal flooding periods of different frequencies (albeit at a lower level) that last for decades. For example, in 2001 a US team led by Stanley Goldenberg showed that the years 1995–2000 experienced the highest level of North Atlantic hurricane activity in the reliable record. Compared with the generally low activity of the previous 24 years (1971–94), the subsequent 6 years saw a doubling of overall North Atlantic heavy-storm activity, a 2.5-fold increase in major hurricanes (with wind speeds of more than 50 m s^{-1}) and a 5-fold increase in hurricanes affecting the Caribbean.

The greater activity largely results from increases in sea-surface temperature that, the researchers note, has associated costs in terms of human well-being and property damage.

In 2008 Richard Allan and Brian J. Soden used satellite observations and model simulations to examine the response of tropical precipitation events to naturally driven changes in surface temperature and atmospheric moisture content. These observations revealed a distinct link between rainfall extremes and temperature, with heavy rain events increasing during warm periods and decreasing during cold periods. Indeed, the observed amplification of rainfall extremes was found to be larger than that predicted by models, implying that projections of future changes in rainfall extremes in response to anthropogenic global warming may have been underestimated previously.

The annual cost (in 1999) for the USA from cyclone damage was of the order of US$5 billion, but this is expected to rise in future years with increased population and wealth. It has been estimated that if the hurricanes of 1925 had happened in the wealthier and more populous 1990s, instead of a few billion dollars, the cost would have been some $75 billion after adjustment for inflation. In 1998, Hurricane Mitch killed at least 10000 people in Central America, while the average economic loss in the Philippines was estimated at some 5% of gross national product (Bengtsson, 2001).

Floods have taken place throughout human history. Noah's flood and similar stories that might date from prehistory appear to relate to the flooding of the Black Sea, which itself arose due to sea-level rise following the last glacial maximum (LGM) and the onset of the current Holocene interglacial. Furthermore, there has always been tension between the human desire to exploit the fertility of the floodplains of the world's greatest rivers and the fact that floodplains (by definition) flood.

In the late 20th century the floods of the upper Mississippi in the summer of 1993, the winter floods in December 1993 and January 1995 of the Rhine and Bangladesh's chronic flood problem all served to illustrate the material costs and costs to human well-being. The 1993 Mississippi flood had an estimated cost of $15 billion but only a small death toll, of 48 people. Following that flood there was concern expressed over the way the river watershed was managed and the flood-control measures that were in place, but irrespectively that flood was due to an extreme weather event following a season of exceptional snow in Iowa and the 1993 precipitation record was the greatest since records began in 1895. Both the 1993 and 1995 floods of the Rhine were also due to extreme weather events and parts of the region in 1993 had three times the average rainfall in December. The cost of the flood in Germany alone was around US$850 million in the money of the day. However, in 1993 lessons were learned so when in 1995 there were floods again, the damage costs were halved. Even so, there was disruption and in The Netherlands some 200000 people were evacuated from their homes for safety. In England and Wales instrumental measurements have shown that between 1765 and 1995 there was a slight decrease in the intensity of winter drought years accompanied by a similar increase in the intensity of years with high winter rainfall, with average rainfall also increasing by about 17% (Osborn et al., 2000). This trend is what is expected in that region with climatic warming. However, in the case of Bangladesh there is no historic meteorological evidence of increased rainfall (although future projections tell a different story) but deforestation in the

Himalayas and isolated tropical storms appear to have caused much of the problem. Bring climate change factors into the picture and in the future flood problems will undeniably be aggravated. If the recent climate events in Bangladesh are part of a longer-term future trend, then arguably this future may have already begun. Indeed, factor in the dimension of sea-level rise and matters will get worse still. As stated, the region is low-lying and has previously seen damage from sea flooding. In the 1970 tropical storm some 300 000 perished when a tidal wave swept across the Ganges and Meghina deltas (Burroughs, 1997). The implications are clear.

With regards to UK climate trends, the autumn 2000 was then the wettest since records began in 1766. The heaviest rainfall that season was across England and Wales, with a total of 489 mm falling between September and November; the most extreme rainfall was in October, which resulted in extensive flooding that damaged 10 000 properties. UK insurance claims arising from that season's floods came to around £1.3 billion. Many of the same areas of southern UK flooded again in early 2003. In the summer of 2004 a flash flood in Boscastle, Cornwall, was caused by around 20 cm of rain falling in just 4 hours. Two rivers burst their banks and a 3 m wall of water passed along the town's high street. More than 150 people had to be airlifted to safety and despite about 50–60 cars being washed away nobody died, but around £50 million of damage was caused. Another extreme event took place in January 2005, when 100 mm of rain fell around Carlisle in Cumbria. So intense was the rain that surface-water drainage could not cope and flooding began quickly before river monitoring (which the UK Environmental Agency then relied on) generated a flood alert. The flood caused local power cuts and interrupted both the terrestrial and mobile phone systems; consequently, some people became trapped in their homes.

The same year, 2005, saw a similar event on the North Yorkshire moors. After a weekend heatwave in which temperatures reached 33.1°C, the night of Sunday 19 June saw rainbursts over the north east of England that triggered regional flooding. North Yorkshire as a whole had the best part of a month's worth of rain in 3 hours. One fast-flowing flood that went through the villages of Thirlby, Helmsley and Hawnby was caused by 2.7 cm of rain that fell in just 15 minutes. The flood was 2 m deep in places and it carried cars along with it. Meanwhile, in river courses torrents washed bridges away. The highest rainfall in the UK on that day took place in Hawarden, Flintshire, in Wales, with 4.3 cm of rain. Two days after the event some 2500 homes were still without electricity. Indicative of the 'clumping' of rainfall patterns Friday 24 June 2005 saw some 2.8 cm fall on RAF Lyneham in Wiltshire: the usual rainfall for June in that area is 3–6 cm.

In 2007 three storms in June and July were of particular significance, the latter two being the immediate cause of the most devastating of Britain's summer floods. A distinguishing characteristic of the late spring and summer of 2007 was the frequency and spatial extent of extreme rainfall events over a wide range of durations. Correspondingly, the previous maximum May–July rainfall total in the 241-year England and Wales series was exceeded by a wide margin; since 1879 no rainfall total in this period has been within 100 mm of that for 2007. Widespread and severe flooding afflicted many river basins in June and July 2007. In some areas – parts of the lower Severn basin, headwater tributaries of the Thames, and Yorkshire and Humberside, in particular – peak river flows exceeded previous recorded maxima by wide margins.

In many river reaches, the design limit of flood alleviation schemes was exceeded; similarly the design capacities of urban drainage systems were exceeded in many areas. An unusual, and very significant, feature of the summer flooding was the high proportion of damage not attributable to fluvial flooding. Around two-thirds of the properties affected (more than 8000 in Hull alone) were inundated as drains and sewers were overwhelmed following the summer storms. More than 55 000 homes and 6000 businesses were flooded and related insurance claims were approaching £3 billion by the end of 2007 (Marsh and Hannaford, 2008).

From a meteorological perspective, the cause of the 2007 floods was jet-stream displacement that affected weather patterns (the jet stream is a fast, eastward-moving ribbon of air at an altitude of around 8–12 km). This displacement was associated with a commonplace but anomalously strong Scandinavian atmospheric circulation pattern (a Rossby-wavelike train of tropospheric anomalies, with a cyclone over the British Isles and a strong anticyclone over Scandinavia). From a climate change perspective, global warming affects jet-stream location (more of which shortly), in addition to increasing ocean evaporation and increasing the atmosphere's ability to hold water: approximately 6–7% more is carried per degree Celscius of warming near the Earth's surface, as determined by the Clausius–Clapeyron equation.

The following year also saw more increased rainfall. August 2008 saw parts of Britain receive rainfall that was well above average. North-west Wales, Cumbria and parts of the Chilterns saw 50% more rain above the long-term August average; south Cornwall, mid-Devon and west Wales had 100% more rain; parts of Northern Ireland 150% and south-east Scotland 200% more rain. This affected the harvest. In addition to some crops being lost, grain needs to have less than 10–15% moisture content to be harvested. Otherwise the crop needs to be dried within 24 hours of harvesting, and this necessitates the use of costly energy. Both losses and extra energy costs raised food prices. Some market garden crops, which thrive in a warm, damp summer, did well.

Then, in September 2009, there were a number of rainbursts. In the first few days of the month there were 94 flood watches in operation in England and Wales and one in Scotland. Some places received more than a month's worth of rainfall in a single day. With soil still wet from August's rain, there was considerable run-off and homes were flooded in parts of the West Midlands. Flash floods hit Yorkshire, Shropshire, Herefordshire and Worcestershire. Northumberland was particularly badly affected with an estimated 1000 properties flooded in Morpeth. In Upton-upon-Severn, which was cut off by floods the previous year, the river levels rose to more than 5 m. At the height of the 2007 floods they were 5.93 m high.

Nearly two million properties in floodplains along rivers, estuaries and coasts in the UK are potentially at risk of river or coastal flooding. Eighty thousand properties are at risk in towns and cities from flooding caused by heavy downpours that overwhelm urban drains: so-called 'intra-urban' flooding. In England and Wales alone more than 4 million people and properties valued at over £200 billion are at risk. Flooding and flood management cost the UK around £2.2 billion each year: the 2004 spend was £800 million on flood and coastal defences, and in 2005–7 the UK experienced an average of £1400 million of damage a year. The UK Foresight Programme's estimation of future flood risks is fraught due to future uncertainties. However, it concludes that all scenarios point to 'substantial increases'. The number of people in the UK at high risk from river and coastal flooding could increase from 1.6 million in 2008 to between 2.3 and 3.6 million by the 2080s. The increase for urban flooding, caused by short-duration events, could increase from 200 000 people today to between 700 000 and 900 000 (Foresight Programme, 2008).

In 2010, the Environment Agency (the UK governmental agency charged, among other things, with flood defence responsibility) published a report, *The Costs of the Summer 2007 Floods in England* (Environment Agency, 2010). Broad-scale estimates made shortly after Britain's 2007 floods put the total losses at about £4 billion, of which insurable losses were reported to be about £3 billion. The Environment Agency study set out to produce a comprehensive monetary estimate of the total economic cost of the 2007 events. It concluded that total economic costs of the summer 2007 floods were around £3.2 billion in 2007 prices, within a possible range of between £2.5 and £3.8 billion. Overall, about two-thirds (£2.12 billion) of total economic costs were incurred by households and businesses. Power and water utilities accounted for about 10% (£0.33 billion) of total costs, and communications (including roads) about 7% (£0.23 billion). Emergency services, involving the police, fire and rescue services and emergency response by the Environment Agency, accounted for about 1% (£27 million) of total economic costs. Damages to agriculture, associated with inundation of more than 40 000 ha, accounted for about 2% (£50 million). Impacts on public health (including school education) accounted for about 9% or £287 million (Environment Agency, 2010).

Of the damage to agriculture, the British pea (*Pisum sativum*) crop was reduced by around a quarter and this was noticed by consumers following the harvest. The next largest producers of peas in Western Europe were in the Benelux countries, which were hit by the same rain systems, so affirming shortages. Consequently, in March 2008 pea prices of the cheapest brands (that do not have inherent high-branding costs in which to absorb basic production fluctuations) were around 70% or more than before the floods. The 2007 rains also prevented farmers spraying some crops with anti-fungal pesticides. This affected crops of potatoes (*Solanum tuberosum*), cabbages and broccoli (both cultivars of *Brassica oleracea*).

On continental Europe the most disastrous floods in recent times took place in August 2002 along the Elbe and Danube rivers. In the Czech Republic some 200 000 had to leave their homes whereas in Germany 3600 patients had to be moved from hospitals threatened by waters. Rough cost estimates for the Elbe flood were US$3 billion in the Czech Republic and over $9 billion in Germany. Flood damage (which is not the same as flooding) of this magnitude had never happened in Europe before. There have been 10 extreme floods in Dresden, Germany, since the 13th century, with water levels peaking between 8.2 and 8.8 m, but the 2002 flood peaked at 9.4 m. However, the previous record flood, in 1845, had a higher flow rate than the flood in 2002. The reason why the 2002 flood was higher despite a lower flow rate was greater river containment as a result of floodplain development. This again exemplifies how non-climate factors can compound extreme weather events. Having said this, the rain that caused the 2002 flood exceeded most previously measured rainfall amounts and intensities, and in the core area of precipitation rainfall levels exceeded the 24-hour record observed in Germany since records began and were close to the possible maximum precipitation (Becker and Grunewald, 2003).

The future Western European picture that is just beginning to emerge is therefore a complex mix of changes in seasonal precipitation (down in summer and up in winter), extreme events (similar frequency but slightly greater intensity) and watershed development (increased floodplain development). This is exemplified by historical analysis of rivers, about which there is substantial historical documentation. Among these are some European rivers. In 2003 German researchers led by Manfred Mudelsee looked at records for the middle Elbe and middle Olde rivers, primarily concentrating on events in the past 500 years, for which the records are more reliable. Discerning long-term trends in monthly run-off was difficult. They found that there was no substantial trend emerging above the natural variability in events. So, the historical analysis did not provide evidence for a climate change-related cause for increased European flooding.

How can this conclusion sit alongside model predictions? Leaving aside model reliability (which is continually improving), historical records look back in time while models look forward. The past has seen just a degree's worth of global warming since the Industrial Revolution, whereas the coming decades are expected to see a B-a-U rise above 1990 temperatures of 2.0–4.5°C (best-estimate all-model scenarios) with overall low and high estimates of 1.4 and 5.8°C, respectively. In short, retrospective reviews and prospective forecasts are not the same. So, the former may show hardly any change while the latter might suggest a definite future change. What this means is that the early 21st century sees us on the cusp of a new climatic regimen. Even so, the actual changes in the future could be quite small, with just marginal excesses over previous instances. Here the problem is that such floods are catastrophic events. Small changes can have big consequences. Matters are fine until river banks are breached (such as with a peak rise of just a few millimetres), and then there is a quick transition from no damage being incurred to considerable damage.

Nonetheless, the above does not answer the question as to whether or not such flood events are likely caused by anthropogenic warming, given – as the IPCC said in 2011 with regards to its then forthcoming 2012 SREX report (cited in Chapter 6) – that a specific event cannot be attributable to anthropogenic warming. In 2011 Seung-Ki Min, Xuebin Zhang, Francis Zwiers and Gabriele Heger, based in Canada and Scotland, looked at the human contribution to rainfall extremes by comparing observations with multi-model simulations. They showed that human-induced increases in greenhouse gases have contributed to the observed intensification of heavy precipitation events found over approximately two-thirds of data-covered parts of northern hemisphere land areas. Indeed, they warned that changes in extreme precipitation projected by models, and thus the impacts of future changes in extreme precipitation, may be underestimated because models seem to underestimate the observed increase in heavy precipitation with warming. Published alongside this paper was another by Pardeep Pall, Tolu Aina, Dáithı Stone and colleagues, based in Britain, Switzerland and Japan, that looked at the 2000 British flood event (the meteorological cause of which was similar to that of the 2007 event). They used a seasonal-forecast-resolution model that better represented the extra-tropical jet stream than lower-resolution counterparts typically used for climate simulations. They also used a river run-off model as that is a better predictor of flooding (the inability to drain land) than precipitation and used actual precipitation data from the autumns between 1958 and 2001. They concluded that whereas the precise magnitude of the anthropogenic contribution remained uncertain, in nine out of ten cases their results indicate that 20th-century anthropogenic greenhouse gas emissions increased the risk of floods occurring in England and Wales in the autumn of 2000 by more than 20%, and in two out of three cases by more than 90%. Together these two papers indicate that the frequency of intense rainfall events is likely to increase with global warming. (See also an accompanying review article by Richard Allan, 2011.)

Finally, increased major floods could well happen in the summer despite European summers becoming drier. Seemingly paradoxically, computer models predict an increase in *intense* summer rainfall with global warming. Instead of the lower rainfall being spread across summer months, there will be a tendency for this precipitation to clump into extreme weather events (Christensen and Christensen, 2002). This is because in a warmer world there is more water vapour in the air. Of course, some air is warmer than other air. When water-laden air (more than is historically anticipated at a given latitude) meets cool air the resulting rainfall will be greater. Additionally, matters can even be worse. Consider two bodies of warm, water-laden air colliding. There is nowhere for the air to go but up. At a higher altitude it cools, again releasing water but even more than in the previous scenario. The consequence of this for flooding is significant. If rainfall is spaced out over weeks then the water has the opportunity to sink into the ground and then into the subsurface geology. If a few weeks' or even a month's worth of rain falls at once, due to such an atmospheric collision, then there is no time for the resulting precipitation to drain away and so it remains on the surface. Problems are compounded if the surface is corrugated with hills and valleys, as this volume of rain will run off the hills into the valleys as fast-flowing floods. Of course, there are parts of the Earth's surface that historically are warmer than temperate latitudes and which also have high rainfall. However, these regions have waterways gouged out by previous rainfall and if they are inhabited the appropriate drainage systems are in place. With climate change, a place that previously only had moderate, well-spaced rain with time becomes exposed to more extreme rainfall regimens; there will not be appropriate drainage systems in place and so flooding results, with property damage and possible loss of life.

Flooding from both sea-level rise and increased precipitation from greater marine evaporation in a warmer planet are expected symptoms of global warming, and this is stated by the IPCC. The UK government's Foresight report, *Migration and Global Environmental Change* (Vafeidis et al., 2011), estimated that in the year 2000 there were some 628 753 970 people (more than 10% of the then global population) in low-lying coastal areas less than 10 m above sea-level: the areas of land prone to both rain-fed flooding and sea-level rise. More than 23% of these were urban dwellers. Table 7.1 lists the population living in such low-elevation coastal zones (LECZs) in selected countries. It is expected that these numbers will rise with population growth in the 21st century.

However, the opposite of an excess of water, drought, is paradoxically also a problem and is also likely in other areas of a warmer world that have greater evaporation

Table 7.1 Size of population in low-elevation coastal zones (LECZs; less than 10 m above sea level) in selected countries (Vafeidis et al., 2011)	
Country	Population in LECZs in 2000
Australia	2 196 520
Bangladesh	63 122 300
Brazil	11 552 800
Canada	1 192 880
China	143 989 000
Denmark	1 455 450
Egypt	25 461 200
Germany	4 585 850
India	63 925 500
Indonesia	39 255 200
Ireland	338 629
Japan	30 193 500
Maldives	290 799
Mexico	5 615 320
Netherlands	11 551 400
New Zealand	472 986
Pakistan	4 556 230
Poland	903 503
South Africa	392 006
Thailand	16 422 700
Tuvalu	9 903
UK	7 071 210
USA	23 366 000

from the land. Further, the human impact of droughts in a more crowded future, and hence one that consumes more water, will be exacerbated. This is yet another example of a non-climate factor exacerbating impacts.

7.3.4 Droughts

As with floods, droughts are not new but have been increasing in their severity. Droughts not only affect natural and human-managed ecosystems (such as agricultural lands) but also damage property. The drying out of soil can damage housing, as exemplified by the 1976 drought in England with the cost of repairing domestic properties exceeding £100 million in the money of the day (Burroughs, 1997).

In 2004 the south of China experienced the worst drought for half a century and low levels of rain continued in 2005 and 2006. Difficult choices had to be made. In autumn 2004, 100 hydroelectric plants were shut down so that water could be diverted to rice fields and farmland, 730 000 ha of which were affected by the drought, including 36 000 ha abandoned due to lack of water. Widespread power cuts resulted. The level of one river, the Dongijang, fell by 80%, which dramatically reduced supplies to 36 million people.

Just as climate change sees monsoons deliver extreme maximum precipitation, so there are minima too. In 2009 India suffered its weakest monsoon for 40 years. Rainfall was at its lowest at the end of the monsoon season since 1972. There were regional differences: the north west had the worst rainfall deficit, at 36%, while the southern part of the country was just 7% below average. India is the world's second biggest producer of rice (*Oryza* spp.), wheat (*Triticum* spp.) and sugar.

Whereas one incident cannot be specifically blamed on climate change, as noted before, it is the pattern of events that is revealing and this instance is the sort of event that one might expect with global warming. The greatest health impacts arising from climate change are those that adversely impact on the increasing global population's ability to feed itself. And then there are the biological secondary effects. For example, in China, from analysis of records for the past 100 years, the years of most severe locust outbreak were in the warm, dry years with warm, dry summers and warm, wet winters. Locust outbreaks damage crops. Both interannual and decadal variability of higher temperature changes have led to the highest locust outbreaks in the past 1000 years (Yu et al., 2009).

Water availability, be it a deficit or a surplus, is one factor that links climate change to food security. This is the subject of the next section, in which we will discuss drought in a little more detail.

7.4 Climate change and food security

7.4.1 Past food security

Climate and food security have historically been central to a society and/or culture's survival (see section 5.2.1). Humanity's food security improved greatly in the latter quarter of the 20th century. This was due to the mechanisation of agriculture and the use of chemical fertilizers and pesticides, which together formed the Green Revolution that dramatically increased agricultural productivity (which is defined as output per unit area). The success of the Green Revolution was evident by the late-20th-century growth in global population (see the beginning of this chapter), which in turn was built on previous population growth. As previously noted, the global population (Figure 7.4a) began to increase following the Renaissance and continued to rise with the Industrial Revolution. The rate of growth (as opposed to population itself) peaked in the 20th century. This determined the size of the population in the following decades (Figure 7.4b), which itself is expected to peak during this century (Figure 7.5). All these extra mouths required, and will require, feeding.

Yet, there has been a price for this success. First, it did not come without its failures. New agricultural practices were not always sustainable. Concerns such as soil erosion, ecotoxicological impacts from chemical inputs and resistance to pesticides have meant that techniques have had to be refined. Furthermore, the post-agricultural component to food supply has come to dominate the economics of food supply so that, for example, in the 1990s in the UK while agriculture contributed to a little over 1% of its GDP, post-agricultural processing and commerce added more than another 2% (see section 7.1.4 on food). This post-agricultural dominance can be seen in virtually all countries. Consequently, post-agricultural commercial concerns have shaped much of the way the non-subsistence proportion of the global population feeds itself. Today people in developed nations are used to eating out-of-season produce and products that have a high energy intensity (again, see section 7.1.4). Irrespective of climate change there is the very important question (worthy of a textbook in its own right) of how the global population of the 21st century will feed itself.

In summary, and remembering the backdrop of growing population, the position as revealed by FAO data to date since the middle of the last century is as follows.

- Annual grain production (which includes wheat, maize [corn] and rice) increased from around 500 million t to around 1800–1900 million t in the 1990s, when it peaked and levelled off. Conversely per-capita grain production has stabilised at around 290–330 kg year^{-1} since the 1970s and indeed the trend has been downwards since a mid-1980s peak.
- Annual meat production (more than 90% of which is pork, poultry and beef) increased from 44 million t in 1950 to 237 million t in 2001, whereas average individual consumption rose from 17.2 kg in 1950 to 39.0 kg in 2002.
- The above has been fuelled by fertilizers, global production of which has increased from 14 million tonnes in 1950 to 120–46 million t in the 1990s.
- The global annual fish catch (as opposed to aquaculture, or fish farming) increased from 19 million t in 1950 to around 92–6 million t in the latter half of the 1990s. However, against a backdrop of growing global population, the annual catch per person despite growth since 1950 has levelled off since the mid-1970s to around 15–17.5 kg. Fortunately the global fish supply has been augmented since the 1980s with aquaculture, the production of which increased from around 7 million t in the mid-1980s to some 36 million t in 2000.

Those in industrialised countries consume more grain than non-industrialised ones via animal feed for the increasing meat in their diet. For example, a North American consumes nearly 10 times as much grain as someone from a sub-Saharan nation. Middle-income countries also showed an increase in per-capita meat consumption during the 20th century. Not irrelevant is the increasing incidence of obesity and diabetes in these countries. Despite this inefficiency in healthy nutrition, globally those facing chronic hunger have decreased from 956 million people (approximately 26% of the global population) in 1970 to 815 million (approximately 13%) in 2002.

Since 1975 politicians in wealthy countries have seemed unconcerned about food supply. After all, if there is a shortage (irrespective of the cause) of one crop harvest (especially if restricted to one region) then their nation will have the money to purchase alternatives on the global food market. In less-wealthy countries matters are more acute but despite this many in the political classes seem more concerned (as do their wealthy nation counterparts) with staying in power than with the under-privileged proportion of their electorate (assuming that their electoral system is not corrupt in the first place). However, should market circumstances, climate change extreme events and fossil fuel economics (itself of climate change policy concern) come together then it is possible for the resulting effect on global food supply to be considerable. This happened in 2007/8.

The year 2007 and first half of 2008 saw food staples on the international market rise sharply in price. The causes were several. First, extreme weather events in 2005–7, including drought and floods, affected major cereal-producing countries. World cereal production fell by 3.6% in 2005 and 6.9% in 2006 before recovering in 2007. Two successive years of lower crop yields in a context of already low stock levels resulted in a constrained supply in world markets. Growing concern over the potential effect of climate change on future availabilities of food supplies aggravated market fears. Second, until mid-2008, the increase in energy prices had been very rapid and steep, with one major commodity price index (the Reuters-CRB Energy Index) more than tripling since 2003. Petroleum and food prices are highly correlated. The rapid rise in petroleum prices exerted upward pressure on food prices as fertilizer prices nearly tripled and transport costs doubled in 2006–8. (Urea, a major component of fertilizer, requires considerable energy to fix nitrogen from the air – and hence fossil fuel in the absence of non-fossil carbon energy – as well as the use of methane as a chemical feedstock for its manufacture, again commonly from fossil fuel. So, fertilizer prices are linked to those of fossil fuel.) Third, the emerging biofuel market is a significant source of demand for certain agricultural crops. The stronger demand for these commodities caused a surge in their prices around the world. And then there were the longer-term drivers of restricted availability of quality agricultural land and population growth (both of which we will return to shortly). All these factors together served to restrict food supply and to drive food prices upwards so that between 2005 and 2008 the FAO food price index doubled (over a 100% increase) from its 1975–2004 average (FAO, 2008). The shortages sparked food riots in 2007–8 in Burkina Faso, Cameroon, Senegal, Mauritania, Côte d'Ivoire, Egypt, Morocco, Mexico, Bolivia, Yemen, Uzbekistan, Bangladesh, Pakistan, Sri Lanka and South Africa. Conversely, if you were to ask anyone in Western Europe or North America as to the biggest event affecting their lives in the latter part of the first decade of the 21st century they would undoubtedly instead refer to the financial market crash of 2008/9.

Here from a human-ecology perspective it must be stressed that the increase in food prices and global unrest of 2008 were due to a real shortage of a physical environmental resource. Conversely, the financial crisis of 2008–9 was 'just' due to the notional revaluation of sub-prime mortgages: no loss of crop, farmland, mine output, fishery or any physical (i.e. 'real' as opposed to 'notional' or abstract) resource. Indeed, the physicality of the houses to which the sub-prime mortgages related remained unchanged. All of which begs the question of what will happen with climate change impacts on environmental resource use later in this evermore populated century?

Leaving aside economic and extreme weather variables, there are two fundamental dimensions undermining the future sustainability of food supply. First, on land much of the most productive ecosystems are already heavily exploited. Much of the most suitable land has already been turned to agriculture while other land has gone to urban sprawl and related infrastructure. Worldwide from 1700 to 1980 crop lands increased by 466.4% from 265 million ha to 1501 million ha. Most of this has been at the expense of forest and woodland, which has declined by 18.7% from 6215 million to 5053 million ha (Turner et al., 1990). At sea the price of human success has been the collapse of fisheries, such as in the North Sea. There have also been effects elsewhere in the marine food web. For example, around Alaska the tripling of the pollack (*Pollachius* spp.) catch between the mid-1980s and the end of the century saw the stellar sea lion (*Eumetopias jubatus*) population that feeds on the pollack collapse by around 90% (Starke, 1999).

The second factor is that the growth in human population, and its modern fossil-energy-intensive culture, appears not to have been sustainable, not just in food-security terms, but also in terms of the sustainable harvesting of ecosystems. For example, non-sustainable fossil fuels are the main source of urea that is used for agricultural fertilizer. Again, regarding fresh water (another essential agricultural input), the growth in demand for water in many parts of the planet is not sustainable. In other words, not only are we quantitatively overexploiting ecosystems so that they collapse, we are also doing this qualitatively, in a non-sustainable way.

However, whereas both these dimensions are real, combating hunger in modern times has been arguably hindered by an enabling factor of economic development. True, while much of the most productive land around centres of population has been exploited, it is often possible to irrigate marginal lands, develop sustainable agricultural systems or buy in food from elsewhere. However, this takes both finance and a sound commercial framework within which to operate. This development dimension is cited by the FAO as being one of the current key factors determining food security. Unfortunately international trade barriers and development agreements have in some instances undermined sustainable development. Whether 20th-century-style economic development will remain in a more crowded future with climate change remains to be seen.

7.4.2 Present and future food security and climate change

As discussed, without future climate change we know (from current human longevity records and birth and death rates) that the global population will continue to rise through to the middle of the 21st century. Given that we are continuing to push for increased food production, what does the future hold? To begin to answer this we need to know the current situation.

The 1996 World Food Summit set the goal to reduce the number of people under-nourished globally by half to no more than 420 million by 2015. Since then the FAO has produced annual reports on *The State of Food Insecurity in the World* (FAO, various dates). In summary, its early reports point to a declining number of people who are undernourished. In 1970 there were approximately 960 million people undernourished. The FAO estimate that in the period 1997–9 this figure declined to 815 million. Of these, more than 95% were in developing nations, around 3% in transitional nations and 1.4% in developed industrial nations. The FAO's 2001 report noted that the (then) current decline in undernourished was 6 million a year, which in the 1990s represented a reduction in the rate of decline in undernourishment compared to the 1980s. In part this might be attributable to the increase in global population (see earlier in this chapter) but nonetheless by 2001 if this reduced trend continued then it would take 60 years to reach the 1996 Summit's target number, which was supposed to be achievable 14 years hence, by 2015. In short, to achieve the original target the annual reduction in undernourished people would have to increase from 6 million to 21 million a year. This target, the 2001 report concluded, was unlikely to be reached without 'rallying political will and resources'. This was prophetic.

The publication of the FAO's 2004 report announced a turning point. The numbers of undernourished people had stopped declining and in fact had increased, while the world population also continued to grow. The 815 million undernourished for the period 1997–9 increased to 842 million for 1999–2001. Consequently the FAO reported that the world was further away from meeting the 1996 World Food Summit target. This could now be reached only if annual reductions were to be accelerated to 26 million per year, more than 12 times the pace of 2.1 million per year achieved on average from the 1996 summit to 2004. The 2004 report also pointed to Africa as the continent with the greatest nourishment problems.

With regards to climate change as a factor affecting food security, the FAO food security reports do not specifically raise this as a concern but do touch upon it tangentially. The closest the 2001 report came to mentioning climate change was when it briefly cited weather extremes – specifically droughts, floods, cyclones and extreme temperatures – as threatening progress towards food security. By the time of the 2004 report the FAO gave weather extremes a higher profile, although developmental concerns (including human conflict) were (rightly) of greater current concern. With reference to both these issues the reports said: 'Many countries that are plagued by unfavourable weather but enjoy relatively stable economies and governments have implemented crisis prevention and mitigation programmes and established effective channels for relief and rehabilitation efforts. But when a country has also been battered by conflict or economic collapse, programmes and infrastructure for prevention, relief and rehabilitation are usually disrupted or destroyed.' The 2004 FAO data demonstrated that the trends in either increasing or decreasing hunger were both related to a nation's growth in GDP, with those countries with the lowest economic growth faring the worst.

The FAO's 2004 food-security report did acknowledge the importance of weather-related events in causing food crises, but it noted that the number of crises caused by weather, and jointly weather and human conflict, had decreased from 86% of crises in the period 1986–91 to 63% for 1992–2004. Conversely problems due to human activities (mainly conflict) had increased. This explains the FAO's current focus away from climate change.

Despite the aforementioned immediate-crisis caveat, from a climate change perspective the 2004 report did highlight the importance of water to long-term food security. It observed that agriculture is by far the biggest user of water, accounting for about 69% of all withdrawals worldwide and more than 80% in developing countries. Reliable access to adequate water increases agricultural yields, providing more food and higher incomes in the rural areas that are home to three-quarters of the globe's hungry people. So, not surprisingly, countries with better access to water also tend to have lower levels of hunger. Drought ranks as the single most common cause of severe food shortage in developing countries. For the three most recent years for which data were available to the 2004 FAO report's authors, weather-related causes of hunger increased year on year, with drought listed as a cause in 60% of food emergencies. Even where overall water availability is adequate, erratic rainfall and access to water can cause both short-term food shortages and long-term food insecurity. Floods are another major cause of food emergencies. Sharp seasonal differences in water availability can also increase food insecurity. In India, for example, more than 70% of annual rainfall occurs during the 3 months of the monsoon, when most of it floods out to sea.

Conversely, farmers who lack irrigation facilities must contend with water scarcity throughout much of the year and the threat of crop failures when the monsoons fail. Where water is scarce and the environment fragile, achieving food security may depend on what has been called virtual water: foods imported from countries with an abundance of water. It takes 1 m^3 of water to produce 1 kg of wheat. Extrapolating from those numbers, the FAO calculated that to grow the amount of food imported by Near-Eastern countries in 1994 would have required as much water as the total annual flow of the Nile at Aswan. In such conditions it may make sense to import food and use limited water resources for other purposes, including growing high-value crops for export. As we have seen (Chapter 6), in a globally warmed world not only can we expect extremes in precipitation (more evaporation of seas, more rain) but also greater evaporation on land and through plant photorespiration. With global warming water availability will become a key issue and, against a backdrop of an increasingly populous planet, a driver of food insecurity.

The 2008 FAO *The State of Food Insecurity* report focused on the 2007–8 food crisis discussed in section 7.4.1. The 2009 report noted that even before the food and economic crises, hunger was on the rise and that the World Food Summit target of reducing the number of undernourished people by half to no more than 420 million by 2015 will not be reached if the trends that prevailed before those crises continue. It also estimated that 1.02 billion people are undernourished worldwide. This represents more hungry people than at any time since 1970 and a worsening of the unsatisfactory trends that were present even before the 2008–9 financial crisis. It reported that this financial crisis was now affecting the poor in poor nations as incomes were not rising and global food market prices remained high: the food problems of 2009 were not due to poor harvests. Conversely, the 2010 report was marginally more optimistic, reporting that the number of undernourished had declined from the levels of the previous 2 years but was still higher than before the 2007–8 food crisis.

The FAO's 2011 report summarised the food context of the two – food and financial – crises, opining that prices are generally expected to rise because continued population and economic growth will put upward pressure on demand, as will the anticipated increased use of biofuels (depending on biofuel policies and the price of oil). On the supply side, if oil prices continue to rise, agricultural production costs will increase, contributing to higher food prices. Natural resource constraints, especially climate change and the limited availability of productive land and water in some regions, pose substantial challenges to producing food at affordable prices. On a more positive note, there remains significant potential for raising crop productivity through new technologies and improved extension, as well as for reducing losses in the supply chain. However, these gains will not materialize without increased investment.

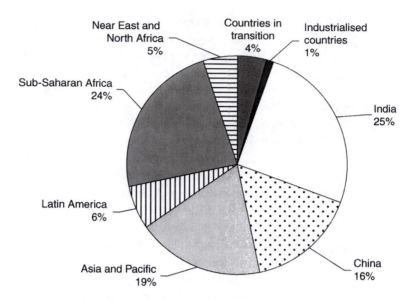

Fig. 7.13 The location of the world's 842 million undernourished people (1999–2001). Based on data from FAO (2004).

There are also compelling arguments suggesting that, in addition to being higher, food commodity prices will also be more volatile in the future. If the frequency of extreme weather events increases, production shocks will be more frequent, which will tend to make prices more volatile. Furthermore, biofuel policies have created new links between the price of oil and the price of food commodities. When oil prices increase, demand for biofuels will increase, thus raising food prices, with the opposite happening when oil prices decrease.

With regards to where food security is currently the greatest problem, as can be seen from Figure 7.13, Africa is the continent bearing the brunt.

Perspectives on the prospects for improved food security extend beyond the FAO. An analysis by Lester Brown of the Worldwatch Institute is not encouraging (Brown, 1995; incidentally, the Worldwatch Institute contributed to the aforementioned 2001 FAO food-security report). Lester Brown noted that two of the four largest grain producers (the USA, China, the former Soviet block and India), which historically in 1950 were either net exporters or virtually self-sufficient, had by 1990 become net importers, with only the USA increasing its production to a level that more than offset the decline in the others. However, over the same period this US surplus was also offset by declines in production by the nine other most-populous countries. Taking into account the anticipated future growth in population it becomes clear that we cannot continue to feed the world as we now do.

Contrary to initial impressions, this is not a doom-and-gloom conclusion. In 1972 the report for the Club of Rome, *The Limits to Growth*, was prophetic (Meadows et al., 1972). The report predicted that resources would run out *unless* we changed the way they were exploited and used. Fortunately, this happened. For example, since its publication the efficiency of copper extraction has improved, so remarkably the spent tailing mounds from previous copper extraction can now be mined, while at the other end of the production cycle material recycling has boomed (see Chapter 8). Such is the current wastage of what we eat, similar improvements in efficiency could happen with global food.

However, if this does happen then Western consumers would have to accept changes. To take just three examples, first, food miles would need to be reduced dramatically to free agri-energy to improve productivity elsewhere, as well as to reduce transition times during which spoiling can take place. This last would mean a return to the seasonal availability of many foods and a reduction in year-round food choice (although not choice across the year). A second possibility might be a change in animal protein and fat consumption, which through much of the 20th century grew ahead of population and which arguably is one of the factors behind obesity-related health impacts that now dominate the health agendas of a number of developed nations. Finally, there would need to be an increase in (sustainable) agricultural intensity. This rise in production per unit area (productivity) in turn means a decrease in biological resources available to wildlife within that agricultural area. There would be biodiversity implications both within and outside of these agricultural areas (see the discussion on agriculture, biodiversity and landscape in section 7.1.5).

Whereas improvements in food-supply efficiency may be a note of hope for our species, if we are to avoid increasing starvation then humanity must feed itself differently 30 years from now compared to today, and probably as differently as it was fed 50 years ago. However, even with this it is hard to see pressures on natural systems decline; indeed, it is almost inevitable that such pressures will increase. Either way, greater resource management worldwide will be required.

The above summary does not emphasise the climate change factors. So, what are the food-security implications of global warming? We have already seen that past climate change has had a detrimental impact on civilisations, such as with the Incas in the New World and the people affected by the Little Ice Age in the Old World. It is therefore difficult to see why future climate change should not have impacts on future food supply. With past change agricultural belts moved and the problems arose because some civilisations did not move with them and exploit new areas. The same problem is likely to affect future climate-related changes in agricultural production. The big difficulty is that the present and future are far more crowded than the past. There are increasingly limited options for opening up land that does not already have a claim on it. A complicating matter is that a significant proportion of the human population is concentrated along land-mass coastlines, some of which will be under threat from climate-induced sea-level rise compounded by occasional intense run-off from the land.

Climate–food models of future global agricultural production vary. Much depends on both which model is used and the agricultural products that will be grown. This is in part because, as with climate models, their resolution is too coarse. Also, because climate and agriculture are so closely related, while climate models are not (yet) sufficiently sophisticated for precision use in regional and local planning, so climate-food models cannot be used for regional and local agricultural planning. Finally, it all depends on what is actually being modelled. A model based on a region's biological productivity and future climate will give different results to a model based on current agricultural output because this last will not take into account how cultivated crops and agricultural practices will alter with climate change.

Having said that, it is possible to look back and see how climate change has impacted on crop production to date and to link that to a model of recent past climate. This is just what US researchers David Lobell, Wolfram Schlenker and Justin Costa-Roberts did in 2011. They found that in the cropping regions and growing seasons of most countries, with the important exception of the USA, temperature trends from 1980 to 2008 exceeded one standard deviation of historic year-to-year variability: that is to say there was considerable climate change over nearly three decades that exceeded annual variability. Global maize and wheat production declined by 3.8 and 5.5%, respectively, relative to what they would have been had there been no climate change. For soya beans and rice, winners and losers largely balanced out. Climate trends were large enough in some countries to offset a significant portion of the increases in average yields that arose from technology, carbon dioxide fertilization and other factors.

Regarding food security, the IPCC (2001a) concludes that while plant growth will benefit from the increased atmospheric carbon dioxide, agricultural belts will shift with climate. This means that farmers in existing agricultural areas will need to change their crops or, in some cases, cease farming or move. The IPCC also note that in areas that become favourable for agriculture there will be new opportunities.

The IPCC raise the concern that extreme weather events will become increasingly problematic. Indeed, examples from food production subsequent to their latest report (2001) already abound. For instance, the Western European heatwave of 2003 scorched enough of the potential wheat harvest to make a noticeable impact on bread prices. Most of the harvest was lost in southern Europe compared to the north. France lost about a fifth of its anticipated harvest whereas Italy and Britain lost more than 10%. Moldova was one of the countries worst hit, losing about three-quarters of its anticipated harvest. In August 2003, Rank Hovis, the UK's largest flour miller, announced that the price of its milling flour had increased from £35 per tonne to around £245. This in turn increased bread prices by around 10–15%. In part this modest rise was due to flour costs being a minor component of bread's retail price. The modest increase was also due to being able to import wheat from elsewhere, as these losses were not similarly reflected worldwide, with the year's global grain (including rice and maize [corn]) harvest being down on expectations by less than 2%. But other crops were affected in Europe; for example, potatoes also increased in price. Another instance was the Russian drought in 2010 that was so severe that some farmland burned and around 155 000 people were needed to combat the fires. In the central Russian republic of Tatarstan alone the drought damaged nearly 800 000 ha of crops. Russia banned wheat exports as domestic prices soared. As Russia was in 2009 the world's fourth largest exporter of wheat, behind the USA, the EU and Canada, the Russian drought even affected the US wheat futures market, in which prices rose by 80% between June and August that year.

A major future challenge for global agriculture, the IPCC predict (2001a), will be the degradation of soil and also water resources. Drier summers in many temperate areas but with bursts of more intense rainfall will necessitate greater water management and the ability to capture downpours for storage. Heavier winter rainfall again will necessitate greater watershed management to minimise soil erosion and flooding.

The IPCC's overall conclusion is that modest climate change that might be expected in the first quarter of the 21st century will result in agricultural winners and losers.

However, more significant annual temperature increases of 2.5°C or more could prompt global food prices to rise. Again, the IPCC note that the greatest impacts will be on the poorer farmers and consumers.

So why would increases in annual temperatures of upwards of 2.5°C have a net detrimental effect? After all, a warmer, carbon dioxide-rich planet should mean that there are some (new) places that will favour agriculture even if some other places become unsuitable. The concern is that higher temperatures shorten the life cycle of many cereals, hastening senescence and so reducing the length of the growing season. The world's staple cereal crops tend to tolerate narrow temperature ranges, which, if exceeded during the flowering stages, can damage production of fertile seeds and reduce yields. Given that global warming is to increase the number of extreme weather events, crop yield is likely to be impaired. This would also be compounded by increases in ozone in the troposphere (which is low down in the atmosphere; Porter, 2005). This is likely to be more of a factor in parts of the globe where there are fewer effective controls on air pollution, such as Asia.

Add climate change concerns to those of global demographics and some feel that we might hit a period of 'peak food' just as there is thought to be a time when there will be 'peak oil' consumption.

Given the almost certain demographic pressures anticipated for the rest of this century, agricultural research needs to become a priority and this needs to be related to high-resolution climate models that include the biological dimensions (which they are now beginning to do). This is worth stressing because many developed nations have been markedly reducing their investment in whole-organism agricultural research due to past successes – reflected in the agricultural productivity (production per unit area) rises – and past European and American food surpluses. Consequently a number of governments have felt that with past success achieved there is little new challenge on the horizon. In the UK governmental investment in policy-driven agricultural research from the department responsible for farming and fisheries has declined in real terms year on year since the mid-1980s to the present (2006/7).

Internationally, one fairly recent initiative to help co-ordinate research into food security and environmental change was the Global Environmental Change and Food Systems (or GECAFS) project. GECAFS was established in 2001 with formal partnerships with the UN FAO and WMO. It ran for a decade and a number of its outputs were published in *Food Security and Global Environmental Change* in 2010, edited by John Ingram, Polly Ericksen and Diana Liverman.

Then there are broader policy concerns beyond those of science research: policies that enable us to reduce the speed of climate change and its extent, and adapt to likely impacts, will help our global society to cope with the warming world of the 21st century.

Finally, to bring sections 7.2 and 7.3 together, it is worth remembering that it is the poor who will be detrimentally affected by climate change: both the poor in wealthy nations and a greater proportion of the populations of less-wealthy nations. Yet it has been the wealthy countries that have over the years added the most greenhouse gas to the atmosphere. Figure 7.14 spatially compares national greenhouse emissions with nations' incidence of four climate-sensitive diseases. It can be seen that developed nations are those responsible for most emissions whereas less-developed southern nations are likely to bear the brunt of changes in these diseases (with the

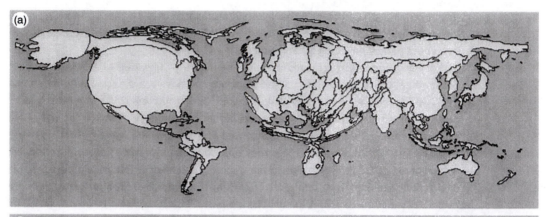

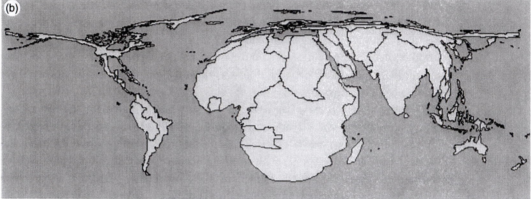

Fig. 7.14 Spatial map comparison of (a) national cumulative carbon dioxide emissions by country for 1950–2000 and (b) the regional distribution of four climate-sensitive health consequences (malaria, malnutrition, diarrhoea and inland flood-related fatalities). Reproduced from Patz et al. (2007), with permission from Jonathan Patz.

notable exception of China, the large population of which masks its lower per-capita emissions).

7.5 The biology of reducing anthropogenic climate change

It is a combination of human short-circuiting of the deep carbon cycle (through fossil fuel burning) and altering of the fast carbon cycle (through land-use change such as deforestation) that is increasing atmospheric carbon dioxide. Similarly, it is possible for us to take carbon cycle actions that reduce climate change, or at least the rate of change. Improving the efficiency with which fossil fuels are consumed and switching reliance towards non-fossil energy resources will be discussed briefly in the next chapter. However, in terms of human-ecology (our own species' demographics and relationship with other species) we might manipulate the fast carbon cycle through land-use, and perhaps marine, management. The aim would be to use the fast carbon cycle to offset carbon dioxide from fossil carbon (part of the deep carbon cycle) by increasing fast-cycle carbon sinks as well as partly replacing dependence on fossil carbon with fast-cycle carbon (or biofuels). However, care would be needed to ensure that we did not build up ecological reservoirs of carbon that would subsequently be released through climate change.

In terms of tapping into the deep carbon cycle, it is broadly estimated that the total carbon released by human action between the onset of the Industrial Revolution in the 18th century and the middle of the 21st century's first decade is around 500 GtC (Allen et al., 2009) and of this broadly two-thirds (over 300 GtC) comes from fossil fuel. This compares with a broad upper estimate of conventional fossil carbon resources (past and the next two centuries) of approximately 1700 GtC. Alternatively, if one includes unconventional fossil reserves (such as tar sands and shale oils), then this increases to a potentially exploitable total fossil carbon inventory of almost 5000 GtC. (And this last excludes carbon fuels that we have yet considered as unconventional resources, such as marine hydrates [clathrates], which the IPCC estimate at 12 000 GtC.) This estimate of approximately 300 GtC for historic fossil releases compares to the forecast total historic *and* future releases (1880–2100), which the IPCC (2001a) use based on their Special Report on Emission Scenarios (SRES), of between 1000 and 2150 GtC for the 21st century. In other words, the IPCC think it likely – depending on various economic futures and compared to the 12 decades up to the end of the 20th century – that 21st-century carbon releases may be double to over six times that released historically.

The aforementioned data give us a rough upper estimate for *average* annual fossil carbon emissions for the 21st century of around 18 GtC year^{-1} for a high Business-as-Usual (B-a-U) scenario (although more is most likely to be released annually at the end of the century than the beginning). For biological sequestration to have a significant impact it is going to have to be extremely effective. So what is its likely potential?

7.5.1 Terrestrial photosynthesis and soil carbon

The B-a-U annual fossil release of 18 GtC for the 21st-century is small compared to the roughly 50 GtC photosynthetically captured by the Earth's terrestrial ecosystems (i.e. excluding ocean productivity) each year *after* allowing for respiration. In theory, therefore, it should be possible to harvest 18 GtC of terrestrial vegetation (say, as wood) each year and bury it (effectively removing carbon from the atmosphere) or burn it as fuel (effectively part-offsetting fossil carbon emissions). This would negate human average annual 21st-century carbon emissions. However, to do this we would need to harvest the equivalent of all new tree growth globally each year and this is clearly impractical. (Note: notwithstanding in addition to the approximately 50 GtC photosynthetically captured there is also a carbon stock from previous years' growth and this last forms the bulk of the carbon stocks depicted in Figure 7.15.)

However, the theoretical possibility of photosynthetically drawing down such a large amount of carbon can perhaps be better understood by looking at terrestrial–atmospheric carbon exchange. Such is the present rate of carbon removal from the atmosphere that at current rates the entire atmospheric volume of carbon in theory might be exchanged within about a decade. Of course, there are mixing problems and just as carbon is drawn down in the form of carbon dioxide it is also respired back as well as returned in other ways and places (such as the oceans), not to mention

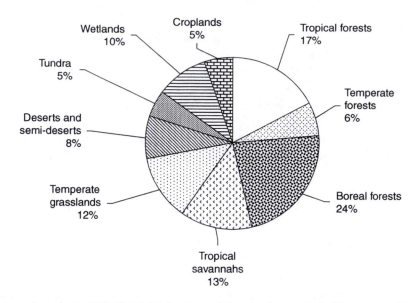

Fig. 7.15 The proportions of some 2.3 Tt of total global carbon stocks in soil and vegetation in different terrestrial biomes. After Royal Society (2001).

turned into other forms of carbon, such as methane. Another theoretical problem of this hypothetical situation would be that actually drawing down that much carbon would serve to draw carbon out of the oceans. This is one reason why the atmospheric residence time of adding a single molecule of carbon dioxide to the atmosphere is effectively between 50 and 200 years (see Table 1.1). Nonetheless, the idea of trapping (sequestering) carbon in biomass and soils is intriguing. Indeed, we know that roughly half the carbon released into the atmosphere through human action each year remains in the atmosphere. The other half is absorbed by vegetation, soils and the oceans, so sequestration already takes place somewhere in the biosphere.

The possibility that forests and soils might be deliberately manipulated to mitigate carbon dioxide emissions was recognised by the UN's Framework Convention on Climate Change (FCCC) in 1992 and the Kyoto Protocol (1997). (The convention and protocol will be discussed further in the next chapter.) Both emphasise natural terrestrial sinks rather than marine ones, largely because land carbon is easier to manipulate and also because of issues of ownership.

There is very roughly 2.3 Tt of carbon in vegetation and soils (and a further 1 Tt in the oceans' surface layers, to which we will return later): this is very much an approximate park figure for discussion purposes (see Figure 1.3). However, carbon is not distributed equally across terrestrial biomes (categories of regional communities of species). This is not only because some biomes contain more carbon per unit area than others but because some biomes are more extensive globally than others. Figure 7.15 provides a proportional breakdown showing in which terrestrial biomes the carbon can be found.

Forests and forest soils contain about 47% of terrestrial-biome carbon and so this form of carbon storage is attracting considerable attention. However, broadly speaking, forests that have been in existence for thousands of years (climax forest communities) tend to be greenhouse-neutral. ('Broadly speaking' because a forest's carbon balance is climate-dependent and will change with temperature and water availability; see Chapters 1, 5 and 6.) It is the establishment of new forests that has a net, short-term effect sequestering carbon from the atmosphere. If you like, this is the opposite of deforestation, which is the second anthropogenic factor contributing to the build-up of atmospheric carbon dioxide (see Table 1.3). This new forest necessity places a severe constraint on forestation as a form of carbon sequestration. Not only is land itself finite, it is also required for other uses. Furthermore, not all the Earth's land is suitable for forestation. The IPCC's second assessment report (IPCC, 1995) estimated that by 2050 some 60–87 GtC could be conserved or sequestered in forests. Compared to the above-estimated mean annual 21st-century B-a-U fossil fuel emissions of 18 GtC, this represents an annual saving of around 1.2–1.7 GtC, or around 6–9% of annual emissions.

Soils can also be managed to increase their carbon content (although we have a lot to learn about soil-carbon mobility and fixation). The potential here might at first appear quite considerable, especially as roughly three times the amount of carbon is stored in soils than as vegetation above ground. Indeed, as noted previously, sub-Arctic soils already hold a considerable volume of carbon (albeit a small proportion of terrestrial carbon in both vegetation and soils). Of the various soil types, peatlands represent a huge store of carbon (see Chapter 4). High-latitude peatlands (which include parts of tundra, boreal and semi-desert soils and wetlands), with between 180 and 455 GtC, represent up to about a third of the global soil carbon pool. Some 70 GtC has been sequestered since the LGM but the LGM was more than 15 000 years ago and the Holocene interglacial is only 11 700 years old, so this represents a small annual sequestration rate, which would be difficult to enhance to meaningful levels. So, despite being a large carbon pool, high-latitude soils do not have a commensurate potential for further sequestration. If anything, there is concern for the opposite. As with forests these high-latitude peatlands and permafrost soils are climate-sensitive. It is quite likely that they will release their store of carbon if warmed. In short, with global warming they could become a carbon source.

One added problem is that, while some high-latitude soil carbon has been sequestered since the end of the last glacial (15 000 years ago), some is very old carbon. In 2008 Canadian geologists led by Duane Froese and John Westgate reported that some ground ice in sub-Arctic Canada is more than 700 000 years old. The implication is that this ice and so indeed the carbon in these soils has survived past interglacials that were warmer than our own to date. If so, then our current warming, which is set to take us to global temperatures above that of even these warmer past interglacials, will release carbon that effectively has been out of the carbon cycle for 700 000 years or more.

Conversely, agricultural soils do have more potential. Soils that are currently used for agriculture are regularly ploughed, so bringing organic carbon to the surface and destroying plant root networks that physically trap carbon compounds. Ploughing also greatly aerates soils, so facilitating oxidation of carbon compounds. On the other hand natural and semi-natural grassland and forest systems are not subject to regular ploughing but additionally provide the soil with carbon compound inputs. The IPCC (2001a) notes that some agricultural land, such as set-aside (which represents some

10% of agricultural land in the EU), could be managed to enhance soil carbon. In total the IPCC's second assessment report (1995) estimated that up to 2050 some 23–44 GtC could be sequestered by agricultural soils, or around 2.6–5% of estimated mean annual 21st-century B-a-U emissions. This excludes the potential saving of fossil fuel emissions (mitigation) from biofuels.

To take one example of management options, an estimate of annual change in soil carbon stocks for Scotland between 2000 and 2009 is -810 ± 89 kt year^{-1}, equivalent to $0.037 \pm 0.004\%$ year^{-1}. It is thought that increasing the area of land-use change from arable to grass has the greatest potential to sequester soil carbon, and reducing the area of change from grass to arable has the greatest potential to reduce losses of soil carbon. Across Scotland, model-simulated changes in soil carbon from carbon-rich soils (C content $>6\%$) between 1950 and 2009 is -63 Mt, compared with -35 Mt from non-carbon-rich mineral soils; losses from carbon-rich soils between 2000 and 2009 make up 64% of the total soil carbon losses. One mitigation option that could be used in upland soils to achieve zero net loss of carbon from Scottish soils is to stop conversion of semi-natural land to grassland and increase conversion of grassland to semi-natural land by 125% relative to the present rate (Smith et al., 2010).

The IPCC's 2001 estimate for carbon mitigation from *both* vegetation and soil management up to the year 2050 was 100 GtC. This is equivalent to 10–20% of fossil fuel emissions estimated for that period. The IPCC (2001a) also notes that 'hypothetically', if all the carbon released due to historic land-use change (forest clearance and such) could be reversed, then projected 2100 atmospheric concentrations (approximately 800 parts per million (ppm) under B-a-U scenarios) could be reduced by 40–70 ppm. This last would be a one-off gain from the finite land area historically affected by land use to release carbon dioxide (see Chapter 5), for once a new forest is fully established it ceases to be a net absorber of carbon.

Managing the terrestrial short-term carbon cycle carbon can offset a small, albeit significant, proportion of carbon likely to be emitted from 21st-century fossil fuel burning. However, it is not risk-free. The more carbon built up in an ecosystem the more the likelihood of possible carbon leakage. Clearly, if an ecosystem is devoid of carbon then there is none to leak. On the other hand, a system rich in carbon, be it through peaty soils or woodland, has plenty of carbon that can escape. For instance, a forest fire can release nearly all the above-ground carbon extremely quickly, whereas ploughing grassland managed to maximise soil carbon can undo much of the benefit in a single season. Leakage can also take place in soils whose carbon pools have adjusted to the comparatively stable Holocene climate of the past 10 000 years or so, but which could well release their carbon with warming anticipated for the 21st century that takes them beyond temperatures seen in this past time (see Chapter 1).

Some research suggests that the state of the Earth system is such that soils' ability to be greenhouse gas sinks will be limited. Carbon dioxide promotes plant growth and plant roots help incorporate carbon into soils. This is all well and good, but there are other greenhouse gases in addition to carbon dioxide. Methane and nitrous oxide emissions from soils are anticipated to increase in a warmer world. In 2011, Kees van Groenigen, Craig Osenberg and Bruce Hungate reported that increased carbon dioxide (ranging from 463 to 780 ppmv) stimulates both nitrous oxide emissions from upland soils and methane emissions from rice paddies and natural wetlands. These emissions are expected to negate some (at least 16.6%) of the climate change mitigation potential previously predicted from an increase in the terrestrial carbon sink under increased atmospheric carbon dioxide concentrations due to increased plant growth. Their argument is that increased carbon dioxide leads to reduced plant transpiration (the evaporation of water from plant surfaces, and leaves in particular due to fewer and/or more closed stomata; see section 6.1.3), which in turn increases soil water content and so promotes the existence of anaerobic microsites in soils. This, together with increasing biological activity, probably stimulated denitrification and consequently nitrous oxide production. Also, the carbon dioxide-induced increase in root biomass may have contributed by increasing the availability of labile carbon, a crucial energy source for denitrification. The carbon dioxide induced stimulation of methane emissions from wetlands and rice paddies was probably the result of higher net plant production, leading to increasing carbon availability for substrate-limited methanogenic microorganisms. The problem is that both methane and nitrous oxide are more powerful greenhouse gases compared to carbon dioxide (see Table 1.2). (See also Knohl and Veldkamp, 2011.) This could be the making of yet another future critical transition resulting in crossing a climate threshold (section 6.6.8) in addition to that due to warming alone.

The potential for ecological carbon leakage should be of concern to policy-makers concerned with climate change. As we shall see in the next chapter, some countries (typically nations that use a lot of fossil fuels) have insisted that it be possible to trade, using permits, the right to emit carbon dioxide for carbon-sequestering ecosystem-management schemes. Trading in such greenhouse permits at best can have a short-term mitigating effect. However, it comes with a risk, and there is the question as to what would happen in this permit game if after a number of decades a forest – created, say, by funding from a fossil fuel station – burned down? Or if its soil carbon was released with global warming? How would the trading permit be paid back and the permit's environmental impact nullified? These questions have not been resolved. Indeed, the IPCC (2001a) warn that 'larger carbon stocks [in ecosystems] may pose a risk for higher CO_2 emissions in the future'. It says that 'if biological mitigation activities are modest, leakage is likely to be small. However, the amount of leakage could rise if biological activities became large and widespread'.

Boreal and tundra soils hold considerable carbon stocks, and of these, as noted in Chapters 1 and 4, high-latitude soils currently have up to a third of the global soil carbon pool. These, as per the IPCC warning, pose a risk for higher carbon dioxide emissions with warming. Already there is some experimental evidence suggesting that warming will result in carbon loss from these high-latitude soils and indeed carbon loss from temperate and tropical soils (see Chapter 1). Yet soil carbon loss may also increase another way as the climate warms.

The amount of carbon held in a soil should really be viewed as a balance between the rate of carbon entering the soil and the rate at which it leaves. Such a consideration should include the living plant dimension, as it is the photosynthetically driven primary productivity (plant growth) that largely draws down carbon, with plants creating roots within the soil and plant remains lying on the soil surface, from which carbon compounds can leach into the soil beneath. Therefore, if something were to hinder primary productivity then the balance of carbon flows would shift. In 2005 it was reported (Ciais et al., 2005) that this indeed happened in Europe during the 2003 heatwave. Then the July temperature in places was up to 6°C above long-term means and rainfall showed deficits of up to 30 cm year^{-1}. This extreme weather event impeded primary productivity and reversed these ecosystems' former net carbon sequestration.

The work (Ciais et al., 2005) was conducted as part of the EU CarboEurope research programme and, although not as detailed an assessment as many ecologists would like, as a partial snapshot it gives cause for thought. The researchers looked at the programme's monitoring of carbon dioxide, water and energy from one grassland and 14 forest sites for 2003 (the heatwave year) and 2002 (the control year). They also analysed crop-harvest data at country level and compared 2003's figures with the 1998–2002 annual averages. Finally, they linked these field data to a sophisticated ecosystem computer model that necessitated supercomputing facilities (which were provided by the French Commissariat à l'Energie Atomique). The results suggested that the 2003 heatwave months resulted in a 30% reduction in gross primary productivity for the year and that carbon release across Western Europe was about 500 GtC. Up to then it had (tentatively) been thought that Europe's soils had slowly been accumulating carbon. This carbon loss roughly equalled some 4 years of such accumulation. What was not known is the effect of this on the following year's carbon balance. However, what is known is that the 2003 European heatwave temperatures, while those of an extreme event early in the 21st century, are destined to be average summer temperatures later in the century (and that heatwave summers then will be correspondingly warmer; see Chapter 6). In short, Western Europe's 2003 carbon loss was not a one-off event.

All of this does not bode well for a system of trading of greenhouse permits that uses carbon sequestration in natural systems, because it assumes that these systems are known and can be taken for granted given a certain management regime in the future. As a way of mitigating fossil emissions, it can at the very best only be as good as however long is the commitment to ecosystem management. ('At the very best' because, as noted, a warmer world facilitates ecosystem carbon release regardless of commitment and present carbon savings could easily turn into future carbon losses). The success of such trading schemes also depends on our knowledge (currently not sufficiently complete) as to agricultural sinks and especially soils, and our ability to monitor them, as well as the schemes' robustness lest they be open to gamesmanship by fossil-energy producers and consumers. Nonetheless, in the 1990s and early 2000s many policy-makers welcomed such permit-trading schemes.

However, whereas permit-trading schemes that rely on soils as carbon sinks may not function properly, carbon sequestration can still play a role in helping (albeit temporarily) reduce the rate at which atmospheric carbon dioxide builds up. This last could help slow the rate of climate change and may (marginally) help both ecosystems and human systems adapt even if a hectare of ecosystem cannot be equated with the consumption of a specific amount of fossil fuel.

7.5.2 Manipulating marine photosynthesis

Compared to a very broad figure of around 50 GtC for annual net terrestrial biological productivity for a biotic carbon stock of 550 GtC, the oceans have a small biotic carbon stock of only around 3 GtC. Much of the 1 Tt of carbon in the oceans that was noted in the previous subsection is in inorganic (non-biotic) forms. The other big carbon difference between the terrestrial and marine biotic reservoirs is that, whereas terrestrial ecosystems have an annual carbon turnover of roughly 20% of the terrestrial biotic stock, the carbon turnover of the marine carbon stock is in excess of 1000%.

In other words, the terrestrial biota represent a large mass of biotic carbon with more modest carbon flows to and from this volume. Conversely, marine biota represent a small mass of carbon with a comparatively very high carbon turnover whereby the carbon ends up in the large inorganic carbon ocean reservoir. Managing the marine carbon cycle to offset atmospheric increases in carbon dioxide from fossil fuel burning depends on this naturally high turnover and indeed this is the attraction of such management.

As noted in section 1.3, there have been attempts to encourage algal blooms with the IronEx experiments. However, while successful in forming blooms, the energy required to make, transport and distribute the iron used to fertilise the ocean more or less balanced the fossil energy that was saved in carbon sequestration. Furthermore, the ecological implications should an IronEx-type scheme be implemented on a large scale are unknown. With regard to marine ecosystem manipulation, the IPCC (2001a) conclude that much remains 'unresolved' and that such schemes 'are not, therefore, ready for near-term application'.

7.5.3 Biofuels

Biofuels arise from living species and are greenhouse-neutral in that they sequester atmospheric carbon dioxide when grown and return it when burnt. However, if the products of biofuel consumption are captured (for example, from the flue of a biofuel power station) and stored, then there is a net transference of carbon from the fast to the deep carbon cycle (see section 8.2.4). This is one way in which carbon dioxide can be usefully sequestered from the atmosphere. Biofuels can also be considered to have a positive effect offsetting fossil carbon emissions when biofuels displace oil, gas and coal consumption. Here the transference of carbon from the deep carbon cycle to the atmosphere is reduced.

There is a continuum of agricultural biological resources from which biofuels may be generated. This continuum is essentially founded within the agriculture sector but also includes forestry. At one end of this spectrum the biofuel resources are derived almost by happenstance from agricultural wastes such as animal slurry (from piggeries, poultry wastes, etc.) and vegetative wastes (such as straw from cereal production). At the continuum's other end, biofuels can be derived from existing agricultural products such as sugars and oils. In between these there are both dedicated and dependent biofuel sources. Dedicated biofuel sources are those whose planting solely provides biofuels and nothing else of economic value. The tropical grass *Miscanthus sinensis*, for example, has no other real value as a crop in the USA or Europe (although it is used as an ornamental plant in the European domestic sector). Conversely, dependent biofuel sources are those that are dependent on another economic gain (commonly food-product-related) from growing the crop. However, the reality is that few biofuel sources fall strictly into either category, even though some may be predominantly one or the other. *Miscanthus*, for instance, is also being investigated as a possible UK crop for fibre production. Birch (*Betula* spp.), willow (*Salix* spp.) and alder (*Alnus* spp.) can contribute to biofuel production while continuing to provide stock for paper and fibre-board production. Since these are broadleaved, woody species, their appropriate cultivation adds substantially to the biodiversity of associated flora and fauna and offers visually attractive amenities during production.

As to the potential contribution of biofuels, taking the UK at the turn of the millennium as an example, if all 500 000 ha of UK set-aside land (farmland deliberately not used for agriculture, equivalent to 3% of the land *within* agricultural holdings) was planted with, for instance, *Miscanthus*, and assuming a median dry-matter production of 15 t ha^{-1} year^{-1} and a broadly typical 40% conversion to oil in energy-content terms, then some 3 mtoe of biofuel would result. This would be equivalent to nearly 4% of UK end-20th century/early-21st-century annual oil demand. If *Miscanthus* oil production could be increased to an optimistic 25 t of dry matter ha^{-1}, then more than 6% of UK oil demand would be met. However, *Miscanthus* does not reach its peak yields until the third or fourth season, even though absolute peak yields for *Miscanthus sacchariflorus* of 44 t of dry matter ha^{-1} year^{-1} have been reported from Denmark. This peak, if possible across all such potential UK *Miscanthus* production, would result in 14 or 15 t of oil ha^{-1} year^{-1}. However, such high yields are highly theoretical and cannot be considered a realistically obtainable national average at this time (Cowie, 2003), although they might with appropriate breeding or genetic modification if environmental concerns could be addressed.

A similar calculation for oilseed rape would take into account the approximately 3.2 t harvested per hectare (in the UK), the 37% of recoverable oil from the rapeseed and the 95% conversion to diesel. This gives us about 0.4 mtoe year^{-1} (0.5% of UK oil demand around the turn of the millennium). The advantages of rapeseed are that the plant's entire life cycle takes place in the UK, but has the disadvantages that methanol (equivalent to about 12% of fuel yield) is required during the diesel-making process; however, this methanol could come from biological sources too. It should be noted that there are rotational constraints on rapeseed production: rapeseed is usually grown one year in five in the UK.

Miscanthus and rapeseed oil are taken as just two examples of how currently unused set-aside land could contribute to make the UK more energy-secure through increasing energy self-sufficiency. However, it may be that some other crops, or crop mix, would be preferable. Climate conditions do not favour *Miscanthus* rhizome production in Europe as they do in the tropics. Nonetheless, rhizome production is possible in the UK (which is warmed by the Gulf Stream) and in some other parts of Europe. Consequently, the examples are illustrative and the production levels of *Miscanthus* and rapeseed oil are not atypical in energy terms of other biofuels.

It should be noted that in addition to the 3% of land within agricultural holdings being set aside, 97% is not. (Of this 24% is rough grazing and 11% unspecified. Indeed, of the 97% of agricultural land that is not set aside, only 25% is used for crops.) In short, there is room for some considerable increases in the land that might be available for biofuels. Assuming, solely for purposes of example, that 6% of current UK oil consumption was met through the use of specific biofuel crops, then there is the additional possibility of a biofuel contribution from dependent sources; that is, crops that are grown for other purposes. Two examples are straw (as mentioned above) or crop trimmings, such as from field vegetables. This last is currently on land covering over eight times the size of set-aside land. Indeed, much of the above-ground biomass that is not harvested for food or export (the waste biomass) from these species could be used for biofuel production. Finally, there are the biofuels available from forestry and woodland, such as from short-rotation coppicing.

In 2002 the Institute of Biology ran a biofuels workshop in London as part of an energy series run also by the Institute of Physics and Royal Society of Chemistry, which was sponsored by the Department of Trade and Industry. Workshop participants, all associated with either the biofuel industry or biofuel research, were asked in advance as to the potential for biofuels to offset UK oil demand. The modal answer was around 20 ± 5% of UK oil demand or about 7% of total UK primary energy consumption. If this estimate is realistic then it represents some 16 million t of UK domestic annual oil consumption and a roughly £1.3 billion gain to the UK's domestic economic output at the point of refinement/production (not point of sale) in terms of crude-oil savings. This 7% figure of UK primary energy goes a long way towards meeting the government's medium-term renewable targets, although the UK will require greater displacement of fossil fuel than that if it is to meet its Kyoto targets. Even so, it is difficult to see any fossil fuel-reducing strategy in the medium-to-long term in the UK that does not include biofuels (Cowie, 2003).

The estimate for a practical biofuel contribution to the UK of some 15–25% of current oil consumption, or 7% of UK primary energy, is not at all unrealistic. True, while a greater proportion may well be feasible, equally it might perhaps incur significant opportunity costs (probably either to strategic food production and/or the environment). The importance of biofuels to the UK would be best appreciated if undertaken as part of a coherent energy strategy, along with improved energy efficiency and changes of behaviour favouring lower energy consumption. The question is whether a UK government would consider a firm drive for biofuels to be a worthy goal in economic, environmental and strategic terms. It is also a question other governments will need to face. The concerns raised by the workshop that the UK was not coherently exploring biofuel options were subsequently echoed by a House of Commons Select Committee report (Environment, Food and Rural Affairs Select Committee, 2003; see also the UK policy case study in section 8.3.3).

Finally, the 2002 Institute of Biology biofuels workshop looked at costs. The range of figures presented did result in an end biofuel cost per barrel that was of the same order of magnitude as that for crude oil (which still has refining costs to bear). Unfortunately the current vagaries of small-scale production, lack of distribution and other factors made a firm costing difficult. Nonetheless, biofuels remain a useful option as part of a suite of measures to counter fossil fuel emissions of carbon dioxide.

Globally, aside from non-commercial local use of biofuels (see section 8.2.3) a little over 1% of commercially traded energy currently comes from biomass. Of all countries, Brazil arguably leads in biofuel production and consumption with biofuels contributing some 20% of its primary energy supply. (However, questions need to be asked as to how much of this comes from biologically sustainable sources that do not contribute net greenhouse emissions from climate change.) While biomass could theoretically make a major contribution to offsetting fossil fuel consumption it is more likely to be used to meet chemical feedstock demand as oil becomes more expensive, especially as there are competing land pressures for food production. It has been calculated that if, for example, the USA were to displace 10% of its car petrol consumption with biofuel then 12% of US cropland would be required, but a further 38% would be needed (bringing the total to nearly 50% of currently available cropland) once the energy requirements for processing, harvesting and transportation are taken into account (Kheshgi et al., 2000). Here the equation – the balance between land for food and/or biofuel and/or chemical feedstock – is a complex one. The world has already seen how in 2007–8 biofuels were a contributing factor to a world food price increase (see section 7.4.1) and in an evermore populous world (see section 7.1.2) the food 'opportunity cost' of biofuels will be high. Indeed, such are the energy factors of growing, harvesting and refining biofuels that their use as fuels (to simply burn) may be limited compared to use as chemical feedstocks. However, where high energy content liquid fuel use has few alternatives there may be a role for biofuels. The obvious example is aviation fuel. The complexity of the policy issues was underlined in 2008 by a Royal Society report, *Sustainable Biofuels: Prospects and Challenges*, which warned of the need to avoid the untold consequences of solving one problem at the expense of another.

If demand for food increases (as it almost inevitably will) and there is also pressure for biofuels, then natural systems may be converted to farmland for biofuel production. Converting forest and grassland to new cropland to replace the grain (or cropland) diverted to biofuels could end up generating more greenhouse gas than the reductions that are created by replacing fossil carbon with biofuel. In 2008 US researchers Timothy Searchinger and Ralph Heimlich used a worldwide agricultural model to estimate emissions from land-use change. They found that corn-based ethanol, instead of producing a 20% saving, nearly doubles greenhouse emissions over 30 years and increases greenhouse gases for 167 years.

Consequently, it is only likely that commercial biofuels could possibly make a small but significant contribution to meeting future energy needs within an early to mid 21st-century window of opportunity (see Table 8.2). This would be before the preference tends to food production, unless in a minority of instances there is a compelling case for biofuel use.

7.6 Summary and conclusions

Human ecology is different from the ecology of any other species in that sentience has enabled a shaping of *H. sapiens*' relationships with other species. This shaping has often been with disregard to possible ecological and environmental side effects (anthropogenic global warming itself being just one instance). This sentient dimension to the shaping of human ecology has been enhanced since the Renaissance and the subsequent Industrial Revolution and this in turn is reflected in the growth in human numbers and the quantity of energy harnessed by our species, both globally and on a per-capita basis. Yet the environmental resources commensurately consumed in the process have impacted, and will continue to impact, markedly on natural systems on which our species continues to depend. Human numbers and projected population growth for the 21st century are such that, *irrespective* of climate change, a more rational and efficient use of resources is required. This can in part be achieved with an understanding of the carbon cycle and related ecology. Climate change policies coincide with more rational and efficient environmental resource use so that irrespective of global warming it would be prudent to pursue such policies (Cowie, 1998). The history of international environmental policies, the possible measures that might be taken to address climate change concerns, and international climate change policies will be summarised in the next chapter.

7.7 References

Allan, R. P. (2011) Human influence on rainfall. *Nature*, 470, 244–5.

Allan, R. P. and Soden, B. J. (2008) Atmospheric warming and the amplification of precipitation extremes. *Science*, 321, 1481–4.

Allen, M. R., Frame, D. J., Huntingford, C. et al (2009). Warming caused by cumulative carbon emissions towards the trillionth tonne. *Nature*, 458, 1163–6.

Anon (2012) Malaria deaths. *Nature*, 482, 137.

Balmford, A., Bruner, A., Cooper, P. et al. (2002) Economic reasons for conserving wildlife. *Science*, 297, 950–3.

Becker, A. and Grunewald, U. (2003) Flood risk in Central Europe. *Science*, 300, 1099.

Bengtsson, L. (2001) Hurricane threats. *Science*, 293, 440–1.

Bentham, G. (1999) Direct effects of climate change on health. In Haines, A. and McMichael, A. J., eds., *Climate Change and Human Health*, pp. 35–44. London: Royal Society.

Bentley, R. W. (2002) Global oil and gas depletion: an overview. *Energy Policy*, 30, 189–205.

BP Economics Unit (2005) *BP Statistical Review of World Energy*. London: British Petroleum Corporate Communications Services.

BP Economics Unit (2011) *BP Statistical Review of World Energy*. London: British Petroleum Corporate Communications Services.

Brown, L. (1995) *Full House: Reassessing the Earth's Population Carrying Capacity*. London: Earthscan.

Burroughs, W. J. (1997) *Does The Weather Really Matter? The Social Implications of Climate Change*. Cambridge: Cambridge University Press.

Cabinet Office (1994) *Sustainable Development: The UK Strategy*. Cm 2426. London: HMSO.

Cardillo, M., Mace, G. M., Jones, K. E. et al. (2005) Multiple causes of high extinction risk in large mammal species. *Science*, 309, 1239–41.

Ceballos, G., Ehrlich, P. R., Soberón, J., Salazar, I. and Fay, J. P. (2005) Global mammal conservation: what must we manage? *Science*, 309, 603–7.

Chichilnisky, G. and Heal, G. (1998) Economic returns from the biosphere. *Nature*, 391, 629–30.

Christensen, J. H. and Christensen, O. B. (2002) Severe summertime flooding in Europe. *Nature*, 421, 805.

Ciais, P., Reichstein, M., Viovy, N. et al. (2005) Europe-wide reduction in primary productivity caused by the heat and drought in 2003. *Nature*, 437, 529–33.

Coates, S. and Cohen, T. (2003) Weird and dangerous: the side-effects of a heatwave. *The Times*, 8 August.

Cohen, J. E. (1995) *How Many People Can the Earth Support?* New York: Norton.

Costanza, R., d'Arge, R., de Groot, R. et al. (1997) The value of the World's ecosystem services and natural capital. *Nature*, 387, 253–60.

Cottrell, F. (1955) *Energy and Society*. Westport, CT: Greenwood Press.

Cowie, J. (1998) *Climate and Human Change: Disaster or Opportunity?* London: Parthenon Publishing.

Cowie, J. (ed) (2003) *Fuelling the Future 3: Biofuels*, 2nd edition workshop report. London: Institute of Biology and the British Council for Crop Protection.

Department of Health (2002) *Health Effects in the UK*. London: Department of Health.

Department of Heath (2004) *West Nile Virus: a Contingency Plan to Protect the Public's Health*. London: Department of Health.

Department of Health and the Health Protection Agency (2008) *Health Effects of Climate Change in the UK 2008: An Update of the Department of Health Report 2001/2002*. London: Department of Health and Health Protection Agency.

Donaldson, C. G., Tchernjavskii, V. E., Ermakov, S. P., Bucher, K. and Keatinge, W. R. (1998) Winter mortality and cold stress in Yekaterinburg, Russia interview survey. *British Medical Journal*, 316, 514–18.

Ehrlich, P. R. and Holdren, J. P. (1971) Impact of population growth. *Science*, 171, 1212–17.

Environment Agency (2010) *The Costs of the Summer 2007 Floods in England*. Bristol: Environment Agency.

Environment, Food and Rural Affairs Select Committee (2003) *Biofuels*. Norwich: House of Commons EFRA Select Committee, Stationery Office.

Food and Agriculture Organization (2001, 2003, 2004, 2008, 2009, 2010, 2011) *The State of Food Insecurity in the World* [various annual reports]. Rome: UN Food and Agriculture Organization.

Food and Agriculture Organization (2002) *FAOSTAT Statistical Database*. Rome: UN Food and Agriculture Organization.

Foresight Programme (2008) *Foresight Future Flooding*. London: Office of Science and Technology.

Froese, D. G., Westgate, J. A., Reyes, A. V. et al. (2008) Ancient permafrost and a future, warmer Arctic. *Science*, 321, 1648.

Goldenberg, S. B., Landsea, C. W., Mestas-Nuñez, A. M. and Gray, W. M. (2001) The recent increase in Atlantic hurricane activity: causes and implications. *Science*, 293, 474–9.

Hallam, T. (2004) *Catastrophes and Lesser Calamities*. Oxford: Oxford University Press.

Hay S. J., Cox, J., Rogers, D. J. et al. (2002) Climate change and resurgence of malaria in the East African highlands. *Nature*, 415, 905–9.

Imhoff, M. L., Bounaoua, L., Ricketts, T. et al. (2004) Global patterns in human consumption of net primary production. *Nature*, 429, 870–3.

Ingram, I., Ericksen, P. and Liverman, D. (eds) (2010) *Food Security and Global Environmental Change*. London: Earthscan.

Intergovernmental Panel on Climate Change (1990) *Climate Change: the IPCC Scientific Assessment*. Cambridge: Cambridge University Press.

Intergovernmental Panel on Climate Change (1995) *Climate Change 1995: the Science of Climate Change*. Cambridge: Cambridge University Press.

Intergovernmental Panel on Climate Change (2000) *Emission Scenarios – IPCC Special Report*. Geneva: World Meteorological Organization.

Intergovernmental Panel on Climate Change (2001a) *Climate Change 2001: Impacts Adaptation and Vulnerability – a Report of Working Group II*. Cambridge: Cambridge University Press.

Intergovernmental Panel on Climate Change (2001b) *Climate Change 2001: the Scientific Basis – Summary for Policymakers and Technical Summary of the Working Group I Report*. Cambridge: Cambridge University Press.

Intergovernmental Panel on Climate Change (2007a) *Climate Change 2007: the Physical Science Basis – Working Group I Contribution to the Fourth Assessment of the IPCC*. Cambridge: Cambridge University Press.

Intergovernmental Panel on Climate Change (2007b) *Climate Change 2007: Impacts, Adaptation and Vulnerability – Working Group II Contribution to the Fourth Assessment of the IPCC*. Cambridge: Cambridge University Press.

Isbell, F., Calcagno, V., Hector, A. et al. (2011) High plant diversity is needed to maintain ecosystem services. *Nature*, 477, 199–203.

Johnson, C. G. and Smith, L. P. (1964) *The Biological Significance of Climatic Changes in Britain*. London: Institute of Biology and Academic Press.

Keatinge, W. R., Donaldson, G. C., Cordioli, E. et al. (2000) Heat related mortality in warm and cold regions of Europe: observational study. *British Medical Journal*, 321, 670–3.

Keilman, N. (2003) The threat of small households. *Nature*, 421, 489–533.

Kerr, R. A. (2004) A few good climate shifters. *Science*, 306, 599–600.

Kheshgi, H. S., Prince, R. C. and Marland, G. (2000) The potential of biomass fuels in the context of global climate change. *Annual Review of Energy and the Environment*, 25, 199–244.

Knohl, A. and Veldkamp, E. (2011) Indirect feedbacks to rising CO_2. *Nature*, 475, 177–8.

LaDeau, S. L., Kilpatrick, A. M. and Marra, P. P. (2007) *West Nile virus emergence and large-scale declines of North American bird populations*. Nature, 447, 710–13.

Lenton, T. and Watson A. (2011) *Revolutions that made the Earth*. Oxford: Oxford University Press.

Lobell, D. B., Schlenker, W. and Costa-Roberts, J. (2011) Climate trends and global crop production since 1980. *Science*, 333, 616–20.

Markham, S. F. (1944) *Climate and the Energy of Nations*, revised edition. London: Oxford University Press.

Marsh, T. and Hannaford, J. (2008) *The Summer 2007 Floods in England & Wales: A Hydrological Appraisal*. Wallingford, Oxford: Centre for Ecology & Hydrology.

Martens, P. (1999) Climate change impacts on vector-borne disease transmission in Europe. In Haines, A. and McMichael, A. J., eds., *Climate Change and Human Health*, pp. 45–53. London: Royal Society.

Maugeri, L. (2004) Oil: never cry wolf – why the petroleum age is far from over. *Science*, 304, 1114–16.

McMichael, A. J. (1993) *Planetary Overload: Global Environmental Change and the Health of Human Species*. Cambridge: Cambridge University Press.

McMichael, A. J. and Haines, A. (1997) Global climate change: the potential effects on health. *British Medical Journal*, 315, 805–9.

McMullan, J. T. Morgan, R. and Murray, R. B. (1977) *Energy Resources*. London: Edward Arnold.

Meadows, D. H., Meadows, D. L., Randers, J. and Behrens, III, W. W. (1972) *The Limits to Growth*. London: Pan.

Meehl, G. A. and Tebaldi, C. (2004) More intense, more frequent, and longer lasting heat waves in the 21st century. *Science*, 305, 994–7.

Min, S.-K., Zhang, X., Zwiers, F. W. and Heger, G. C. (2011) Human contribution to more-intense precipitation extremes. *Nature*, 470, 378–81.

Mooney, H., Cropper, A. and Reid, W. (2005) Confronting the human dilemma. *Nature*, 434, 561–2.

Mudelsee, M., Börngen, T. and Grünewald, U. (2003) No upward trends in the occurrence of extreme floods in central Europe. *Nature*, 425, 166–9.

Murray, C. J. L., Rosenfeld, L. C., Lim, S. S., Andrews, K. G. et al. (2012) Global malaria mortality between 1980 and 2010: a systematic analysis. *Lancet*, 379, 413–31.

Myrskylä, M., Kohler, H.-P. and Billari F. C. (2009) Advances in development reverse fertility declines. *Nature*, 460, 741–3.

National Health Service (2004) *Heatwave Plan for England: Protecting Health and Reducing Harm from Extreme Heat and Heatwaves*. London: Department of Health.

Nature News (2004) Reform of land use urged as floodwaters rise across Asia. *Nature*, 430, 596.

Osborn, T. J., Hulme, M., Jones, P. D. and Basnett, T. A. (2000) Observed trends in the daily intensity of United Kingdom precipitation. *International Journal of Climatology*, 20, 347–64.

Pall, P., Aina, T., Stone, D. A. et al. (2011) Anthropogenic greenhouse gas contribution to flood risk in England and Wales in autumn 2000. *Nature*, 470, 382–6.

Palmer, T. (2006) Making the paper: how climate data are helping to predict malaria outbreaks in Africa. *Nature*, 439, xi.

Parliamentary Office of Science and Technology (2008) *Ecological Networks*. No. 300. www.parliament.uk/parliamentary_offices/post/pubs2008.cfm.

Patz, J. A., Gibbs, H. K., Foley, J. A., Rogers, J. V. and Smith K. R. (2007) Climate change and global health: quantifying a growing ethical crisis. *EcoHealth*, 4, 397–405.

Phalan, B., Onial, M., Balmford, A. and Green, R. E. (2011) Reconciling food production and biodiversity conservation: land sharing and land sparing compared. *Science*, 333, 1289–91.

Pierce, J. L., Meyer, G. A. and Jull, A. J. T. (2004) Fire-induced erosion and millennial-scale climate change in northern ponderosa pine forests. *Nature*, 432, 87–90.

Porter, J. R. (2005) Rising temperatures are likely to reduce crop yields. *Nature*, 436, 174.

Raleigh, V. S. (1999) World population and health in transition. *British Medical Journal*, 319, 981–3.

Roberts, I., Stott, S. and the Climate and Health Council Executive (2010) Doctors and climate change. *British Medical Journal*, 341, c6357.

Rogers, D. J. and Randolph, S. E. (2000) The global spread of malaria in a future, warmer world. *Science*, 289, 1763–5.

Rojstaczer, S., Sterling, S. M. and Moore, N. J. (2001) Human appropriation of photosynthesis products. *Science*, 294, 2549–51.

Royal Society (2001) *The Role of Land Carbon Sinks in Mitigating Global Climate Change*. Policy document 10/01. London: Royal Society.

Royal Society (2008) *Sustainable Biofuels: Prospects and Challenges*. Policy document 10/08. London: Royal Society.

Schroter, D., Cramer, W., Leemans, R. et al. (2005) Ecosystem service supply and vulnerability to global change in Europe. *Science*, 310, 1333–7.

Searchinger, T., Heimlich, R., Houghton, R. A. et al. (2008) Use of U.S. croplands for biofuels increases greenhouse gases through emissions from land-use change. *Science*, 319, 1238–40.

Simon, J. (1996) *The Ultimate Resource 2*. Princeton, NJ: Princeton University Press.

Smith, J., Gottschalk, P. and Bellarby, J. et al. (2010) Estimating changes in Scottish soil carbon stocks using ECOSSE II application. *Climate Research*, 45, 193–205.

Stanwell-Smith, R. (1999) Effects of climate change on other communicable diseases. In Haines, A. and McMichael, A. J., eds., *Climate Change and Human Health*, pp. 55–69. London: Royal Society.

Starke, L. (ed) (1999) *Vital Signs: 1999–2000*. London: Earthscan.

Sustainable Development Commission (2006) *The Role of Nuclear Power in a Low Carbon Economy*. London: Sustainable Development Commission.

TEEB (2009) *Climate Issues Update*. Brussels: European Communities.

Thomson, M. C., Doblas-Reyes, F. J., Mason, S. J. et al. (2006) Malaria early warnings based on seasonal climate forecasts from multi-model ensembles. *Nature*, 439, 576–9.

Turner, II, B. L., Clark, W. C., Kates, R. W. et al. (1990) *The Earth as Transformed by Human Action: Global and Regional Changes in the Biosphere over the Past 300 Years*. Cambridge: Cambridge University Press.

Turner, W. R., Brandon, K., Brooks, T. M. et al. (2007) Global conservation of biodiversity and ecosystem services. *Bioscience*, 57(10), 868–73.

UN Population Division of the Department of Economic and Social Affairs (2010) *World Population Prospects: The 2010 Revision*. New York: United Nations Secretariat.

Vafeidis, A., Neumann, B., Zimmermann, J. and Nicholls, R. J. (2011) *Migration and Global Environmental Change: MR9: Analysis of land area and population in the low-elevation coastal zone (LECZ)*. London: Foresight.

van Groenigen, K. J., Osenberg, C. W. and Hungate, B. A. (2011) Increased soil emissions of potent greenhouse gases under increased atmospheric CO_2. *Nature*, 475, 214–17.

Vitousek, P. M., Mooney, H. A., Lubchenco, J. and Melillo, J. M. (1997) Human domination of Earth's ecosystems. *Science*, 277, 494–9.

Whitlock, C. (2004) Forest, fires and climate. *Nature*, 432, 28–9.

World Health Organization (1999) *El Niño and Health: Report of the Task Force on Climate and Health*. Geneva: World Health Organization. [Summarised in WHO Fact sheet 192 (2000)].

World Health Organization (2001a) *Climate and Health*. Fact sheet 266. Geneva: World Health Organization.

World Health Organization (2001b) *Water – Too Much or Too Little – The Foremost Cause of Natural Disasters*. WHO feature no. 203. Geneva: World Health Organization.

World Health Organization (2004) *Surge in Demand Leads to Shortage of Artemisinin- based Combination Therapy for Malaria*. Media release WHO/77. Geneva: World Health Organization.

World Health Organization (2005) *Ecosystems and Human Well-Being: Health Synthesis*. Geneva: World Health Organization.

World Health Organization (2006) *WHO Announces Pharmaceutical Companies Agree To Stop Marketing Single-Drug Artemisinin Malaria Pills*. News release WHO/23. Geneva: World Health Organization.

World Health Organization (2008) *The Impact of Climate Change on Human Health*. Statement WHO/5. Geneva: World Health Organization.

World Health Organization (2011) *World Malaria Report 2011*. Geneva: World Health Organization. http://apps.who.int/malaria/world_malaria_report_2011/en/index.html.

World Health Organization/Food and Agriculture Organization (2004) *Minimizing Food-borne Illness Requires Co-ordination Right up the Food Chain*. Joint news release WHO/FAO/71. Geneva/Rome: World Health Organization/Food and Agriculture Organization.

Worldwatch Institute (2003) *Vital Signs 2003–2004: the Trends that are Shaping Our Future*. London: Earthscan.

Yu, G., Shen, H. and Liu, J. (2009) Impacts of climate change on historical locust outbreaks in China. *Journal of Geophysical Research*, 114, D18104.

Sustainability and policy

Having an understanding of climate change is one thing, but relating it to real-world development is another. In the latter half of the 20th century it became apparent to politicians that human impacts on the environment were sufficiently detrimental that they undermined the sustainability of human well-being, and hence environmental quality. 'Human well-being' is a catch-all term relating to material and cultural standards as well as quality of life.

Many of these terms, while having a clear meaning to Western politicians and policy-makers, have no strict definition or individual basis of quantitative indexing in the strict scientific sense, although in some instances attempts have been made. Other terms have been used so much by the media that they are often used in policy-making, although human ecologists are often more precise. For example, 'carbon footprint' is misleading as it generally does not include biofuels (which *are* carbon-based) and it also seemingly relates to the spatial concept of ecological footprint as opposed to a quantitative dimension of carbon mass. Academic comment on misleading terms and usage has reached the highest impact-factor journals (for example, see Hammond, 2007). A better term is fossil carbon burden, or fossil burden for short. Indeed some terms (such as medieval climatic optimum, climate tipping point and fertility that were discussed in earlier chapters, and zero carbon, which will be mentioned in section 8.5.1) are not only misleading but mean different things to different people. In 2009, with regards to the term 'carbon neutral' the UK government's Department of Energy and Climate Change even held a formal consultation on the term's usage (Department of Energy and Climate Change, 2009).

'Sustainability' itself, as we shall see, does have a specific definition that is enshrined in international agreements, and sustainability is affected by climate change. The history of what politicians mean by sustainability, and how current climate change issues affect it, are central to developing climate change and human ecology from topics of academic interest to those of application. The history of sustainability and climate change also provides a valuable lesson as to how long it takes to recognise a problem scientifically and to get it into policy discussions, before generating legislation and international agreements and then achieving a policy goal with tangible results.

Table 8.1 Summary of Stockholm's declaration of principles

1. Human rights should be asserted, and apartheid, colonialism, etc., should be condemned.
2. Natural resources must be safeguarded.
3. The Earth's capacity to produce renewable resources must be maintained.
4. Wildlife must be safeguarded.
5. Non-renewable resources must be shared and not exhausted.
6. Pollution must not exceed the environment's capacity to clean itself.
7. Damaging oceanic pollution must be prevented.
8. Development is needed to improve the environment.
9. Developing countries therefore need assistance.
10. Developing countries need reasonable prices for exports to carry out environmental management.
11. Environment policy must not hamper development.
12. Developing countries need money to develop environmental safeguards.
13. Integrated development planning is needed.
14. Rational planning should resolve conflicts between environment and development.
15. Human settlements must be planned to eliminate environmental problems.
16. Governments should plan their own appropriate population policies.
17. National institutions must plan development of states' natural resources.
18. Science and technology must be used to improve the environment.
19. Environmental education is essential.
20. Environmental research must be promoted, especially in developing countries.
21. States may exploit their natural resources as they wish but must not endanger others.
22. Compensation is due to states thus endangered.
23. Each nation must establish its own standards.
24. There must be cooperation on international issues.
25. International organizations should help to improve the environment.
26. Weapons of mass destruction must be eliminated.

8.1 Key developments of sustainability policy

8.1.1 UN Conference on the Human Environment (1972)

The 1972 Stockholm Conference on the Human Environment was the first of a series of UN conferences that arguably developed international environmental policy significantly. It was attended by 113 nations, but due to a dispute on the status of East Germany the USSR and all the Eastern European nations (many of which now are considered Central European), bar Romania, refused to participate.

The idea for the conference had been around since the end of the 1960s and Sweden played a leading role in much of the preparatory work. Sweden was at the time suffering the effects of acid rain and so realised that many environmental problems could only be tackled by international action. Not surprisingly, the theme of Only One Earth arose again and again throughout the conference.

Three things emerged from the conference: a proclamation, a list of principles (Table 8.1) and an action plan.

The proclamation is memorable for one sentence included by the head of the Chinese delegation, Tang Ke: 'We hold that of all things in the World, people are the most precious'.

The declaration consisted of 26 principles in all. The UN Environment Programme (UNEP) that arose as a consequence of the conference later adopted some of these. The UNEP, you will recall (Chapter 5), was to be the UN agency under the auspices of which (along with the UN World Meteorological Organization; WMO) the Intergovernmental Panel on Climate Change (IPCC) was established 16 years later. Of relevance to anthropogenic climate change the declarations included nations must be free to exploit their own resources, no nation should endanger others, compensation is due if they do, each nation should plan its own population policy, non-renewable resources must be shared by all and environmental concerns must not infringe human rights. Today anthropogenic climate change is being caused mostly by the developed nations and this is endangering others, through both climate change and sea-level rise. Fortunately, for a number of the participating states, the UN conference's principles are not enshrined in international law, for if they were there would undoubtedly be severe economic repercussions. Conversely for others, at the sharp end of anthropogenic climate change, the principles provide no protection.

The conference's action plan consisted of 109 recommendations covering, albeit broadly and loosely, all areas of environmental concern. Eighteen recommendations (numbers 1–18) related to human settlements, 51 recommendations (19–69) to natural resources, 16 recommendations (70–85) to pollution generally, nine recommendations (86–94) to marine pollution, seven recommendations to education and culture (95–101) and eight recommendations (102–109) to environment and development. Climate change was covered mainly by the general pollution recommendations and much of the science to address this was either reviewed or conducted by the WMO following the conference. However, recommendation 70, which called for countries to consult with other nations before doing anything that might affect the climate, has been an abject failure. This was recognised by many (well before the 1990 IPCC report), including the International Institute for Environment and Development, whose briefing a decade on from Stockholm said that 'provisions for this have become *seriously unstuck*' (the italics represent the source's own emphasis; Clarke and Timberlake, 1982). In July 1980 the UNEP wrote to all nations submitting the provisions for co-operation in weather modification. A year later only nine replies had been received, of which six were simply acknowledgements of receipt of correspondence; just three accepted the proposals.

The decade following the conference saw the development of some climate models that were basic by today's standards. It has to be remembered that at that time many large computers used by industry and academia had *less* computer power than today's home PCs. Indeed, up to 1982 PCs were extremely rare in homes and prior to 1980 they were virtually unheard of domestically. Those that became available early in the 1980s had memory capacities that were measured in kilobytes, not the gigabytes of today. Similarly, the models that academics used were very elementary. Nonetheless, back then most models predicted a rise of between 1.5 and 3°C for a doubling of carbon dioxide (Clarke and Timberlake, 1982), which compares quite favourably with the IPCC's first assessment in 1990 (IPCC, 1990). All of this demonstrates that it takes a leap of orders of magnitude in climate-model complexity and data processing to realise relatively modest improvements in detailed output and reductions in uncertainty.

8.1.2 The Club of Rome's Limits to Growth (1972)

In 1968, 30 individuals – scientists, educators, economists, humanists, industrialists and civil servants – gathered in the Accademian dei Lincei in Rome. It was the first meeting of the Club of Rome and they were there to discuss the present and future predicament of humans.

Jay Forrester of the Massachusetts Institute of Technology (MIT) had devised a computer model of the global human population and its resource consumption, which incorporated not just population levels, but industrial output, food availability, pollution and finite resource levels. Again, by today's standards it was a very simple model. Then in 1972, the same year as the Stockholm Conference, an MIT team led by Dennis Meadows produced a report based on runs of the Forrester World Model for the Club of Rome. This was *The Limits to Growth* report (Meadows et al., 1972).

The Limits to Growth report made headline news in many countries and in turn this spawned numerous documentaries and books. Yet it also came under severe criticism from a number of academics and was misunderstood by many. In essence it reflected the idea of the Reverend Thomas Robert Malthus that arithmetic (exponential) growth in population and resource consumption tends to outstrip additive (linear) growth in the ability to harvest resources (Malthus, 1798), so that eventually checks come into play that reduce population. These checks include famine, pestilence, disease and war, but also include 'moral restraint'. *The Limits to Growth* report argued that human population growth and per-capita resource consumption cannot continue to increase indefinitely, so that if we were to avoid some sort of collapse then 'entirely new approaches are required to redirect society towards goals of equilibrium rather than growth'.

The Limits to Growth's many critics point to the fact that the decades since the report have not shown the collapse in resource use and society that were predicted. This is certainly true. To take just one example from the report, copper was predicted to run out in 36–48 years' time and gold in just 11–29 years from when the report was written, in 1972. This meant that by today gold should have run out while copper would be on the point of being exhausted. This has not happened, so clearly, the critics say, *The Limits to Growth* analysis must be wrong.

However, such criticisms completely miss the point. *The Limits to Growth* report pointed to a mismatch between resource consumption and supply that would come about unless circumstances changed, hence the need for new approaches. Indeed, in part what we got were new approaches. To continue with the examples of copper and gold, the situation regarding the development and use of these resources has improved. Change has taken place. Within a decade of the publication of *The Limits to Growth*, copper mining and ore-refining techniques had improved so much that the spent tailing mounds from past copper mining were themselves being mined routinely for copper. Similarly, technological improvements had been made that enhanced the extraction of gold and also the uses to which gold had been put changed; indeed, its importance as an international basis for monetary exchange diminished, so freeing the metal for alternative use.

Twenty years on from the original report its authors published a follow-up (Meadows et al., 1992) that included a rebuttal of many of the criticisms levelled at them. This included pointing out that they did not seek to provide 'a warning about the future [that] is a prediction of doom' but 'a warning . . . to follow a different path'. They also emphasised that they were not making a judgement as to whether 'growth' per se was all good or all bad but that what was required was 'sustainable development'. The 20-year follow-up pointed out that the future held a range of options, but that considerable change was still required in the management of a number of key resources (including fossil fuels, water and food supply). They also pointed out that at the time of the UN's Stockholm Conference the number of nations with environment ministries numbered no more than 10, whereas by 1992 it was around 100. In short, the follow-up report urged for continued efforts to develop sustainably. To be fair to the critics of *The Limits to Growth*, the first report did have (rightly or wrongly) an alarmist tone, whereas the 20-year follow-up was more measured. However, in equal fairness to the authors, the first report did attract considerable media attention, something at which the 20-year follow-up was not nearly so successful.

8.1.3 World Climate Conference (1979)

Several international conferences on climate and global warming were held in the 1970s following Stockholm. Arguably one of the most significant was the 1979 World Climate Conference organised by the WMO. It concluded, echoing Stockholm, that the carbon dioxide problem deserved the most urgent of attention by the international community of nations. Nonetheless, following the conference, little discernable action was taken by politicians.

8.1.4 *The World Conservation Strategy* (1980)

In 1980 the International Union for the Conservation of Nature (IUCN), the World Wildlife Fund (WWF) and the UNEP came together to produce the *World Conservation Strategy*. It was a useful document, slim and to the point, that outlined the central tenets of sustainable development from an ecological perspective. It took the Stockholm Conference's recognition of economic endeavours and environmental conservation as being two sides of the sustainable development coin, but went on to cite key principles and current priorities to realise sustainable development. Its three core principles were:

- to maintain ecological processes and life-support systems,
- to preserve genetic diversity and
- to ensure the sustainable utilisation of species and ecosystems.

The strategy warned, similar to *The Limits to Growth*, that the planet's ability to support its growing human population was being undermined through destruction of (biotic) resources. Importantly, the strategy stressed the fundamental need for 'genetic diversity' and this was a precursor to the popular concept of 'biodiversity' that was not formalised (strictly in policy terms) until the 1987 Brundtland Report (see section 8.1.6). However, it was the marrying of biological conservation ideas (and not just broad environmental concerns, as with Stockholm) with development issues that made the report stand out. This fact, and that it sprang from three major international bodies, meant that it attracted significant press coverage worldwide. For example, in the UK *The Times* ran a major feature on aspects of the strategy each day of the week of its launch, while the final programme of David Attenborough's *Life on Earth* TV series, devoted to the future of wildlife and humans, specifically cited the *Strategy* and its conclusions.

There was in 1991 a follow-up to the *World Conservation Strategy* called *Caring for the Earth: a Strategy for Sustainable Living* (IUCN, WWF and UNEP, 1991). Yet despite a simultaneous launch in several capital cities (the one in the UK had the Duke of Edinburgh as the lead speaker) it did not capture the attention of the media as did the original *Strategy*. The reasons for this are not entirely clear. However, *Caring for the Earth* was longer than – and hence lacked the punch of – the *World Conservation Strategy*. It also lacked a proper summary and key points: these had focused the reader's attention in the original. This may have been because *Caring for the Earth* was subject to extensive consultation and revision but either way it failed to make anywhere near as big an impact as the *World Conservation Strategy*.

Yet, while the *World Conservation Strategy* made the headlines, and even caught some political attention, it had little immediate effect on the ground, although it did pave the way for future policy documents. Had the *Strategy* been properly implemented then each nation would produce its own national strategy and implement it. A few countries did develop their own national response on paper but fewer still went on to implement these. The UK was one such. The Prince of Wales, three governmental agencies and two non-governmental bodies as well as the WWF launched its national plan, The Conservation and Development Programme for the UK, in the summer of 1983. Its ten-point action plan was never implemented. Yet the idea that an international policy document might spawn national strategies was one that caught on. As we shall see, more formal international climate and biodiversity policies of the 1990s (with political leadership from the UN as opposed to being led by non-governmental organisations), among others, also called for individual national strategies. Here such strategies, or national action plans, were developed.

8.1.5 The Brandt Report: Common Crisis North-South (1980)

In the same year as the *World Conservation Strategy* a report came from a commission of 18 politicians of international repute (some of them former premiers), led by West Germany's Willy Brandt. The report (Brandt Commission, 1980) was primarily concerned with the growing 'Third World' problem (or 'developing nations' as they are more commonly referred to today). The North and South in the title referred to the fact that most developed nations were in the north while those developing or undeveloped were in the south. The report was concerned that the divergence between the north and south was increasing.

The second report of the Brandt Commission, *Common Crisis* (1983), pointed out that the problem of wealth divergence affected both the north and the south and that both depended on each other. The south depended on the north for financial institutions and technology, while the north depended on the south for raw materials and tropical and semi-tropical cash crops.

Whereas the *World Conservation Strategy* looked at global problems from the perspectives of biological conservation and human ecology, the 1980 Brandt report looked at developmental problems from a political perspective. As such it focused on international trade and monetary agreements. However, pointedly it did mention the need 'to integrate conservation with development'. Also, like the *World Conservation Strategy*, the report attracted considerable media coverage. Again like the *Strategy*, because politicians actively in power at the time did not produce the report, Brandt's impact on the ground was muted, although it did help establish a tone of political thought. Three years later (1983) the follow-up report was very much along the same lines.

8.1.6 The Brundtland, World Commission on Environment and Development Report (1987)

The World Commission on Environment and Development was the first body to produce an international environment and development report that had real political credibility. Not only was it created as a consequence of a UN General Assembly resolution (38/161), but it was chaired by the premiere of Norway, Gro Harlem Brundtland, and had on it those who were either currently active senior politicians or who worked for major organisations such as the Organization for Economic Cooperation and Development (OECD), and leading scientific organisations and environmental agencies.

In effect the Brundtland report, called *Our Common Future* (World Commission on Environment and Development, 1987), was a synthesis of both the *World Conservation Strategy* and the 1980 report of the Brandt Commission. On the one hand it recognised the fundamental need to conserve biological resources and have sustainable development, but on the other it was aware that institutions, financial systems and international economic relations needed a radical overhaul.

With regards to biology and (separately) climate change it made some key (for international policy-makers at the time) observations and recommendations. On the biological front it called for a 'Species Convention' and the preservation of genetic diversity. It recognised the need to preserve ecosystems and have sustainable development. As such it re-iterated the arguments of the *World Conservation Strategy*. The Brundtland report has been cited as one of the first popularisers of the term biodiversity, although the term actually used in the chapter on species and ecosystems and in the report's summary is biological diversity. (Walter G. Rozen, of the US Academy of Science, used the term biodiversity in 1985 in the run up to the National Forum on Biodiversity held in Washington DC the following year; later the ecologist Edward O. Wilson used it as the title of his 1988 book.)

As for energy, the Brundtland report recognised the need for energy for economic growth and human well-being but also that much was being used inefficiently and that global warming and acid rain were major problems to be avoided. It called for improved energy efficiencies and alternative energy sources, and a 'safe, environmentally sound, and economically viable energy pathway that will sustain human progress into the distant future'. It said that this was 'clearly an imperative'.

However, the report was not clear as to how this might be achieved. This was largely because no country had a significant long-term energy strategy (although a few – such as France and Norway – had short- and medium-term plans that were being implemented).

8.1.7 United Nations' Conference on the Environment and Development: Rio de Janeiro (1992)

In 1992 in Rio de Janeiro the UN held a Conference on Environment and Development (known as UNCED for short). It was convened on much the same basis as the Stockholm Conference but unlike Stockholm had two distinct foci: biodiversity and climate change. Indeed, at the conference two international conventions were signed relating to each of these subjects. The biodiversity convention was in essence the realisation of recognition in international policy terms of the *World Conservation Strategy*'s and the Brundtland report's aspirations of the importance of biodiversity. The climate-related outcome was the UN Framework Convention on Climate Change (FCCC).

In the run up to UNCED over a score or so of nations had declared targets for the reduction of carbon emissions. The EU had a target of carbon emissions to be no higher than they were in 1990 by the year 2000, whereas the UK's target was emissions no higher than 1990 by 2005 and a reduction in all greenhouse gases by 20%. Not all these targets, and those of nearly all other nations, were met. The UK did manage to reduce its carbon emissions to below 1990 by 2004 and had even greater success with its total greenhouse emissions; however, as we shall see below, these savings owed much to a switch to natural gas (especially for electricity generation), with its higher energy content per carbon atom than coal or oil. Meanwhile, prior to UNCED the USA had no official targets, although it did sign up to the FCCC.

The convention lacked any clearly binding, short-term commitments to reducing greenhouse emissions. What it did was to establish a process for negotiating further measures and introduced mechanisms for transferring finance and technology to the less-developed nations. One of the UNCED's successes was that in addition to the nations that already had adopted some form of greenhouse policy, the USA signed the convention.

The UNCED had its failures too. These included the removal of much of the potential from the FCCC's original draft but also proposals with climate implications outside of the FCCC. For instance, the UNCED Convention on Biodiversity was not signed by the USA. This convention would have increased the financial value of standing tropical forests, in that there would have been equitable sharing (between developed and less-developed nations) of the benefits that flow from using genetic resources (of which tropical forests abound). President George Bush Snr, perceiving that signing would have a cost for the USA in employment terms, and with an eye on the forthcoming presidential election, declined to sign. Another failure with greenhouse implications was a third convention on Sustainable Forestry. Here the international gulf between the developed and less-developed nations was so great that before the UNCED began the convention had been watered down to a set of so-called Forest Principles.

The UNCED was a summit at which nations took two steps forward and one step back: progress was made only in that nations agreed that progress needed to be made. Nonetheless, there was a framework for development. The next major step down this road from a climate perspective was taken at Kyoto in 1997.

8.1.8 The Kyoto Protocol (1997)

The World Climate Conference was held in Kyoto, Japan, in December 1997 and was a direct result of the UN's FCCC. The problem during Kyoto was that Russia and, arguably more importantly, the USA, dragged their feet. Russia did so because its economy was behind that of Europe and the USA, and although it sought to develop its industrial base had gone into recession after the 1988–90 devolution of many of the former Soviet Bloc nations. Russia did eventually sign up in 2004. The USA was noticeably unenthusiastic and had much at stake, as North America had less than 5% of global population yet around 24% of global emissions at the turn of the millennium. Here the problem was not so much that President Clinton's administration did not accept the urgency of climate change issues, but that the administration was gridlocked between the Senate and Congress. There was little point in the USA agreeing to policies internationally that would not be ratified in its own country. However, the USA did call for carbon trading between nations, whereby efficient users of carbon could trade permits with inefficient users. It also called for trading to take into account a nation's planting of forests. The problem with this last, as noted in Chapter 7, is that whereas new forest growth does sequester carbon, as do some soils, and quantifying this is difficult. There are also problems in ensuring that carbon so sequestered remains sequestered.

The Kyoto Protocol set an international target for there to be a reduction of emissions (of all anthropogenic greenhouse gases) by 5% of the 1990 level (the FCCC benchmark year, although a 1995 base year was allowable for some nations' accounting) by the Protocol's first implementation period of 2008–12. Nationally, though, commitments to savings made by individual nations varied. The OECD nations generally had to make the most savings (an exception being Australia and some of the smaller nations) and the UK and Germany agreed to make greater savings so as to lessen the burden on some other EU states. The UK's target at the time was to reduce its greenhouse emissions by 12.5% below the 1990 base year by 2008–12. This is greater than the collective EU target (by the then 1997 EU states) of 8%: individual EU nations contributed to this collective goal by agreement subject to their economic circumstances.

Importantly, the Protocol established three 'flexible mechanisms' to achieving mechanism reductions: Joint Implementation (JI), the Clean Development Mechanism (CDM) and international emissions trading. Together these allow countries to achieve part of their legally binding targets by actions in other countries that are party to the FCCC. Emissions trading enables countries which achieve greater reductions than needed to meet their own target to sell the surplus to other countries. JI and CDM allow countries with targets to receive a credit for project-based activities that reduce emissions in other nations. Even so, under the Protocol, these three mechanisms must be in addition to action to curb emissions taken by a nation itself, with such domestic action contributing a significant amount to the country's total effort to meeting its target.

The problem with the Protocol is that the enthusiasm for it of individual nations differs. Although the Kyoto conference took place at the end of 1997, Russia did not sign up to it until 2004. Worse, the USA, which is responsible for about a quarter of global emissions, withdrew from the Protocol in 2001 on the grounds that it would harm its economy, even though it negotiated hard for carbon trading and for carbon sinks to be taken into account. (Currently, early 2012, the US has not ratified the Kyoto Protocol, but is a signatory.) Further, even many of the nations that did sign up to Kyoto had failed to make any savings by 2011. Not surprisingly then, globally carbon emissions have risen and this is reflected in increased atmospheric accumulation of carbon dioxide. Nonetheless, the Kyoto Protocol to date represents the most serious international policy attempt to curb global warming. Its first commitment period ends in 2012 and the UN has been holding a number of preliminary meetings to develop the Protocol beyond that date and also to encourage nations that are not on board to join. The first of these was held in Montreal in November 2005 (of which more shortly, in section 8.1.10).

So what effect would the Kyoto Protocol have on the typical range of IPCC Business-as-Usual (B-a-U) scenarios for the 21st century? If it were implemented by itself with no further carbon emission cuts then the 21st century would see global emissions reduced to 5% less than those of 1990 by 2012 and then remain steady. Under such a scenario one might presume that the developed nations would continue with further cuts (beyond 5%) to offset the increase in emissions from developing countries: the remainder of the 21st century would then see a static level of emissions at 5% less than those in 1990. At the end of the century global warming would be delayed by just a few years (optimistically, 2–5 years) compared to most of the IPCC's 2000 B-a-U-type scenarios. To actually stabilise atmospheric concentrations at 420 parts per million (ppm) CO_2 (50% above pre-industrial levels of 270–80 ppm) and to avoid even further warming beyond what this will incur, the IPCC say we need to cut emissions to 50% of 1990 levels by 2050. Conversely, more drastically, if we were to have stabilised the atmospheric concentration at its 1990 level of around 353 ppm, then the IPCC (1990) state that we would have needed to have made an immediate cut of carbon emissions of 60–80%. Yet by 2010 the year average atmospheric concentration was 388.5 ppm[1]. In short, the Kyoto Protocol by itself (assuming its goals are fulfilled) is just a beginning in an attempt to manage fossil carbon emissions. Much depends on the continuing Protocol negotiations.

Having said all of this, it is important to note the assumption in the above conclusion, namely that the Protocol, within its own terms, succeeded: many nations agreed to sign up to climate goals. This is a start, but it would necessitate further development so that emissions could be meaningfully curbed to reduce (if not eliminate) anthropogenic climate forcing. So, is the Protocol really succeeding in terms of this broader goal? Two-thirds of a decade on from Kyoto, and halfway to 2010

---

[1] A point of pedantry, but sharp-eyed readers of the literature might wonder: 388.5 ppm is the 2010 average taken from globally averaged marine surface data given by the National Oceanic and Atmospheric Administration Earth System Research Laboratory (NOAA/ESRL) website. See Dr Pieter Tans, www.esrl.noaa.gov/gmd/ccgg/trends. Often quoted is the Mauna Loa average for 2010 of 389.8 ppm.

(the mid year between 2008–12), the 2004 global carbon dioxide emissions figures were 26% higher than in 1990, and subsequent annual emissions were even higher still (not withstanding a temporary dip in emissions due to the 2008/9 global financial recession). Consequently, only an undaunted optimist could consider Kyoto to be working. Yet it should not be dismissed. Kyoto was undoubtedly a spur for the limited switching of energy away from fossil carbon that is occurring. Had this not taken place then emissions would have been, and would be, even higher still and so make subsequent reductions that much harder. Furthermore, those Kyoto countries that have developed more non-fossil energy sources may well find future economic benefits in being more independent from fossil fuels and this could act as a spur to further, more ecologically meaningful, cuts in the longer term.

8.1.9 Johannesburg Summit: UNCED+10 (2002)

The UN World Summit on Sustainable Development in Johannesburg was also known as UNCED+10 and this last more accurately reflects its purpose. This was to chart the progress made since the UNCED in Rio de Janeiro in 1992 and to take things forward. Alas, despite the presence of many premier politicians and G8 leaders, UNCED+10 was hardly considered a success. It transpired that on the ground little progress had been made since 1992. Further, the future commitments were not binding and the wording was loose. This was also reflected in those issues that have a major focus on biology, human ecology and climate change.

Food security had deteriorated so it was agreed at Johannesburg that this must be improved and that the numbers suffering from hunger in Africa should be halved by 2015. It also had a goal that 'on an urgent basis and where possible by 2015, [to] maintain or restore depleted fish stocks to levels that can produce maximum sustainable yield'. With regards to the UNCED Biodiversity Convention, it was recognised that little progress, if any, had been made. Those at Johannesburg resolved to 'achieve by 2010 a significant reduction in the current rate of loss of biological diversity'. This signalled that they accepted that biodiversity was going to continue to deteriorate and so only planned to slow the rate at which this happened. By 2005, with just half a decade to the deadline, progress on even this modest goal had yet to show signs of success. With regards to human mortality it reaffirmed a goal the UN made at the turn of the millennium (the Millennium Development Goal) to 'reduce by 2015 mortality rates for infants and children less than 5 by two-thirds, and maternal mortality rates by three quarters, of the prevailing rate in 2000'. It also affirmed to reduce HIV prevalence in the most affected countries in people of 15–24 years of age by 25% and globally by 2010. This too, as of 2006, appeared to be difficult to realise as deaths from AIDS (for all ages) had increased from under a third of a million a year in 1990 to nearly 3 million in 2000. Today, 2012, HIV/AIDS in many nations (especially some of those with weak economies) continues to be a growing problem.

The commitments of UNCED+10 to climate change were limited, primarily because politicians were leaving this to the development of the Kyoto Protocol taking place at the same time (and which itself was having its own difficulties, especially with the recent loss of the USA from negotiations). However, energy did feature, and here the energy goals, though lacking in specificity, were in line with Kyoto goals. These included diversification of global supply and increasing the contribution from renewable energy, tackling energy poverty, freeing markets and removing harmful subsidies that might exacerbate the greenhouse effect and improving energy efficiency. The UNEP, though, did launch a new initiative, called the Global Network on Energy for Sustainable Development, to promote research and energy-technology transfer to developing nations. Meanwhile, the EU announced a US$700 million partnership initiative and the USA said that it would invest up to $43 million in 2003. However, a fair proportion of these monetary pledges were already earmarked for international aid, and not all the monetary pledges were met in full. UNCED+10 never lived up to the promise of the UNCED itself.

8.1.10 2002–2007

Subsequent to UNCED+10 among the most significant international climate change policy developments was the 2005 G8 Summit in Gleneagles, Scotland. This had two principal items on its agenda: Africa and climate change. However, the communiqué that was released was a watered-down version of the earlier drafts, saying that fossil fuels 'contribute a large part to increases in greenhouse gases associated with the warming of our Earth's surface'. This carefully crafted sentence (and rest of the communiqué) left considerable room for interpretation and political manoeuvrability but was viewed by some as signalling that US President George W. Bush was beginning to recognise the science behind the issue. The communiqué was accompanied by a list of 38 action points including the promotion of low-energy vehicles and systems.

In essence, whereas Kyoto imposed carbon limits, the US view (and that of the G8 communiqué) was that although climate change was a problem, carbon limits were not essential, but rather the ability to switch to a low-fossil economy. The USA sought to increase the economic activity produced per unit of carbon dioxide released into the atmosphere. However, Kyoto did score in one area. The G8 meeting also committed to the rapid adequate funding of the Kyoto CDM, whereby rich countries or industries can buy rights to emit carbon dioxide provided that they supply the technology and expertise to reduce emissions in developing countries. The success or not of national and international climate policies over subsequent years will play their part in the next UN environment and development conference that was set for 2012.

At their Gleneagles summit the G8 countries also agreed to a new dialogue with the leaders of China, India, Brazil and other emerging economies. Together they were to hold dialogue climate meetings annually up to 2008, when they were to report to the Japanese G8 Presidency. Even though some of the countries involved are not part of the Kyoto agreement, the goal of these meetings is to try to achieve real progress in stabilising greenhouse gas emissions before the Kyoto 2012 deadline.

The 2005 G8 summit also saw international discussion on the further development of the Kyoto Protocol in Montreal: this might be considered the opening discussions in the aforementioned dialogue. The meeting consisted of two sets of parallel talks. One track of discussion consisted of nations who had signed up to the Kyoto Protocol. They agreed to further future discussions for greater carbon emission cuts and to find ways to help developing countries beyond 2012, up to when the original Kyoto Protocol relates. The other track consisted of parties who had signed the 1992 UN FCCC in Rio de Janeiro. Importantly these included nations like the USA, who had signed the FCCC but not the subsequent Kyoto Protocol (and who had made it clear at the abovementioned G8 summit earlier that year that they could not at that time join the Protocol). These talks with non-Kyoto nations enabled non-Kyoto countries to participate in climate-related policy discussion and keep the door open for them to potentially join the Kyoto nations later.

A pessimistic perspective might be that the 2005 Montreal meeting achieved little, with no clear decision for further cuts in greenhouse gas emissions beyond 2012. Optimistically it did commit to future discussions that could potentially lead to ecologically meaningful cuts in greenhouse gases.

In 2007 the IPCC published their fourth assessment (AR4; IPCC, 2007a, 2007b, 2007c, 2007d). See section 5.3.1.

8.1.11 The run-up to Kyoto II (2008–2011)

The 13th UN FCCC Conference of the Parties (COP) meeting was held in Bali, Indonesia in 2007. It concluded with the Bali Roadmap, or Bali Action Plan, as a basis for how future negotiations could progress.

The 14th COP meeting was held in Poznań, Poland in 2008. There was an agreement of principles (only) of funding to help less wealthy nations adapt to climate change and transfer to a low-fossil carbon economy.

Given the lack of real progress in 2007 and 2008, it was hoped that the COP 15 UN climate summit meeting at Copenhagen in December 2009 would see considerable developments, especially as the Kyoto Protocol agreement was due to end in 2012. US President Barack Obama replaced President George W. Bush at the beginning of 2009 and Obama professed concern over anthropogenic climate change, unlike George Bush Jnr who for much of his presidency did not accept that humans were responsible for changing the global climate. Indeed, Barack Obama actually attended the Copenhagen COP meeting and helped broker an accord (the Copenhagen Accord) with major greenhouse gas emitters and leading emerging countries. It included a recognition of the need to limit temperature rises to less than 2°C and promises to deliver US$30 billion of aid for developing nations over the next 3 years. The agreement outlined a goal of providing $100 billion a year by 2020 to help poor countries cope with the impacts of climate change. It also includes a method for verifying industrialised nations' reduction of emissions. However, the Accord was not adopted by the 193-nation conference (as it was considered an external document) but delegates did formally agree to take note of it. Finally, it should be appreciated that the measures that had been proposed earlier in Copenhagen, but which failed to make it into the Accord, were not enough to prevent 2°C warming. The best that could be said of it was that it was a meaningful starting point upon which to build.

The following year's COP 16 summit was held in Cancún, Mexico. Just prior to the 2010 meeting, the Canadian Senate voted against a greenhouse emissions bill that would have called for a reduction in the country's emissions by 25% of its 1990 level by 2020. The bill had previously passed through Canada's House of Commons in 2009 but, even if it had cleared the Senate, Canada would have had a smaller reduction target than some other nations, including the EU, whose policy it was to cut emissions by 30% of 1990 levels by 2020.

At Cancún itself Japan, China and even the USA were not supportive of the cuts necessary to curb warming to 2°C above pre-industrial levels. However, it was agreed to establish a Green Climate Fund (see COP 14 discussion, above) of US$100 billion by 2020 – initially using the World Bank as a trustee – to help protect developing nations from the worst effects of climate change and transfer their economy to one of low-fossil carbon reliance. There was also a framework to pay some nations not to engage in deforestation (Reducing Emissions from Deforestation and Forest Degradation, or REDD) and a formal recognition that deeper emission cuts were necessary. However, there was no actual ruling to enforce deeper emission cuts and no mechanism was developed to negotiate for these nor anything on the status of Kyoto II (the anticipated successor to the 1997 Kyoto Protocol, which only covered the period up to the end of 2012).

The 2011 COP 17 meeting was held in Durban, South Africa. The USA kept a low profile during the meeting (President Obama was facing a presidential election in 2012 for his second term and a politically significant proportion of the US electorate do not accept that anthropogenic climate change is taking place). China and India were two countries determined to be allowed to develop their economies without undue fossil carbon constraints: they wanted the developed OECD nations to make the most contributions to global emission cuts. A roadmap was proposed largely by the EU, the Alliance of Small Island States (AOSIS) and the Least Developed Countries (LDC) bloc: together the EU, AOSIS and LDC bloc represented a formidable number of nations to sway decisions. However, by the conference's supposed end no overall agreement as to how to work towards Kyoto II had been decided. Some delegations from small, less-developed countries then left, but the COP continued for nearly a day and a half beyond its official end date. In the end an agreement was passed: the Durban Platform.

There was also some progress on REDD and a management framework was adopted for the Green Climate Fund. At the Durban COP meeting it was also agreed to establish a technical working party to examine the question of whether agriculture's greenhouse impact should be considered as a separate section within the UN FCCC.

However, the Durban Platform did not have embedded in it the actions purportedly necessary to keep warming below 2°C (although it should be said, as will be noted later, that some scientists are of the view that by Durban the world was already on track to exceed 2°C warming and some opined that the Durban track would lead to 3.5°C warming by 2100). The agreement called for talks on 'Kyoto II' (or whatever it will be called) to be completed by the 2015 COP 21 summit and to come into force by 2020. As the Kyoto Protocol was due to expire at the end of 2012, the Durban Platform included a provision to extend the Protocol by between 5 and 8 years, and the exact date was scheduled to be determined at the next COP meeting in 2012, which will be COP 18, held in Qatar. Just after the 2011 COP meeting an editorial in the journal *Nature* summarised the position: ' . . . for climate change itself it [Durban] is an unqualified disaster' (Anon, 2011). Meanwhile, since 1990 (the oft-used policy benchmark year) the clear trend has been one of increasing greenhouse emissions, hence rising atmospheric carbon dioxide concentrations, with no sign of stabilisation let alone decline. Also, since 1990 the climate science has continued to provide more detail as to the impact of greenhouse gases on the global climate. It seems as though the action required, as underpinned by the science, and the action delivered by international policy-makers are on diverging tracks.

8.2 Global energy sustainability and carbon

An overview of the history of human energy-resource production and consumption is given in section 7.2. Figure 7.9 shows how the growth in global energy consumption roughly quintupled in the latter half of the 20th century and that this energy provision was by far dominated by fossil fuel consumption. For this reason alone, it is easy to see how any greenhouse impact from the addition of carbon dioxide became more marked at the century's end. It also suggests – should such trends continue – that climate change impacts will continue to increase. However, in terms of possible resource consumption and availability (as opposed to biological impact), how likely is it that we could continue trends in a B-a-U way?

The IPCC have in part answered this with their B-a-U scenarios, but these only cover the 21st century. As we have seen, in geological and biological terms a century is insignificant. So, what are the prospects for continued profligate fossil carbon consumption given the likely exploitable geological base?

The Kyoto Protocol calls for a reduced overall level of carbon emissions but, to make the math easier, and allowing for just some increase in consumption by developing nations, let us assume that carbon emissions stabilise in the near future. How long can we continue to consume fossil carbon at near-current rates? From Chapter 7 we can see that we have over two centuries of coal, and while oil and gas will run out long before then (at present rates of consumption), future economically recoverable fossil reserves (i.e. reserves not *currently* economically recoverable) could extend the potential fossil fuel era at current levels by another century or so, to over 300 years. Indeed, there are a number of resources, such as low-grade coal and orimulsion (a high-sulphur water/oil mix), that could be harnessed if the price were high enough to cover the extra processing costs.

Currently (using data from 2011) we are releasing over 8 Gt year^{-1} of fossil carbon into the atmosphere, of which around 4 Gt of carbon (GtC) accumulates. Of course there is also the deforestation factor to take into account. Deforestation, though, will not last at the current rate for 200 years as the forests will have gone by then, so this dimension of land use may not be a carbon source in the long term. Then again, there are other carbon sources and the possible atmospheric accumulation from soils as the boreal climate warms (and that of other areas with high soil carbon). There is also the confounding factor of reduced ocean absorption once the oceans have absorbed a critical amount of carbon dioxide. So, making long-term estimates is a complex business.

Nonetheless, a-back-of-an-envelope calculation (which, if good enough for Arrhenius is good enough for us) suggests that if total fossil fuel consumption continues as it is now (1995–2005) for up to two (or more) centuries, to use up our currently commercially available reserves (and without using some form of carbon-capture technology), then we will easily increase the atmospheric concentration of carbon dioxide to over three or four times pre-industrial levels, as opposed to around 33% above pre-industrial levels, which is where we are today, early in the 21st century. This carbon emission, if continued for two or more centuries at current levels (through tar sand and orimulsion exploitation, etc.), is likely to raise global temperatures by easily over 6°C and so have a palaeo-analogue earlier in the Cenozoic than a Pliocene palaeo-analogue. This last itself might be expected somewhere towards the end of the 21st century with just the 2007 IPCC's anticipated 2–5.4°C rise. Whether enough of this 6°C or more will by then sufficiently penetrate the oceans to disrupt methane hydrates is not known with certainty (yet), but it is almost certain to release some carbon in soil, so compounding warming. For the palaeobiologist the nightmare scenario is that we might release enough carbon directly (through fossil fuel and land use) and indirectly (due to soil carbon and possibly methane hydrates) to incur an extinction event. This may sound dramatic, and it is, but remember equally that we are already in the midst of a human-induced (anthropogenic) extinction event through biodiversity loss that is distinct from extinction by anthropogenic climate change. The majority of the population do not perceive this as serious because the human species continues to flourish, as do the major commercial species on which we depend.

It is important to remember that the above discussion is based on rates of emission in 2001. As was noted in the previous chapters, the twin drivers of increasing per-capita energy demand and increasing global population synergistically conspire to drive up emissions beyond the 2001 level. (Indeed, in 2007 it looks like this is happening and will continue to do so unless we switch our energy dependence away from carbon; clean-carbon technologies notwithstanding.) In short, should emissions continue to rise this increase in mean global temperature of 6°C or more will happen long before two centuries are up.

However, if a major (6°C+) climate-driven event were to occur, what would be the consequence? Well, we can with a fair degree of confidence rule out turning our Earth into a Venus-like planet. Yet this does not mean that we can be excused from striving for greater energy sustainability that relies less on fossil carbon, or from adopting carbon-mitigation strategies. The palaeo-analogue of continued substantial atmospheric fossil carbon accumulation is likely be the carbon isotope excursions associated with the Initial Eocene Thermal Maximum/Palaeocene–Eocene Thermal Maximum (IETM/PETM) and, worse, the Toarcian event (see Chapters 3 and 6).

Should we see a future event analogous to the IETM (let alone the Toarcian event) then this would have a profound effect on most of the planet's ecosystems. As it stands, the IETM did have a considerable ecological effect some 55 mya, even though globally much biodiversity was maintained (the marine extinction notwithstanding). Yet this maintenance of global biodiversity only took place because of biome (hence ecosystem) shift. However, such an ecological response would not be analogous to the situation today. To begin with, present-day ecosystems arose (up to the so-called present Anthropocene era) in a cooler Pliocene and ice-age Pleistocene of alternating glacials and interglacials: our current interglacial climate is already at the warm end of the Quaternary spectrum of global temperature in which present ecosystems evolved. So, further warming of more than a few degrees takes us into a new temperature realm for these ecosystems. Additionally, as mentioned in previous chapters, current human ecosystem management to our species' own ends means that options for natural ecosystems to shift and evolve with climate are restricted due to human fragmentation of the landscape, which was not the case back in the Eocene. Furthermore, the ecological disruption of the IETM resulted in subsequent opportunities for new evolutionary paths, this again would be limited in the human-dominated global environment. Certainly the diversification and expansion of primates began around the time of or shortly after the IETM, so in theory at least there could be evolutionary opportunities. It would also be a little ironic if our species – one current end-point of this primate diversification – triggered a climatic event analogous to the IETM that arguably marked the beginning of a new diversification for some other species.

The ecological disruption of an IETM-type event today would in all probability be more severe (because of our current ecosystem management, which itself is resulting in biodiversity loss). Equally, as with the IETM itself, there would be winners and losers. One of the losers would be humanity – more specifically our global culture – for although we as a species would probably not be completely wiped out, many of our 7 billion-plus population (and growing) would face dislocation, as parts of the planet become too hot, and famine, unless we radically adapted our grain crops (including rice) that currently dominate human calorie provision.

So, how realistic is this nightmare scenario? This picture may well be a longer-term one than the IPCC's 100-year scenarios. Equally, as the IPCC warn, being wary of surprises, depending on natural systems' critical thresholds the scenario might be manifest sooner. There has so far been only informal discussion about this among researchers but, to date, no analysis has been published in a high-impact, peer-reviewed journal. Nonetheless, globally, the IPCC (1990) say, we need to reduce carbon emissions by some 50% of their 1990 value by 2050 to stabilise CO_2 at 50% above pre-industrial levels (which, we should note, would probably result in a low probability of an eventual IETM-type event). This in turn necessitates both cutting fossil carbon emissions per capita today and planning for even greater per-capita reductions for tomorrow's larger population. In addition, 22 years after the first IPCC report we have singularly failed to halt growth in emissions. Therefore, it would be foolish to dismiss this IETM-analogue nightmare scenario out of hand.

As things stand, irrespective of whether or not this worst-case scenario comes about, we are facing the very real prospect of there being climate change to a regime not seen at least since the Pliocene 3 mya or more. As such we are likely to lose much of the fauna and flora that have evolved over this time and see some major changes in the location and type of many of today's ecosystems.

So, what then are the theoretical prospects of reducing fossil emissions? This question relates more to the abiotic aspects of human ecology rather than biology per se and has been covered elsewhere (such as in Cowie, 1998a and by the IPCC Special Report on Emission Scenarios [SRES] scenarios; IPCC, 2000) and so is only very briefly summarised in the following subsections and only for the short term to 2050.

8.2.1 Prospects for savings from changes in land use

As noted towards the end of the last chapter, the IPCC's second assessment report (IPCC, 1995) estimated that by 2050 some 60–87 GtC could be conserved or sequestered in forests. Compared to the above-estimated mean 21st-century B-a-U fossil fuel emissions of 18 GtC year^{-1}, this represents an annual saving of around 1.2–1.7 GtC, or around 6–9% of annual emissions. More optimistic estimates (a number are cited in Cowie, 1998a) increase this saving to over 2 GtC per annum. Due to uncertainty, the 2007 IPCC Working Group III assessment (IPCC, 2007b), which deals with mitigation, gives a wide range for possible 21st-century carbon mitigation from forestation of approximately 100–450 GtC (much depends on the models used and the economic value placed on carbon mitigation). However, even this degree of saving would not last indefinitely, for once forests have become firmly established (reached their climax community) they become (climate change itself notwithstanding) carbon-neutral. Nonetheless, as a fossil carbon-mitigation strategy it has the potential to make a significant contribution for at least the first half of this century. A round-figure equivalent to an average figure of 2 GtC year^{-1} for a major global forestation strategy to 2050 is not unreasonable but, because of the aforementioned reasons, unlikely to be of significant long-term benefit.

Of course, as the IPCC 2007 Working Group III and the 2007 synthesis report point out (IPCC, 2007a, 2007b), forestation has additional benefits than just carbon mitigation. One example is biodiversity conservation.

8.2.2 Prospects for savings from improvements in energy efficiency

Since the early 1970s there has been a considerable body of work looking at the potential for improvements in energy efficiency that, of course, have direct implications for reducing energy use and hence possible fossil carbon demand and emissions. Much of this energy-efficiency work was originally prompted by the 1973 oil crisis and some of this research is referred to in the USA and UK case study sections that follow (section 8.3; and again many are cited in Cowie, 1998a). Prior to 1973 the predominant policy view on both sides of the Atlantic was that a nation's energy consumption was proportional to its GDP. After the oil crisis this changed as it was recognised that different types of economic activity consume different amounts of energy per unit of economic turnover. In other words, people did not want to consume energy per se but to receive the goods and services that can be realised by harnessing energy resources.

It is never possible to achieve a theoretical 100% of energy transference from fuel to user application due to the laws of thermodynamics and physics, so it is always theoretically possible to improve energy efficiencies of generation, transfer and use. Indeed, because historically efficiencies have been so low, there has been an overall historical trend of improved efficiencies. The situation is further complicated because over time consumers have found new ways to use energy. For example, two decades ago (in the early 1980s) few Western homes had home computers, no homes had DVD players and only a few had video recorders (VHS had only just become established as the surviving format out of four vying for the commercial privilege). Then again, household sizes change (Chapter 7) and that also affects energy efficiencies or energy consumed per capita.

Thermodynamically, energy efficiency from a heat engine or generator can be mathematically expressed in the following equation:

$$\text{Carnot efficiency} = \frac{\text{The heat converted into work}}{\text{Initial heat intake}} = \frac{T_2 - T_1}{T_2}$$

T_1 is heat discarded in degrees Kelvin and T_2 is heat input in degrees Kelvin. Efficiency trends for generation have improved. Energy generation that relies on a boiler or the thermal expansion of gases is restricted by the Carnot cycle that relates the temperature (in Kelvin) of the burning fuel to the temperature of the outgoing coolant (see above). Because of engineering constraints it is difficult to have commercial boilers operating much above 550°C (823 K), whereas coolant temperatures (in temperate countries) tend to be a little above the average environmental temperature and so are around 10–20°C (283–293 K). This gives a theoretical maximum practical efficiency of around 65%. This applies to both fossil- and nuclear-power generation. Such early 20th-century power stations had an efficiency of around 25% while at the century's end, in the more developed nations, efficiencies of new generating plants had risen to 30–37%. One trick around this limitation is not to transform the fuel energy solely into another form (electricity) but to give the station's heat output direct to the consumer. So, instead of the station inefficiently generating electricity for consumers' electric heaters, heat from the plant goes straight to their houses. This is the theory behind combined heat and power stations. Another way around this limitation is not to use a boiler at all but some alternative way of transforming energy. Hydroelectric plants capture the energy of falling water while solar photovoltaic cells capture that of the Sun and both do so with far higher efficiencies than power stations. However, there are other limitations, here, respectively, the availability of water and sunshine, as well as limitations of cost. These costs can change with time, as the downward trend in the cost of photovoltaic cells demonstrates.

Aside from the cost of new, more-efficient generating plants, there is the cost (and indeed the carbon-energy cost) of discarding older, less-efficient plants. To take an extreme illustration, consider what might happen if a new technology became available tomorrow that doubled efficiency in construction and running in terms of both financial costs and energy costs (in terms of oil and cement). It would actually probably be less efficient in both these terms to discard recently built plants that used the less-efficient, old technology if the old plants had not been running long enough to pay off their construction costs (in terms of energy and money).

There is therefore a very real time lag involved in introducing new energy-efficient technology. Regardless of development time, new efficient technologies, even if theoretically practical, cannot be introduced instantaneously.

This same time-delay factor also applies to introducing energy-consuming technology with improved efficiency. A motor car needs to have paid off its construction energy and finance costs before it is replaced by a more efficient car. True, we only tend to think of such factors in financial terms, but the principle applies equally to the energy costs of technology, be it the energy used to refine (or recycle) the metal or the actual carbon in the said technology (plastics and so forth).

Again, as with energy generation, efficiency trends in consumption have also tended to improve, especially since 1950, and progress was marked in the years following the 1973 and 1980 oil crises (see Figure 8.1a). However, these efficiency gains in consumption have been confounded in a number of ways.

First, we actually might use the improved efficiency savings to maintain consumption comparable to old inefficient levels rather than lower consumption itself. This is known as Jevons' paradox (or Jevons' effect) because it was first articulated by

(a)

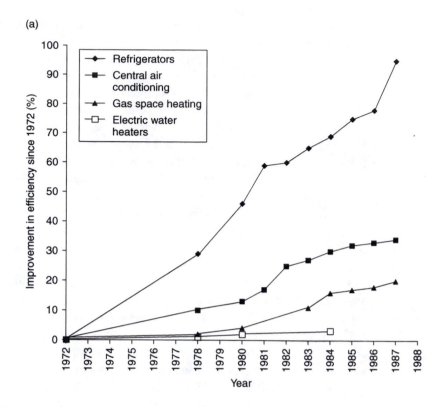

(b)

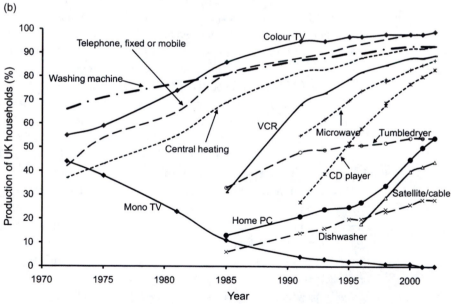

Fig. 8.1 (a) The improvements in energy efficiency of some household electrical equipment between 1972 and 1987. Note that this does not mean, for instance, that by 1987 refrigerators were 90% efficient, rather that their typical efficiency had improved by 90% over their efficiency in 1972. Data based on Schipper and Hawk (1991). (b) Proportion of UK households with selected appliances (Office of National Statistics, 2002).

William Stanley Jevons in his 1865 book *The Coal Question*. The classic example is space heating. Over the years improved household insulation has reduced the loss of warmth from homes. However, the trend has not been one of reducing the energy that nations use for domestic space heating. Leaving aside the question of reduced household occupancy (Figure 7.3), home owners have tended to recoup the energy savings by being able to have warmer homes. This maintaining of dwellings at higher temperatures offsets potential savings. Similarly, whereas there has been an overall trend in improved fuel efficiency of cars (in terms of fuel consumption/unit of car mass per kilometre travelled) so this has been more than offset by the distance travelled per year by car owners. (This is one reason why the green lobby is against construction of new motorways as it leads to increased car use, hence the need to build even more motorways; Tyme, 1978).

Second, we might devote our increased affluence (see the previous chapter) to using energy in less efficient ways. An example might be the use of less-efficient (in terms of common occupancy energy per passenger kilometre), larger sports utility vehicles (or SUVs) than traditional smaller and lighter family cars. This trend was manifest in developed nations early in the 21st century where some SUVs began to be used solely in urban settings and not off-road, for which they were notionally designed. The motivation here is largely psychological in that SUVs are sometimes perceived as conferring status for the driver/owner.

These two trends of improved efficiency and increased consumption serve to offset each other. Indeed, in some cases they can cancel each other out. Here an example would be the energy used in watching television in the UK. In the 1950s televisions were still a luxury and very rare but the number of households in developed or industrialised nations with television sets was increasing. For example, in the UK in 1955 television sets consumed about 830 GWh. By 1970 more households had television sets and the energy consumed that year was around 5000 GWh. However, during that decade semiconductor technology was being introduced to replace the old thermo-electronic valves and so the electricity consumption per set began to decline. In the late 1970s colour televisions began to replace the former black-and-white models (see Figure 8.1b) and energy consumption per set rose once more. However, by 1984, despite almost universal use in the UK, these sets too had become more efficient and so annual consumption decreased almost to the 1970 level (Evans and Herring, 1989). The 1990s saw an increase in the number of television sets in the UK and with the advent of video (and later still DVD) increased total UK energy consumption not just through viewing but also by the sets themselves being left on standby rather than being turned off (in addition to, of course, the ancillary video and DVD equipment). Domestic equipment left on standby accounts for approximately 10% of total domestic electricity consumption in the UK today. Finally, the new generation of high-definition, wide-screen plasma sets consume almost five times as much electricity as the sets they are replacing.

This pattern of counteracting trends – of improved efficiency on one hand and higher-use, additional new energy-consuming technology on the other – is common to virtually all areas of energy use, although on balance one tends to dominate over the other so that they rarely negate each other exactly. For these reasons, since the 1970s to the early 2000s electricity consumption by UK domestic appliances and lighting increased at a rate of around 2% a year (Parliamentary Office of Science and Technology, 2005).

Finally, there has been one misunderstanding about energy efficiency that has re-surfaced in the literature over the decades. This is that energy efficiency per se is a substitute for energy generation. This is not strictly true. While it is eminently important to improve energy consumption and production efficiencies, improved effi-ciency by itself is meaningless: *something* needs its efficiency improving. Arguments that the *bulk* of energy investment should be directed towards improving energy effi-ciency are not practical. Investment will *always* be required first for energy production (although hopefully this will be efficient production). Calling for investment to be almost exclusively directed to energy conservation will fail because a nation needs power production as well as consumption (Greenhalgh, 1990). Given this, and the offsetting twin trends in improved energy efficiency and increased consumption, it is unwise to consider improved energy efficiency by itself as a way of countering increased greenhouse gas emissions. Instead, a package of measures is required, of which increased efficiency of energy production and consumption is but one. The others include low- (or zero-) fossil options such as nuclear and renewable energy, and carbon-capture technologies as well as pricing. These are summarised briefly in the following subsections.

The bottom line with regard to the savings that might be realised through energy efficiency – should a broad quantified estimate be required – might amount to around 20% of consumption over 1990 efficiencies by 2025. (For a more detailed overview see Cowie [1998a], which is in line with the IPCC [1990, 2001a] and the IPCC emission scenarios [IPCC, 2000].) We will return to this broad estimate later in the context of other potential savings that might accrue from possible energy strategies, and, with regards to efficiency, low-energy strategies.

8.2.3 Prospects for fossil carbon savings from renewable energy

Renewable (flow) energy resources are currently (using 2010 data) contributing to 6–7% of global commercially traded energy (BP Economics Unit, 2011). (This is up $\approx$1% since 2002, reported in this book's first edition.) It excludes wood and other biofuels gathered locally and consumed. (Estimates for early-21st-century global non-commercial local-biomass energy production are around 12 600 TWh year^{-1} [Royal Society of Chemistry, 2005], or about 1000 million t of oil equivalent, or mtoe, which itself is close to about 10% of commercially traded energy.) Among renewable commercially traded fuels, hydroelectric power (HEP) dominates (at about 6–8% of global commercial energy traded in terms of coal-plant equivalents). However, there are also wind, wave, geothermal and solar (thermal and photoelectric) forms of renewable energy. These are not biological in nature and will not be described in this book save to say that, of these, wind is the renewable energy currently being significantly developed but which in the middle of the 21st century's first decade (2006) only contributed around 0.5% of global commercial energy.

Biofuels are biological, though, and a summary account appeared in the previous chapter: they make a small but significant global contribution. Biofuels not only have the potential to offset additions of fossil carbon to the atmosphere but, when

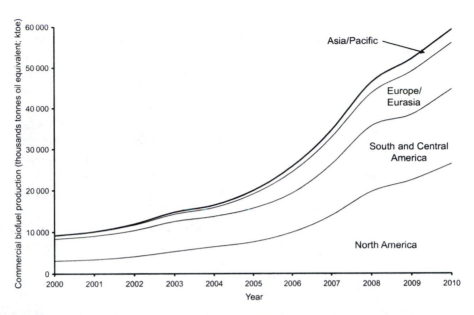

Fig. 8.2 Commercial biofuel production 2000–10 by world region (BP Economics Unit, 2011). *Note:* most of the other energy graphs in this book refer to million tonnes of oil equivalent (mtoe) whereas this figure uses thousands of tonnes of oil equivalent (ktoe).

combined with the carbon-capture technology described below, can take carbon out of the atmosphere.

Commercial biofuel production (excluding local gathering of wood, etc.) increased dramatically (by around 500%) in the first decade of the 21st century (see Figure 8.2). Much of this growth has taken place in North and South America as well as Europe. It is thought to have competed with food production, and hence contributed to the 2007–8 food crisis (see section 7.4.1). Nonetheless, despite this growth, by 2010 biofuels were contributing less than half of 1% of commercial energy production. This means that non-commercial biofuel use (mainly rural local harvesting of wood from forests, burning animal dung, etc.) is substantially greater than its commercial counterpart.

As with fossil fuel consumption and nuclear power, renewable energy production does have environmental impacts, but these are largely unrelated to climate – hence are not covered in detail in this text – and the exceptions to this only have a local (not global) impact. An example might be those hydroelectric schemes that create a substantially increased surface area of water over that of the previous river course, such as with the Iron Gates scheme on the River Danube between Romania and Serbia. The valley in which the scheme is located has acquired a more Mediterranean (as opposed to its formerly continental) climate.

Developing renewables, indeed novel non-fossil carbon energy-generating techno-logy, necessitates both research and then deployment. Both these have costs. While the former may be financed by a combination of industrial and governmental research investment, deployment needs to be actively encouraged with financial investments so that the new technology can compete on a level playing field with established (hence cheaper) fossil carbon technology. Here, one of the best proven incentives are what are known as 'feed-in tariffs' whereby renewable energy producers are subsidised with an enhanced payment for the electricity they generate for national grids. The first feed-in tariff was created by President Carter in 1978 with the US National Energy Act. By 2011 some 50 nations have feed-in tariffs.

For feed-in tariffs to be accepted by private generators it is important for private financial investors to be assured that agreed feed-in tariffs will remain for those who sign up to them and not removed in the future. (Although tariffs may at some stage be closed to new entrants.) These tariffs can be paid for in a number of ways, but principally these are taxation of energy users, taxation of fossil energy consumption (carbon tax) or taxation of fossil fuel production, or a combination thereof. Feed-in tariffs need not continue for new energy production investors once the renewable technology has matured and can compete without them. For example, according to the Renewable Energy Association (the trade body for UK renewable energy producers) in their evidence to a 2011 House of Commons Energy and Climate Change Committee Select Committee enquiry, in Britain in 2010 photovoltaic solar power seems to be near this point.

There is a substantial body of literature as to the potential of various renewable energy resources to be found. My previous review (Cowie, 1998a) and the renewable potential for the year 2025 is in line with the proportions of total non-fossil energy contributions from a number of the IPCC (2000) scenarios including the principal scenarios within its A1, A2, B1 and B2 groups. One reason for this is that the various IPCC scenarios see their greatest divergence in emissions after 2025 through to 2100. Another reason is that the most economically attractive sites that lend themselves to renewable energy will be exploited first, which means that there is greater uncertainty associated with estimates further into the future. Finally, because bulk exploitation of renewable energy other than HEP is still in its infancy there is uncertainty associated with these technologies. Broad working upper and lower estimates for potential non-fossil renewables for the year 2025, should there be a drive to increase renewable energies, are depicted in Table 8.2. These also straddle the International Energy Agency's (2004) renewable potential forecast for 2030. The low estimate of 1296 mtoe roughly equates with a little over 1 GtC year^{-1} whereas the high estimate of 1960 mtoe equates with around 1.7 GtC year^{-1}. (For comparison, in 2000 some 8 GtC was released globally from fossil fuel.)

8.2.4 Prospects for carbon-capture technology

Improving energy efficiency is one way to increase the economic effectiveness of fossil energy resources without increasing fossil consumption and developing renewable energy is another. However, as indicated in the previous sections, both these ways have their limitations in contributing to a low-fossil energy strategy. With enough coal for over two centuries of global consumption the question arises of whether there is a way of continuing to use fossil carbon energy sources yet reduce carbon emissions to the atmosphere. The comparatively new idea of carbon capture and storage (CCS) has recently come to the fore.

Table 8.2 The potential contribution of renewable energy to our energy supply in 2025. Source: Cowie, 1998a, and checked to see if data are compatible with IPCC SRESs. Further, the table's total low/high estimates of 1295–1960 mtoe for 2025 fit the International Energy Agency's (2004) estimate of some 2000 mtoe possible from renewable energy for the year 2030. The mtoe data are based on the assumption that the fossil fuel it displaces would have been burned in a power station with an efficiency of around 35%

| 2025 Energy resource | Potential contribution (mtoe) | |
	Low estimate	High estimate
Existing (2000) HEP	600	600
New HEP	300	500
Photovoltaic cells	5	45
Wave	20	220
Tidal	50	125
Wind	50	90
Geothermal	20	40
Biomass	250	340
Total	**1295**	**1960**
Of which is new	695	1360

The idea behind CCS is that instead of fossil fuel consumption short-circuiting the deep carbon cycle (Chapter 1) through the burning of geological deposits of carbon-releasing carbon dioxide to the atmosphere, the carbon is captured and stored geologically. It could also apply to the burning of biofuels, in which case carbon from the fast carbon cycle is transferred to the deep carbon cycle as opposed to the other way around. This means that there would be a net reduction in atmospheric carbon dioxide.

There are other storage options but geological storage is both the most viable and has the least environmental risk (although the technology is currently far from proven). Other options include precipitating the carbon out as carbonate, so in effect creating limestone, or pumping it into the deep ocean. The former is not yet commercially viable. The possible environmental impact of the latter on abyssal ecosystems is unclear, as is its stability (especially if the oceans warm up).

The capture of carbon dioxide takes place directly after the fossil fuel, or even biofuel, is burnt, by taking it out of the exhaust flue of a power station and pumping it into geological formations. The use of depleted oil and gas fields, or alternatively deep saline aquifers (water-laden porous rock formations), are the principal options. There are advantages with the former. Primarily, oil and gas fields have a proven stability and storage ability or tightness, whereas saline aquifers do not. Tightness is an important factor as the carbon must remain trapped for longer than the time in which fossil fuel is used at biosphere-affecting levels. Given that this last is likely to be between four and eight centuries, and given that the half life of carbon dioxide in the atmosphere is of the order of a couple of centuries, ideally the geological structure's tightness needs to be such that the carbon dioxide is secure for over a thousand years. One problem is that exploratory drills and oil- and gas-extraction processes serve to undermine tightness. Well-capping technology is reasonably developed but may need optimising. Estimates vary, and further work needs to be done, but optimistically it is thought that leakage could be 0.004–2.4% on a 1000-year horizon. Further, some 0.03% leakage might take place during transportation and injection (Department of Trade and Industry, 2003). Less optimistically, the *IPCC Special Report on Carbon Dioxide Capture and Storage* (IPCC, 2005) estimates that the technology could be reasonably assumed to be capable of storing 80–99% for 100 years and some 60–95% for 500 years.

There is another key advantage to using oil and gas fields, which is that the actual process of storing carbon dioxide can be used to help extract remaining oil and gas. Pumping in carbon dioxide at the base of one end of the field drives remaining oil and gas further away and upwards. This is known as enhanced oil recovery (EOR). It is particularly attractive because it helps extend an oil or gas field's economic life and offset the cost of the carbon-capture exercise. It also means that the field provides more energy, which is itself important as the carbon dioxide from power-station flue gases needs to be compressed, and then pumped from the power station before being injected into the geological structure. This takes energy and so can offset (in energy terms) the utility of carbon capture.

Estimate costs per tonne of carbon dioxide emissions reduced by carbon capture vary both due to the uncertainties associated with the new technology and because much depends on the locations of the geological fields and the power stations. Storage costs without EOR could be about £30–90 t^{-1} (2003 prices) and with EOR significantly less. (For comparison it is interesting to note that the price of carbon on the EU trading scheme in the latter half of 2005 was around £22 t^{-1}.) It should be noted that if EOR was always economically viable then the oil companies would always be using it. That they do not means that governments must provide a long-term economic and regulatory framework to ensure the process's economic viability.

Although it is early days, the suggestion is that fossil fuel power stations with carbon-capture methods have costs comparable with nuclear power and offshore wind energy. The Policy Exchange, a British think tank, in 2008 estimated carbon storage reduction costs for a small-scale pilot plant to be £70 tCO_2^{-1} (£257 tC^{-1}) but if carbon capture was rolled out in a large scale then costs might be just £30 tCO_2^{-1} (£110 tC^{-1}) and incur an annual extra cost per household of around £60: this would make carbon capture and storage coal-fired electricity plants roughly as expensive as wind energy. It also estimated that fitting CCS to UK plants could cut emissions by 20% by 2020 and global emissions by between 28 and 50% by 2050. (The 2007 Stern review estimated that CCS has the potential to contribute up to 20% of global CO_2 mitigation by 2050. Furthermore, to achieve stabilisation at 550 ppm without CCS will increase costs by more than 60%.) However, because carbon capture is not immediately economic, the prospects for wide-scale adoption of the technology in the short term are unrealistic. What is more easily possible is to ensure that any new fossil fuel power stations are built as 'capture-ready' so that CCS can be added later. This is particularly important for less-developed nations, where future fossil fuel consumption is likely to rise markedly. A number of nations, such as India and China, are already thinking about CCS and are participating in the Carbon Sequestration Leadership Forum. Other than a few exceptions, developed Western nations have yet to fund CCS in developing nations but this is a likely option in the future.

It is important to note that there are limitations to CCS technology other than those already cited. Chief of these is that, as currently envisioned, CCS systems only apply to power stations where the amount of carbon dioxide being captured is of sufficient quantity for CSS to be practically feasible (even if it is still not generally economical without some support). This limits our ability to reduce carbon emissions.

The argument against CCS is that it facilitates continued reliance on fossil fuels. There is also a sustainability argument in that in the long term (many decades and centuries), as fossil fuels dwindle, other longer-lived energy resources will need to be exploited. The nuclear option is currently one of the few technically proven, greenhouse-friendly options available, although it has its own associated environmental concerns. Because of these some green campaigners view CCS as the preferable investment option.

The counter argument to this is that because carbon capture is predominantly related to fossil fuel consumption, and because consumption of finite fossil fuels is not sustainable in the long term (beyond a couple of centuries as our main energy resource), fossil fuels' (non-CCS) climatic impact, although very significant, will be eventually curtailed. CCS could therefore have a limited role in providing time while we transfer to a sustainable, long-term energy strategy. Nonetheless, if considerable quantities are to be stored then the tightness of the geological formations used needs to be greater than 60–95% for 500 years, as the IPCC's 2005 special report estimates, if we are to completely prevent a long-term climatic impact.

The IPCC (2001a) SRES scenarios for the 21st century assume 1000–2200 GtC consumption and one might presume approximately double this if we included the 22nd century. This last greatly exceeds the estimated carbon release that triggered the IETM/PETM. This thermal maximum (and indeed the earlier Toarcian event) lasted well over 100 000 years. Consequently, from a palaeo-environmental perspective if CCS is used to store hundreds or thousands of gigatonnes of carbon then significant leakage needs to be minimal over this kind of 100 000-year time frame if a similar thermal event is to be avoided. (It is interesting to note how this perspective compares with that for storage of nuclear waste: see section 8.2.5.) Upon further technology appraisal it may be that such storage tightness is not achievable. However, this does not mean that the technology is without value. It would appear from both the Toarcian and Eocene carbon isotope excursion (CIE) events that the build-up of atmospheric carbon took tens of thousands of years. We know that the biosphere coped with both these events (ecosystem disruption, biome latitude migration, greatly increased global precipitation and erosion notwithstanding). What we do not know, even loosely, is how the biosphere will react to at least 1000 GtC injected into the atmosphere in a century. Even if carbon-capture technology could help reduce our 21st- (and 22nd-) century anthropogenic impact to 'only' that of an Eocene event, it would confer major benefits. It would give more time for the biosphere to adjust and contribute to minimising human impacts. The IPCC's 2005 report on carbon capture cites the security (longevity) of storage sites as one of the unknowns requiring further investigation.

The IPCC's 2007 Working Group III assessment (IPCC, 2007b), which assessed climate mitigation, estimated the total potential for reducing carbon emissions for the entirety of the 21st century to be between about 500 and 1000 $GtCO_2$, or between 136.4 and 273 GtC.

8.2.5 Prospects for nuclear options

Nuclear fission is a non-biological energy resource and so will not be described in this book other than to note its possible contribution to a sustainable energy strategy. Nor does nuclear power pose a serious threat to the global commons in the way that fossil fuel carbon release does. As with renewable energy there are potential local impacts and among these the potential for radiological ecotoxicology is in the fore, especially (for politicians) if humans are affected. The biggest radiological burden on natural systems does not come from the power stations themselves (other than accidents primarily arising out of poor design, such as the disaster at the Chernobyl nuclear power plant in 1986) but the creation of radioactive isotope-enriched fuels, fuel reprocessing and waste disposal. For example, in the UK the greatest radiological burden imposed on natural systems comes from the Sellafield processing and reprocessing plant (formerly Windscale), which greatly exceeds the sum total of radiological burden from all British nuclear power stations. Even so, this is not so great as to be considered a public health risk. What is of concern is the question of the long-term disposal of radioactive waste. In both the UK and the USA (and virtually all other nuclear countries) this question has not been resolved and both nations have been engaged in lengthy investigations and consultations about viable options. No country has yet fully resolved the question of high-level radioactive waste but some, such as Finland, have excavated repositories for low- and intermediate-level waste. The major concern is that high-level waste depositories need to be kept isolated from the biosphere for many thousands of years. This, by definition, will involve an unproven method, and hence incur a risk that needs to be anticipated if it is to be minimised. There is also a cost involved. Not surprisingly, this dimension to the nuclear option makes the question of nuclear fission power in a number of countries politically sensitive.

Chernobyl and carbon impact timescales compared

The worst nuclear disaster was at the Chernobyl power plant in the Ukraine in 1986, involving a power plant of inherently unstable design. (None of that type has been commissioned since and 21st-century reactors cannot undergo such a runaway reaction.) The UN estimates that after 100 years only plutonium and americium isotopes are of concern. While the amount of americium-241 (^{241}Am) originally released was very small, the total activity of ^{241}Am will increase with time due to the decay of plutonium-241 (^{241}Pu). It will reach a peak after 72 years, following which it will slowly decline. After 320 years, the total activity of ^{241}Am will be the highest of all the remaining radionuclides (UNSCEAR, 2008). Although plutonium isotopes and

[241] Am will persist perhaps for thousands of years, their contribution to human exposure is low. In 2005, the UN estimated that a total of up to 4000 people could eventually die of radiation exposure (UN Department of Public Information, 2005). The necessity of a concrete sarcophagus surrounding the plant itself is of most concern as it will have to remain for around 100 000 years or more. There is currently (2012) restricted access to the 30 km zone around Chernobyl but beyond this human life largely goes on as normal.

By comparison the carbon isotope excursion (CIE) events of the Toarcian (section 3.3.7) and early Eocene (section 3.3.9) also lasted around 100 000 years or more and were global in extent: glacials also last a similar magnitude of time. It is thought that the Toarcian and Eocene CIE events may be analogous to the climate change we might incur should Business-as-Usual emissions continue this century. The conclusion is that the greenhouse-induced global change caused by current human action, and the local impact of the most severe nuclear power disaster, both have a similar duration. So, in these terms, which is worse and/or avoidable? The debate will no doubt continue.

Although the nuclear waste problem has yet to be solved by most countries with nuclear energy there is a far greater understanding at the moment of the concerns and related science than for carbon capture. Whereas there is not space in this text to go into the issues in any depth, it is interesting to note the timescales involved and compare these with carbon burial (carbon capture; see section 8.2.4). One scheme is that proposed for a repository in the Yucca Mountains in the USA. In 2001 the Environmental Protection Agency (EPA) issued standards for the radiological dose one would receive near the proposed repository of 15 milli-rems year^{-1} for 10 000 years. However, a court ruling in July 2004 found that the EPA, by limiting its compliance assessment to 10 000 years, had not heeded the recommendations (as it is statutorily obliged to do) of the National Academy of Sciences. They had earlier concluded that 'peak risks might occur tens to hundreds of thousands of years or even further into the future'. In response to this court ruling the EPA issued a consultative statement for public comment in August 2005 that there be a two-tiered protection regime: a maximum dose of 15 milli-rems year^{-1} for the first 10 000 years and 350 milli-rems year^{-1} for up to 1 million years thereafter (assuming some leakage by then; Carter and Pigford, 2005). This means that the civil-engineering considerations for leakage for nuclear power and carbon capture (to avoid carbon leakage contributing to an Eocene-like warming event) are in timescale terms broadly similar.

In terms of growth rate, nuclear power was the most rapidly developed of the major energy resources in the 20th century. That it does not involve fossil carbon (the moderate use of carbon in gas-cooled reactors' excepted) makes it very attractive from both climate-impact and energy-sustainability perspectives. In 1975 the global production of nuclear power was just less than 100 mtoe (assuming the oil was burnt in a modern station with 38% efficiency) and this rose to over 600 mtoe by 2004, so contributing around 6.5% of global commercially traded energy (BP Economics Unit, 2005) and 5.2% (626.2 mtoe) in 2010 (BP Economics Unit, 2011). However, because of the aforementioned environmental concerns the growth of nuclear power capacity (as opposed to supply) has slowed since the Chernobyl disaster in 1986. Nonetheless, of the non-fossil carbon resources it is the only one that has seen significant growth and contribution to the global energy budget in the latter half of the last century. It is therefore the only non-fossil energy resource that can demonstrably contribute to meaningful cutting of future carbon emissions without a decline in energy supply in this century.

However, from the perspectives of environmental science and human ecology the question of whether we should adopt nuclear fission as part of a move towards a sustainable-energy strategy depends on the balance of biological risk between nuclear and fossil fuels. What should not be an issue in climate terms, although early in the 21st century it has featured in much public discussion, is whether energy investment is made in either nuclear or renewable energy. Such a question simply removes investment from the development of non-fossil carbon energy resources. The key question is whether nuclear investment should be included in investment that displaces fossil carbon consumption. Given the time it is taking to develop significant amounts of renewable capacity, it is very difficult (though not entirely impossible) to dismiss the nuclear option. Indeed, some of the public may be surprised at this human-ecology and environmental science perspective, which differs markedly from that of the green movement.

From a political and sociological perspective, there are other (non-scientific) problems of nuclear power to do with proliferation: that of nations acquiring nuclear military capability through civil nuclear programmes. However, these might be surmounted if it is possible to keep nuclear power generation quite separate from the other parts of the nuclear epicycle (of fuel enrichment, processing and disposal). This is especially important if the generation of nuclear power takes place in less stable countries.

It is not often noted, but as with finite fossil fuels there are limits to nuclear fission that depend on the finite supply of fissile uranium. As such, some fission can be considered a finite or fund resource. The size of this fund depends on the rate of use, which has seen periods of rapid growth up to the 1970s and which slowed markedly after 1986 (and Chernobyl). Consequently the estimates for the global supply of uranium vary from over three centuries at broadly current use to less than a few decades assuming that uranium consumption grows to meet some 40% of anticipated global electricity demand by around 2030. There is, though, a way in which nuclear physics can be used to extend this resource's lifetime. This is by using a more energetic reactor to produce more nuclear fuel from certain isotopes. These reactors are called fast breeder reactors; 'fast' because they actively depend on an energetic type of neutron (a fast neutron) and 'breeder' because they breed fuel through fast-neutron interaction with appropriate isotopes in the reactor's core. However, because the process relies on high-energy fast neutrons, the resulting high thermal energy produced needs to be carried away quickly from the core and so typically such reactors use liquid metal (sodium) coolant to conduct the heat away. This presents significant engineering challenges that make such reactors inherently more expensive than non-breeder reactors. Nonetheless, fast breeder reactors would extend the global uranium resource by some 4–6-fold.

The other problem with any major global use of nuclear power to offset greenhouse emissions is that of nuclear proliferation. Even without breeder reactors it is possible to manipulate conditions in reactors and to increase reactor-core dwell times to produce transuranic elements such as plutonium that can then be used in warheads. Further, those countries with nuclear-epicycle capability (primarily, isotope enrichment) for enriched nuclear-fuel production can produce materials for nuclear warfare. Such considerations, though beyond the scope of this book, are of global sociopolitical consequence. Indeed nuclear war, no matter how limited, is not without its own biological impact.

Nuclear fusion is the process by which light elements are built up to more nuclear-stable heavier elements (the most stable being iron) with the release of energy. As such it is a counterpart to fission, whereby heavy elements are split into more nuclear-stable lighter elements (again the most stable being iron). Fusion is the process that takes place in stars (including the Sun). The attractions of fusion include that far more energy per light atom (such as hydrogen) is released than with an atom of fissile uranium or plutonium. Also, whereas fission converts the fuel to radioactive isotopes, fusion can convert hydrogen to helium without the production of long-lived radioactive isotopes. However, both fusion and fission do lead to the transmutation of their reactors' non-fuel structural components to radioactive isotopes. (In fusion, it is the neutral fast neutron which is not contained by the magnetic 'bottle' that does the transmuting.) So, both fusion and fission produce waste, albeit of markedly different natures, and it is considered that a fusion reactor would result in far less radioactive waste per unit of energy produced than fission.

The other problem with fusion is that it is technically difficult in engineering terms to contain a plasma of hydrogen in a magnetic bottle at a sufficient temperature and density for fusion to take place, let alone to extract the heat on an ongoing basis. The first serious proposal for harnessing fusion came in 1946 with the Thomson and Blackman toroidal design: a toroidal has no ends from which a plasma may leak. Famously the 1957 Zero Energy Thermonuclear Assembly (ZETA) was predicted to be the forerunner of electricity that would be too cheap to meter. It failed. Over subsequent decadesthere has been increased plasma density, temperature and duration in various designs, including the Russian Tokamak and the Joint European Torus (based in Culham, England). The next major experimental fusion reactor that is hoped to demonstrate sustained net energy output will be expensive, at over US$10 billion, so necessitating international co-operation. It will be the International Experimental Thermonuclear Reactor (ITER), hosted in France. The final decision on constructing this was made in 2005 and it is thought that the first ITER plasma should be possible by 2015. However, just as underfunding of agreed budgets slowed fusion research between the 1980s and early 21st century, so bureaucracy and administrative hurdles slowed the construction of the ITER and added further funding difficulties.

In terms of energy sustainability and carbon the big attraction of fusion is 2-fold. First, it imposes a low greenhouse burden. (There will always be some greenhouse burden, if only because of the carbon dioxide generated in the cement-making process, even if this is reduced through carbon capture.) Second, compared to human energy needs there is an abundance of hydrogen and its more rare isotope, deuterium. If all humanity's commercial energy production came from deuterium fusion (at current levels) then theoretically there would be enough fuel for well over one billion years. Indeed, should fusion become sufficiently economical (i.e. only marginally cheaper than oil's real-term price for much of the 20th century) then, aside from the issue of radioactive waste, a real environmental concern might become the thermal waste from profligate energy production and consumption. (For example, consider far more

Table 8.3 Potential B-a-U carbon emissions against maximum likely possible savings for the year 2025. Non-fossil energy includes renewable energies and fission (hence is higher than the estimates for solely non-nuclear renewables given in Table 8.2)

Carbon source or sink	Human-related CO_2 emissions/savings, in GtC year^{-1}		
	2000	2025	B-a-U 2025 with savings
Total human activity	8	14	(savings not realised) 14 GtC
Decline in deforestation and some reforestation	–	0	Save ≈2 GtC
Improvement in energy efficiency	–	0	Save ≈20%
Offset with non-fossil energy	-	0	Save ≈2.2%
Total	**8**	**14**	**7 (low)–8.5 (high)**

pronounced current urban thermal islands.) However, we still have at least decades to go before the first commercial fusion reactor. ITER has to prove itself theoretically and then a fully commercial reactor needs to be designed and produced along with a fusion epicycle plant to produce fuel and process waste. Fusion is therefore unlikely to make a significant contribution to meeting global energy before the mid-21st century. It is, though, a worthy long-term goal and an option that arguably we cannot afford to ignore.

8.2.6 Overall prospects for fossil carbon savings to 2025

Such are the errors of estimation it is not worthwhile presenting an overall picture for potential carbon dioxide emissions for the present (taken here as 2000) and those possible for the near future (taken as 2025) to greater than one or two significant figures. Table 8.3 below depicts the overall current and near-future carbon scenarios with the latter considered both in a B-a-U way (i.e. continuing as we did in the 1990s without adopting any greenhouse strategies) and what might realistically be done with greenhouse measures. Note the total carbon emissions include all those associated with human activity and so include fossil fuels, cement manufacture and land-use change.

The second and third columns of Table 8.3 tie in with many of the IPCC SRESs. In part this is a cheat on my part because most of the IPCC's scenarios do not diversify until the second quarter of the 21st century onward: after then, things not only get difficult to summarise but the outcomes and scenarios are so varied that any attempt to include them would have limited meaning. However, although calculated independently, the final column of savings happens to approximate broadly the IPCC's B1 SRES optimistic scenario, especially the high with-savings total. The low with-savings total also ties in with the extreme low-fossil SRESs. Although this scenario could in theory be realised, it would be foolish to assume that it will occur in practice, for the global political and economic will for this needs to be considerable.

It should be noted that the non-fossil contribution includes elements of both renewable energies and nuclear fission. It may be possible to invest solely in non-nuclear renewable energy but as some of this would be new unproven technology (or established technology operating in untried circumstances), individual nations' policymakers have to decide whether to go down the nuclear route. Globally speaking, from the 1970s to the Chernobyl incident in 1986 nuclear-generating capacity was growing at between 25 and 30 GW year^{-1}. After Chernobyl capacity typically grew between 1 and 5 GW year^{-1}. However, this belies a more complex picture revealed by the number of nuclear reactors starting to be constructed. These peaked *before* the Three Mile Island nuclear plant incident in 1979 and declined markedly after it (Worldwatch Institute, 2003). There is little doubt that these two incidents undermined global confidence in nuclear power. Consequently, the contribution nuclear is likely to make in the future is uncertain, although the potential is there.

Nonetheless, even the broad-brush picture that Table 8.3 portrays provides a couple of illuminating points as to our species' near-future relationship with the carbon cycle and climate. First, without taking any action to combat carbon emissions and to continue as we are, anthropogenic carbon emissions are set to approximately double between 2000 and 2025. Second, even with the adoption of major greenhouse measures (developing non-fossil fuels, enhancing energy efficiency, reducing deforestation and engaging in some reforestation) we will barely stabilise emissions.

The IPCC 2001 report, based on the then-peer-reviewed science, is quite clear as to what a B-a-U future without greenhouse-combating measures will look like. By 2025 global temperatures will have increased by 0.75–0.9°C above 1990 temperatures and by the century's end temperatures will have risen by 1.4–5.8°C. The IPCC's 2007 report did not substantially change this assessment. At the high end this would take us well beyond Quaternary climates and the likelihood of the IPCC's warning of climate surprises becomes very pertinent (see Chapter 6). With greenhouse measures by the end of the century the temperature is still likely to exceed anything seen during our own time (that of *H. sapiens*) and especially the past 11 700 years (the Holocene). Our world will change.

8.3 Energy policy and carbon

It is the human influence on the carbon cycle that is changing our world, above and beyond the direct impacts of our species' physical presence. Of course these impacts are affecting the global carbon cycle through land-use change, but beyond this it is fossil fuel emissions that are by far the dominant factor (see Table 1.3). Consequently no exploration of the human ecology (the way our species relates to others) of climate change can be complete without looking at the energy dimension and energy policy in carbon (and non-carbon) terms. This itself is a huge subject worthy of a number of books in its own right.

However, it is perhaps worth looking at five case studies. The first relates to a country whose inhabitants emit the most carbon when taken as a combination of a per-capita basis *and* as a nation: the USA. (Although of course if looking purely at a per-capita basis there are smaller nations that consume more.) The second is about Canada. The third is the UK, whose citizens along with a number of their Western European peers (and a number of other developed nations such as Japan), emit a little over half that of their US counterparts. Then, we briefly examine India and China. Their citizens each release less carbon than North Americans and Europeans but because they are developing and because India and China's populations are 1.1

and 1.3 billion, respectively (CIA, 2008), together they represent around 38% of the Earth's total population and consume (in 2004) 18.7% of fossil carbon produced globally. It is these countries that will generate much of the new climate forcing expected this century. Finally we look at Australia and New Zealand as examples of two developed nations with smaller populations.

All told, in 2010 the USA and China accounted for just over 40% the planetary fossil carbon consumption (BP Economic Unit, 2011) and coincidentally together had nearly the same proportion of the global population. However, the average US citizen individually consumes more energy than their Chinese and Indian counterparts (see Figure 7.1).

Turning away from purely carbon concerns, energy policy will seem – to some – to have little to do with human ecology. Indeed, in the past many countries had little regard for the environmental impact of energy-resource utilisation. This has slowly changed, at least among the most developed (OECD) nations. Nonetheless non-fossil energy-policy decisions do affect fossil energy use (consider France's nuclear energy policy in the latter half of the 20th century). And even fossil energy-policy decisions can affect the degree of a nation's anthropogenic climate impact (consider the late 20th century's so-called dash for gas). Both non-fossil and fossil energy-policy decisions have implications for environmental sustainability. These then are the reasons why anybody trying to understand the biology of climate change needs to have an appreciation of the energy-policy implications for human ecology.

Each of the nations in the case studies that follow has a fossil carbon and energy graph (Figures 8.3–8.9, below) that conveys its carbon energy relationship for at least the last three decades. Each depicts the nation's fossil fuel energy consumed (the total of gas, oil and coal) in million tonnes of oil equivalent (mtoe) and the fossil fuel energy produced. If the production line (dashed lines) is below the consumption line for fossil fuels (thin solid lines) then the nation is not managing its needs self-sufficiently (this is just one, limited, aspect of sustainability). The separation between these two lines indicates the amount of imports required. Conversely, if the fossil fuel production line is above the consumption line then the nation is a net exporter of fossil fuel. The thick solid lines denote the nation's total energy requirements including non-fossil carbon energy. Therefore, the gap between that and the line for fossil fuel consumed denotes the amount of non-fossil energy that the country produces (mainly nuclear power and HEP but also including wind and other renewable energy). This gap can be viewed (the finite nature of global uranium notwithstanding) as indicative of the proportion of sustainable energy that a nation has in its energy mix.

8.3.1 Case study: USA

As noted in the last chapter, the USA, with less than 5% of the planet's population, generates nearly a quarter of the Earth's anthropogenic greenhouse gases. In part this is due to the individual high energy consumption of US citizens. Such per-capita consumption has been growing steadily since the Industrial Revolution and in no small part is due to the nation's indigenous resources: initially of wood and then coal and oil. However, the finite nature of US oil became apparent as long ago as the late 1940s.

If any single thing in the latter half of the 20th century could have shaken US citizens from profligate energy consumption and demonstrated the need for a successful energy policy then it should have been the 1973 oil crisis. At that time oil from the Organization of the Petroleum Exporting Countries (OPEC) increased from US$1.77 to $7.00 per barrel at a time when nearly half of the USA's annual oil demand (818 mtoe) was met by imports. However, the 1973 crisis did focus some political attention on energy concerns. Then the western hemisphere's energy gap (i.e. the gap between production and consumption) was some 9 million barrels per day, or around 13% of global refinery throughput.

In 1973 US President Ford proposed a series of plans to ensure that by the end of the decade the USA would not have to rely on any source of energy beyond its own. These proposals were packaged in what the President called Project Independence 1990. However, by 1990, 40% of US primary energy needs were still being met by imported oil. With the benefit of hindsight few today might be surprised at this failure. Back in the 1970s some energy-policy experts had their doubts. David Rose was one such, and in January 1974 he condemned US energy policy as unrealistic and unworkable without the 'application of Draconian measures' (Rose, 1974). Indeed, the question energy-policy experts were asking at the time was whether the President's policy ideals would be formally adopted and implemented by Congress. So, how did matters turn out?

In 1990 the USA produced some 417 million t of oil and 561 mtoe of coal. Yet because the USA (only) consumed 481 mtoe of coal it had a coal surplus of some 80 mtoe (17%). US oil consumption was 782 million t, resulting in an oil deficit of 365 million t. Natural gas consumption, 486 mtoe, just exceeded production, by 23 mtoe (or 5%). In total in 1990 the USA saw fossil carbon consumption exceed production by 21% or 308 mtoe. In terms of overall fossil sustainability (combining oil, gas and coal production and consumption figures) the USA has been importing increasingly more fossil carbon to meet its needs. Notwithstanding the climate dimension, given the finite nature of fossil reserves this fossil carbon trade deficit is clearly not sustainable (see Figure 8.3).

Today the US fossil self-sufficiency situation has deteriorated further. In 2010 the USA consumed the following primary fossil resources: 850 mtoe of oil, 524.6 mtoe of coal and 621 mtoe of gas, a total of 1995.6 mtoe. In contrast, it produced 339.1 mtoe of oil, 552.2 mtoe of coal and 556.8 mtoe of gas, a total of 1448.1 mtoe. In short, the fossil deficit between 1990 and 2004 of US fossil carbon consumption over production more than doubled, increasing to just over 50% of its fossil fuel production level. This growing fossil deficit has been a hallmark of the USA's energy-resource profiles throughout the post-World War II years.

The finite nature of US fossil energy resources was predicted by energy experts as long ago as the late 1940s. As mentioned, the 1973 oil crisis served to highlight the problem. Today, with the benefit of hindsight, we can see that President Ford's Project Independence 1990 failed. Indeed, just one and a half decades after Rose's 1974 condemnation of US energy policy, Raymond Seiver, in an introduction to the 1980 reprint of Rose's paper, noted that 'when Rose's article was written, in late 1973, there was no USA energy policy. In mid-1979, there was still none. How long will we wait?' (Seiver, 1980).

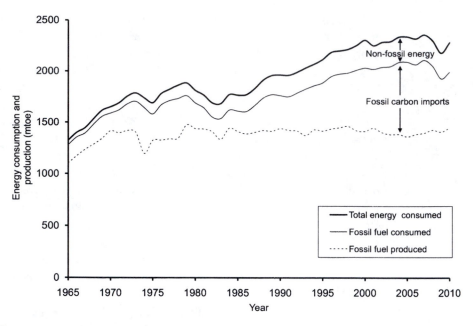

USA's fossil carbon domestic production and consumption trends. US domestic fossil carbon production remained broadly constant since 1975 but overall fossil consumption has increased. Further, non-fossil energy has been increasingly relied upon to meet the nation's total energy needs.

Regrettably a decade later, in the early 1990s, the target decade of President Ford's initiative, the answer remained effectively unchanged. This despite the US Energy Policy Act in 1992 that encouraged renewable energies and natural gas (with its higher energy content per carbon atom than coal or oil), and the provision of more research and development as well as the extension of tax credits. Even so, the US Department of Energy estimated that by 2000 annual fossil savings would be of the order of just 2.5% of the USA's 1993 primary energy consumption. Such a saving, though significant, is small. As it was, the Act ignored car efficiency and lacked major economic incentives proposed by energy-efficiency groups (Eikeland, 1993).

As can be seen from Figure 8.3, since then trends have still continued more or less the same: US fossil consumption has grown, although its domestic production has remained static so that the US fossil deficit has increased commensurably. More recently, in 2001 the National Energy Policy Development Group under the chairmanship of Vice-President Dick Cheney reported back to President George W. Bush, who had instructed it to assess America's energy position and policy at the outset of his administration's first term. Then the President had said, 'America must have an energy policy that plans for the future, but meets the needs of today. I believe we can develop our natural resources and protect our environment.' The report, entitled *National Energy Policy*, highlighted the need for some 38 000 miles of new gas pipelines along with 255 000 miles of distribution lines as well as increasing oil-refinery capacity, and a significant upgrading of the nation's electrical transmission grid (National Energy Policy Development Group, 2001). It estimated that over the next 20 years the nation's oil consumption would increase by 33%, natural gas by

50% and electricity by 45%. Yet it noted that both domestic oil and gas production were failing to grow to meet increases in demand. Conversely it observed that in terms of its economy's energy intensity – energy consumed per US$ of GDP – it was becoming more effective, in that since 1980 it was getting more dollars per unit of energy consumed. Above all, it called for increased energy production by not just renewable energy and nuclear power, but all fossil fuels including natural gas. This last was particularly interesting in terms of energy sustainability because estimates for the ratio of US reserves to (economic) production for gas were then of the order of just a decade. However, the Policy Development Group anticipated this criticism by calling for a vigorous domestic fossil energy exploration programme, part of which would be in an opened part of Alaska's Arctic National Wildlife Refuge. It also called for the USA to build strong relationships with its overseas energy providers, namely the Middle East. In short, it assumed that the nation's fossil carbon demand needed to increase along with increases in alternative energy so as to meet continued growth in energy consumption. In 2005 the essence of the report was enshrined in an Energy Bill, for a brief summary of which see Kintisch (2005).

In light of the above, it was not surprising that George W. Bush's administration also refused to sign the Kyoto Protocol. This meant that as a nation it continued not to have any targets for cutting fossil carbon use. 'As a nation' because some US city mayors and state governors have introduced local targets. In 2005 nine north-eastern US states – Connecticut, Delaware, Maine, Massachusetts, New Hampshire, New Jersey, New York, Rhode Island and Vermont – planned to cap some 600 of their power stations' carbon dioxide emissions at roughly 150 million t (about that year's level), starting in 2009. Further, between 2015 and 2020 those states plan to reduce power station emissions by 10%. The grouping represents about 3% of global fossil energy emissions, although some states have pulled out and rejoined, and at the end of 2011 New Jersey pulled out. In addition, two eastern Canadian states (Quebec and New Brunswick) have observer status in the initiative. The proposal came after 2 years from the Regional Greenhouse-gas Initiative. Alongside this there is the New England Governors/Eastern Canadian Premiers Climate Change Action Plan, which calls for a reduction in greenhouse gas emissions to 10% below 1990 levels by 2020. (This compares with the EU aim to reduce emissions to 20% below 1990 levels by 2020.)

Also in 2009, California (which as a state represents 12.19% of the US population and 13.26% of its GDP) pledged to reach 1990 levels of emissions by 2020. Although these US regional policy goals were welcome, and contrast sharply with US national policy of George W. Bush (less so with Barack Obama), they are only policy goals. As we shall see, and has been noted in the earlier subsection on the Kyoto Protocol (section 8.1.8), policy goals are one thing, but realising them quite another. Notwithstanding this, even if these states, or any country for that matter, did manage to lower their carbon emissions, care is needed to ensure that there has been no carbon leakage. Carbon leakage takes place when an area supposedly has reduced its emissions but in fact has ceased high-carbon-emitting activities and imported goods and/or energy the production of which elsewhere involved high-carbon emissions. (This is not to be confused with carbon leakage from soils, to which some IPCC reports refer.) In short, the savings the US regions are laudably proposing need to be genuine savings without carbon leakage.

In addition to the aforesaid Regional Greenhouse Gas Initiative (RGGI), there is the Midwestern Greenhouse Gas Reduction Accord. Launched on 1 January 2009, the RGGI was the first mandatory market-based US cap-and-trade programme to reduce greenhouse gas emissions.

Regarding the science underpinning the policy the 2006 report *Temperature Trends in the Lower Atmosphere* from the US Climate Change Science Program also signalled a change in how the USA perceived climate change science. Up to 2006, the Bush administration appeared reluctant to accept the IPCC's conclusions. This report, the first of a number from the Climate Change Science Program, recognised some of the apparent discrepancies of past climate data (such as from satellites). It concluded that 'the most recent versions of all available data sets show that both the surface and troposphere have warmed, while the stratosphere has cooled' (Climate Change Science Program, 2006). This is what one would expect with a world warming due to the effects of increased greenhouse gases. Although the report fell short of explicitly spelling this out, it did call for more research and assessment of climate-related data. The importance of this report is not that it said anything new as far as many US climate-related scientists were concerned, but that it carried weight with US policy-makers and so represented the beginnings of a shift in the way President Bush's administration was likely to view climate science.

As a nation, in the first years of the 21st century the USA was very much on the same energy track as it had been for more than 50 years. It continues to increase its fossil consumption and enables this through fossil energy imports and so, on a per-capita basis, continues to make a disproportionately large contribution to global carbon emissions. In terms of Kyoto, by 2004 the USA had increased its fossil energy consumption by over 18% above 1990 (the Kyoto base year). The 2005 US administration had policies in place that would allow for emissions to rise by 30% over the nation's 1990 emissions by the target window of 2008–12.

President Barack Obama replaced George Bush Jnr at the beginning of 2009 and Obama professed concern over anthropogenic climate change. He created the White House Office of Energy and Climate Change Policy. As noted earlier, Barack Obama actually attended the Copenhagen COP meeting and helped broker an accord (the Copenhagen Accord) with major greenhouse gas emitters and leading emerging countries. However, he has not ratified the Kyoto Protocol as it is unlikely that he could get approval for ratification from Congress and the Senate. He did try to introduce an emissions trading bill – the American Clean Energy and Security Act – but this did not get through the Senate. However, as of 2011 the mayors of over 400 US cities with a combined population of 60 million (around 20% of US) agreed that their constituencies should aim to realise the principles of the Kyoto Protocol.

8.3.2 Case study: Canada

Canada was a ratified signatory to the Kyoto Protocol, and in 2007 Canada's Kyoto Protocol Implementation Act (KPIA) received Royal Assent. This bound Canada to reduce its emissions as per its Kyoto obligations. Among other things it stipulated that the Government produce an annual report (KPIA Plans) detailing Canada's progress in meeting its Kyoto obligations. This it did through Environment Canada.[2] Canada's emission reduction target was that of 6% below its 1990 levels over the 2008–12 period. The first KPIA report, *A Climate Change Plan for the Purposes of the Kyoto Protocol Implementation Act 2007* (Environment Canada, 2007), despite allowing greenhouse emissions to rise until 2009, presented the goal of reductions thereafter. However, it was already clear that it would be unable to meets its 6% reduction on the 1990 Kyoto target. So in 2007 the Canadian Government announced a new target to cut its then current (2006) emissions by 20% by 2020. As such Canada was the first nation to publicly abandon its Kyoto target without leaving the Protocol.

In 2008 the Canadian Government produced the report *From Impacts to Adaptation: Canada in a Changing Environment 2007* edited by Donald S. Lemmen and colleagues (Lemmen et al., 2008). It was a scientific assessment published by Natural Resources Canada and Environment Canada. It concluded that climate change impacts could be discerned in every region of Canada, that there would be new risks and opportunities for Canadian infrastructure, that Canadian communities were vulnerable to extreme weather events, that Canada's potential adaptive capacity was high and that barriers to adaptation need to be addressed. As such, science and policy in Canada were largely chiming.

In the 2008 Speech from the Throne, the Government announced its commitment to work with provincial and territorial governments, and other partners, to develop and implement a North America-wide cap-and-trade system for greenhouse gases. The Canadian Government also re-committed itself to a national objective of a 20% absolute reduction in greenhouse gas emissions from 2006 levels by 2020. Its third KPIA plan (Environment Canada, 2009) noted that while 73% of Canada's electricity is generated by non-emitting sources such as HEP, nuclear and wind; the Government had set an objective of raising this to 90% by 2020. The Government was to also take action to reduce emissions from the transportation sector by regulating carbon dioxide emissions from cars and light trucks, and that energy conservation is critical and reducing Canada's emissions must also involve changes in how Canada consumes and conserves energy in the country's homes and offices. However, buried in an annex at the report's end was a table showing which of Canada's sectors (residential, commercial, transportation, etc.) would see emission reductions each year through to 2012. From this it is clear that Environment Canada expected the large industrial emitters sector to suddenly make major reductions in 2010 with a jump in reduction of just 0.6 Gt CO_2-equivalent in 2009 to 40.9 Gt CO_2-equivalent the following year. (Remember, CO_2-equivalent relates to the standardisation of all greenhouse gases – such as methane and nitrous oxide – to carbon dioxide based on global warming potentials as described in Chapter 1.) Such a jump in reductions here (let alone elsewhere in the plan) was at best optimistic. The sector sources of Canada's greenhouse gas emissions are given in Figure 8.4a.

---

[2] Environment Canada is the Canadian Government's Department of Environment. (In UK terms it is the equivalent of Department of Environment, Food and Rural Affairs [DEFRA] combined with the Environment Agency.)

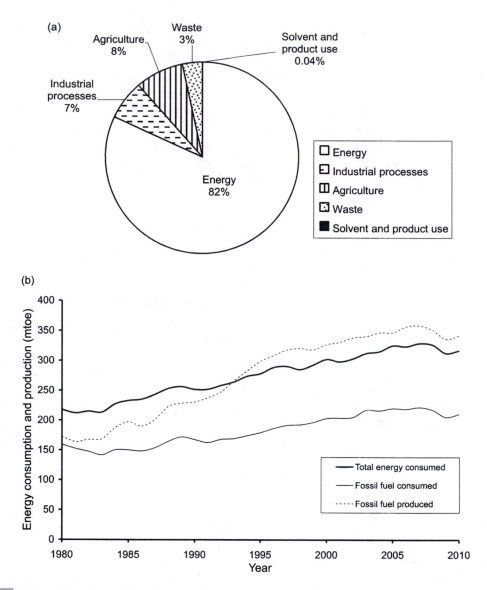

(a) Sectoral breakdown of Canadian greenhouse gas sources in 2009 (Environment Canada, 2011). (b) Canada's fossil carbon domestic production and consumption trends.

Against this backdrop Canada's fossil energy consumption (the principal source of the nation's emissions) increased from over 167 mtoe in the 1990 benchmark year to over 221 mtoe in 2007: see Figure 8.4b. (Remember that although it is not the only greenhouse gas in terms of warming the planet, carbon dioxide is the most important gas – see Figure 1.2 – and here the dominant source of anthropogenic carbon dioxide is fossil fuel combustion in Canada's energy sector.) This was a 32% increase in fossil fuel use and not a 6% reduction, even if that would be needed in CO_2-equivalents across all Canada's greenhouse emissions. Canada's fossil fuel consumption declined a little in 2009 in response to the global economic recession, but in 2010 it rose again. Canada was not going to be able to fulfil its Kyoto obligations.

In December 2011, just following the COP meeting in Durban (see section 8.1.11), Canada formally withdrew from the Kyoto Protocol. Canada will, however, continue with international negotiations regarding Kyoto's successor. Meanwhile, as with the USA, on a regional level in Canada there are initiatives to combat greenhouse emissions.

With regards to long-term potential greenhouse issues, Canada has extensive oil sands. These are more expensive and energy intensive to processes (which itself releases considerable greenhouse gas). As the world supply of cheap oil diminishes and oil prices rise, Canada has the potential to be a major oil producer. As the IPCC's scenarios are largely based on conventional oil futures, and less on extensive future harnessing of unconventional fossil fuels such as tar sands, there is potential for Canada, along with other nations producing expensive unconventional fossil fuel, to contribute to global warming beyond that which it forecasts later this century.

8.3.3 Case study: UK

The average UK citizen produces just over half as much carbon dioxide as their US counterparts (see Figure 7.1), but far less as a nation, as its population is just less than a quarter of the USA's. In fossil fuel terms the UK consumes (in 2011) around 200 mtoe compared to the US consumption of 2300 mtoe. As such, in fossil carbon terms it is (loosely speaking) broadly representative of Western European nations, although it uses slightly more fossil fuel per capita than most others. By 2009 in Europe only Germany as a nation consumed more fossil carbon than the UK. The UK is also roughly similar in fossil carbon terms to Japan and so can be considered to be part of a group of nations that represent a halfway house between our US case study and the large developing nations.

Perhaps the single reason why the UK is less fossil carbon-intensive per capita than the USA is due to oil tax. Notwithstanding this, the USA is far larger than the UK and so expends more energy on transport, and historically it has a higher proportion of energy devoted to air conditioning. Factors such as historically poor car fuel efficiency have also played their parts in determining the intensity of fossil fuel use. Automobile petrol has throughout the post-World War II period been cheaper in the USA than the UK by, very broadly (allowing for exchange-rate fluctuations), some 38% (see Table 8.4).

The UK saw considerable fuel switching throughout the 20th century and towards the century's end it strove for energy sustainability (meaning, in this case, self-sufficiency), albeit with only limited success as far as the very early 21st century is concerned.

Following World War II UK energy policy (such as it was) focused on recovery from the war's rationing years and then in the 1960s in reducing acid emissions from coal burning (mainly in domestic homes and then in electricity power stations). However, it was the 1973 oil crisis that prompted much debate over energy sustainability, possibly more so than in the USA, which had not seen its imports threatened during the war, as had Britain. In the years that followed 1973 the UK Department of Energy produced a number of Energy Papers, number 22 of which specifically dealt with national energy policy (Department of Energy, 1977). It defined the energy-policy

Table 8.4 Prices for car petrol in the money of the day in mid-1992 and mid-2005 (prior to the 2005 sharp rise in crude oil). Note: fluctuations in the exchange rate mean that caution is advised when making strict comparisons and hence an allowance of around ±5% is advisable. Nonetheless, this price comparison between nations is fairly representative				
Year	Automobile fuel	Price in UK (London)	Price in USA (east)
1992	1 US gallon	£1.75 ($3.24)	$1.30 (£0.74)
1992	1 UK gallon	£2.10 ($3.88)	$1.56 (£0.84)
2005	1 US gallon	£3.50 ($5.95)	$2.04 (£1.20)
2005	1 UK gallon	£4.20 ($7.14)	$2.45 (£1.44)

objective as securing the nation's energy needs at the lowest cost in real resources, but consistent with national security, and environmental, social and other objectives. It highlighted needs to develop guidelines for the future expansion of coal, to ensure that reactor technology was sound for post-1990 expansion if required, to have regard for the long-term role of North Sea oil and gas, to secure all cost-effective savings of energy and to ascertain the viability and potential of renewable energy. It also noted that 'There [was] an argument for devoting some of the benefits to come from North Sea oil to investment in energy production and conservation'. This energy paper was particularly important as it preceded a governmental energy-policy Green Paper. (UK Green Papers are discussion documents and usually accompany a public consultation as a precursor to a White Paper[3] that formally outlines forthcoming national policy and legislative intent.) The subsequent 1978 Green Paper (Department of Energy, 1978) was a landmark in UK energy policy for it represented the next most significant UK energy-policy official statement since the previous White Paper on fuel policy in 1967.

The 1978 Green Paper outlined the criteria for the formulation of a possible long-term energy strategy and covered much of the ground presented in the aforementioned 1977 Energy Paper. It addressed each of the primary fuels (gas, oil, coal, HEP and nuclear). It concluded that there was an urgent need to allow for the dwindling potential of global oil reserves to sustain economic growth and hence the need to switch to other fuels. It said:

> This change was in no way comparable with previous changes in the predominant source of World energy – from wood to coal, from coal to oil – since these changes were not in general dictated by an insufficiency of supply of the old fuel but took place as and when the new fuel appeared more advantageous. By contrast the change away from oil will be an enforced change and subject to the constraints in timing. We may have to accept that the available alternatives will be less convenient, more expensive or less environmentally attractive than the use of oil. Careful planning will be required if we are to effect the transition successfully and without damage to the global economy. (Department of Energy, 1978)

---

[3] A White Paper is a formal policy document of the government. Conversely, a Green Paper is a government policy discussion document.

In terms of manifesting the 1978 Green Paper's policy aspirations, UK politicians have both succeeded and failed. They succeeded in that UK oil consumption largely stabilised between 1979 and 2004 to 74–84 mtoe (1984 and its coal-industry problems excepted). It failed because there was no long-term decline in oil consumption. Indeed, spending on cutting fuel use through energy-efficiency measures, for example, was cut: the Department of Energy energy-efficiency budget fell from £24.5 million in 1985 to £15 million in 1990.

In 1982 the UK Secretary of State, Nigel Lawson, liberalised the energy market and made it competitive, which was to be governmental policy through to the 1990s. Since then UK Conservative governments have privatised gas, electricity and coal. Electricity privatisation was a success in that it met the policy goals of its time and reduced prices. However, the cost was in cutting the strategic development of new supply, closing the former Central Electricity Generating Board's research laboratories and shortening the industry's longer-term perspectives. This was to lead to supply concerns in the early years of the 21st century.

Yet UK politicians were concerned about global warming issues from fossil carbon consumption, and prior to the 1992 UNCED (see section 8.1.7) the UK was for a while leading Europe in calling for a carbon tax. (The problem was that unilateral action by a single country would disadvantage that country economically as it would be burdened with higher energy costs while its competitors would not.)

Following the 1992 UNCED, in 1994 the UK published *Climate Change: The UK Programme* (HMSO, 1994), which set out its proposals to combat anthropogenic climate change. The UK was the first country to fulfil this obligation made under the UNCED FCCC. It featured several measures that were considered by some as weak, and implementing it was not helped by the closure of the UK Department of Energy (which was then subsumed within the Department of Trade and Industry). However, *The UK Programme* did explore the aforementioned possibility of a carbon tax (as did some other Western European countries at the time). To date this has not come to pass. However, a tax (value-added tax) was added to energy sold to consumers. Originally it was proposed to be at a level of 17.5% but this proved so unpopular that it was implemented at a reduced rate. Indeed, it continued to be lowered so that by 1997 it was 8% and by 2005 just 5%. Another concern with this energy tax was that it did not discriminate between fossil carbon and non-fossil carbon energy sources. The UK strategy strongly relied on voluntary measures and what was known as the 'dash for gas': natural gas (methane) having a higher energy content per carbon atom than either coal or oil and so being more greenhouse friendly.

During the mid-1990s it became clear that the original *Climate Change: The UK Programme* was not enabling the UK to meet its climate policy goals and so other measures were introduced. First there was a levy on electricity from fossil fuel power stations. However, because of EU Charter restrictions this could not continue beyond 1998. Second, there was a series of Non-Fossil Fuel Obligations (NFFOs) that guaranteed the sales of electricity from a variety of non-fossil fuel sources. These too were criticised as weak and the third NFFO in 1995 related to just 1% of UK electricity production.

In 1997 the Labour Party took office in the UK. It published a number of policy documents, including *Climate Change: The UK Programme* (Department of

Environment, Transport and the Regions, 2000), which is not to be confused with the 1994 document of the same title. It affirmed the UK's obligation under the Kyoto Protocol (unlike the USA), stating the UK's commitment to reducing greenhouse gas emissions to 12.5% below 1990 levels by 2008–12, and stating its 'domestic goal' of a 20% reduction in carbon dioxide by 2010. The means by which this was to be achieved included a continued move to gas-fired power stations ('subject to cost'), increased contribution from renewable energy and a target to double combined-heat-and-power stations. (Such stations have a lower efficiency of electricity production but as waste heat is used to warm buildings the overall energy efficiency from fuel consumption to total energy – electricity and heat – delivered is greater.) It also formalised the UK commitment to the Kyoto measures of Joint Implementation (JI) and the Clean Development Mechanism (CDM). Climate-related criticisms of the new *UK Programme* included the cost restrictions on renewable energy and that it side-stepped the question of replacing existing UK nuclear plants approaching the end of their generating life. Nonetheless, its aspirations to address greenhouse gas emissions were commendable.

In 2003 the same government published the UK's first energy White Paper for over two decades. The *Energy White Paper: Our Energy Future – Creating a Low Carbon Economy* was published under the auspices primarily of the Department of Trade and Industry, and the Department for Transport and Department of Environment, Food and Rural Affairs (DEFRA) (2003). It was one of a number of Government and Parliamentary documents that used the phrase 'low-carbon economy'. Technically this was misleading as there was still an emphasis on switching from coal to gas (which contains carbon) and encouraging biofuels (which also contain carbon).

The 2003 *Energy White Paper* was the first energy White Paper to have as one of its core themes climate change concerns. The other core themes were energy diversity (for security purposes), maintaining a competitive energy market and tackling energy poverty (to prevent the poor from being unable to afford energy). Laudably – from a climate change policy perspective – it recommended, above and beyond its existing Kyoto and domestic goals, that the UK should reduce carbon dioxide emissions from the current (2003) levels by 60% by 2050. It also affirmed commitment to the policy goals of *Climate Change: The UK Programme* (2000) regarding renewable energy and combined-heat-and-power stations. However, on cost grounds it did not commit to any nuclear build, although it did 'keep the option open'. Another problem area identified by the White Paper was that of transport, especially road transport, and the dominance of the use of cars for personal transport. It said that the solutions lay with hydrogen and biofuels.

The White Paper's greenhouse-emission aspirations were lauded, although a number of commentators and policy analysts were not so confident that the paper had identified the means to deliver these. Climate policy-related criticisms included that it side-stepped the question of nuclear build (given the forecast decline in UK nuclear power which meant that a modest increase in renewable energy offset nuclear as a non-fossil carbon fuel rather than fossil fuel), and that the road-transport goals could not be fulfilled by domestic biofuel production (where even optimistic forecasts were a power of 10 too low for required levels) and hydrogen (which is an energy vector and not a fuel, as it takes energy to make the hydrogen in the first place). Such criticisms reflect those by others that the UK was not striving as it might to develop alternatives (for instance, see section 7.5.3). The other criticism, made less vocally at the time, was that the UK was leaking carbon (in the economic as opposed to the ecological sense). Since the 1980s it had been shutting down some of its energy-intensive industrial processes such as steel manufacture and so was importing such energy-intensive goods instead. This meant that the carbon emissions were being generated overseas.

Over the 25 years up to 2005 the UK has struggled to reduce its reliance on fossil imports and – for much of this time, from the mid-1980s onwards – to address greenhouse gas emissions. Since 2005 and the decline in North Sea oil production, the UK has not been self-reliant on fossil fuels (see Figure 8.5a). It has had some success in that total energy consumption has risen less than the USA's. Further, in a broad sense, it managed to roughly equate fossil fuel consumption with domestic fossil production. Again this was unlike the USA, but it did rely heavily on its North Sea oil and gas reserves, production of which peaked around the turn of the millennium. However, longer-term energy-policy concerns focus on the slow development of non-fossil carbon renewables as well as the lack of a nuclear strategy. Given the absence of the former, addressing the latter in some way was becoming an issue. Here, in Parliamentarian minds, concern over nuclear power related to public perceptions, and the uncertain long-term environmental impact of nuclear waste, counted against the technology. (Here it is perhaps worth being reminded of the comparability of past natural carbon [CIE] releases into the biosphere of the Eocene and Toarcian with that of the timescale needed to secure nuclear power's radioactive waste, as noted in section 8.2.5. Up to this time neither policy-makers nor the public were making such comparisons between the environmental impact of carbon emission and that of radioactive waste.)

By 2005 energy-policy concerns over UK security of supply, sustainability (mainly meaning medium-to-long-term national self-sufficiency) and climate came together. This was perhaps most strongly articulated by the UK learned community in a 2-day discussion symposium, 'Challenges & Solutions: UK Energy to 2050', followed by a half-day conclusion launch by a clutch of learned and professional societies led by the Geological Society, including the Royal Society of Chemistry, Institute of Physics, Institution of Civil Engineers and the Energy Institute and with the support of UK research councils. Only the absence of biological learned societies prevented nearly all of the major disciplines being represented. One of the main conclusions was the need for government to seriously consider a new generation of nuclear power stations (the current stations then coming to the end of their generating lifetimes). A month later, on 29 November 2005, Prime Minister Tony Blair announced an energy review based on widespread consultation that included the question of whether the UK should embark on a new nuclear programme. This review was due to be completed in 2006. Before then, though, there were two other documents published that throw an intriguing light on the use of science to underpin UK climate and energy policy and to meet UK climate-policy goals.

The first of these was a consultation launched in December 2005 on *Proposals for Introducing a Code for Sustainable Homes* (produced by the Office of the Deputy Prime Minister, 2005). The idea was that the Government was to introduce a code of practice that would apply to all government-funded home construction and be

(a)

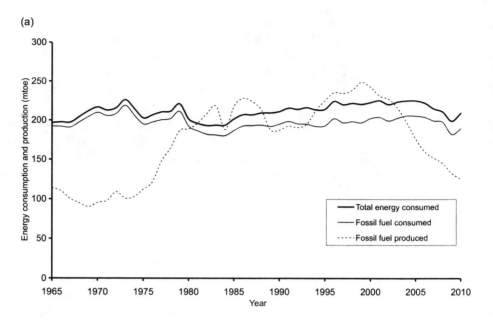

(b)

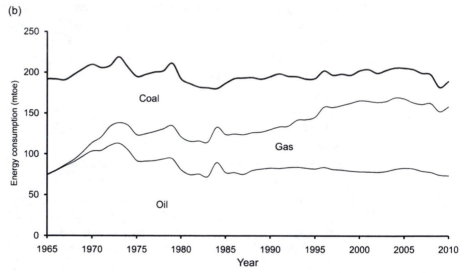

Fig. 8.5 (a) UK's fossil carbon domestic production and consumption trends. (b) UK's fossil carbon consumption broken down by fuel type.

voluntary for private builds. This code would help ensure that new homes were more environmentally sustainable and, importantly, impose a smaller carbon burden. But how committed was the Government in involving the UK science community in achieving this policy goal? The document's introductory section (page 7) stated that 'sustainability is not an exact science and is still developing', which is a fair comment. What perplexed some involved in science policy was that the appended list of those to whom the consultation was sent did not include the chartered bodies for biology, chemistry, geology or physics, or the Royal Society (Britain's 'academy' of science), even though the document appeared to recognise the need for science to underpin sustainability. The consultation document itself said that the proposed code was to include the need for homes to have a lifetime in excess of 60 years. Yet the proposed code was that the homes considered to be most sustainable would be heated by natural gas. Another concern from the greenhouse perspective was that the lowest level of the proposed code was the same as the then existing legal energy-conserving requirements. In short, builders, should they wish, needed not to do anything more than they had been doing already in terms of energy efficiency.

The reason for citing this policy document is that it is illustrative of one of the main reasons why successive UK governments since 1990 have had difficulty in meeting climate-policy goals. The second document actually recognised this difficulty and it was the Governments' *Climate Change: The UK Programme* (Department of Environment, Food and Rural Affairs, 2006; this was the successor to the *Climate Change: The UK Programme* 2000 and 1994 policy documents). Although some of the UK Government's 2006 policy documents had yet to be published the 2006 *UK Programme* did predict some of their content. For example, it stated that the then forthcoming code for sustainable homes would raise carbon standards above the mandatory level. Its forecast for UK electricity generation depicted a constant contribution from nuclear power between 2000 and 2020 that clearly suggests that policy-makers were actively considering building new nuclear reactors to replace old ones scheduled to be decommissioned. A number of policy analysts consider that the reason for *The UK Programme* 2006 document anticipating forthcoming carbon-saving measures was because *The UK Programme* 2006 itself predicted that the Government's previous goal to reduce by 2010 carbon dioxide emissions by 20% (over the 1990 base level) would not be achieved: instead, a new forecast of a 15–18% reduction was given. Nonetheless, the Government remained guardedly optimistic that its original long-term aspirations were still in theory achievable. In 2007 the UK Government produced another policy White Paper, *Meeting the Energy Challenge: A White Paper on Energy* from the Department of Trade and Industry (2007; the government department then responsible for energy). It sought to address the twin problems of climate change and energy security and, as with the 2003 energy White Paper, without causing fuel poverty. It proposed that the UK foster an international framework to tackle climate change and provide legally binding carbon dioxides targets. These last would be delivered by increased use of renewables, improved efficiency of energy production and use, and carbon capture and storage.

The UK created a new government department in 2008 to tackle the twin issues of climate change and energy with the Department of Energy and Climate Change (DECC). As part of this reform the government also created an independent Committee on Climate Change that was to set legally binding emission reduction targets and advise on other climate change policy concerns. One of its first decisions was that the UK should aim to cut its 1990-level emissions by 80% by 2050. In 2009 the Government published a White Paper called *The UK Low Carbon Transition Plan* via the Office of Public Sector Information (2009) in an attempt to identify the policies needed for this 80% cut. However, in 2010 the Committee on Climate Change warned that a step change in reducing UK emissions was still required as reductions in 2009 were largely due to the effects of the 2008/9 global financial recession.

One problem the UK has is that while its emissions are slowly reducing, its dependence on fossil fuel has barely reduced from the long-term 1965–2010 mean (see Figure 8.5b). How can this be: a reduction of emissions yet near-constant dependence on fossil fuel? The answer to this illusion of carbon dioxide emission reduction is 2-fold. First, the UK has increased its gas consumption at the expense of oil and (mainly) coal. Methane gas (CH_4) has four high-energy carbon–hydrogen bonds per carbon atom whereas coal and oil have less than two. So natural gas provides more energy when burnt (oxidised) to form carbon dioxide than either coal or oil. The problem is that natural gas remains a finite fossil fuel and so has all the energy security problems associated with oil in addition to its use as a fuel being a source of greenhouse gas. What is needed is for Britain to increase its non-fossil fuel energy use and it is not doing this with any significance: nuclear energy is set to decline in the 2010s as no new reactor programme has been instigated since the Central Electricity Generating Board was privatised in 1989, and renewable energy, though growing in the UK, provides (2011) less than 3% of the nation's energy. Second, the UK had been reducing its industrial and manufacturing energy consumption. Yet British citizens were still purchasing manufactured goods but these were being imported and overseas manufacturers were emitting the carbon dioxide instead: this is economic carbon leakage (as opposed to ecological carbon leakage). In short, while the UK's stated goal of reducing greenhouse emissions is admirable and the realisation of this seemingly manifest (albeit modestly), the reality is that the UK has not put in place a long-term energy strategy reducing reliance on fossil fuel or ensuring its population's consumption of goods whose manufacture is not reliant on fossil fuel. Having said that, it is further along the road (again albeit modestly) to reducing its reliance on fossil fuel compared to nations such as those in North America.

In 2009 a policy move was made to begin to address this with the White Paper *The UK Renewable Energy Strategy* (H. M. Government, 2009). It promoted the security of UK energy supply, reducing its overall fossil fuel demand by around 10% and gas imports by 20–30% against what they would have been in 2020. It also aimed to provide opportunities for the UK economy with the potential to create up to half a million more jobs in Britain's renewable energy sector, resulting from around £100 billion of new investment. Whether these ambitious policy aspirations will be realised remains to be seen.

In 2010 Britain's Climate Change Committee published its *The Fourth Carbon Budget: Reducing Emissions Through the 2020s* (Climate Change Committee, 2010). It called for a 2030 target to reduce emissions by 60% relative to 1990 levels (46% relative to 2009 levels). Interestingly, its perspective was one of carbon dioxide equivalents (which is the more accurate way of looking at emissions; see Chapter 1) and to consider land-use change and ecological sinks. (The problems with the latter are difficulty with accurate measurement and ecological leakage, especially as encouraging carbon sequestration by ecosystems can make them prone to be carbon emitters with further warming and, as warming is expected over the next two centuries, this is a likelihood.) In May 2011, the newly elected Conservative-Liberal coalition government accepted *The Fourth Carbon Budget* in full.

From a purely policy-analysis perspective it is also worth noting that, reminiscent of its earlier contribution to the discussion of a possible European carbon tax, the UK has taken an international lead to encourage all nations to invest together in addressing greenhouse gas concerns. This was not a purely altruistic action. British politicians did not want to make their own nation's markets uneconomical through unduly high energy prices (that is, significantly higher than Western European nations) to pay for renewable development. In short, the UK is in favour of multilateral action to address greenhouse emissions and so it only undertakes limited unilateral action. The two leading UK Parliamentary parties have maintained this approach for over two decades.

Economics lies at the heart of politics, and also at the heart of climate change policy issues. That successive UK governments have recognised climate change as a problem since the 1980s, yet have failed since 1990 to set in train strategic policies to meaningfully reduce UK emissions of greenhouse gases, exemplifies political weakness as far as the environment is concerned. To help address this, in July 2005 the Government (through the Treasury's Chancellor) asked the former World Bank Chief Economist Sir Nicholas Stern to conduct an independent review of the economics of climate change. The review was taken forward jointly by the Cabinet Office and Her Majesty's Treasury. It reported both to the Prime Minister and the Chancellor on 30 October 2006 with an electronic version placed on the Internet, and a mass-market paper version published in 2007 (Stern, 2007).

Its conclusions were that climate change will incur real costs, both in terms of reducing the causes (reducing greenhouse emissions) and in terms of dealing with its impacts. Stern warned that costs could be as high as 20% of the world economy. However, by taking 'strong action' now we could 'avoid the worst impacts'. Action could cost as little as 1% of the world economy if emissions were reduced. However, delay would exacerbate matters. Importantly, Stern's report said:

> Action on climate change will also create significant business opportunities, as new markets are created in low-carbon energy technologies and other low-carbon goods and services. These markets could grow to be worth hundreds of billions of dollars each year, and employment in these sectors will expand accordingly. The World does not need to choose between averting climate change and promoting growth and development. Changes in energy technologies and in the structure of economies have created opportunities to decouple growth from greenhouse gas emissions. Indeed, ignoring climate change will eventually damage economic growth. (Stern, 2007)

This argument was not new. Some scientists and the green lobby had noted previously that climate change would have severe costs. Further, some had said that adopting low-fossil carbon measures would be advantageous through freeing an economy from its dependence on dwindling finite fossil resources. Indeed, both these arguments were detailed together nearly a decade earlier in *Climate and Human Change: Disaster or Opportunity?* (Cowie, 1998a). What was new from Stern was that this message had come from an independent, yet Government-commissioned, review led by a senior economist. What remains to be seen is whether such measures will be adopted, especially as at the time of publication the UK was still struggling to keep on track to meet its targets to reduce emissions.

With regards to any climate-stabilisation target, the UK view has been influenced by EU policy. Since 1996 EU policy documents have frequently referred to a carbon dioxide stabilisation level of around 550 ppm, which (depending on the greenhouse gas mix) has been associated with a 2°C rise above pre-industrial temperatures, although the EU itself has no official target. One reason for this figure is that European crop yields are expected to begin to fall when warming reaches 2°C. More recently evidence suggests that to ensure that the temperature does not rise above this 2°C limit levels will need to be kept below 400 ppm. As carbon dioxide levels in 2011 were around 390 ppm and rising at about 1.5 ppm per year it is virtually certain that the 400 ppm limit will be reached in a few years' time and, given that carbon dioxide levels are expected to rise faster as the century progresses, the 550 ppm limit will be reached around, or shortly after, the middle of the 21st century. The UK's current aspirational target (an 80% cut in carbon dioxide emissions below 1990 levels by 2050) is consistent with the UK's overall contribution to global emission reductions required for stabilisation at 550 ppm. Because curbing UK and global emissions is proving so difficult it therefore looks as if the 2°C limit will be exceeded.

8.3.4 Case study: China and India

China and India have many similarities in terms of energy and greenhouse emissions. Further, on the spectrum of energy used by citizens, they are at the opposite end to North America in the amount of energy that their respective citizens use, being currently low energy per capita in contrast to North America's high energy per capita. However, representing some 38% of the global population China and India both make up for their low per-capita energy consumption in high capita numbers (CIA, 2008). Further, as both are developing rapidly and so increasing their per-capita energy consumption, these nations are likely to dominate emissions from before the middle of the 21st century. This is especially inevitable because, in addition to their per-capita energy consumption increasing, their respective populations are also growing. Of the two countries, India is likely to see proportionally more population growth, for unlike China (whose population has a comparatively low fecundity of 1.77) India has an average of 2.76 children born per woman (CIA, 2008). This is despite China having the larger population, with some 1.33 billion (2008 estimate), or just over 20% of the global population, and is due to that nation having had a policy that actively discourages parents from having a second child.

Of the two, China has the higher per-capita energy consumption: over 1 t of oil equivalent a year in 2004 (inclusive of minority non-carbon fuels) but growing rapidly to around 1.8 t in 2010, compared to an Indian citizen, who consumed just over a third of a tonne of oil equivalent in 2004 but which has also increased to 0.46 t per person in 2011 (again, inclusive of non-carbon fuels). The per-capita fossil carbon emissions in 2002 and 2009 from energy are given in Figure 7.1. Of China's domestic energy consumption only around 6% comes from non-fossil carbon sources. Consequently China makes a considerable contribution to global anthropogenic carbon emissions and as a nation is currently (2012) the largest carbon dioxide emitter, having overtaken the USA in 2007. India and China, with some 38% of the global population, together accounted for nearly 19% of global human carbon emissions from fossil fuels in 2004

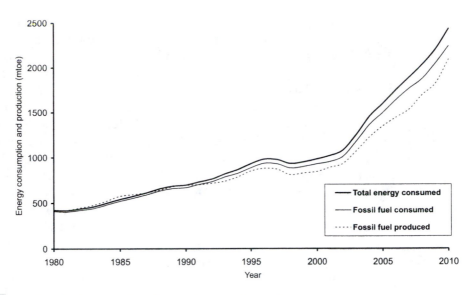

Fig. 8.6 China's fossil carbon domestic production and consumption trends.

and this rose to 26% (which itself had increased) in 2010 (BP Economics Unit, 2005, 2011).

China's fossil fuel consumption, its fossil fuel deficit (domestic production less consumption) and total energy consumption (fossil and non-fossil consumption) have steadily increased over the past quarter of a century, with an exception around the turn of the millennium when its coal production and domestic consumption declined. This dip was in part caused by a decrease in coal supply and also a move towards increased electrification combined with the transference of electricity regulation from the former State Power Corporation. This rise in energy consumption has fuelled its economy, and its GDP quadrupled between 1978 and 2004 and has continued to rise since. Its major energy-intensive industries include iron and steel making, aluminium refining, cement manufacture and coal mining.

As can be seen from Figure 8.6, China has gone from being a marginal exporter of fossil fuel prior to the early 1990s to a marginal (currently slowly increasing) fossil fuel importer. It does produce (hence consume) a small amount of non-fossil energy. This carbon and energy picture over the past 30 years is shown in Figure 8.6. The small amount of non-fossil energy can be seen in the difference between the total energy consumed and the total fossil fuel consumed, and this non-fossil contribution has been up to now clearly a minority energy resource. Both China's growing energy demand, and its small but increasing fossil deficit, have caused increasing governmental unease that has been detected internationally. This last is not a trivial point as China had very secretive governments in the latter half of the 20th century. However, China's recent opening up of its economy internationally has revealed particular governmental concern over the nation's growing dependence on imported fossil carbon (largely in the form of oil) and it is known that China is seeking to improve its energy security.

In November 2002 the Chinese Government announced an oil strategy. It included new fuel-efficiency standards for cars and lorries sold there. The first phase of the standards came into effect in 2005 and the new standards ranged from 38 miles US gallon^{-1} for the lightest cars to 19 miles US gallon^{-1} for heavier trucks. In 2008, the standards will increase to 43 and 21 miles US gallon^{-1}, respectively.

On a broader economic basis, China aims to quadruple its GDP from 2000 to 2020 while only doubling energy use. It will be a challenge to address its aforementioned energy security, and social welfare and environmental, concerns. It markedly contrasts with the years up to 2000 when energy use had been rising faster than GDP. As for the future, China's energy demand has been greatly increasing and is expected to grow at about 5.5% per year to 2020. By 2030, the International Energy Agency predicts the country will account for one-fifth of the total global annual energy demand. At the 2005 Kyoto next-phase meeting in Montreal, China declared an intent to double its use of renewable energy to meet 15% of its electrical demand by 2020.

There is no one body responsible for energy in China; the government's energy responsibilities are shared mainly between the State Development Planning Commission, the State Economic and Trade Commission and the Ministry of Land and Natural Resources. They respond to a set of objectives provided by China's Parliament, the National People's Congress. These objectives are set down in national Five-Year Plans. In China's tenth Five-Year Plan (2001–5) it aims to increase the share of HEP and natural gas in its energy budget. Its commitment to nuclear power is considerable and it hopes to more than treble this. So currently there is much emphasis on the construction of gas pipelines and terminals, as well as many more hydroelectric schemes and nuclear plants. However, the impression given overseas is that China's priorities lie with economic development first and environmental and climate concerns second.

In 2006 the International Energy Agency (IEA) published a report, *China's Power Sector Reforms: Where to Next?* It noted that every 2 years China adds as much power-generation capacity as the total in France or Canada. Also, the country is now the biggest electricity consumer in the world after the USA and its needs are still growing. It concluded that China needs to strengthen its institutional and governance framework to use energy far more efficiently. Importantly, in greenhouse gas and climate change terms, this is needed to tackle the environmental consequences of coal, which fuels 70% of China's electricity. To date China has adjusted some of its internal energy prices but much still needs to be done if it is to minimise its greenhouse impact. Development (as opposed to sustainable development) appears to be the overriding priority. However, in 2007 China issued reports on both climate change and its energy policies, *China's National Climate Change Programme* (National Development and Reform Commission People's Republic of China, 2007) and *China's Energy Conditions and Policies* (its first white paper on energy; Information Office of the State Council of the People's Republic of China, 2007): the former naturally had a climate change emphasis but, importantly, so did the latter. That year China's Standing Committee of the 11th National People's Congress, the nation's top legislative body, approved a resolution to actively deal with climate change. It stated that China will continue to participate constructively in international conferences and negotiations on climate change, and advance comprehensive, effective and sustained implementation of the international convention and its protocol. But it also said that China 'as a developing country' will firmly 'maintain the right to development', and opposes 'any form of trade protectionism disguised as tackling climate change'. Developed nations, in China's view, should 'take the lead in quantifying their reductions of emissions' and honour their commitments to 'support developing countries with funds and technology transfers'.

In 2009, China invested US$34.6 billion in the clean energy economy: nearly double the USA's total of $18.6 billion. Furthermore, China's growth in installing renewable generation grew by 79% over the 5 years to the end of 2009. This is compared with 40% growth in Australia, 31% in India, 30% in the UK, 24% in the USA and 18% in Canada (Pew Charitable Trusts, 2010).

China's 12th Five-Year Plan in 2011 (2011–15), in a first for its Five-Year Plans, included greenhouse gas emission 'reductions', so reversing the trend of previous decades. However, the same plan also included economic growth targets to increase GDP by 7% annually on average. As such the plan called for a radical decoupling of economic growth from fossil carbon consumption. The notional 'reductions' proposed were not as they might first have appeared. The country viewed its emissions not so much as the amount of gas released but emissions per unit of GDP, with the aim of reducing carbon dioxide emissions per unit of GDP by 17% from 2010 levels by 2015 and to reduce energy consumption per unit of GDP by 16% from 2010 levels by 2015. If China's economy was to grow by 7% annually over the 5-year period then the total economic compound growth would be 40%, and if it succeeded in reducing emissions per unit of GDP by 17% from 2010 levels by 2015 then its emissions would still be 16% more than they were in 2010.

China planned to realise these goals by (1) improving the efficiency of its fossil fuel plants, (2) expanding its nuclear programme and (3) further developing the use of renewable energy sources. With regards to the latter, in addition to large HEP schemes it planned to increase its share of the global solar photovoltaic (PV) module market by 10%, making it responsible for 59% of the world's solar PV industry. The Five-Year Plan also included a goal to increase the area of forest cover by 12.5 million ha by 2015.

Although renewables, as elsewhere in the world, only contribute a minority share to China's energy generation mix, this early 21st-century growth in the nation's non-fossil energy sector can only be welcomed by those concerned about greenhouse emissions.

India has less than a quarter China's national energy consumption. Yet, like China, it has seen considerable growth in both consumption and production, albeit at a far slower rate and over a longer time period, since the late 1970s (see Figure 8.7). Also like China, it has been almost exactly self-sufficient in gas, which it uses as a minority primary fuel and of which it has (according to 2010 reserve/production [r/p] ratios) around three decades' worth of reserves. Further, again as with China, India mainly relies on coal but it is not quite self-sufficient and so is a net coal importer. Combined with its oil deficit, this means that India has had a significant, and growing, fossil deficit for more than the past 25 years. It has just under the estimated size of economical coal reserves of China but, because it produces (mines) so much less, it has a higher estimated r/p ratio, of over two centuries. This means

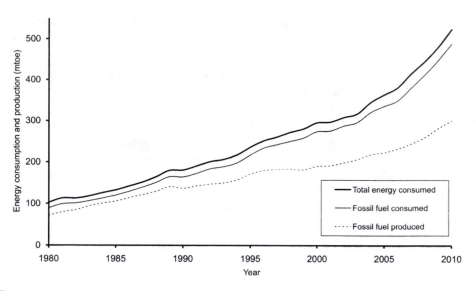

Fig. 8.7 India's fossil carbon domestic production and consumption trends.

that India, should it wish, could sustain its increasing coal production for much of the 21st century.

With regards to India's oil deficit, it produces just a third of the oil it consumes. Its oil consumption (in energy terms) is equivalent to just over half of its coal consumption and around four times its gas consumption. As with China, India has some non-fossil energy (represented in Figure 8.7 by the difference between total fossil fuel consumed and the total energy consumed). This non-fossil contribution is increasing but not nearly as fast as its fossil energy demand and in turn its domestic production of fossil fuel is increasing at a slower rate than consumption. Therefore, proportionally speaking, India has a greater fossil sustainability problem than China in that the gap between what it can produce domestically and its consumption is already increasing steadily.

With regards to energy policy, India's energy-policy formulation and implementation is split between several ministries: one for each fuel, including electricity, and also one for energy planning. Since the 1980s India's policy has barely changed and this is reflected in the way it produces and consumes energy and fossil carbon (Figure 8.7). Foreign investment, especially from Japan in the mid-1990s, has enabled the nation to develop its coal resources. However, around the turn of the millennium it was considered that the lack of a comprehensive energy policy had been a barrier to significant further investment and so the government published its *Hydrocarbon Vision 2025* report. This states that India should revise its foreign ownership regulations for refinery operations to allow 100% foreign ownership but be more restrictive of foreign control of petroleum trading. It also established a goal of India supplying 90% of its own petroleum. However, attempts to increase domestic oil production have had minimal success since the mid-1980s while demand has increased. So, like China, India has a net oil deficit but, unlike China, India's domestic production has effectively remained static since the mid-1980s.

In common with many other developing countries whose economies in the 1990s and 2000s were seeing considerable growth, part (although not all) of their increase in carbon emissions is due to carbon gain from developed nations. This is the opposite of carbon leakage. Such rapidly developing countries are tempted to take on energy-intensive industrial processes to make goods or raw materials for export to developed nations.

On 30th June 2008, Prime Minister Manmohan Singh released India's first *National Action Plan on Climate Change* (from the Prime Minister's Council on Climate Change, 2008), which outlined existing and future policies and programmes to address climate change: both mitigation and adaptation. The plan identified eight core 'national missions' covering the period up to 2017. These were related to solar power, improving energy efficiency, sustainable homes, water, Himalayan ecology, forestation, agriculture and strategic climate change knowledge.

With regard to 21st-century carbon emissions, neither China nor India are capable of sustainably accessing fossil carbon inside their own borders to fuel their intended ongoing decades of economic growth. This means that they will become increasingly subject to external fossil limitations, other than (for much of this century) for coal. In theory, should oil petroleum imports become sufficiently expensive, both India and China could synthesise liquid fuel from coal. Both China and India have a policy of increased fossil energy consumption to drive growth. However, should the developed (OECD) nations provide low-emissions technology at the same cost as the necessary investment in conventional energy technology (what economists call at no extra 'opportunity cost'), then they would explore this option. What they are not prepared to do is to halt their economic development.

This technology transfer from the developed nations is enshrined in the Kyoto Protocol. Participation of developing countries in carbon-emission mitigation is not specifically quantified by the Protocol, although it is a goal: India and China were not therefore required to limit their carbon emissions by 2012. However, the commitment to mitigate per se was conditional on the fulfilment of developing countries' own commitments to transfer to developing nations the enabling finance and technology. By 2012 this had not happened to the degree necessary to curb increases in emissions, let alone the rate of increase in emissions so far this century (see Figures 8.6 and 8.7).

Here some (not all) politicians in OECD developed nations, such as those from North America and Europe, argue that until countries like India and China curb their emissions there is little point in OECD curbing emissions. This view is a minority political view at the national level in Europe but sufficiently strong in North America to, so far, prevent the USA and Canada from committing themselves to binding emission reductions. One implication of this is that as North America and China together produce over half of global emissions, and less-developed nations are increasing their fossil carbon consumption, currently (two decades on from the first IPCC assessment in 1990) we are still on a Business-as-Usual emissions track.

Although European politicians (at national level) have policy aims of reducing emissions, neither they nor their North American, nor their Chinese or Indian, counterparts have closely examined the question of economic carbon leakage. Not all the economic growth derived from fossil fuel consumption in either India or (and, indeed, less so in) China has translated into material wealth for its citizens. India and China have provided developed nations with cheap goods manufactured with the energy from fossil carbon. If you like, China and India are gaining fossil carbon emissions from other developed nations that (having declined their industrial base) are economically leaking fossil carbon from their economies to China and India. If developed nations were not buying goods from China and India then not only would China and India's emissions be lower but the developed nations' emissions would be higher (through having to manufacture the goods themselves). The IPCC defines (economic fossil) carbon leakage as 'the increase in carbon dioxide emissions outside the countries taking domestic mitigation action divided by the reduction in the emissions of these countries'. It also focuses more on oil-exporting countries than the fossil carbon used to generate internationally traded goods. The IPCC 2007 assessment covers the question of economic carbon leakage in its Working Group III report on mitigation (IPCC, 2007b) but notes that 'critical uncertainties' remain as to its assessing the scale of the problem.

The traditional view has been that this international shift in industry and manufacture is all part of globalisation and necessary to bring living standards in China and India up to those in developed nations. However, a view borne of understanding the drivers of emissions globally is different. In 2008 Glen P. Peters and Edgar G. Hertwich from the Norwegian University of Science and Technology presented the following analysis. They found emissions of over 5.3 $GtCO_2$, or 1.44 GtC, embodied in international trade flows (increase this by a further 7.5% if the energy in the international transportation of goods is taken into account): this compares with the average annual global fossil fuel emissions in the 1990s of 6.4 GtC. From a global climate change perspective, they argue, it is more desirable to have production occur where it is environmentally preferable and then trade the products internationally. That is to say, it would be better for Europe to manufacture its own goods as its power stations and industrial processes are more efficient than countries like China. Another approach to reduce the impact of trade on climate policy, they say, is to adjust emission inventories for trade. Currently, emission inventories are *production-based* and this causes a separation between a country's consumption and the global production system. Arguably, it is this separation that causes the competitiveness concerns in the Kyoto Protocol. With *consumption-based* emission inventories consistency is returned between a country's consumption (which occurs domestically) and the production system required for the consumption (which occurs globally). Consumption-based inventories eliminate carbon leakage and encourage mitigation to occur where the costs are lowest.

8.3.5 Case study: Australia and New Zealand

Australia's energy and fossil energy sustainability has been one of self sufficiency throughout much of the latter half of the 20th century. Indeed, since 1980 its energy balance has become one of significant export compared to its domestic needs (see Figure 8.8). In part because of this – Australia has a strong fossil fuel lobby – climate change has been a politically contentious issue. Australia's emissions per person are among the highest in the world and roughly comparable to those of US and

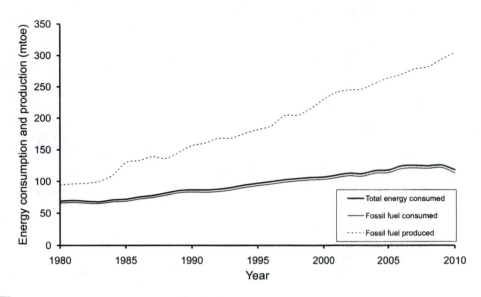

Fig. 8.8 Australia's total energy consumption, its fossil carbon domestic production and consumption.

Canadian citizens, hence twice those of most European countries (see Figure 7.1). Nonetheless, despite high per-capita fossil carbon use and much political debate over greenhouse issues, in 1998, under Prime Minister John Howard, Australia created the Australian Greenhouse Office, that was then the world's first such governmental agency. Then at the end of 2007 the Department of Climate change was established, and subsequently renamed the Department of Climate Change and Energy Efficiency. Australia aimed to have a national emissions trading scheme operating after the end of the first commitment period of the Kyoto Protocol (ending in 2012). Yet, perhaps not surprisingly, progress has been difficult and in 2010 implementation of Australia's Carbon Pollution Reduction Scheme was delayed due to lack of political support until after the Kyoto period ended (in 2012). Australia had a coalition government which meant that legislation either already had to have broad political support to be implemented or support had to be gained through arduous politicking. Here climate change science and politics met in the form of a report by the Climate Commission (from the Department of Climate Change and Energy Efficiency). This was in association with the Commonwealth Commission. The report, *The Critical Decade: Climate Science, Risks and Responses* (Climate Commission, 2011), drew on two science reviews of climate change, by Professor Ross Garnaut (2008, 2011). Nonetheless despite political hurdles, in 2011 Australia narrowly passed a carbon tax bill, the Clean Energy Bill, that would apply to around 500 large industrial emitters. This saw a levy of AUS$23 per tonne of carbon dioxide emitted and this will increase above inflation for 3–5 years, after which a broader emissions-trading scheme is envisioned. The aim was that by 2020 emissions would be reduced to 5% below levels seen in 2000, an overall saving of some 160 million t of carbon. Additionally, the government increased Australia's long-term emissions-reduction target from a 60% cut below 2000 by 2050 to an 80% reduction, and this despite Prime Minister Julia Gillard's pre-2010 election pledge not to introduce a carbon tax. Currently (2012)

(a)

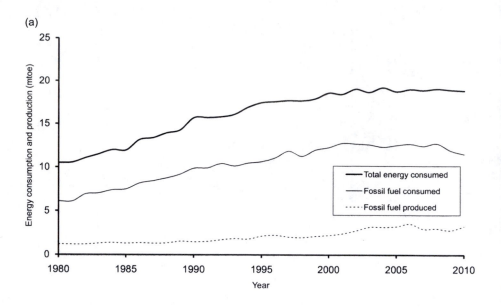

(b)

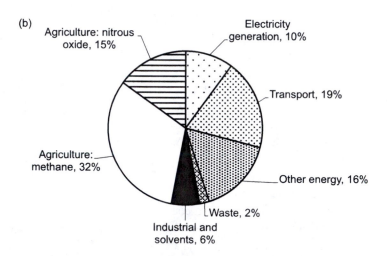

Fig. 8.9 (a) New Zealand's fossil carbon domestic production, consumption and overall (including non-fossil) energy for the past three decades. (b) New Zealand's greenhouse gas emissions by sector in 2008 (Ministry for the Environment, 2011).

the opposition parties say that they will campaign on removing these measures should they win the general election, which is at present scheduled for 2013.

New Zealand, of all this book's case study nations, is the one least dependent on fossil fuel (see Figure 8.9). New Zealand is not a big greenhouse emitter either as a nation or on a national per-capita basis. Nearly 40% of its energy comes from HEP and renewable energy sources, and it is proportionally least dependent on fossil fuel of all the nations discussed above. This is in no small part because historically it has relied heavily on fossil fuel imports, proportionally more so than any other nation examined in this chapter, although in gross terms its imports have been smaller.

New Zealand has many advantages when it comes to transitioning to a low-fossil carbon economy. These include a low population in relation to its land area, food sustainability and the potential for increased wind and HEP renewable resources. However, its population is set to grow and its agriculture intensify.

New Zealand is somewhat different to the other case study nations in this chapter, which need to curb their fossil carbon emissions if they are to significantly reduce their overall greenhouse gas emissions. Sources of emissions in these nations are similar to the global proportion of greenhouse gases emitted (see Figure 1.2b). Conversely, less than half of New Zealand's emissions (in terms of 100-year global warming potential; see section 1.2) in 2008 were from fossil energy use (carbon dioxide), and a minor contribution was made by industry and solvents (CFCs, etc.). However, 47% were in the form of methane and nitrous oxide from agriculture (see Figure 8.9b): New Zealand's agriculture is an important greenhouse emissions sector. As the country's agricultural industry intensifies, there will be pressure to increase nitrogenous inputs, which will hence lead to increasing nitrous oxide emissions. In 2007 the Agribusiness and Economics Research Unit at Lincoln University reported that New Zealand agriculture has such low energy and nitrogen inputs that the UK had 34% more greenhouse gas emissions per kilogram of milk solids and 30% more per hectare than New Zealand for dairy production, *even* including the emissions resulting from shipping to the UK (Saunders and Barber, 2007). Can New Zealand continue to be such a greenhouse-efficient agricultural producer with higher inputs in a more populated mid-21st-century world?

In August 2009 New Zealand announced an emissions-reduction target for 2020 of 10–20% below 1990 levels. The 2020 target is conditional upon an effective global agreement, including appropriate commitments by developed and developing countries, and rules relating to land use and forestry and carbon markets that are important to New Zealand (Ministry for the Environment, 2011). Then in March 2011 Climate Change Minister Nick Smith announced the Government's long-term target of a 50% reduction in New Zealand greenhouse gases emissions from 1990 levels by 2050.

New Zealand, starting in 2002, tried to introduce a carbon tax but, following a review of climate policy in 2005, did not introduce it. In 2008 it introduced the NZ Emissions Trading Scheme but the timetable for different economic sectors to enter the scheme is spread over a number of years: agriculture is not scheduled to enter it until 2015. The scheme (in common with a number of other trading schemes in different countries) has been criticised for the number of free emission permits allocated.

So far in this chapter we have looked at carbon drivers and carbon trends that have affected the anthropogenic manipulation of the carbon cycle. This manipulation has in one sense been inadvertent because it was not intended, only the harnessing of energy resources. These drivers have shaped our species' manipulation of the biosphere's carbon cycle, and will in all likelihood continue to do so in the future. So, given current carbon trends in developed (currently high-carbon-emitting) and developing (potential future high-carbon-emitting) nations, can we now manage our effect on the carbon cycle with a view to reducing our forcing of the climate, for example as the Kyoto Protocol intended? Also, can we also do this to meet sustainability concerns, in the economic sense, given the finite nature of comparatively easy-to-access economic fossil carbon? (Once again, note that resource sustainability and climate change concerns coincide.) Indeed, what are our prospects for successful carbon management?

8.4 Possible future energy options

8.4.1 Managing fossil carbon emissions: the scale of the problem

As noted, to have stabilised the global climate at 1990 levels around the turn of the millennium a reduction in human greenhouse gas emissions of 60–80% would have been needed (Chapter 5). Indeed, in order not to exceed an atmospheric carbon dioxide concentration of 50% above pre-industrial levels – 420 ppm – then we would have needed to start reducing emissions in the 1990s, or make greater cuts later, so that they would be reduced by 50% of their 1990 value by the year 2050 (IPCC, 1990). One question that repeatedly arose from audiences of the lectures I gave subsequent to this book's first edition (2007) was what were our chances of keeping warming to below 2°C above the pre-industrial temperature? (Note that 2°C above pre-industrial levels – or 1.2°C above the Earth's 2006–7 temperature – takes the Earth's global temperature above that seen any time in the current, Quaternary, ice age of glacial and interglacial cycles of the past two million years and this is considered the safe limit for warming.) I therefore wrote an online essay to address this question in 2009 (Cowie, 2009). My conclusion then was that 'it seems *very likely* (without a really major change in global human behaviour) that we *will exceed* our 'safe' 2°C above pre-industrial level'. Subsequently a number of others (who are arguably more informed than I) have come to the same conclusion (United Nations Environment Programme, 2010; Levin and Bradley, 2010; Rogelj et al., 2010; Friedlingstein et al., 2011).

The rate at which we are adding carbon to the atmosphere is so fast, and the amounts sufficiently great, that there has not been enough time for the Earth system to adjust and ultimately return to the pre-industrial state. It is therefore not so much a problem of reducing the rate and amount of emissions released in any one year, but keeping the cumulative total of carbon added since industrialisation to a specific limit. By 2009 humans added roughly half a trillion tonnes of carbon, 1000 GtC (or 1.83 t of CO_2) to the atmosphere. In 2009 a key analysis was conducted by primarily British scientists, and one German, who concluded that the total addition of a trillion tonnes of carbon (3.67 trillion t of CO_2) to the atmosphere since industrialisation would ultimately (a century later) see the Earth warm by 2°C (Allen et al., 2009). This conclusion was simultaneously echoed by a separate international team of Europeans (Meinhausen et al., 2009). Clearly, as we are currently adding over 7 GtC a year (see Table 1.3) and rising then, *should* current trends continue, it is likely that we will exceed this trillion-tonne limit in a few decades' time. Such is the magnitude of the reductions needed to keep within this limit, it is exceedingly unlikely that these limits will not be breached, especially given the current drivers and trends outlined in this chapter.

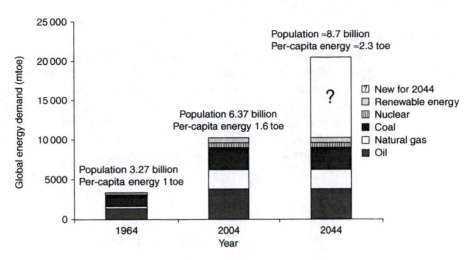

Fig. 8.10 Past, present and likely future global energy demands, for 1964, 2004 and 2044.

Exactly how great is this problem of managing fossil carbon emissions? To get an idea of the scale of the problem we can look at the energy demands of the past 40 years, the present and the likely demands of the future. Energy demand is the main contributor to atmospheric carbon dioxide, the principal anthropogenic greenhouse gas. To date, how we have met this energy demand?

Forecasting the future is not possible with any certainty but it is possible to make some reasonable assumptions to underpin a scenario. In this case the scenario is depicted in Figure 8.10. Three fundamental assumptions are made. One is that future global population growth follows the median UN forecast but, as the 2010 revision gave a higher mid-21st-century population estimate, this assumption based on the earlier UN estimate is more conservative (UN Population Division of the Department of Economic and Social Affairs, 2002, 2010). Second, future energy demand is a middle IPCC SRES (IPCC, 2000) and so close to many of the IPCC scenarios including those used in the 2007 assessment report. Finally, at least part of the energy contribution in 2044 will be similar to that in 2004. In other words, that we will be consuming at least a similar amount of coal, generating at least as much from HEP, etc., plus additional energy to allow for the economic and population growth that will take place between now and then.

In short, the assumption is that both global per-capita energy consumption and population will grow over the next 40 years up to 2044 but that this growth will be *less* than that which actually took place since 1964. Under this scenario global energy consumption approximately doubles to 2044. Indeed, the International Energy Agency in 2005 forecast that global energy consumption would be even higher: it forecast an energy increase between 2005 and 2030 of over 52%. So, the assumption of 50% growth by 2044 can be considered a conservative one.

Given the aforementioned facts, then if fossil emissions in 2044 are to be held at 2004 levels then annually we need to find a similar amount of non-greenhouse-impacting energy to the total energy used in 2004. The options, of course, are renewable energy, nuclear power and fossil fuels with carbon capture. Figure 8.10 visually encapsulates the scale of the problem by suggesting the likely amount of new fuel capacity required. This is not a trivial quantity.

The climate-related question here is whether fossil-related emissions can be kept at present levels or whether new fossil consumption (without carbon capture) will take place. When addressing such questions it is important to note that the limited ability of trees and soils to absorb carbon (assuming that a successful global sequestration programme is possible) can only at best (given likely boreal soil carbon release) offset a comparatively small proportion of any extra greenhouse emissions (see Chapter 7). Consequently, the biological implication is that we do run a considerable risk of increasing atmospheric carbon dioxide above 420 ppm (50% above pre-industrial levels). Indeed, unless radical action is taken we will exceed this 420 ppm limit, so incurring consequential biological and human ecological impacts.

8.4.2 Fossil futures

Virtually all the multinational oil companies and energy consultants forecast (as much as anyone can forecast) that oil production outside of the 11 nations of OPEC will peak around or before 2015: in 2004 they met 60% of international demand. After 2015 OPEC nations, especially its Middle Eastern members, will have to meet the vast majority of international oil demand. OPEC production itself might peak as early as 2018 or as late as 2025 (Kerr, 2005). So by 2044 the Middle East will in effect have strategic control of the global oil supply. This brings us to considering fossil energy considerations beyond our 2044 scenario. The one piece of good news is that both the UN low and medium global population forecasts see the population begin to stabilise; indeed, the low scenario sees the beginning of a decline (see Figure 7.5). Only the UN high-population scenario sees continued growth. Conversely, other than coal, *currently economically recoverable* reserves of oil and gas are likely to be largely depleted. This means either that the price of these fuels will rise markedly in real terms due to scarcity (as per the law of supply and demand) or the price of these fuels will rise due to the exploitation of *currently* uneconomic reserves. In short, by the middle of the 21st century it is almost certain that we will be paying significantly more in real terms for these fuels. This will make currently uneconomical technologies such as many types of renewable energy and nuclear power more economical, though far more expensive in real terms than those commonly utilised today. They will also make the use of unconventional fossil fuel resources, such as orimulsion and tar sands, more economical (and here the greenhouse problem is that there is a sizeable reserve of such fuels). These price rises will herald the end of the post-Industrial Revolution era of cheap fossil energy with implications for the utilisation of non-fossil energy resources. Importantly, though, it will not necessarily significantly impact on the further perturbation of the carbon cycle or climate due to our past, existing and near-future fossil carbon-related activities.

In 2005 an independently researched report was published that had been undertaken for the US Department of Energy. The study for the report was led by Robert L. Hirsch and so it is commonly known as the Hirsch Report (Hirsch, 2005). It noted that many estimates for the date at which peak conventional oil production will take place are before 2025. It also said that to most easily survive the economic disruption that will follow the peak, policies to both facilitate production and curb oil consumption (through improved efficiency and the use of other sources of energy) would need to be instigated two decades in advance; that is, now.

On a purely economic basis real-term price increases in natural gas and oil will again make liquid- and gas-fuel conversion from coal more attractive. However, there will be economic repercussions in the 22nd century as economic reserves of coal, if heavily exploited, will themselves become depleted. Consequently, in the very long term (from a human perspective), all fossil- (including coal-derived) fuel costs will be subject to a real-term marked rise. On a climate basis, without some form of carbon capture (which at this time is not easy to envisage), extensive use of coal as predicted over such a scale of the coming century will mean that climate change is likely to follow the higher IPCC forecasts into the 22nd century (which has not to date been the subject of principal IPCC focus). What it also means is that by then the human perturbation of the carbon cycle will be of a comparable order to that of the early Eocene carbon isotope excursion (see sections 3.3.9 and 6.6.4), and hence increase the likelihood of setting in train some similar climatic event.

8.4.3 Nuclear futures

In part because of the above economics, and in part because it is a low- (although not entirely zero-) fossil carbon resource, nuclear fission is likely to become increasingly more attractive in the future. Although there are, and will likely continue to be, concerns over radioactive waste and potential nuclear proliferation, it is difficult not to see many countries adopting nuclear power, or increasing their existing nuclear-generating capability by the middle of the century. The scale of fission development (and its nature, such as developing or not fast-breeder reactors) will influence the rate at which high-grade uranium ore reserves are depleted. At some stage (somewhere from the middle of the century, but it is unclear when), as high-grade uranium ore deposits are depleted, the cost of fission will rise. Increased fission costs would be bearable if cheaper fossil fuel ran out and/or if fossil fuel combined with carbon capture was at least as expensive. This would enable lower-grade uranium ore be extracted and processed.

Ore extraction and processing currently has a carbon cost in that oil is used to transport the ore and energy is used to process it and to isotopically enrich the fuel. This processing and enrichment energy in our high-fossil-using present day currently necessitates some fossil carbon consumption (although nuclear power still offsets more carbon than it consumes; see Chapter 7). However, in a potentially low-fossil future (with high use of nuclear or renewable energy) this nuclear-fuel-manufacturing energy could conceivably come from renewable energy and nuclear power itself: such are the complexities that make energy futures difficult to predict. Currently part of this carbon cost is offset by nuclear electricity generation displacing electricity generation that would otherwise have been undertaken by fossil fuels. However, if lower-grade uranium ore is mined, then more oil is used in its transportation and more energy for processing and enrichment. Of course, if we were by then in a largely renewable- and alternative-energy future then conceivably fossil carbon would not be required to contribute to a nuclear infrastructure.

Increased future (beyond the mid-21st century) fission and fossil costs are likely to make fusion economically attractive: assuming, of course, that the current International Experimental Thermonuclear Reactor (ITER) project demonstrates fusion viability[4]. It is unlikely to be cheap, hence there are fossil carbon-release implications, but *if* fusion does prove technically viable, and economic in a more energy-expensive end of the 21st century, then fusion combined with a hydrogen economy could be a long-term, climate-friendly hope. There would, though, still be a cost in the form of radioactive waste and we would still have to live with the climate impacts already in train and also likely due from whatever carbon emissions are to come. Finally, like fission, fusion, with its fast-neutron flow, also has nuclear-proliferation consequences for any nation with sufficiently advanced technology.

8.4.4 Renewable futures

The aforementioned higher price of future energy will make the majority of currently expensive renewable energies increasingly attractive and so they are an option for nations that can afford to develop these resources. Large-scale hydroelectric schemes are the cheapest of large commercial energy sources. However, there are a limited number of sites available for developing these without incurring (currently?) unacceptable environmental impacts, not to mention socio-economic consequences. Currently (using 2010 data), HEP accounts for some 775.6 mtoe, which is about 6.4% of commercially traded global fossil- and non-fossil energy consumption (BP Economics Unit, 2011). It may well be possible to double this contribution by our 2044 snapshot date in terms of tonnes of oil equivalent but not in terms of its percentage contribution to the global energy market. This means that new HEP development will not be a factor that keeps energy prices down. Furthermore, HEP in itself does not offset fossil consumption for personal transport. Were there enough of it then it could theoretically be possible via a hydrogen economy but other forms of energy generation will be required for that option to work.

That currently expensive renewable energies will become economically attractive in the future, not to mention climate change concerns, will mean that these technologies will become increasingly developed. Yet because most renewable energies, such as solar, wind and tidal power, exploit a low-density energy resource they take up considerable land or sea-surface area. This in turn confers a significant local environmental impact. If renewable energies were extensively developed worldwide then this impact would be extended and begin to have a global effect. In addition, given the existing human domination of the terrestrial landscape and the current impacts that this has (most notably the post-Industrial Revolution increase in Holocene biodiversity loss), renewable energies are likely to compound current impacts and so not be the totally environmentally benign energy resources that they are currently popularly believed to be.

---

[4] Remember, fusion programmes have not recently had the full R&D funding that had originally been planned. For example, the 1976 US Energy Research and Development estimates and the funding anticipated by the US Magnetic Fusion Engineering Act were not met. By 2003 only half the funding anticipated by this act was made available (Nuttall, 2008). And the current IETR programme has ongoing bureaucratic hurdles and costs.

Having said the above, there is scope for a significant contribution from a mix of renewable energies. Further, there is scope for considerable microgeneration, by solar and wind power, or on individual dwellings in the urban setting. Here cost and individual motivation are factors that will affect implementation.

8.4.5 Low-energy futures

Low-energy strategies are those that maximise energy efficiency and conservation and so, in theory, have lower energy demand. It is important to be aware of the difference between energy efficiency and energy conservation: improved energy conservation means, for example, taking fewer car journeys, while improved energy efficiency means going further per unit of fuel consumed. As noted earlier in this chapter, there have been continual improvements in the efficiency of energy generation by power stations that use boilers. Mileage (distance travelled per unit of fuel) of the most efficient family cars has also improved. However, variations in family preference regarding type of car and in use mean that there is still much wastage, so improvements are still possible.

As also previously noted (see section 8.2.2) there are other behavioural difficulties with low-energy strategies. Improvement in home insulation resulted in some homeowners preferring warmer houses to lower energy bills. Another example from the UK was the 1990s fad for adding glass conservatories to the sides of houses. These were thought to help reduce energy bills by providing a green-housed thermal buffer. However, many home owners decided to treat conservatories as extensions to the all-year living space and so heated these in winter, increasing energy loss.

Yet, low-energy strategies are, in terms of reducing energy-driven climate impact, a necessary part of the energy equation. Nobody in normal circumstances would consider carrying water in a leaky bucket, and even if leaky buckets were the only type available one would surely choose a bucket with the fewest holes. However, people do not tend to think like that when it comes to energy. Nonetheless, using energy efficiently is an important part of reducing energy consumption, and hence lowering greenhouse gas emissions.

Low-energy strategies are most effective when included in settlement planning and building construction rather than when introduced later. In the developed world ensuring that planning and construction standards are such that energy efficiency is greatly enhanced is undermined by a lack of what the industry calls a level playing field. If firms and commercial interests are not obliged for their products to meet the same specific minimum energy-efficiency standards then a competitor is able to save on efficiency measures and so undercut the prices of suppliers who have paid to meet standards. Consequently, legally determining efficiency standards and monitoring these standards in the market are both essential if maximum energy efficiencies are to be realised. Again, human psychology is a factor. Many consumers prefer to pay less for a product today even if over several years they may pay more through higher energy bills. Surveys have also shown that managers in the workplace often see less value in addressing energy-efficiency concerns as they save little money (and often cost more) in the short term, even if long-term savings are realisable and ongoing. I have addressed this in more detail elsewhere (Cowie, 1998a).

Looking at energy efficiency and the prospect of low-energy strategies on an international basis, historically the developed nations used to export their old and inefficient power-generating plants, and technology which used that power, to less-developed and developing nations. This is a cousin of carbon leakage, whereby developed nations do not produce, but import, products that are likely to result in high carbon emissions in their manufacture. For this reason the Kyoto technology-transfer mechanisms are important to help developing nations develop more cleanly (efficiently). Without such technology transfer developing nations, whose overriding priority is growth, will develop in a more energy-wasteful way, so raising emissions more than they otherwise would have from such economic growth.

It is therefore generally agreed that what are required, be it domestically or internationally, are highly energy-efficient and low-energy-consuming market controls and standards that are monitored and enforced. Within this regulatory regime the free market would be allowed to operate. However, as mentioned earlier in this chapter, such measures have to go hand in hand with greenhouse-friendly energy-production strategies.

8.4.6 Possible future energy options and greenhouse gases

Given the above, and the scale of the likely future energy demand to meet an evermore energy-intensive and growing human population, it is a virtual certainty (barring some cataclysm) that fossil fuel consumption will continue to increase beyond present values as we approach the middle of the 21st century (and beyond); see Figure 8.10. Carbon capture, even if the technology works on the necessary timescale, is only applicable (as far as can be currently foreseen) to large, point-source carbon dioxide emitters, namely carbon-fuel power stations (be they fossil or biofuel) and carbon dioxide-emitting industries such as cement manufacturers, and here only really effectively if they are located near, or within pipeline reach of, the appropriate geological strata to act as a carbon store. Consequently, for the medium-term future continuing to use fossil fuels for the most part means continuing to emit carbon dioxide and at levels higher than at present.

Nobody can predict the future with certainty, which is why the IPCC have their various futures from the Special Report Emissions Scenarios (SRESs). Not surprisingly, given the aforementioned range of pressures for increased carbon dioxide emissions, all the SRESs forecast increased emissions until 2050 and only a minority predicted reductions after that. However, such is the scale of reduction needed to stabilise climate change, let alone return the global temperature to its 1990 level, that all the SRES forecasts result in global warming and none see carbon dioxide at less than 500 ppm through to the end of the 21st century. See how this compares with the discussion of climate and carbon dioxide levels in the earlier section on the Kyoto Protocol (section 8.1.8).

The alternatives to fossil fuel power – renewable energy and nuclear – are currently more expensive than fossil fuel. Market regulation (different to that currently in place) will therefore be required. Having said that, some countries may adopt civil nuclear power for other (military-related) reasons. The future therefore seems to be one of increased energy prices. Increased prices impact inherently most on the poor and poorer nations. This in turn will have biology-related impacts on poor-nations' population health and longevity as well as reducing their ability to pay for climate change impacts. Even so, such is the magnitude of likely energy demand, even before the middle of this century (see Figure 8.10), that the required growth in non-fossil energy generation will be unprecedented. This is not impossible but, as with climate change itself, it is new territory.

With regards to the developing and developed nations, carbon emissions and economic growth have been the subject of some discussion about compromise, and indeed it can be found in the Kyoto Protocol. This is the idea of contraction and convergence. The developed, wealthy nations would switch to a low-fossil, high-energy-efficiency economy and so contract or reduce their carbon emissions. Meanwhile, the developing nations would be allowed some leeway and so increase emissions. In this way the developing and developed nations, on a per-capita basis, would see their respective levels of emissions converge. The problem at the moment (since Kyoto) is that the wealthy developed nations *en masse* are not contracting emissions but increasing them, and the developing nations are similarly raising emissions. At the moment (at least) our species is firmly on a high-emission IPCC scenario track and, despite international rhetoric by the Kyoto nations, there seems little prospect of this changing (indeed, 5 years on from this book's first edition the high end of the high-emission track seems more likely). Climate change and climate change impacts are therefore not only inevitable but could be more towards the extreme ends of the range discussed in this book.

The consequence of this is that not only will we have to adapt to our currently slightly warmer world but we will need to make greater adaptations to an even warmer world in the mid-21st century and beyond. Of importance to policy-makers, the question is almost certainly no longer one of expenditure on either adaptation *or* greenhouse-emission mitigation, but expenditure on both, or paying for the consequences (namely the range of climate impacts and fossil fuel scarcity).

8.5 Future human and biological change

Anthropogenic (human-induced) climate change is undoubtedly taking place. We can measure greenhouse gas concentrations in the atmosphere and their respective heat-absorbing properties in the laboratory. We have detected how their atmospheric concentrations have fluctuated in the past: for example, between significant climate (glacial and interglacial) swings. We know that other factors also affect the global climate, within certain margins of error, depending on the factor being considered. So, the case for human activity causing climate change is virtually conclusive, even if the exact detail is unclear (Chapter 1).

What is a little less certain is whether we are in the process of triggering a biosphere event like the Initial Eocene Thermal Maximum/Palaeocene–Eocene Thermal Maximum (IETM/PETM). There is less evidence for this and considerable uncertainty as to the environmental thresholds that would have to be crossed to initiate such an event. Only in recent years has there been concern that the amount of carbon being

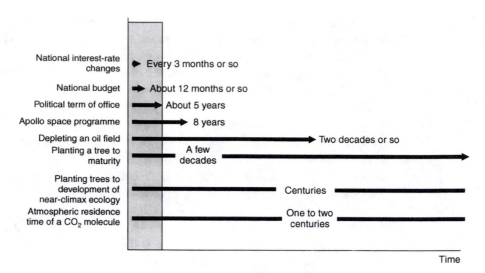

Fig. 8.11 Time horizons of various human activities, natural processes and atmospheric carbon dioxide. The political time horizon is primarily dictated by the length of term of office, which is a few orders of magnitude less than many energy-related natural processes; this time horizon (the shaded area in the figure) is also the typical period of *direct* policy involvement (responsibility). Reprinted from Cowie (1998a).

released into the atmosphere is of a similar magnitude to that involved in the Eocene event. Yet this similarity exists and was noted by the IPCC's 2007 assessment report[5]. Computer models as yet do not include the necessary biospheric components needed to simulate such events with sufficient detail/confidence, but they probably will in time. However, the evidence that does exist seems to point to an Eocene-maximum-type possibility. This suggests that future global warming impacts are likely to be very significant, especially in the long term from the point of view of humans.

Here the language and perceptions of scientists and policy-makers varies. Geologists commonly view matters in thousands to tens of millions of years and longer. Such timescales do not figure in policy-makers' operational perceptions. For them 2 or 3 years is short term and a decade or more is long term. Even the timescale of short-term climate science processes tends to be greater than that which policy-makers usually consider (see Figure 8.11).

Reconciling these two sets of perceptions is not easy. Consequently biosphere scientists need to recognise that their long-term concerns will not be readily (if at all) appreciated by policy-makers and so the scientists will need to adjust their presentations of science to politicians. Yet climate change and its associated impacts will be very significant on *both* these levels.

On the decade level we will see significant change and impacts. For example, over the coming years parts of the Earth (be it parts of the USA or the UK or South Africa) will become markedly drier, with profound economic and ecological effects. Over the coming couple of decades past weather extremes, such as the heatwaves experienced

[5] The IPCC AR4 (2007) can be accessed from the IPCC website and chapter 6 of the Working Group I *The Physical Science Basis* report (IPCC, 2007c), entitled Palaeoclimate, is freely available in both HTML and PDF formats. See the report's pp. 442–3.

in 2003 by people in London and Paris, will be normal for the summer. Heatwaves at that time will be that much warmer and their associated biological impacts greater (Chapters 6 and 7).

On the 1000-year level we will see several metres of sea-level rise. Large parts of Greenland will become ice-free. Year-round sea ice at the North Pole will have long gone (in fact we will lose end-summer Arctic sea ice in just a few decades' time) and we will see carbon cascade between various biospheric pools via the atmosphere, which will maintain global temperatures even if the global economy has by then switched away from fossil carbon. This will persist for many thousands of more years unless some means of sequestering atmospheric carbon is found. Over this longer timescale the oceans will rise scores of metres. All of Greenland and parts of Antarctica will become ice-free.

As said, many thousands of years, let alone 1000 years, hardly figures in the thinking of politicians or civil servants. Yet 1000 years, very nearly the time from the Doomsday Book to the present, is in human terms just 40 generations!

As I have previously argued, if environmental scientists, or green lobbyists (two quite distinct groups), wish to focus people's minds then they should learn from biology (Cowie, 1998a). Here the biggest bio-psychological imperative is associated with reproduction. Our bioclade (the sum total of life from a common primordial nucleic acid ancestor to date) has had over 3.5 billion years (Chapter 3) of Darwinian evolution dictating successful reproduction, enabling the selective passing on and evolution of genes. We are therefore programmed to care for our offspring. The intergenerational arguments of a climatically changed Earth, and the exhaustion of cheap fossil carbon fuel over the coming decades, need to be brought to the fore. Anyone born today will witness many of the climate change impacts discussed in this book and will almost certainly see the peak in consumption of global oil and gas as well as witnessing their decline to scarcity. This is not a shock statement but one of simple virtual certainty and something very pertinent to all parents.

Yet, if to help get the message across we recognise the biological imperative to nurture offspring then equally we should recognise other biological instincts that serve to counter moves to curb greenhouse gas emissions. While it is difficult on moral grounds to expect fast-developing nations (such as China and India) with low standards of living to forgo economic growth (and hence the energy needed to drive such growth), recognition of biological imperatives that encourage profligate resource consumption needs to be accepted. Affluent North Americans and Western and Central Europeans consume energy (and fossil carbon) not for essentials (basic warmth, clean water and so forth) but for luxuries (the SUV in an urban setting, overly warm homes and such). From a biologist's perspective this too can be viewed as part of the reproductive imperative and can be seen in many species. It is the driver behind the spectacular tail-feather displays of male peafowl such as the blue peacock (*Pavo cristatus*) or the large antlers of male fallow deer (*Dama dama*): the examples are legion. These signal to females that the males have the necessary genes to secure resources for such displays and so are worthy mates. The problem is that it can seem to be a never-ending race. As soon as one individual raises the stakes to a level above those of his or her peers, so the peers are forced to raise their own stakes if they are to compete successfully.

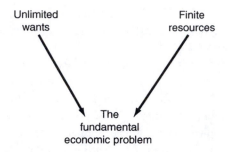

Fig. 8.12 The fundamental economic problem.

In humans such resource-control signals include being able to command resource consumption so as to afford the big car or the latest fashions. It leads to what economists call the fundamental economic problem (see Figure 8.12). Countering it is possible by changing perceptions of what is desirable. Humans can do this, being both a sentient and a social species, but just because it is possible does not necessarily make it easy.

So how is it possible to get around the fundamental economic problem? In biology and in the wild such displays have their limitations. Too large a set of display feathers, or antlers, can impede flight from predators. There are balances. With humans such checks and balances include limitations of access to resources (availability), which is where sustainability comes in. Failure to curb resource consumption leads to overconsumption and this applies to resources common to life on the planet. These common resources are principally the main biogeochemical sources and sinks in the biosphere and overconsumption leads to what is known by environmental scientists as the tragedy of the commons (Hardin, 1968). Global commons include the oceans and atmosphere. The overconsumption comes about through competition up to and beyond the sustainability limit. A simple model of grazing sheep on common land illustrates how this comes about. Each sheep herder is motivated to get the most grass for the flock. That the common can become overgrazed is a secondary consideration, for if a herder takes some sheep away from the common then sheep from other herders can have that grass. The result is that herders keep their sheep on the common, which then becomes overgrazed. The parallels between this and perturbing the fast carbon cycle, hence atmospheric carbon concentrations, should be clear. What the tragedy of the commons illustrates is the difficulty in modifying motivations so as to preserve commons. This brings us back to employing other powerful biological drivers such as the desire to nurture offspring (the next generation).

8.5.1 The ease and difficulty of adapting to future impacts

It is now clear that at best we might marginally slow global warming in the 21st century; it is most unlikely that it will be stopped, and virtually unrealistic to consider the Earth's system returning to the climatic regimen of the early 20th century. Consequently 21st-century policies addressing climate change, in addition to any mitigation, are going to have to either include adaptation or pay for the consequential costs of climate impacts.

Leaving aside temperature, in terms of human global distribution, sea-level rise is arguably one of the factors that will incur both impacts and adaptation costs. Worldwide some 70% of the global population live in coastal areas or near rivers. A significant proportion of these could potentially be affected by sea-level rise and/or possible flooding due to increased (seasonal?) precipitation. The IPCC (2001b) estimate that in Western Europe alone in 1990 some 25.5 million people were exposed to coastal flood risk and that of these roughly one in 1000 actually experienced flooding. Assuming no adaptation the IPCC consider that between half as much again to 90 times that number might be affected by the 2080s (see also section 7.3.3 and Table 7.1). The degree of uncertainty reflects the combination of uncertainty in the proportion of the population with a degree of flood-proofing and the uncertainty in the IPCC's forecasts. However, either way this represents a significant cost.

Adapting to climate change-related impacts, such as flooding, does not mean reducing risks or flooding to zero. This is for two reasons. First, to eliminate all risks and flooding would be the most expensive option and is likely to be too expensive for society/communities to condone. Second, not all flooding does significant damage or incurs a permanent cost. For example, you may be aware of a street near you that becomes difficult to traverse after a heavy sustained downpour but then drains comparatively quickly. There are also land areas that are naturally prone to, and might even benefit from, occasional flooding. Marshes and other wetlands are obvious examples. Finally, extreme flooding for some will be a very rare event and proofing everyone against these may be too expensive. In these cases 'soft' measures (such as allowing fields and parks to flood) as opposed to physical infrastructure (flood barriers) may be more economical. It is a question of assessing costs and risk.

There are two key problems of assessing costs and risks with climate change. First there is the fundamental assumption that the risk and cost assessments are accurate. The 2005 example of Hurricane Katrina and New Orleans is illustrative. The 1998 proposed protection plan *Coast 2050* had a potential cost of some US$14 billion. It did not get implemented as nobody was prepared to pay; the risk or hazard was not considered to be worth the cost. It was a gamble that such an event would not happen. Yet following Katrina the economic losses were in excess of $125 billion. (We will return to this shortly.) The second problem is that climate change is by definition dynamic and not static: past experience cannot be used as a guide to the future. For example, most of the New Orleans levees breached by Katrina were designed to deal with floods that occur once every 30 years or so. Now, notwithstanding whether this itself was an appropriate judgement of the degree of protection required, given the change in strength and frequency of extreme weather events due to global warming (Chapter 6), basing safety decisions on past criteria will be unreliable whereas including forecasts of future change would be prudent. That there is and will be global warming is not a gamble, even if the exact rate and extent of this warming are not exactly certain.

One of the countries that has historically spent the most on sea protection per length of coastline is The Netherlands. Quite simply this is because it is a low-lying country, with some 60% below sea level, and it is also a wealthy Western European nation. In addition, some 70% of its gross national product is earned in its low-lying areas and this provides a certain financial motivation. Indeed, the country is due to spend nearly $1 billion on a programme called ARK (which translated from the Dutch stands for Adaptation Programme for Spatial Planning and Climate) to generate a comprehensive agenda to deal with climate change across several sectors of society and economy. As with other low-lying areas, the Netherlands faces a number of climate adaptation questions. First, how to identify and cope with increased risk. Second, on what timescale will the adaptations (identified from answering the first question) need to be implemented: when does acting later become too late? Third, what are the comparative costs and benefits of alternative strategies? Such adaptations are not just restricted to enhancing conventional sea defences (dykes, etc.) For example, the 2003 summer was the hottest in Europe for some 500 years but it is thought that it will represent a normal summer in the middle of the century. During the 2003 summer the low discharge of the Rhine resulted in seawater seepage into the ground water that in turn affected agricultural and horticultural activities. One option therefore will be new canals to bring in extra fresh water. However, with low river flow securing an extra volume of fresh water is likely to prove difficult. Another option therefore would be to increase the area of land devoted to freshwater reservoirs. However, this will take land out of alternative use and income generation. A compromise therefore may be to build the reservoirs but to house large floating rafts, or hydrometropoles, on them to provide an area for, say, greenhouses (Kabat et al., 2005).

Of course, just as overall in a warmer world there will be more ocean evaporation and hence precipitation, so on land there will be more evaporation and some areas will see significant water shortage. So, global warming will incur costs to address water shortages and costs of surplus (floods). Even some wealthy countries, where politicians recognise climate change, ignore the full implications of the costs associated with such change, especially if such costs fall outside of government departments whose remit relates to climate and energy matters. The UK provides a typical example. Although the UK Government has plans to build homes for some 500 000 people, mainly in the area of Milton Keynes and north Kent, it ignored many of the major implications of climate change, about which its own agencies warned (see Hulme et al., 2002). In 2005 it launched a public consultation into *Proposals for Introducing a Code for Sustainable Homes* that would be used in the construction of these new (and any other government-funded) homes (Office of the Deputy Prime Minister, 2005). However, the proposals did not take into account the broader landscape and regional issues for the areas in which the homes were to be built. (Nor were the independent UK learned scientific – biology, chemistry, geology and physics – societies on the initial consultation distribution list.) Yet the areas in which nearly all the new build was to take place are already experiencing water shortages in the form of lowering water tables, and stream and river flow, and are forecast to see less rainfall with 21st-century climate change. Further, there were (at the time) no plans to enable water to be routinely transferred from parts of England and Wales predicted to see an increase in rainfall. Even the Kyoto-supporting government of one of the planet's leading economies was ignoring costs. And indeed, when the Code for Sustainable Homes was implemented by the Department for Communities and Local Government in 2006 it was voluntary and did not raise minimum standards. Although it did bring all the sustainability criteria for dwelling construction into one document it was hardly the 'step-change in sustainable home building practice' its subtitle suggested.

Part of the problem is the lack of understanding as to what sustainability means, the willingness to invest (it costs money even though there are long-term financial benefits) and the use of inappropriate terminology (see the beginning of the chapter). In case you thought that this problem of nomenclature is trivial, with regard to this last, in 2008 the UK's Department for Communities and Local Government conducted a consultation on the *Definition of Zero Carbon Homes and Non-Domestic Buildings*.

As indicated, climate-adapting measures have a substantial price tag. Even if the cost of adaptation is far smaller than the economic activity protected, such costs can only be paid if there is the finance both available and in place. In 2003 the international news media expressed surprise that the wealthiest nation had not protected itself from, and then was slow to address, Hurricane Katrina. As mentioned above this was because the estimated $14 billion was not available to implement a protection scheme even though the resultant cost from failing to make this investment was some nine times greater. Clearly the USA had the economic resources to make that investment but, for whatever reason (let's call it policy-implementation inertia) it did not. In Britain the government has been increasingly aware of the need to adapt to climate change and, for example, in 2011 its environment department (DEFRA) produced the report *Adapting to Climate Change: Helping Key Sectors to Adapt to Climate Change* (Department of Environment, Food and Rural Affairs, 2011). However, the proportion of the global population living in wealthy countries is a distinct minority (indeed, so is the proportion of wealthy within these countries). Indeed, Britain recognised that climate change has far broader policy implications and included a number of references to it in the H. M. Government's 2010 defence White Paper *The National Security Strategy*.

An example of a far less-wealthy nation to be significantly affected by sea-level rise is Bangladesh. Sea-level rise of just 100 cm (which is predicted by virtually all the IPCC SRESs for the 22nd century) could potentially flood nearly 30 000 km^2 of Bangladesh, representing over 20% of its land area. This will expose 14.8 million people, or 13.8% of its current population, to dislocation. The problem is compounded because the nation is not only one of the most densely populated but most of its population relies either directly or indirectly on site-dependent agriculture: over a third of Bangladesh's GDP comes from agriculture (IPCC, 2001b). Yet despite the threat of floods, its lack of wealth is impeding it from developing strategies to address future sea-level rise. This is evident if only because it currently suffers from regular monsoon flooding that it is equally unable to address.

This problem of the poor adapting to climate change is not restricted to sea-level rise but includes, as has been mentioned (Chapter 7), the human-ecological aspects of climate change, especially with regards to health and food security. Yet, as has been emphasised in this chapter, it is both the wealthy and those nations striving to develop that are dominating greenhouse gas emissions. So, while adapting to climate change will be easier (more affordable) for the wealthy, it is not an option available to (or within control of) the poor. Further, it is an unpaid environmental cost (or environmental externality) imposed by the wealthy and those accruing wealth.

If the poor will have a difficult time adapting, then what of natural systems? Semi-natural systems of woodland and agriculture will at least see management, as they have a recognised economic value. Of course, the most vulnerable components of such semi-natural systems will be those species and ecosystems within the semi-natural biome that have no immediate economic value attached to them, such as wildlife species (see section 8.5.3).

As for agricultural system regions themselves, by the decades around the end of the 21st century a good proportion globally will have to either change significantly due to warming and/or be relocated due to decreases in precipitation. Figure 8.13a depicts the principal regions in the world devoted to agriculture (both crops and pasture). Figures 8.13b and 8.13c are a computer model depiction of where increases and decreases in precipitation are expected in a world that is 4°C warmer, as might be expected under Business-as-Usual towards the end of the 21st century compared to its beginning. These show that whereas some parts of the world will see more precipitation (as expected in a warmer world with more evaporation from the oceans), in parts of the North American, Europe, South Africa and Australia agricultural areas will see reductions in rain. When looking at these model outputs (in addition to the usual caveats and caution that is required when looking at models) it is important to note that these represent annual averages and that seasonal changes will be more marked (a larger area of the world is likely to see a summer deficit). Also, in a warmer world plants will need more precipitation due to the increased evaporation (see section 6.6.6, and the discussions in section 6.1.3 on evapotranspiration and notwithstanding the section 6.1.3 discussion of stomata reduction in a carbon dioxide-rich world). This increased evaporation means that areas of a globally warmed world that receive the *same* precipitation as today could well be regions of agriculture water *deficit* in the warmer future. In short, comparing Figure 8.13a with 8.13c does not reveal all the current agricultural areas likely to be adversely affected through a decline in water around the end of the 22nd century. Of course, there will be new areas of agricultural opportunity in a 4°C warmer world, such as in parts of Canada, Siberia and South America. Taken together, a significant proportion of the world's food production systems will need to relocate and those that do not will need to adapt to the new temperature regimen with new crops and farming techniques. Such adaptation will be easier for those capable of making the necessary investments: the wealthy.

Mapping the likely impact of climate change on existing agricultural areas is currently being undertaken. Figure 8.13 only provides an initial indication of areas of likely risk. However, shortly after this book's second edition is published, the Inter-Sectorial Impact Model Intercomparison Project (ISI-MIP) is due to deliver its first results and so worth looking up on the Internet. This venture is largely being co-ordinated by the Potsdam Institute for Climate Impact Research in Germany.

As noted in Chapter 7, even natural ecosystems with no apparent immediate economic worth have a value, in that their ecosystem function enables biosphere processes on which we rely to take place. In theory it may be possible to attach a value to these ecosystems for performing these functions. However, acting practically on such environmental economic theory is not an affordable option for poor nations and, without assistance, their options for adaptation remain limited.

Finally, climate change often results in a significant step change as the system goes through a critical transition and the climate crosses a threshold. This is because many of the factors relating to climate forcing (see Chapter 1) and impacts (Chapter 6)

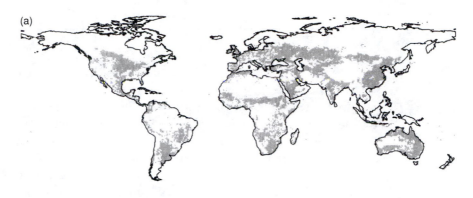

(a)

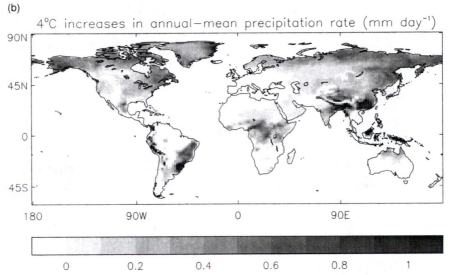

(b)

4°C increases in annual−mean precipitation rate (mm day⁻¹)

0   0.2   0.4   0.6   0.8   1

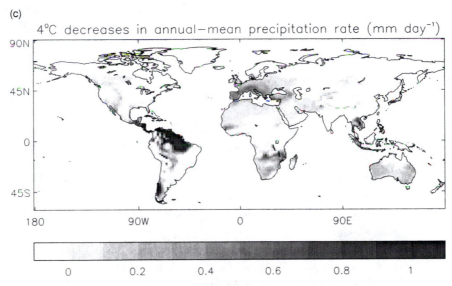

(c)

4°C decreases in annual−mean precipitation rate (mm day⁻¹)

0   0.2   0.4   0.6   0.8   1

Fig. 8.13 (a) Crop and pasture regions of the world. Adapted with permission from IPCC (2007c), figure 2.15, Cambridge University Press. (b) Regions of the world likely to see an increase in precipitation in a 4°C world. (c) Regions of the world likely to see a decrease in precipitation in a 4°C world. Panels b and c kindly provided by the UK Meteorological Office in 2011. For technical details as to the HadGEM2-ES model used for the CMIP5 centennial assumptions see Jones et al., 2011. © Crown Copyright 2011, the Met Office.

interact with varying positive and negative feedbacks until a new equilibrium is reached (section 6.6.8). Again, wealthy nations can afford to adapt or relocate climate-sensitive economic activity. This is not an option for poorer countries. For example, near-subsistence communities in the sub-Sahara can do little in the face of dune-stabilising vegetation dying due to climate change, and hence the consequential dune movement and covering by sand of other vegetation.

In short, the difference in the ability of societies to adapt to climate change is both wealth-related and likely to exacerbate wealth differences. The question of difficulty in adapting to climate change then becomes one of managing international development. This is an entire topic in itself, but the struggle that the global community has had in meeting the UN's millennium goals (primarily to reduce poverty and increase human well-being and environmental sustainability) is a testimony to how successfully it will manage international development to reduce climate change impacts on human well-being.

8.5.2 Future climate change and human health

As future climate change takes the planet beyond the climate regime of the Holocene (Chapter 5) a number of regions that have been inhabited in past centuries will experience new climate-related health impacts. Later in the 21st century, as the global climate regime becomes one not experienced since our species evolved, these will become more pronounced. That this will occur when the planet's population is more than four times greater than it has been at any time since our species evolved up to 1900 will serve to compound impacts on human well-being, and hence health. Global warming and this human demographic maximisation are related because it is human activity that is driving atmospheric concentrations of greenhouse gases beyond previous Quaternary levels (Chapter 7).

As with other aspects of climate change and human interactions, there will be winners and losers. Again, the wealthy are likely to fare far better than the poor, and again the disparity between the two will increase. An ironic illustration is that of atmospheric particulates (a cause of respiratory impairment) that are commonly associated with industrial and urban areas. In the past this association was universal but since the late 20th century urban atmospheric particulates have declined markedly in OECD countries. Yet atmospheric particulates have served to provide a countering force through global dimming (Chapter 5). If they decline elsewhere as developing countries progress with the cleaning of their industries and economies, then this brake will be lessened and global warming enhanced. In short, one set of human health problems will be exchanged for another.

Finally, generally speaking (and this is very generally speaking), the environment of a warmer and wetter world will tend to favour terrestrial pathogenic microbes. However, there are still many unknowns. For example, how will domestic and urban water use change? Will climate change and associated ecosystem changes affect microbial mutation rates and successful speciation? How would an increase in the tempo of extreme weather events and natural disasters affect infectious disease occurrence? (McMichael, 2005). Combining such climate developments with the forthcoming human demographic maximisation, and the continued post-Industrial Revolution trend in increased international transportation, the likelihood is that health risks will increase.

8.5.3 Future climate and human-ecology implications for wildlife

As we have seen, climate and climate change have discernable, indeed marked, impacts on species and biological systems. Consequently, through a variety of ways (a number of which were covered in Chapter 2) we have managed to discern many of the principal global climatic regimes and changes of the past couple of hundred million years, although, of course, we know far more about relatively recent climate change than that in distant deep time. The picture that each tool provides is necessarily incomplete. However, taking evidence together enables the bigger picture to emerge, and it is still emerging with ongoing research. This picture is coherent as both the palaeo and current evidence is very much corroborative (Chapters 3 and 4), which in turn lends confidence to the conclusions drawn.

It is therefore certain that current climate change influences (Chapter 5) and likely future change (Chapter 6) would, without the presence of humans, cause species to migrate and ecosystems to transform. Ecotones (the boundary zones between habitats) will be a focus of profound biological change: one that in no small part reflects the changes to take place within bordering ecosystems. Yet, in all likelihood, without humans terrestrial species would largely survive if (as we expect will happen) a similar amount of carbon were added over the next century to the atmosphere to that added since the Industrial Revolution (of course, there would be a genetic fingerprint; Chapter 4). Calcareous marine species, though, would be susceptible to extinction through increased ocean acidity, as happened with the Initial Eocene Thermal Maximum (Chapter 3). Other than this, most groups and families of species would survive and the biosphere would recover. However, this is theoretical. The reality is that we do currently have a climbing population of over 7 billion humans that already is exerting considerable pressure on natural systems (Chapter 7), even without additional impacts due to anthropogenic climate change. So, whereas wildlife has adapted to the series of considerable glacial–interglacial climatic fluctuations over the past two million years (Chapter 4), and even major carbon cycle perturbations earlier still, the past is not the present. First, the current anthropogenic perturbation of the carbon cycle is far more rapid than previous carbon excursions (consider the Eocene thermal maximum), as is the commensurate rapidity of global climate change (even if in the past some regional climate change has been as fast). Second, competition with humans has seen the landscape fragment: a landscape through which species adapting to change would have to migrate.

That we do have anthropogenic climate change in addition to an extinction already taking place (Barnosky et al., 2011) will only serve to enhance threats to species. Biodiversity is therefore set to continue to decline through the 21st century (Pereira et al., 2010). That current human pressures include the fragmenting of many natural terrestrial systems, and management and conversion of others to semi-natural systems, will impede species' ability to adapt to change. Such negative factors tend to interact synergistically. For example, a forest subject to climate change is one thing: species may migrate. Alternatively, a forest that becomes a large island surrounded by highly managed land becomes a nature reserve. However, impose significant climate change on isolated wildlife reserves, from which species cannot migrate, and extinctions virtually become inevitable.

That this anthropogenic climate change will, beyond the 21st century, take us to a global climate regime not seen since before the Quaternary means that species that have evolved since then, such as in the Pleistocene (or even the Pliocene), will in all likelihood be most adversely affected. This will be especially so for those species accustomed to the cooler end of the global climatic spectrum. Indeed, this can already be seen with some boreal and sub-polar species.

We are therefore going through an extinction event of a magnitude only last exceeded by the Cretaceous/Tertiary extinction 65.5 mya (Chapter 3 and also Barnosky et al., 2011). Yet, unlike that event, much of which took place rapidly, the current extinction arguably began when (some consider) the Anthropocene began with the extinction of Eurasian megafauna; that is to say, an extinction over thousands of years. This extinction is continuing today slowly but steadily and the size of the species affected is getting smaller. Future climate change will exacerbate matters.

Can this extinction be halted? In all probability, no. Can it be slowed and can some species be conserved? In all probability, yes, although it is impossible to estimate how many species that will be.

If we are to minimise extinctions then, aside from investment in biological conservation and management, investment will be needed to continue improving our understanding of biology and climate interactions. Computer models of global climate are now increasingly including biological dimensions and they are also becoming increasingly detailed. However, much still needs to be done. To date much of our present-day climate change understanding (hence computer modelling) comes from our knowledge of Holocene and glacial climates discerned from actual biological and geological evidence. Notwithstanding this, we have still much to learn as to how carbon moves between various biosphere sources and sinks. Filling this knowledge gap is fundamentally important, especially as the 21st century will see the Earth move into a climatic regime not seen for a few million years. Nonetheless, it is possible, before the end of this century's first quarter, that in some areas ecological biogeographic mapping will be sufficiently detailed and the climate models sufficiently accurate, and we will be able to foresee where and what species are best likely to flourish. This will greatly facilitate not just wildlife management but management of natural systems, especially agricultural and forestry systems, and hence generate a financial return on such investment in research, especially as we have a burgeoning human population to feed. In short, there is a very real role for whole-organism biological, ecological and biosphere expertise to address such issues for the foreseeable long-term future.

8.5.4 Reducing future anthropogenic greenhouse gas emissions

Curbing anthropogenic emissions of carbon dioxide, as the principal greenhouse gas, is the main priority (though not the sole one; Chapter 1). Yet such is human ecology (Chapter 7), and especially human energy requirements, that reductions are proving difficult. Even if the Kyoto Protocol was implemented globally in full then the curbing of our current growth in emissions will only amount to saving a few years' worth of warming and climate forcing by 2100 (Chapter 8). It is therefore difficult to envisage, at least from our early-21st-century perspective, that international political measures will have any significant effect in halting global warming. The best that may perhaps be hoped for is that measures above and beyond the Kyoto Protocol might begin to meaningfully slow matters down.

Given this, is it worth investing in the curbing of carbon emissions? The future may be one of continued global warming. Reducing its rate would certainly help both natural and human systems to adapt as well as humans facilitate natural adaptation, but there is (from a purely human materialistic and well-being perspective) another motivation for cutting fossil carbon emissions. This stems from two facts. First, we know that fossil carbon reserves are finite. Second, we know that the human demand for energy to power our technological global society is also certainly set to increase (barring some global human catastrophe). Consequently the need to curb emissions should not be perceived as an economic threat, but an opportunity to develop alternative ways of generating energy and using it more efficiently. Nations that succeed most in doing this will be more protected from fossil fuel shocks and will see more economic benefit per unit of energy consumed and so be more competitive globally (Cowie, 1998a). This economic benefit will also help protect us from some of the adverse effects of climate change and help conserve wildlife. It is up to us, and nobody else, to seize this opportunity. In short, distinct from purely climate change arguments, there is a raft of (largely sustainability based) reasons for reducing our consumption of fossil carbon. That these chime perfectly with climate change arguments for reducing greenhouse gas emissions should be positively welcome (Cowie, 1998b). Yet as global greenhouse gas emissions deviate from the Kyoto path such a welcome is far from being practically realised, even if a few nations espouse the rhetoric.

The null-hypothesis approach is equally compelling. Suppose some nations do not adapt to a low-fossil carbon economy. Well, before the end of the 21st century these nations will be competing for dwindling reserves of oil and gas in a world with a far larger population. One of these factors alone – be it the larger population or the dwindling supply – would drive up the price of fossil fuel; both factors together will have a synergistic effect. This is due to what economists call the law of supply and demand. Economists have long known that if supply decreases but demand stays the same then price rises. They also know that if demand increases and supply stays the same the price also rises. Economists equally recognise a worse situation, that of supply decreasing while demand simultaneously increases. This prospect is the one which prudent national leaders need to take into account when framing policies to develop their respective nations' economies. The conclusion on purely economic grounds, irrespective of climate change's biological and environmental impacts, is that renewable and alternative energy (arguably including nuclear power) need to provide the vast bulk of society's energy needs.

From a human ecological perspective, if human societies are to be environmentally sustainable then one of the fundamental factors is the need to rely on sustainable energy and not the planet's finite fund of fossil photosynthetic energy.

8.5.5 A final conclusion

One final conclusion (before Appendix 4 and matters for further consideration): it is actually more an observation, and it is a personal one at that. Yet as the author of this broad review and appraisal, it is something I feel entitled to make after some 300 000 words and 90 or so diagrams. If we are to comprehend and address climate change concerns and the implications for species, including our own, then this understanding needs to come from science. It will not come from anywhere else. Science in many nations, including a number of highly developed ones, is not as valued as it might be and in some instances is positively under attack. Indeed, even where it is supported it is increasingly being constrained. No Western university today would allow a scientist a few years out to write a book like Newton's *Principia*. Work to develop bold new ideas outside of science institutions (such as a monk growing peas) are also an anathema (so much for any modern Mendel influencing genetics). Yet if we are to see our global society survive the coming century (albeit transformed) without considerable suffering, let alone denuded of many biological and ecological assets, then we are going to have to change the way we relate to both resource use and (scientific and technological) knowledge. These are equally formidable, but not impossible, challenges. If we fail... well, someone else can narrate doom and disaster, at least as far as humans and many wildlife species are concerned. As for the rest of the biosphere, it will remain. Of that I am very confident.

8.6 References

Allen, M. R., Frame, D. J., Huntingford, C. et al. (2009) Warming caused by cumulative carbon emissions towards the trillionth tonne. *Nature*, 458, 1163–6.

Anon (2011) Editorial –The mask slips: The Durban meeting shows that climate policy and climate science inhabit parallel worlds. *Nature*, 480, 292.

Barnosky, A. D., Matzke, N., Tomiya, S. et al. (2011) Has the Earth's sixth mass extinction already arrived? *Nature*, 471, 51–7.

BP Economics Unit (2005) *BP Statistical Review of World Energy*. London: British Petroleum Corporate Communications Services.

BP Economics Unit (2011) *BP Statistical Review of World Energy*. London: British Petroleum Corporate Communications Services.

Brandt Commission (1980) *North-South: a Programme for Survival*. London: Pan Books.

Brandt Commission (1983) *Common Crisis – North-South: Co-operation for World Recovery*. London: Pan Books.

Carter, L. J. and Pigford, T. H. (2005) Proof of safety at Yucca Mountain. *Science*, 310, 447–8.

CIA (2008) *CIA: The World Factbook-2008*. Washington, DC: Central Intelligence Agency.

Clarke, R. and Timberlake, L. (1982) *Stockholm Plus Ten: Promises, Promises? The Decade since the 1972 UN Environment Conference*. London: Earthscan – International Institute for Environment and Development.

Climate Change Committee (2010) *The Fourth Carbon Budget: Reducing Emissions Through the 2020s*. London: The Committee on Climate Change.

Climate Change Science Program (US) (2006) *Temperature Trends in the Lower Atmosphere*. Washington DC: U.S. Climate Change Science Program.

Climate Commission (2011) *The Critical Decade: Climate Science, Risks and Responses*. Commonwealth of Australia: Canberra.

Cowie, J. (1998a) *Climate and Human Change: Disaster or Opportunity?* London: Parthenon Publishing.

Cowie, J. (1998b) The other reasons for cutting carbon dioxide emissions. *Science in Parliament*, 55(4), 21.

Cowie, J. (2009) *Can we Beat the Climate Crunch?* Concatenation Science Communication. http://science-com.concatenation.org/archive/can_we_beat_the_climate_crunch.html.

Department for Communities and Local Government (2006) *Code for Sustainable Homes: A Step-change in Sustainable Home Building Practice*. Wetherby: Communities and Local Government Publications.

Department for Communities and Local Government (2008) *Definition of Zero Carbon Homes and Non-Domestic Buildings*. London: Department for Communities and Local Government.

Department of Energy (1977) *Energy Policy Review – Energy Paper 22*. London: Department of Energy and HMSO.

Department of Energy (1978) *Energy Policy: a Consultative Document*. Cmnd 7107. London: HMSO.

Department of Energy and Climate Change (2009) *Consultation on the Term 'Carbon Neutral': Its Definition and Recommendations for Good Practice*. London: DECC.

Department of Environment, Food and Rural Affairs (2006) *Climate Change: The UK Programme*. London and Norwich: DEFRA and Stationery Office.

Department of Environment, Food and Rural Affairs (2011) *Adapting to Climate Change: Helping Key Sectors to Adapt to Climate Change*. London: DEFRA.

Department of Environment, Transport and the Regions (2000) *Climate Change: The UK Programme*. London and Norwich: DETR and Stationery Office.

Department of Trade and Industry (2003) *Review of the Feasibility of Carbon Dioxide Capture and Storage in the UK*. London: DTI.

Department of Trade and Industry (2007) *Meeting the Energy Challenge A White Paper on Energy*. Cmnd 7124. Norwich: Stationery Office.

Department of Trade and Industry, Department for Transport and Department of Environment, Food and Rural Affairs (2003) *Energy White Paper: Our Energy Future – Creating a Low Carbon Economy*. Cmnd 5761. Norwich: Stationery Office.

Eikeland, P. O. (1993) US energy policy at a crossroads? *Energy Policy*, 21, 987–98.

Evans, R. D. and Herring, H. P. J. (1989) *Energy Use and Energy Efficiency in the UK Domestic Sector up to the Year 2010*. London: Department of Energy and HMSO.

Environment Canada (2007, 2009, 2011) *A Climate Change Plan for the Purposes of the Kyoto Protocol Implementation Act*. Quebec: Environment Canada. (Different annual assessments and projections.)

Friedlingstein, P., Solomon, S., Plattner, G.-K. et al. (2011) Long-term climate implications of twenty-first century options for carbon dioxide emission mitigation. *Nature Climate Change*, 1, 457–61.

Garnaut, R. (2008) *The Garnaut Climate Change Review: Final Report*. Cambridge: Cambridge University Press.

Garnaut, R. (2011) *Garnaut Climate Change Review –Update 2011. The Science of Climate Change*. Update paper 5. www.garnautreview.org.au/update-2011/garnaut-review-2011/garnaut-review-2011.pdf.

Greenhalgh, G. (1990) The comforting illusion of energy conservation. *Atom*, 393, 6–7.

Hammond, G. (2007) Time to give due weight to the carbon footprint issue. *Nature*, 445, 256.

Hardin, G. (1968) The tragedy of the commons. *Science*, 162, 1243–8.

Hirsch, R. L. (2005) *Peaking of World Oil Production: Impacts, Mitigation, & Risk Management*. Washington DC: Department of Energy.

H. M. Government (2009) *The UK Renewable Energy Strategy*. Cmnd 7686. Norwich: Stationery Office.

H. M. Government (2010) *A Strong Britain in an Age of Uncertainty: The National Security Strategy*. Cmnd 7953. Norwich: Stationery Office.

HMSO (1994) *Climate Change: The UK Programme*. Cmnd 2427. London: HMSO.

House of Commons Energy and Climate Change Committee (2011) *The UK's Energy Supply: Security or Independence?* Vol. 1. London: Stationery Office.

Hulme, M., Turnpenny, J. and Jenkins, G. (2002) *Climate Change Scenarios for the United Kingdom: The UKCIP02 Briefing Report*. Norwich: Tyndall Centre.

Information Office of the State Council of the People's Republic of China (2007) *China's Energy Conditions and Policies*. Beijing: Information Office of the State Council of the People's Republic of China.

Intergovernmental Panel on Climate Change (1990) *Climate Change: the IPCC Scientific Assessment*. Cambridge: Cambridge University Press.

Intergovernmental Panel on Climate Change (1995) *Climate Change 1995: the Science of Climate Change*. Cambridge: Cambridge University Press.

Intergovernmental Panel on Climate Change (2000) *Emission Scenarios –IPCC Special Report*. Geneva: World Meteorological Organization.

Intergovernmental Panel on Climate Change (2001a) *Climate Change 2001: the Scientific Basis – Summary for Policymakers and Technical Summary of the Working Group I Report*. Cambridge: Cambridge University Press.

Intergovernmental Panel on Climate Change (2001b) *Climate Change 2001: Impacts, Adaptation and Vulnerability – a Report of Working Group II*. Cambridge: Cambridge University Press.

Intergovernmental Panel on Climate Change (2005) *IPCC Special Report on Carbon Dioxide Capture and Storage*. Geneva: IPCC.

Intergovernmental Panel on Climate Change (2007a) *Climate Change 2007 Synthesis Report*. Geneva: IPCC.

Intergovernmental Panel on Climate Change (2007b) *Climate Change 2007: Mitigation of Climate Change – a Report of Working Group III*. Cambridge: Cambridge University Press.

Intergovernmental Panel on Climate Change (2007c) *Climate Change 2007: the Physical Science Basis – Working Group I Contribution to the Fourth Assessment of the IPCC*. Cambridge: Cambridge University Press.

Intergovernmental Panel on Climate Change (2007d) *Climate Change 2007: Imapcts, Adaptation and Vulnerability – a Report of Working Group II*. Cambridge: Cambridge University Press.

International Energy Agency (2004) *Key World Energy Statistics*. Paris: IEA.

International Energy Agency (2005) *World Energy Outlook 2005*. Paris: IEA.

International Energy Agency (2006) *China's Power Sector Reforms: Where Next?* Paris: IEA.

International Union for the Conservation of Nature, World Wildlife Fund and United Nations Environment Programme (1980) *The World Conservation Strategy*. Gland: IUCN.

International Union for the Conservation of Nature, World Wildlife Fund and United Nations Environment Programme (1991) *Caring for the Earth: a Strategy for Sustainable Living*. London: Earthscan.

Jevons, W. S. (1865) *The Coal Question*. London: Macmillan.

Jones, C. D., Hughes, J. K., Bellouin, N. et al. (2011) The HadGEM2-ES implementation of CMIP5 centennial simulations. *Geoscientific Model Development*, 4, 543–70.

Kabat, P., Vellinga, P., van Vierssen, W., Veraat, J. and Aerts, J. (2005) Climate proofing the Netherlands, *Nature*, 438, 283–4.

Kerr, R. A. (2005) Bumpy road ahead for World's oil. *Science*, 310, 1106–8.

Kintisch, E. (2005) US Energy Bill promises some boosts for research. *Science*, 309, 863.

Lemmen, D. S., Warren, F. J., Lacroix, J. and Bush, E. (2008) *From Impacts to Adaptation: Canada in a Changing Climate*. Ottawa: Government of Canada.

Levin, L. and Bradley, R. (2010) *Comparability of Annex 1 Emission Reduction Pledges*. Washington, US: World Resources Institute. www.wri.org.

Malthus, T. R. (1798) *Essay on the Principle of Population as it Affects the Future Improvement of Society With Remarks on the Speculation of Mr Godwin, Mr Condorcet and Other Writers*. Reprinted in Himmelfarb, G., ed. (1960) *On Population*. New York: Modern Library.

McMichael, A. J. (2005) Environmental and social influences on emerging infectious diseases: past, present and future. In McLean, A. R., May, R. M., Pattison, J. and Weiss, R. A., eds, *SARS: a Case Study in Emerging Infections*. Oxford: Oxford University Press.

Meadows, D. H., Meadows, D. L., Randers, J. and Behrens, III, W. W. (1972) *The Limits to Growth*. London: Pan Books.

Meadows, D. H., Meadows, D. L. and Randers, J. (1992) *Beyond the Limits: Global Collapse or a Sustainable Future*. Godalming and London: World Wide Fund for Nature and Earthscan.

Meinhausen, M., Meinhausen, N., Hare, W. et al. (2009) Greenhouse-gas emission targets for limiting global warming to 2 6°C. *Nature*, 458, 1158–62.

Ministry for the Environment (2011) *Minister's Position Paper: Gazetting New Zealand's 2050 Emissions Target*. Wellington: Ministry for the Environment.

National Development and Reform Commission People's Republic of China (2007) *China's National Climate Change Programme*. Bejing: National Development and Reform Commission People's Republic of China.

National Energy Policy Development Group (2001) *National Energy Policy*. Washington, DC: National Energy Policy Development Group.

Nuttall, W. J. (2008) *Fusion as an Energy Source: Challenges and Opportunities*. London: Institute of Physics.

Office of the Deputy Prime Minister (2005) *Proposals for Introducing a Code for Sustainable Homes*. London: Office of the Deputy Prime Minister.

Office of National Statistics (2002) *Living in Britain 2002*. Norwich: Stationery Office.

Office of Public Sector Information (2009) *The UK Low Carbon Transition Plan*. Norwich: Stationery Office.

Parliamentary Office of Science and Technology (2005) *Household Energy Efficiency*. POST Note 249. www.parliament.uk/parliamentary_of.ces/post/pubs2005.cfm.

Pereira, H. M., Leadley, P. W., Proença, V. et al. (2010) Scenarios for global biodiversity in the 21st century. *Science*, 330, 1496–1501.

Peters, G. P. and Hertwich, H. G. (2008) CO_2 embodied in international trade with implications for global climate policy. *Environmental Science & Technology*. 42(5), 1401–7.

Pew Charitable Trusts (2010) *Who's Winning The Clean Energy Race? Growth, Competition and Opportunity in the World's Largest Economies*. Washington DC: Pew Charitable Trusts.

Policy Exchange (2008) *Six Thousand Feet Under: Burying the Carbon Problem*. London: Policy Exchange.

Prime Minister's Council on Climate Change (2008) *National Action Plan on Climate Change*. New Delhi: Government of India.

Rogelj, J., Meinshausen, M. et al. (2010) Copenhagen Accord pledges are paltry. *Nature*, 464, 1126–8.

Rose, D. J. (1974) Energy policy in the US. *Scientific American*, 230, 20–9. Reprinted in Siever, R., ed. (1980) *Energy and Environment: Readings from Scientific America*. San Francisco: Freeman & Co.

Royal Society of Chemistry (2005) *Chemical Science Priorities for Sustainable Energy Solutions*. London: Royal Society of Chemistry.

Saunders, C. and Barber, A. (2007) *Comparative Energy and Greenhouse Gas Emissions of New Zealand's and the UK's Dairy Industry*. Agribusiness and Economics Research Unit Research Report 297. Lincoln, New Zealand: AERU.

Schipper, L. and Hawk, D. V. (1991) More efficient household electricity use. *Energy Policy*, 19(3), 244–62.

Seiver, R., ed. (1980) *Energy and Environment: Readings from the Scientific American – Energy Policy in the US*. San Francisco: Freeman & Co.

Stern, N. (2007) *Economics of Climate Change: the Stern Review*. Cambridge: Cambridge University Press.

Tyme, J. (1978) *Motorways Versus Democracy*. London: Macmillan Press.

UN Department of Public Information (2005) *Press Release DEV/2539 IAEA/1365 SAG/394: Chernobyl – The True Scale of the Accident*. New York: Department of Public Information – News and Media Division.

United Nations Environment Programme (2010) *How Close Are We To The Two Degree Limit?* Information note to the UNEP Governing Council/Global Ministerial Environment Forum.

UN Population Division of the Department of Economic and Social Affairs (2002) *World Population Prospects: The 2002 Revision.* New York: United Nations Secretariat.

UN Population Division of the Department of Economic and Social Affairs (2010) *World Population Prospects: The 2010 Revision.* New York: United Nations Secretariat.

UNSCEAR (2008) *Sources and Effects of Ionizing Radiation: Volume II Annex D Health effects due to radiation from the Chernobyl accident.* New York: United Nations Scientific Committee on the Effects of Atomic Radiation.

Wilson, E. O. (1988) *Biodiversity.* Harvard, CT: Harvard University Press.

World Commission on Environment and Development (1987) *Our Common Future.* Oxford: Oxford University Press.

Worldwatch Institute (2003) *Vital Signs 2003–2004: the Trends that are Shaping Our Future.* London: Earthscan.

Appendix 1. Glossary and abbreviations

Glossary

The following are just some of the main biological, geological and climate terms used in this book. It is not a complete list but instead focuses on words used in more than one chapter. Although the terms are common to biology, geology or climatology, they are provided to help readers from one discipline understand passages of this book that relate to another discipline.

abyssal (zone)	The zone of greatest ocean depth, frequently defined as deeper than 1000 m.
aerosol	A colloidal system such as a mist or fog in which the dispersal medium is a gas. In climatology the gas is the atmosphere.
albedo	The fraction of light reflected from a surface. It includes the proportion of solar energy reflected from part of the Earth's surface or from the planet in its entirety.
Allerød and Bølling	A warm phase lasting over 1000 years towards the end (deglaciation period) of the last glacial and prior to the Holocene.
allopatric	Species having completely separate geographical distributions (*see also* sympatric).
anthropogenic	Human-generated.
biome	The largest ecological category of a biogeographical region (such as tundra or tropical rainforest).
Bond cycle	A cycle of waxing and waning of the Arctic glacial ice sheet.
boreal	The northern continental biogeographic zone of short summers and long winters.
Broecker circulation	A global circulation of ocean water, part on the surface and part at abyssal depths, sometimes referred to as the thermohaline circulation (THC). One component is the meridional overturning circulation (MOC).
Calvin cycle	The light-independent phase of photosynthesis that follows the capture of light and synthesis of two key molecules (ATP and NADPH). Its products include

three-carbon sugar phosphate molecules, which the plant uses to make more complex carbohydrates.

Carnivora — Flesh-eating mammals.

dendrochronology — Reconstruction of past climate change from tree rings.

ecotone — A transitional zone between two habitats (e.g. forest bordering grassland).

Eemian — The previous interglacial (in some texts referred to as the Ipswichian, Sangamon or Eem-Sangamon interglacial) that occurred some 125 000 years ago.

endosymbiont — Organism(s) living inside the cell of another, in a symbiotic (close and mutually beneficial) relationship.

Eocene — A 22 million-year epoch beginning 55.8 mya (*see* Appendix 2).

eukaryote — An organism with cells that contain a membrane-bound nucleus and other structures (organelles) also surrounded by a membrane. Animals and plants have eukaryotic cells. *See also* prokaryote.

eustatic — Changes in global sea level due to changes in the size of the Earth's ocean basins, as well as in local sea level due to tectonic movement.

evapotranspiration — The total loss of water from soil by evaporation and by the transpiration of plants growing in the soil.

fecundity — The number of offspring produced by an individual or breeding pair.

Fennoscandia — The north-western part of the Eurasian continent and commonly used in association with the glacial ice sheets covering the north of that continent.

fertility — The ability to reproduce (irrespective of realisation). (Note: some human demography texts do not distinguish between fertility and fecundity. Ecologists do, and this text uses the ecological definition to avoid confusion.)

folivorous — Leaf-eating.

forcing — A factor that affects the climate, warming or cooling it. Climate-forcing factors include aerosols and greenhouse gases.

glacial — A cool period within an ice age with expanded ice caps and lower sea levels. However, often in common parlance glacials are known as ice ages and this can lead to some confusion.

global warming potential — The global warming over a *specific* time period resulting from the addition of a greenhouse gas compared to the same mass of carbon dioxide.

Gondwana (Gondwanaland) — The supercontinent in the southern hemisphere that existed 200 mya but which has since fragmented.

Heinrich events — Short periods of glacial Arctic ice rafting.

Holocene	The current interglacial, which began 11 700 years ago (*see* Appendix 2).
Holocene climatic optimum	A warm period (as opposed to a geological period) of time some 8000–5000 years ago.
hyperthermal	A geologically abrupt (that is, centuries to thousand years in initiation) time of global warming leading to a warm period that then lasts for scores of thousands of years. Typically associated with a carbon isotope excursion (CIE; see section 3.3.7).
ice age	A period of time with glacials.
interglacial	A warm period in an ice age between glacials.
interstadial	A cool period within a (cool) glacial.
Ipswichian	*See* Eemian.
Iron Age neoglaciation	A minor glaciation that occurred 4500–2500 years ago.
Laurentide	Commonly used to describe the glacial ice sheet over North America.
Little Ice Age	A short cooling episode particularly associated with, but not restricted to, the northern hemisphere approximately around AD 1550–1850.
medieval climatic anomaly	A short warming episode around approximately AD 1100–1300, also known as the medieval warm period or (not preferred) the medieval climatic optimum.
medieval warm period	*See* medieval climatic anomaly.
metazoan	Multicelled animal.
mitochondrion	An organelle within a cell that consumes carbohydrate and releases energy. Contains DNA that is inherited through the female line.
nunatak	A mountain peak poking through an ice sheet.
Oligocene	The geological epoch lasting between 34 and 23 mya (*see* Appendix 2).
opportunity cost	An economist's term for the lost opportunity of employing one investment (be it cash or a resource) option over another (this book considers resource options). For example, some land used to grow biofuel would have an opportunity cost of not being available to grow food. Both options can also be quantified in product and cash terms.
Pangaea	A supercontinent existing in the deep geological past that subsequently fragmented into Gondwana to the south and Laurasia to the north.
Páramo	A biome type in the tropical Andes, below the snow line and above the tree line (it is also used to describe similar environments elsewhere at and close to tropical latitudes).

Phanerozoic	The eon since the beginning of the Cambrian (542 mya), with multicelled species (*see* Appendix 2).
phenology	The study of plant and animal development in response to the seasons.
photoperiodism	The response of an organism to the day/night cycle as well as its changes in the course of the year.
photosynthesis	The (sun) light-powered process by which plants (including algae) synthesise carbohydrates from carbon dioxide and water.
Pliocene	The epoch that occurred 5.3–1.8 mya (*see* Appendix 2).
pluvial	A geologically short time of high rainfall that is usually associated with Milankovitch or other climatic cycles, such as those related to Heinrich events.
production	The biomass fixed over a period of time (usually a year), expressed in terms of either mass or energy. Primary production is the production due to photosynthesis (also called gross primary production). Net primary production is the production due to photosynthesis after respiration losses are taken into account. Secondary production relates to the production of species living on photosynthetic species. Productivity relates to production in a unit area over a period of time (again, usually a year).
prokaryote	A simple cell that has no internal membrane-bound structures or nucleus. Bacteria and blue-green algae have prokaryotic cells. *See also* eukaryote.
pyroclastic rocks	Rocks formed by materials thrown out by explosive volcanic eruptions.
Quaternary	The period (in the geological sense) of time from the present to 1.806 mya (as provisionally defined by the International Commission on Stratigraphy in 2008) that saw a number of deep glacials and interglacials (*see* Appendix 2). (It may shortly be redefined to 2.588 mya.)
refugium	An ecological area where species can survive surrounding or nearby environmental change (plural: refugia).
Sangamon	*See* Eemian.
Snowball Earth	The Earth covered with snow and ice, either nearly or completely.
speciation	The processes by which new species evolve.
stoma	A minute opening on the leaf of a plant (usually on the underside) through which gas and water exchange takes place (plural: stomata).
stromatolite	A fossilised mat of algae.

sympatric	Species occupying the same, or overlapping, geographical areas (*see also* allopatric).
tundra	The treeless polar regions with permanently frozen subsoil (although surface soil may be thawed for some parts of the year).
Younger Dryas	A short, cool period at the end of the last glacial, 12 900–11 600 years ago, before the start of our current (Holocene) interglacial.
zonation	The occurrence of distinct bands of species across an environmental gradient such as a seashore or mountainside.

Abbreviations

ACC	Antarctic Circumpolar Current
ACR	Antarctic cold reversal
BAP	Biodiversity Action Plan
B-a-U	Business-as-Usual scenario
Be	Bering, a unit of water volume (equivalent to 10^3 km^3). *See also* Sv (water flow). Not to be confused with Be, the symbol for the element beryllium.
bya	billion years ago
C	carbon
^{12}C	Most common isotope of carbon with an atomic weight of 12. Preferred by the enzymes involved in photosynthesis.
^{13}C	Rare isotope of carbon, discriminated against by photosynthetic enzymes.
^{14}C	Rare unstable (radioactive) form of carbon with a half-life of 5715 years. Used in carbon dating.
C_3 (plants)	Plants that use the older photosynthetic pathway with three carbon atoms in the first product during carbon dioxide assimilation.
C_4 (plants)	Plants that use the more recently evolved photosynthetic pathway capable in intense sunlight (hence low latitudes) of taking more carbon dioxide out of the atmosphere than C_3 photosynthesis.
Ca	calcium
CBD	UN Convention on Biological Diversity (1992)
CCD	calcite compensation depth
CCM	carbon-concentrating mechanism
CCS	carbon capture and storage
CDC	US Centers for Disease Control and Prevention
CDM	Clean Development Mechanism
CEH	UK Centre for Ecology and Hydrology (formerly ITE and the NERC River Laboratory)
CET record	Central England Temperature record

CFB (event)	continental flood basalt (event)
CH_4	methane
CIE	carbon isotope excursion
CO_2	carbon dioxide
CO_2-eq	carbon dioxide equivalent
D	deuterium or heavy hydrogen (^2H) with an atomic weight of 2
DECC	Department of Energy and Climate Change (UK/England and Wales)
DEFRA	Department of the Environment, Food and Rural Affairs (England and Wales; formerly DETR and MAFF)
DETR	(former) Department of the Environment, Transport and the Regions (England and Wales; formerly the Department of the Environment) and now DEFRA
DIC	dissolved inorganic carbon
DoE	(former) Department of Environment (England and Wales), now DEFRA
ECM	electrical conductivity measurement (here applied to ice cores)
ENSO	El Niño Southern Oscillation
EOR	enhanced oil recovery
EPICA	European Project for Ice Coring in Antarctica
EU	European Union
FAO	UN Food and Agriculture Organization
FCCC	UN Framework Convention on Climate Change (1992)
G5	The emerging economic powers: China, India, South Africa, Mexico and Brazil.
G8	Group of Eight, consisting of Canada, France, Germany, Italy, Japan, the UK, the USA and Russia.
GDP	gross domestic product
GECFS	Global Environmental Change and Food Systems
GRACE	Gravity Recovery and Climate Experiment (satellite mission)
GSA	Great Salinity Anomaly
GtC	gigatonnes of carbon
GWh	giga-Watt hour
GWP	global warming potential
H	hydrogen
ha	hectare (metric unit of area; 100 m $\times$ 100 m $= 10^4$ m^2, or approximately 2.47 acres)
HEP	hydroelectric power
HMSO	Her Majesty's Stationery Office (now just the Stationery Office)
IETM	Initial Eocene Thermal Maximum (also known as the PETM)
IETR	International Experimental Thermonuclear Reactor
IPCC	Intergovernmental Panel on Climate Change
ITCZ	Intertropical Convergence Zone
ITE	UK Institute for Terrestrial Ecology (now CEH)
IUCN	International Union for the Conservation of Nature (and Natural Resources; now also known as the World Conservation Union)
JI	Joint Implementation

KPIA	Kyoto Protocol Implementation Act (Canada)
kya	thousands of years ago
LAA	leaf area analysis (used to approximate rainfall)
LGM	last glacial maximum (approximately 20 000 years ago)
LMA	leaf margin analysis (used to approximate temperature)
MAFF	(former) UK Ministry of Agriculture, Fisheries and Food (now part of DEFRA)
MAR	mass accumulation rate
MCA	medieval climatic anomaly (also known as the medieval warm period)
MIT	Massachusetts Institute of Technology
MOC	meridional overturning circulation
MONARCH	Modelling Natural Resource Responses to Climate Change
MPT	Mid-Pleistocene Transition
mtDNA	mitochondrial DNA
mtoe	million tonnes of oil equivalent ($= 42$ PJ $= 42 \times 10^{15}$ J $= 12$ TWh $= 12 \times 10^{12}$ Wh; excluding station efficiency factors)
MWP	medieval warm period (also known as the medieval climatic anomaly)
mya	millions of years ago
NAO	North Atlantic Oscillation
NERC	UK Natural Environment Research Council
NFFO	Non-Fossil Fuel Obligation
NOAA	US National Oceanic and Atmospheric Administration
O	oxygen
^{16}O	The most common form (isotope) of oxygen with an atomic weight of 16.
^{18}O	Rarer and heavier isotope of oxygen.
OECD	Organization for Economic Cooperation and Development (Austria, Australia, Belgium, Canada, Czech Republic, Denmark, Finland, France, Germany, Greece, Hungary, Iceland, Republic of Ireland, Italy, Japan, Luxembourg, Mexico, New Zealand, Netherlands, Norway, Poland, Portugal, Slovakia, South Korea, Spain, Sweden, Switzerland, Turkey, UK and USA)
OPEC	Organization of the Petroleum Exporting Countries (Iran, Iraq, Kuwait, Qatar, Saudi Arabia, United Arab Emirates, Algeria, Libya, Nigeria, Indonesia and Venezuela)
PETM	Palaeocene–Eocene Thermal Maximum (also known as the IETM)
PMP	possible maximum precipitation
ppb	parts per billion
ppbv	parts per billion by volume
ppm	parts per million
ppmv	parts per million by volume
ppt	parts per trillion
r/p ratio	(fossil fuel) reserve/production ratio
Si	silicon
SO	Stationery Office (formerly HMSO)
SRES	IPCC Special Report on Emission Scenarios

Sv	Sverdrup. A unit of volume of water transport (flow) used exclusively in oceanography; 1 Sv $= 10^6$ m^3 s^{-1}. See also Be (water volume).
tC GWh^{-1}	tonnes of carbon per giga-Watt hour
TGS	thermal growing season
toe	tonnes of oil equivalent
UKCIP	UK Climate Impacts Programme
UN	United Nations
UNCED	UN Conference on Environment and Development (1992)
UNEP	United Nations Environment Programme
USGCRP	US Global Change Research Program
UV	ultraviolet (light)
W	Watt, the SI unit of power ($= 1$ Js^{-1})
WAIS	West Antarctic Ice Sheet
WHO	World Health Organization
WMO	World Meteorological Organization
WWF	World Wildlife Fund (now World Wide Fund for Nature)

Appendix 2.   Biogeological chronology

Age (mya)	Eon	Era	Period	Epoch	Major events	Time beginning (mya)
0.0117	Phanerozoic	Cenozoic / Neogene	Quaternary	Holocene	Current interglacial (part of which is sometimes called the Anthropocene)	0.0117 (11 700 years ago)
2.588				Pleistocene	Series of interglacials and glacials	2.588
5.332			Tertiary	Pliocene		5.332
23.03				Miocene	35–15 mya, 3–4°C warmer than present but cooling	23.03
33.9		Palaeogene		Oligocene	Gets cooler from Eocene. Antarctic ice from 34 mya	33.9
55.8				Eocene	Early Eocene warm	55.8
65.5				Palaeocene		65.5
99.6		Mesozoic	Cretaceous	Upper	End-Cretaceous extinction	99.6
145.5				Lower		145.5
199.6			Jurassic		CO_2 levels begin a slow, variable trend down to Quaternary levels	199.6
251			Triassic		CO_2 levels rise to ≈5 times present level	251

Date	Precambrian / Eon	Era	Period	Sub-period	Event	Date
299	Palaeozoic		Permian		End-Permian extinction	299
318.1			Carboniferous	Pennsylvanian	CO$_2$ levels low	318.1
359.2				Mississippian		359.2
416			Devonian		CO$_2$ levels peak at over 22 times present level	416
443.7			Silurian			443.7
488.3			Ordovician			488.3
542			Cambrian		'Age of invertebrates'	542
635	Proterozoic (Precambrian)	Neoproterozoic		Ediacaran		635
850				Cryogenian	Snowball Earth II	850
1000				Tonian		1000
1200		Mesoproterozoic		Stenian		1200
1400				Ectasian		1400
1600				Calymmian		1600
2500		Palaeoproterozoic			Snowball Earth I (around 2.2 bya, possibly 2.4–2.3 bya) O$_2$ photosynthesis begins to increase	2500
Start of Precambrian not defined	Archaean				Fossil evidence of bacteria and algae	4000 (for start of Archaean)

The dates given are those designated by the International Commission on Stratigraphy (2008). The spellings are English. The breakdown of nomenclature includes commonly used terms such as 'Quaternary' which the ICS has made a move to discontinue, but debate goes on. Yet because of common use and for student ease (especially by non-geologists) such terms are retained in this chart.

Appendix 3. Calculations of energy demand/supply and orders of magnitude

Calculations of energy demand/supply

From Chapter 7 onwards there are a number of depictions of global energy supply and demand. The following criteria have been applied in constructing these graphs of energy statistics.

Where statistics are given in millions of tonnes of oil equivalent (mtoe), quantities of nuclear and renewable energy (such as hydroelectric power) are expressed in terms of the oil equivalent as if the energy was generated by an oil-fired electricity-generating plant with an efficiency of around 33% up to 2000 and then 38% thereafter. Consequently, it is possible to make some comparison and to see how these non-fossil energy sources (could) displace oil sources.

Historic proportional energy-supply estimates prior to 1950 are taken from Morgan and Murray (1976) and early carbon emissions were assessed from a variety of sources including from the Carbon Dioxide Information Analysis Center (CDIAC) of the US Department of Energy and also the Woods Hole Research Center in Massachusetts, USA. Carbon emissions from energy for Figure 7.10b came from the International Energy Agency's *CO$_2$ Emissions from Fossil Fuel Combustion 2011*. Estimates for fossil fuel from the 1980s and 1990s were also cross-checked against the US Government's Energy Information Administration in autumn 2004 and spring 2005 (for this book's first edition), as well as the *BP Statistical Review of World Energy* (1990, 2000, 2005, 2011) and adapted where necessary, as per previous power-station efficiency assumptions. The Worldwatch Institute's *Vital Signs* series (2003), which in turn draws on additional sources, was used as a further check on statistics post-1950. The UK Department of Trade and Industry's *Energy Trends* (also known as Digest of UK Energy Statistics or DUKES) was similarly used as a check for UK data. Any misinterpretation of data sources is the author's responsibility. Data for this book's second edition were accessed online in the latter half of 2011.

Key websites that have energy and/or carbon data:

- British Petroleum, www.bp.com
- Carbon Dioxide Information Analysis Center, http://cdiac.ornl.gov
- International Energy Agency, www.iea.org
- UK Department of Trade and Industry, www.dti.gov.uk
- US Energy Information Administration (EIA), www.eia.gov

Data from 1950 onwards relates strictly to commercially traded fuels and does not include locally gathered or produced domestic fuels in less-developed nations, such as burning of rural wood or animal dung. The proportional estimates prior to 1900 for the UK and USA in graphs featuring such historical dimensions may include some from such categories, although the elements of change in patterns of energy consumption depicted in this book prior to the 20th century are thought to capture the essential features. Consequently any use of these energy graphs should reference this book.

- 1 t (metric) of oil $= 1.1023$ short tons $= 256$ gallons (imperial) $= 308$ US gallons
- 1 mtoe (million tonnes of oil equivalent) produces around 4 TWh $= 4 \times 10^{12}$ Wh in a power station of about 33% efficiency
- 1 mtoe, when burnt, releases approximately 4.25×10^{13} Btu (British thermal units) or 4.48×10^{16} J
- A machine doing work consuming energy at the rate of 1 Js^{-1} consumes 1 W each second. If this continues for an hour then (60 s $\times$ 60 min) 1 kWh of energy is consumed, which is equivalent to 3600 J.

Orders of magnitude

Other than fossil fuel energy content (above), other non-biological units and nomenclature are based on Baron (1988).

Factor	Prefix	Symbol
10^1	deca	da
10^3	kilo	k
10^6	mega	M
10^9	giga	G
10^{12}	tera	T
10^{15}	peta	P
10^{18}	exa	E
10^{21}	zetta	Z

Sources

Baron, D. N. (1988) *Units, Symbols and Abbreviations: a Guide for Biological and Medical Editors and Authors*. London: Royal Society of Medicine Services.

BP Economics Unit (1990, 2000, 2005, 2011) *BP Statistical Review of World Energy*. London: British Petroleum Corporate Communications Services.

Morgan, R. and Murray, R. B. (1976) *Energy Resources and Supply*. London: Wiley Interscience.

Worldwatch Institute (2003) *Vital Signs 2003–2004: the Trends that are Shaping Our Future*. London: Earthscan.

With this book's first edition (2007) Cambridge University Press kindly afforded a couple of pages as Appendix 4 to allow a brief summary of the IPCC's 2007 assessment (AR4) at the book's final page-proof stage: that assessment literally came out a couple of months before this book's first edition was published. This time the IPCC's 2013 assessment (AR5) will come out several months after this edition sees light of day and so I now use this appendix in a different way.

In the course of lectures and encounters following this book's first edition I have invariably been asked a number of questions as to my personal thoughts, as opposed to recounting the climate science and policy developments. Other than commenting on likely prospective research and policy analysis avenues, in the main I have shied away from answering, especially as most people have expected me to be predictive: what will happen to such and such in coming decades? And of course nobody can predict the future. Having said that, I do have some personal thoughts on climate science and policy. Given that there has been interest in my own take beyond that of appraising the literature, I now make a couple of points in this short appendix, quite separate from the main body of the text: I do not wish to contaminate my earlier (hopefully) sober review of the science and policy with wilder personal musings.

First, on the policy front, I feel sincerely that there is a pressing need for the way nations account for fossil carbon emissions to be more rigorous. It is not just that international aviation and shipping emissions are excluded, but the problem of economic carbon leakage between nations. The way the figures are currently calculated purely relate to a nation's *in situ* greenhouse gas emissions. To take the UK as an example, carbon dioxide equivalent (CO_2-eq) greenhouse gas emissions have – according to the way they are currently calculated – been in decline since 1991–2 to the present (2011 being the last full year of data at the time of writing). Yet the UK imports manufactured goods and also imports very roughly half its food. Fossil energy and fertiliser (part fossil-based) are used to provide these. Of course, some of this currently unaccounted fossil carbon will be offset by UK exports, but only a proportion: the UK is by far a net importer of unaccounted carbon. When looking at UK emissions, and those of other individual nations, what is needed are additional metrics to take this carbon leakage into account. Nations need to know how much greenhouse gas emission their citizens are causing through their 'total net emissions'; that is, imports less exports. Of course, care needs to be taken to avoid double accounting but the bottom line is that a further metric still is needed: the net total fossil carbon used to generate a pound (or dollar), which is effectively total net emissions divided by GDP. This is a measure of the economy's fossil carbon efficiency.

Some may say that additional metrics will make life too complex. This is a specious counter argument. We are already used to multiple metrics for many commonly reported dimensions to modern life. For example, we are all used to multiple metrics to assess a nation's economy, such as its GDP, trade deficit, rate of currency exchange and employment level.

There is currently some discussion over the need for additional emissions metrics. Indeed, as this book goes to press (2012), the UK's all-party House of Commons Select Committee on Energy and Climate Change is holding an enquiry into Consumption-Based Emissions Reporting, with the Select Committee's conclusions due out at the end of the year. The good news is that the country's Parliamentary Select Committees sometimes produce incisive reports. The bad news is that many Select Committees' conclusions can be bland when it comes to a politically controversial topic such as this one, and sometimes they miss the point, especially where enquiries are on a technical topic, again as is this one. Furthermore, even if the Select Committee produces a bold report, the UK government is not obliged to adopt its conclusions (though it is obliged to respond to the points the committee makes) and, of course, other nations need not take any notice at all.

Nonetheless, proper accounting is vital if decisions are to be evidence-based: the null position being that the decision-making process is ill informed. Does anyone need reminding that the 2008–9 global financial crash was due to lack of accounting with due diligence, when sub-prime mortgages were incorrectly valued? Or that the Eurozone crisis that began in 2011 was caused by at least one nation's 'creative' accounting of its national debt?

The importance of proper accounting for emissions is not just because accounting provides the evidence and the evidence needs to be sound (and this should be reason enough), but because the aforementioned financial crises were based on *notional* resources, not genuinely physical ones. In both these cases it was notional value (based on currency) at the root of the problems and not a *real* resource: it was not the failure of an agricultural region, or collapse of a mine, that caused the problem: crops carried on growing regardless of stock market performance. Conversely, climate change *does* affect agriculture (ask the farmers in Australia's Murray-Darling basin), extreme precipitation does affect the physicality of homes (ask the residents of England's Boscastle), sea-level rise does reduce land area (talk to people from Tuvalu) and the intensity of hurricanes is changing (a concern to those in the Caribbean and south-east USA), to take just a few examples. This puts the difficulties of the 2008–9 and 2011–12 financial crises into perspective when compared with the real-world pressures on society that are arising out of climate change. Yet if we fall foul of purely human constructs of notional value, what will happen when real physical resources are diminished?

The second point that I feel is worthy of consideration is one that will be extremely challenging for my ecologist colleagues: the need to consider species and ecosystem translocation as part of global conservation *and* climate mitigation. Not only are some species threatened by climate change, and so arguably could do with a helping hand to colonise new areas, but ecosystems are one of the key ways the Earth system (or biosphere) is regulated. Past major carbon isotope excursions (CIEs) see carbon cascade (often – at least in part, if not in the main – biologically mediated) from one biosphere reservoir to another, be it soil carbon, oceans, wetlands and so forth, via the atmosphere. Reducing atmospheric carbon is difficult but in part might be achieved through the translocation of ecosystems based on climate projections together with Earth system considerations. This will not involve just forestation, but requires us to assist climate change-driven ecosystem change and succession. This is in contrast to maintaining ecosystems in their present locations in the manner of their *current* natural potential, which is the basis for much current ecosystem management (conservation). This is using biological systems to manage the cascade of carbon across the CIE event on which we are now embarked (irrespective of the triggering of any initial Eocene analogue) so as to help regulate the atmospheric carbon dioxide. It must be emphasised that this is not referring to forestation per se as we could plant many forests today only to see some of them release their carbon tomorrow as the world warms and biomes shift.

However, species translocation is not only controversial in itself (for some very good reasons, such as the unforeseen ecological impact of invasive species): some current ecosystems that might be considered for such management have international conservation status. Yet can we continue as we are and should we ignore ecological ways to manage the global carbon cycle? Such questions have so far largely been avoided but perhaps we need to start seriously considering them. Eventually, as climate change progresses, it is likely that policy-makers will start asking such questions and ecologists will be expected to provide the answers. Surely now is the time to start the debate in earnest. Here, the learned ecological societies across the world are among those best placed (but not the sole players) to facilitate this dialogue.

These are just two personal thoughts that I present to you for consideration. It is likely that a number, if not many, reading this book are either already professionally in place to take forward discussion of one or other of these points, or will be so in the future (for instance, today's students[1]). I leave it up to you.

---

[1] To facilitate any online discussion I will post this appendix as a page to which you can link. Just search online for this appendix's title.

Index

CPSIA information can be obtained at www.ICGtesting.com
Printed in the USA
LVOW11s1352220913

353520LV00003B/3/P